PHILIP'S

ATLAS OF THE UNIVERSE

PHILIP'S

ATLAS OF THE UNIVERSE

PATRICK MOORE

CONTENTS

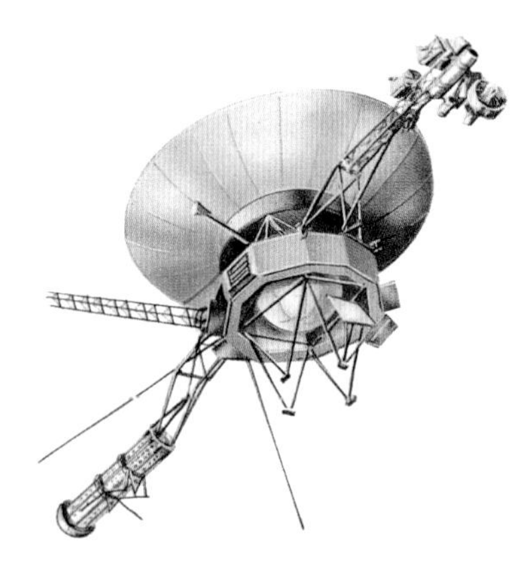

***Front cover**: Nebula in galaxy M33, photographed with the Wide Field Planetary Camera 2 from the Hubble Space Telescope.*

***Half-title page**: Partial solar eclipse of 13 November 1993, seen at X-ray wavelengths from the Japanese spacecraft Yohkoh.*

***Opposite title page**: The Eagle Nebula – a region of star formation, photographed from the Hubble Space Telescope.*

First published in 1994
by George Philip Limited
an imprint of Octopus Publishing Group Ltd
Reprinted 1995, 1996, 1997 (new edition)

This edition published in 2001
by Chancellor Press
an imprint of Bounty Books, a division of Octopus Publishing Group Ltd,
2-4 Heron Quays, London, E14 4JP

A CIP catalogue record for this book is available from the British Library.

Produced by Toppan (HK) Ltd
Printed in China

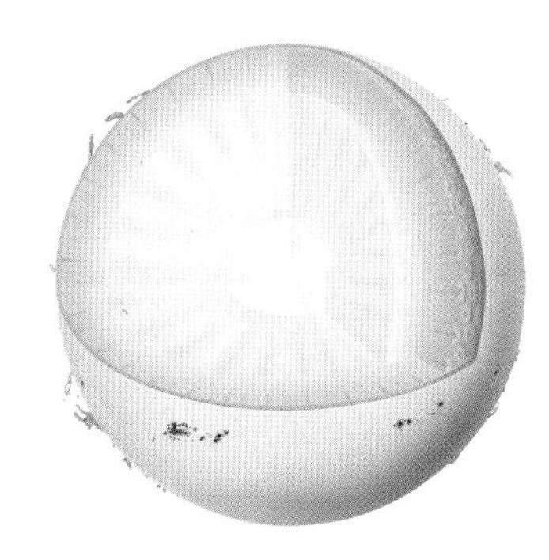

FOREWORD

If there is a field of human enquiry that is pre-eminent in exciting mankind it is surely astronomy. What is out there in the cosmos? How did the universe start? Where is it heading? What will be the fate of the Earth? Is anyone else out there – and if so, how should we react to them? These are some of the great questions that we should all like to have answered. Of course, practising astronomers are involved in solving small parts of these puzzles, but most people – and the many amateur astronomers lead the way – provide considerable stimulus by their interest.

This *Atlas of the Universe*, relating as it does to all these questions, attempts the impossible – and succeeds! Not only does it contain, as its name implies, maps of the heavens, but it covers a wealth of related detail, all fascinating material which will provide excitement, and knowledge, for readers of all ages.

Just as the universe has evolved and, on a very much shorter time scale, mankind's knowledge of it, so has this particular *Atlas*. The first edition was written in 1970 and the most recent in 1988; in his Introduction to the latter, Patrick Moore wrote that 'it will require further drastic revision before 2000'. In fact, as we see, progress in astronomy has been so rapid, and dramatic, that 'before 2000' is already with us. Thus, the new *Atlas* contains maps and data from the very latest space probes and the increasingly complex ground-based telescopes. The information is not in an indigestible form, however. Patrick Moore – a master in the art of synthesis and simplification, yet without compromising on accuracy – has marshalled everything magnificently, with the result that the book is a work of art as well as a work of science.

I recommend this *Atlas* to all who have an interest in the heavens – and in our own place in it.

Arnold Wolfendale.

PROFESSOR SIR ARNOLD WOLFENDALE, FRS
ASTRONOMER ROYAL, 1991–94

◄ ***Supernova in the Large Magellanic Cloud****, 24 February 1987. The Tarantula Nebula is the spidery red object, and the bright supernova may be seen below and to the right of it. Visible to the naked eye, it reached its peak brightness on 20–21 May, before beginning a slow decline.*

INTRODUCTION

When I wrote the first edition of *The Atlas of the Universe*, in 1970, the great astronomical revolution was just beginning. Electronic devices had started to take over from photographic plates, and computers had become a real force even though they were very crude compared with those of today. Space research was in full swing: men had already landed on the Moon, probes had been sent out to the nearer planets, and the first astronomical observatories were in orbit round the Earth.

Since then a great deal has happened. Great new telescopes have been built, allowing us to explore the far reaches of the universe; new theories have forced us to change or even abandon many of the older ideas, even if we have yet to solve fundamental problems such as that of the origin of the universe itself.

The progress of space research has been rather less smooth. There have been spectacular triumphs, such as the controlled landings on Mars, the missions to Halley's Comet and, of course, the epic flight of Voyager 2 to the outer planets; but there have also been delays and disappointments, together with a few tragedies. The Space Shuttle took much longer to develop than had been expected, and neither do we yet know what will happen to the former Soviet programme following the sudden collapse of the USSR.

Undoubtedly there will be further problems during the next few decades, but all in all the outlook remains bright. There are still people who question the value of the space programmes, but the cost of a planetary probe does not seem excessive when compared with that of, say, a nuclear submarine, and there are many benefits to mankind: for example, medical research is now closely linked with astronautics.

All this development has meant a drastic revision of the text of the *Atlas*. In effect it is new, and little of the last edition remains apart from the general plan. There is also a major difference between this *Atlas* and others. We are used to the superb, highly-coloured images produced by the world's greatest telescopes, but in general the colours are added to help in scientific analysis. Obviously I have included some of these false-colour pictures here, but I have concentrated upon things which can actually be seen by an observer who is adequately equipped. This is not always possible (for example, no Earth-based telescope can show surface details upon the satellites of Jupiter or Saturn), but I have kept to my rule as far as I can.

More new information has been obtained even since the last edition of this Atlas. For this latest reprint, I have made some further amendments and additions to bring the text up to date in January 2000.

Patrick Moore

PATRICK MOORE

◀ ***Part of the Cygnus Loop*** *supernova remnant, taken with the Wide Field and Planetary Camera on the Hubble Space Telescope on 24 April 1991. The Cygnus Loop marks the edge of a bubble-like, expanding blast wave from a supernova outburst – a colossal stellar explosion which occurred about 15,000 years ago.*

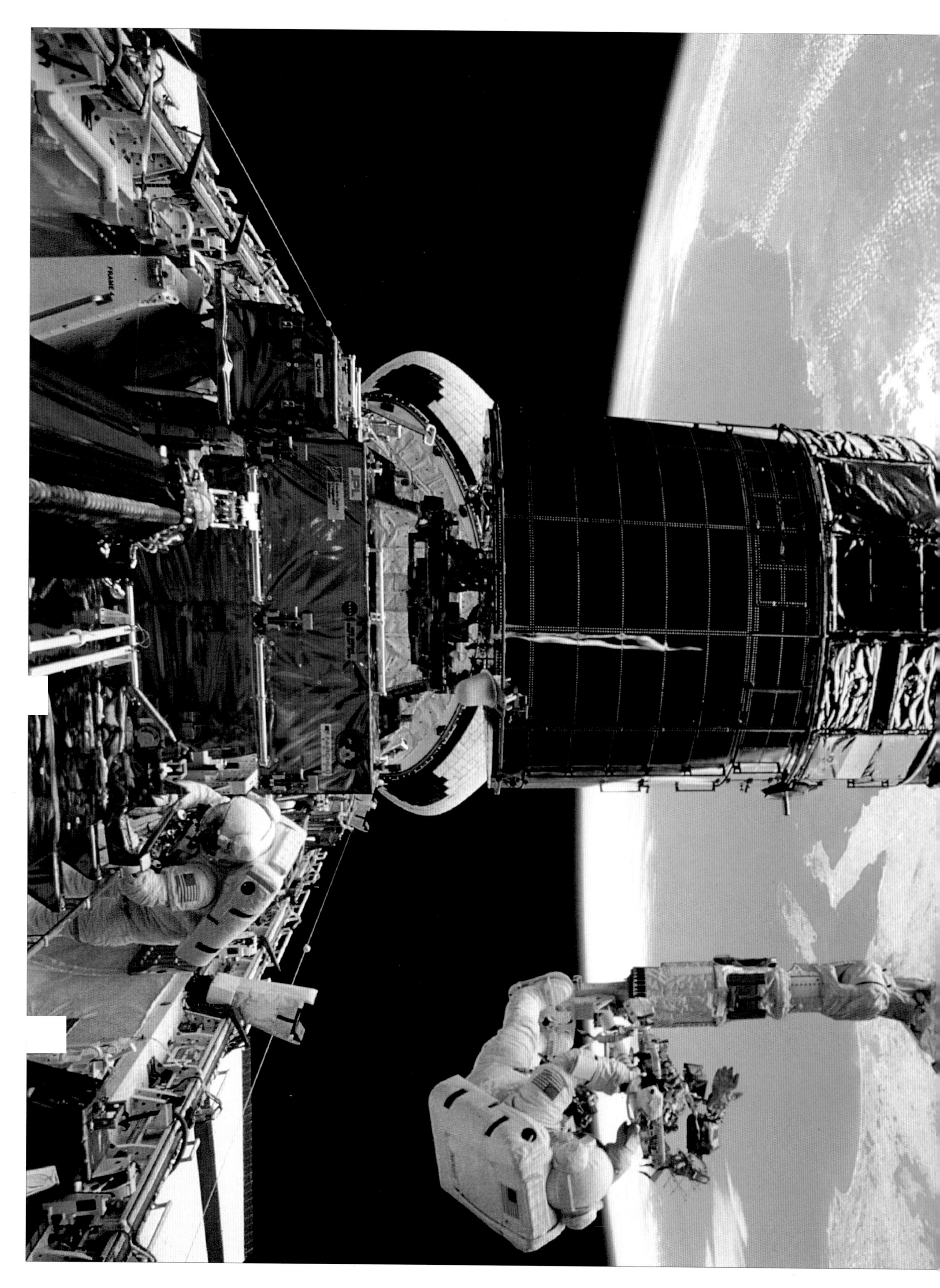

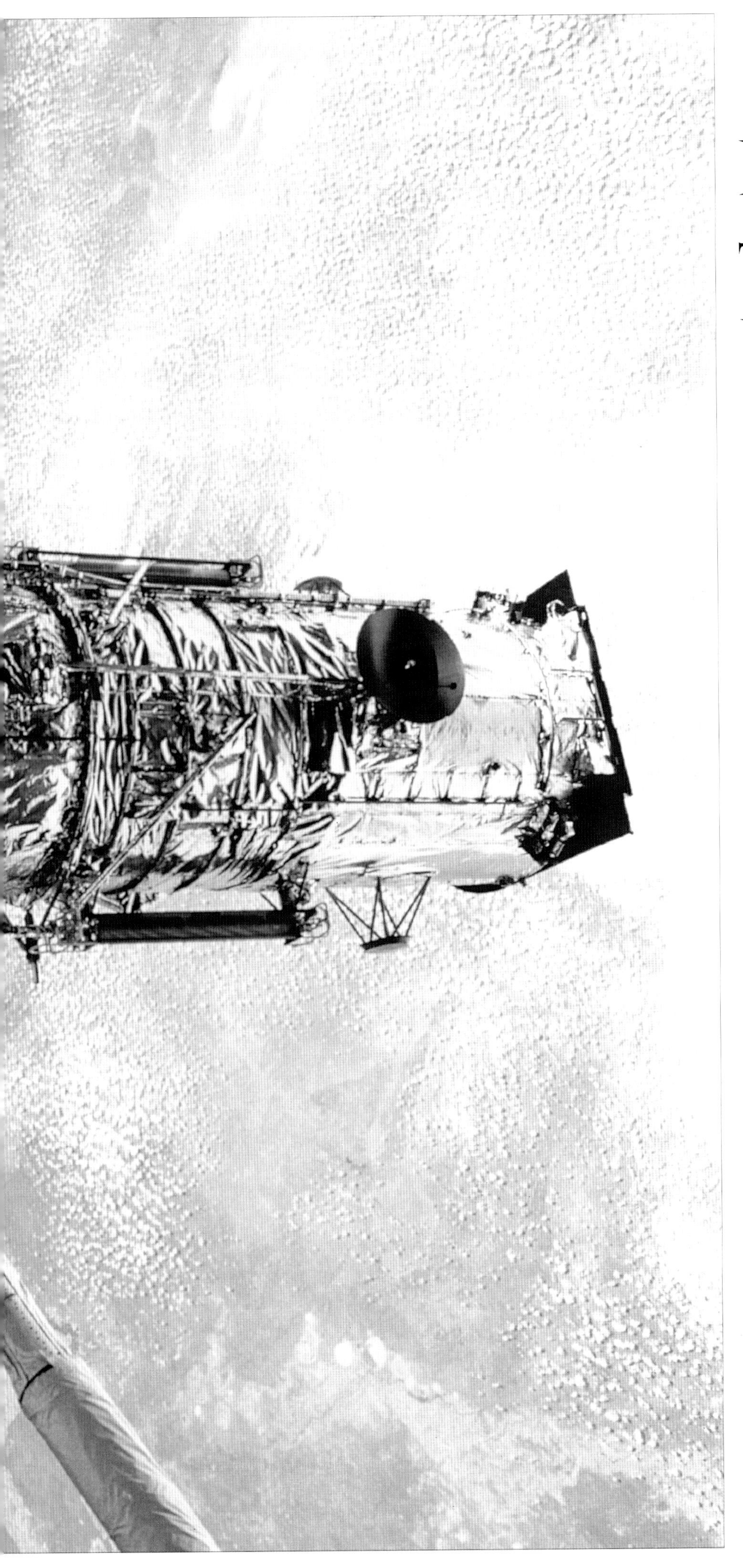

EXPLORING THE UNIVERSE

◀ ***The Hubble Space Telescope*** *was found to have a faulty mirror after it had been launched. A 'repair mission' was carried out in December 1993, so that the telescope is even better than had been originally hoped. Here astronauts F. Story Musgrave and Jeffrey A. Hoffman are shown working during the mission.*

Astronomy through the Ages

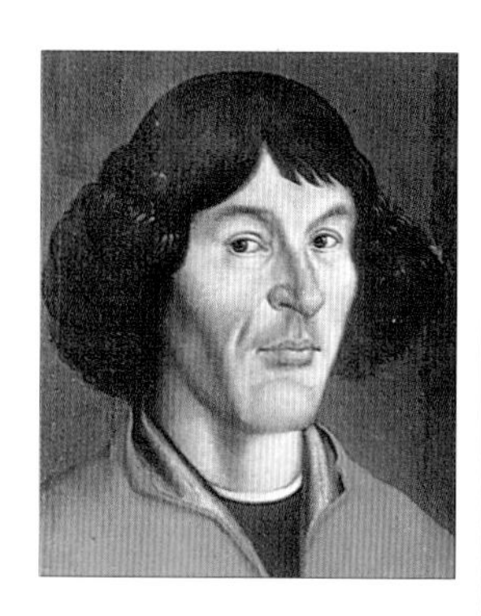

▲ **Copernicus** – *the Latinized name of Mikołaj Kopernik, the Polish churchman whose book,* De Revolutionibus Orbium Coelestium, *published in 1543, revived the theory that the Earth is a planet moving round the Sun.*

▲ **Galileo Galilei**, *the pioneer telescopic observer, was also the real founder of the science of experimental mechanics. He lived from 1564 to 1642; in 1633 he was brought to trial, and condemned for daring to teach the Copernican theory.*

▲ **Isaac Newton** *(1643–1727), whose book the* Principia, *published in 1687, has been described as the 'greatest mental effort ever made by one man', and marked the true beginning of the modern phase of astronomy.*

Astronomy is certainly the oldest of all the sciences. Our remote cave-dwelling ancestors must have looked up into the sky and marvelled at what they saw there, but they can have had no idea what the universe is really like, or how vast it is. It was natural for them to believe that the Earth is flat, with the sky revolving round it once a day carrying the Sun, the Moon and the stars.

Early civilizations in China, Egypt and the Middle East divided the stars up into groups or constellations, and recorded spectacular phenomena such as comets and eclipses; a Chinese observation of a conjunction of five bright planets may date back as far as 2449 BC. Probably the earliest reasonably good calendars were drawn up by the Egyptians. They paid great attention to the star Sirius (which they called Sothis), because its 'heliacal rising', or date when it could first be seen in the dawn sky, gave a reliable clue as to the annual flooding of the Nile, upon which the whole Egyptian economy depended. And, of course, there is no doubt that the Pyramids are astronomically aligned.

The first really major advances came with the Greeks. The first of the great philosophers, Thales of Miletus, was born around 624 BC. A clear distinction was drawn between the stars, which seem to stay in the same positions relative to each other, and the 'wanderers' or planets, which shift slowly about from one constellation to another. Aristotle, who lived from around 384 to 325 BC, gave the first practical proofs that the Earth is a globe, and in 270 BC Eratosthenes of Cyrene measured the size of the globe with remarkable accuracy. The value he gave was much better than that used by Christopher Columbus on his voyage of discovery so many centuries later.

The next step would have been to relegate the Earth to the status of a mere planet, moving round the Sun in a period of one year. Around 280 BC one philosopher, Aristarchus of Samos, was bold enough to champion this idea, but he could give no firm proof, and found few supporters. The later Greeks went back to the theory of a central Earth. Ptolemy of Alexandria, last of the great astronomers of Classical times, brought the Earth-centred theory to its highest state of perfection. He maintained that all paths or orbits must be circular, because the circle is the 'perfect' form, but to account for the observed movements of the planets he was forced to develop a very cumbersome system; a planet moved in a small circle or epicycle, the centre of which – the deferent – itself moved round the Earth in a perfect circle. Fortunately, Ptolemy's great work, the *Almagest*, has come down to us by way of its Arab translation.

Ptolemy died in or about the year AD 180. There followed a long period of stagnation, though there was one important development; in AD 570 Isidorus, Bishop of Seville, was the first to distinguish between true astronomy and the pseudo-science of astrology (which still survives, even though no intelligent person can take it seriously).

The revival of astronomy at the end of the Dark Ages was due to the Arabs. In 813 Al Ma'mun founded the Baghdad school, and during the next few centuries excellent star catalogues were drawn up. In 1433 Ulugh Beigh, grandson of the Oriental conqueror Tamerlane, set up an elaborate observatory at Samarkand, but with his murder, in 1449, the Baghdad school of astronomy came to an end.

The first serious challenge to the Ptolemaic theory came in 1543 with the publication of a book by the Polish churchman Mikołaj Kopernik, better known by his Latinized name Copernicus. He realized the clumsiness and artificial nature of the old theory could be removed simply by taking the Earth away from its proud central position and putting the Sun there. He also knew there would be violent opposition from the Church, and he was wise enough to withhold publication of his book until the end of his life. His fears were well founded; Copernican theory was condemned as heresy, and Copernicus's book, *De Revolutionibus Orbium Coelestium* (*Concerning the Revolutions of the Celestial Orbs*) was placed on the Papal Index. It remained there until 1835.

◄ **An orrery**, *made in 1790; the name commemorates the Earl of Cork and Orrery, for whom the first orrery was made. The Sun is represented by a brass ball in the centre. Around it move the three innermost planets, Mercury, Venus and the Earth; an ingenious system of gears makes the planets move round the Sun in the correct relative periods, though not at the correct relative distances. The Moon's orbit round the Earth is inclined at the correct angle. When the mechanism is moved, by turning a handle, the planets revolve round the Sun and the Moon revolves round the Earth. The Zodiacal signs are shown around the edge of the disk.*

Ironically, the next character in the story, the Danish astronomer Tycho Brahe, was no Copernican. He believed in a central Earth, but he was a superbly accurate observer who produced a star catalogue which was much better than anything compiled before. He also measured the positions of the planets, particularly Mars. When he died, in 1601, his work came into the possession of his last assistant, the German mathematician Johannes Kepler. Kepler had implicit faith in Tycho's observations, and used them to show that the Earth and the planets do indeed move round the Sun – not in circles, but in ellipses.

Kepler's Laws of Planetary Motion may be said to mark the beginning of modern-type astronomy. The first two Laws were published in 1609, though the change in outlook was not really complete until the publication of Isaac Newton's *Principia* almost 80 years later. Meanwhile, the first telescopes had been turned towards the sky.

◀ ***Stonehenge*** *is probably the most famous of all 'stone circles'. It stands on Salisbury Plain, and is a well-known tourist attraction! Contrary to popular belief, it has nothing to do with the Druids; its precise function is still a matter for debate, but it is certainly aligned astronomically. It has, of course, been partially ruined, but enough remains to show what it must originally have looked like.*

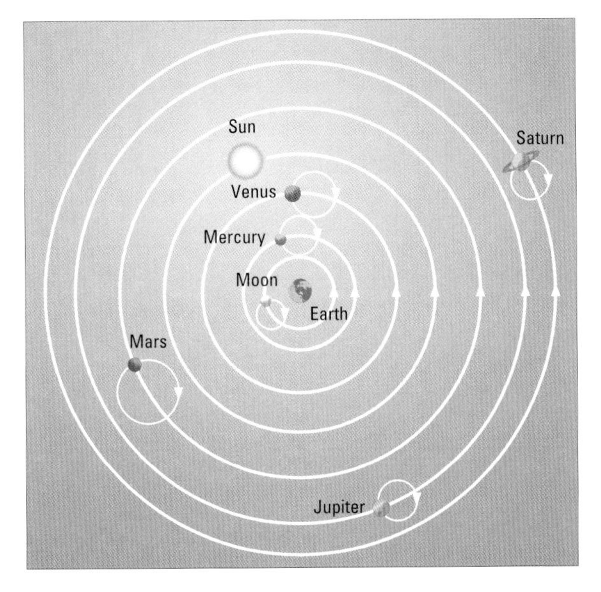

◀ ***The Ptolemaic theory*** *– the Earth lies in the centre of the universe, with the Sun, Moon, planets and stars moving round it in circular orbits. Ptolemy assumed that each planet moved in a small circle or epicycle, the centre of which – the deferent – itself moved round the Earth in a perfect circle.*

▶ ***The Copernican theory*** *– placing the Sun in the centre removed many of the difficulties of the Ptolemaic theory, but Copernicus kept the idea of circular orbits, and was even reduced to bringing back epicycles.*

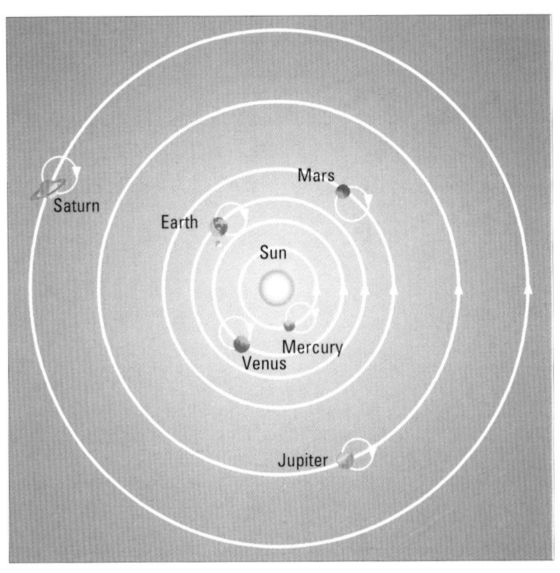

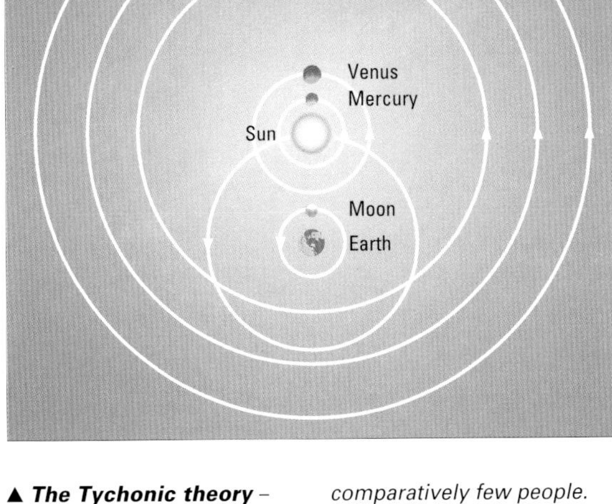

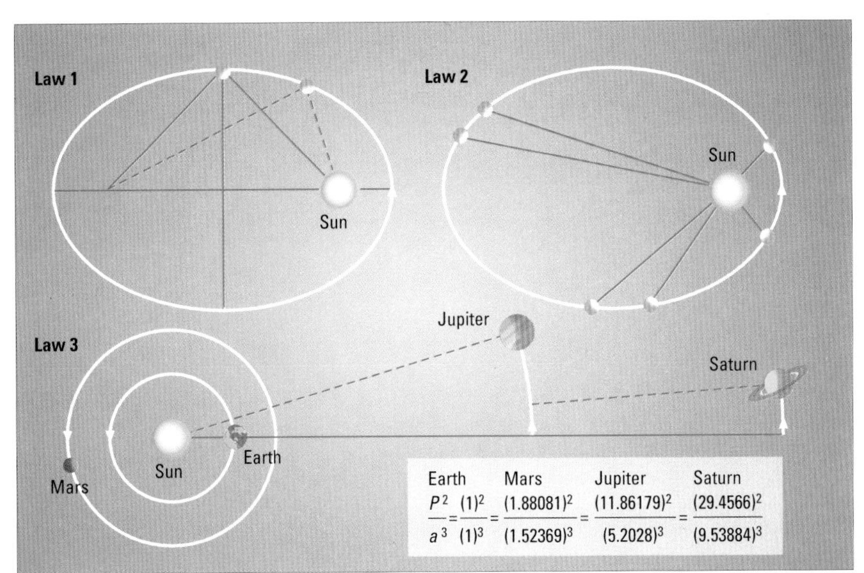

▲ ***The Tychonic theory*** *– Tycho Brahe retained the Earth in the central position, but assumed that the other planets moved round the Sun. In effect this was a rather uneasy compromise, which convinced comparatively few people. Tycho adopted it because although he realized that the Ptolemaic theory was unsatisfactory, he could not bring himself to believe that the Earth was anything but of supreme importance.*

▲ ***Kepler's Laws:***
Law 1 A planet moves in an ellipse; the Sun is one focus, while the other is empty.
Law 2 The radius vector – the line joining the centre of the planet to that of the Sun – sweeps out equal areas in equal times (a planet moves fastest when closest in).
Law 3 For any planet, the square of the revolution period (p) is proportional to the cube of the planet's mean distance from the Sun (a). Once the distance of any planet is known, its period can be calculated, or vice versa. Kepler's Laws make it possible to draw up a scale model of the Solar System; only one absolute distance has to be known, and the rest can then be calculated.

Telescopes and the Stars

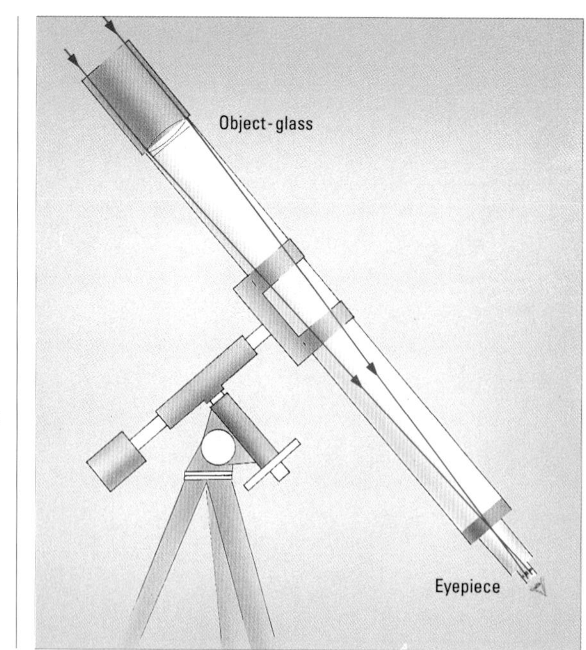

▶ ***Principle of the refractor**. The light from the object under observation passes through a glass lens (or combination of lenses), known as an object-glass or objective. The rays are brought to a focus, where the image is enlarged by a second lens known as the eyepiece or ocular.*

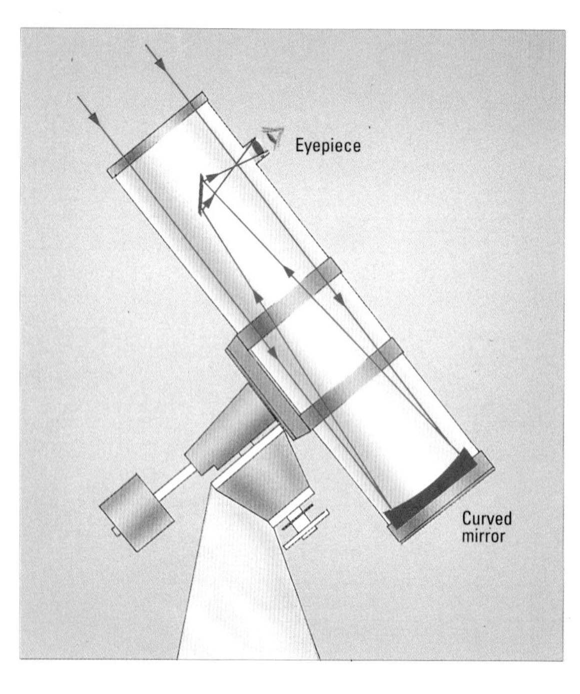

◀ ***Principle of the Newtonian reflector**. The light passes down an open tube and falls upon a curved mirror. The light is then sent back up the tube on to a smaller, flat mirror placed at an angle of 45°; the flat directs the rays on to the side of the tube, where they are brought to focus and the image is magnified by an eyepiece.*

Nobody can be sure just when telescopes were invented, but there is strong evidence that Leonard Digges, in England, built a workable telescope in or around the year 1550. Apparently it used both a lens and a mirror; we do not know exactly what it looked like, and there is no firm evidence that it was ever turned skywards.

The first telescopes of which we have definite knowledge date back to 1608, and came from Holland. During 1609 Thomas Harriot, one-time tutor to Sir Walter Raleigh, drew a telescopic map of the Moon which shows recognizable features, but the first systematic observations were made from 1610 by Galileo Galilei, in Italy. Galileo made his own telescopes, the most powerful of which magnified 30 times, and used them to make spectacular discoveries; he saw the mountains and craters of the Moon, the phases of Venus, the satellites of Jupiter, spots on the Sun and the countless stars of the Milky Way. Everything he found confirmed his belief that Copernicus had been absolutely right in positioning the Sun in the centre of the planetary system – for which he was accused of heresy, brought to trial in Rome, and forced into a hollow and completely meaningless recantation of the Copernican theory.

These early 17th-century telescopes were refractors. The light is collected by a glass lens known as an objective or object-glass; the rays of light are brought together, and an image is formed at the focus, where it can be magnified by a second lens termed an eyepiece.

Light is a wave-motion, and a beam of white light is a mixture of all the colours of the rainbow. A lens bends the different wavelengths unequally, and this results in false colour; an object such as a star is surrounded by gaudy rings which may look pretty, but are certainly unwanted. To reduce this false colour, early refractors were made with very long focal length, so that it was sometimes necessary to fix the object-glass to a mast. Instruments of this kind were extremely awkward to use, and it is surprising that so many discoveries were made with them. A modern objective is made up of several lenses, fitted together and made up of different types of glass, the faults of which tend to cancel each other out.

Isaac Newton adopted a different system, and in 1671 he presented the first reflector to the Royal Society of London. Here there is no object-glass; the light passes down an open tube and falls upon a curved mirror, which reflects the light back up the tube on to a smaller, flat mirror inclined at 45 degrees. The inclined mirror reflects

▶ ***The altazimuth mounting**. The telescope can move freely in either altitude (up and down) or azimuth (east to west). This involves making constant adjustments in both senses, though today modern computers make altazimuth mountings practicable for very large telescopes.*

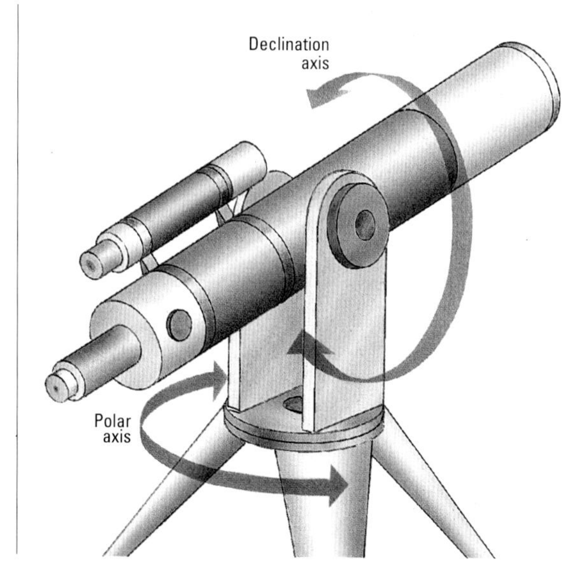

▶ ***The equatorial mounting**. The telescope is mounted upon an axis directed towards the celestial pole, so that when the telescope is moved in azimuth the up-or-down motion looks after itself. Until recently all large telescopes were equatorially mounted.*

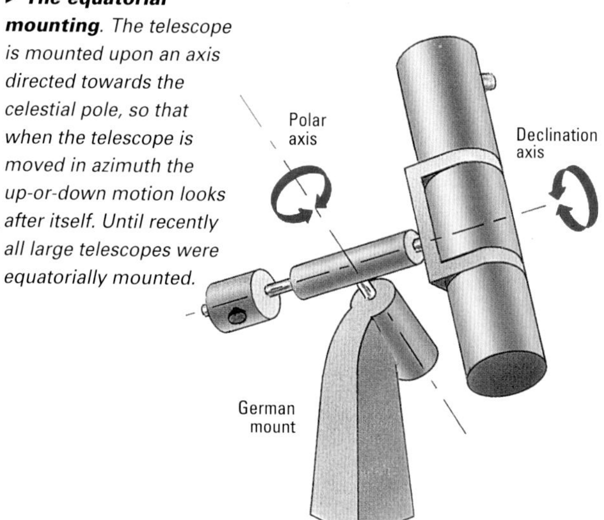

the light to the side of the tube, where an image is formed and enlarged by an eyepiece as before. A mirror reflects all wavelengths equally, so that there is no false colour problem. Newtonian reflectors are still very popular, particularly with amateur astronomers, but there are other optical systems such as the Cassegrain and the Gregorian, where the light is reflected back to the eyepiece through a hole in the centre of the main mirror.

Newton's first reflector used a mirror only 2.5 centimetres (1 inch) in diameter, but before long larger telescopes were made. In 1789 William Herschel, a Hanoverian-born musician who lived in England, built a reflector with a 124.5-centimetre (49-inch) mirror, though most of his work was carried out with much smaller instruments. Then, in 1845, came the giant 183-centimetre (72-inch) reflector made in Ireland by the third Earl of Rosse, who discovered the spiral forms of the star systems we now call galaxies. The Rosse reflector remained the world's largest until the completion of the Mount Wilson 2.5-metre (100-inch) reflector in 1917.

Admittedly the Rosse telescope was clumsy to use, because it was slung between two massive stone walls and could reach only a limited portion of the sky. Moreover, a celestial object moves across the sky, by virtue of the Earth's rotation, and the telescope has to follow it, which is not easy when high magnification is being used. In 1824 the German optician Josef Fraunhofer built a 23-centimetre (9-inch) refractor which was mechanically driven and was set up on an equatorial mount, so that the telescope rides the axis pointing to the pole of the sky; only the east-to-west motion has to be considered, because the up-or-down movement will look after itself. Until the development of modern-type computers, all large telescopes were equatorially mounted.

The late 19th century was the age of the great refractors, of which the largest, at the Yerkes Observatory in Wisconsin, USA was completed in 1897. The telescope has a 1-metre (40-inch) object-glass, and is still in regular use. It is not likely to be surpassed, because a lens has to be supported round its edge, and if it is too heavy it will start to distort under its own weight, making it useless. Today almost all large optical telescopes are of the reflecting type, and are used with photographic or electronic equipment. It is not often that a professional astronomer actually looks through an eyepiece these days. The modern astronomer observes the skies on a computer or TV screen.

▲ ***Herschel's 'forty-foot' reflector*** *was completed in 1789. The mirror was 124 cm (49 inches) in diameter, and was made of metal; there was of course no drive, and the mounting was decidedly cumbersome. The optical system used was the Herschelian; there is no flat, and the main mirror is tilted so as to bring the rays of light directly to focus at the upper edge of the tube – a system which is basically unsatisfactory.*

◀ ***The Rosse reflector.*** *This telescope was built by the third Earl of Rosse, and completed in 1845. It had a 183-cm (72-inch) metal mirror; the tube was mounted between two massive stone walls, so that it could be swung for only a limited distance to either side of the meridian. This imposed obvious limitations; nevertheless, Lord Rosse used it to make some spectacular discoveries, such as the spiral forms of the galaxies. The telescope has now been fully restored, and by early 2000 will again be fully operational.*

▲ ***The Yerkes refractor.*** *This has a 101-cm (40-inch) object-glass. It was completed in 1897, due to the work of George Ellery Hale, and remains the largest refractor in the world; it is not likely that it will ever be surpassed, because a lens has to be supported round its edge, and if too heavy will distort, making it useless.*

Observatories of the World

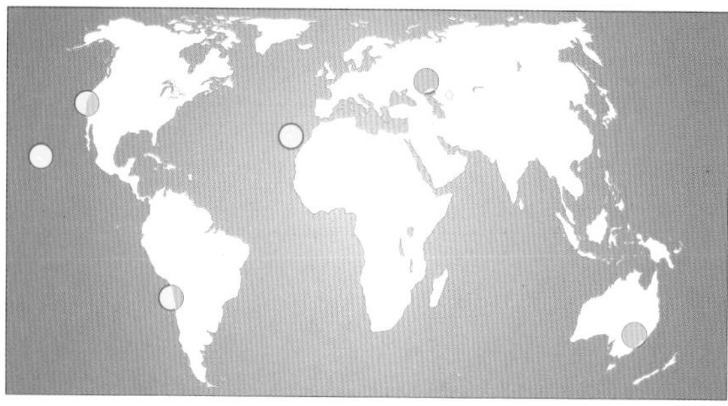

◀ ***Observatory sites****. There are major observatories in all inhabited continents. The modern tendency is to establish large new observatories in the southern hemisphere, partly because of the clearer skies and partly because some of the most significant objects lie in the far south of the sky.*

It is the Earth's atmosphere which is the main enemy of the astronomer. Not only is it dirty and unsteady, but it also blocks out some of the most important radiations coming from space. This is why most modern observatories are sited at high altitude, often on the tops of mountains, where the air is thin and dry.

Of course, this is not always possible. For example there are no high peaks in Australia, and the observatory at Siding Spring, near Coonabarabran in New South Wales, lies at an altitude of less than 1150 metres (3800 feet), though this does at least mean that it is easily accessible (provided that one avoids driving into the kangaroos which roam the Warrumbungle range; the animals have absolutely no road sense!). Another modern hazard is light pollution, which is increasing all the time. The Hooker reflector at Mount Wilson in California was actually mothballed for some years during the 1980s because of the lights of Los Angeles, and even the great Palomar reflector, also in California, is threatened to some extent. Another indifferent site is Mount Pastukhov, where the Russians have erected a 6-metre (236-inch) reflector. The altitude is just over 2000 metres (6600 feet) but conditions are not

▶ ***Dome of the Anglo-Australian telescope****. The AAT is a 3.89-m (153-inch) reflector, completed in 1965 but still among the world's best. Though it is not at high altitude, the seeing conditions are good. Also at Siding Spring is the UK Schmidt telescope, as well as various smaller instruments; the radio telescope at Parkes is within easy driving distance.*

▼ ***Dome of the William Herschel telescope at La Palma****. It has a 4.2-m (165-inch) mirror. It is sited on the summit of Los Muchachos, an extinct volcano in the Canary Islands, at an altitude of 2330 m (almost 8000 feet). The Isaac Newton Telescope is also on Los Muchachos; it has a 256-cm (101-inch) mirror, and was transferred to La Palma in 1983.*

▶ ***Domes at La Silla****. La Silla, in the Atacama Desert in Northern Chile, is an exceptionally good observational site and a number of large telescopes have been set up there. This photograph was taken from the catwalk of the Max Planck 2.2-m (87-inch) reflector. The La Silla facility is the main observing station of the ESO (European Southern Observatory), which has its headquarters at Garching, near Munich in Germany.*

good, and the site was selected only because there are no really favourable locations in the old USSR.

Against this, the mountain observatories are spectacular by any standards. The loftiest of all is the summit of Mauna Kea, the extinct volcano in Hawaii, at well over 4000 metres (13,800 feet). At this height one's lungs take in only 39 per cent of the normal amount of oxygen, and care is essential; nobody actually sleeps at the summit, and after a night's observing the astronomers drive down to the 'halfway house', Hale Pohaku, where the air is much denser. There are now many telescopes on Mauna Kea, and others are planned. Almost equally awe-inspiring is the top of the Roque de los Muchachos (the Rock of the Boys), at La Palma in the Canary Islands. The altitude is 2332 metres (7648 feet), and it is here that we find the largest British telescope, the 4.2-metre (165-inch) William Herschel reflector. The 'Rock' is truly international; La Palma is a Spanish island, but there are observatories not only from Britain but also from Scandinavia, Germany, Italy and other countries. Another superb site is the Atacama Desert of Northern Chile, where there are four major observatories: La Silla (run by the European Southern Observatory), Cerro Tololo and Las Campanas (run by the United States), and the new observatory for the VLT or Very Large Telescope, at Cerro Paranal in the northern Atacama. The VLT will have four 8.2-metre (323-inch) mirrors working together, and will be much the most powerful telescope ever built. By mid-1999 two of the mirrors, named Antu and Kueyen, were already sending back superb images.

A modern observatory has to be almost a city in itself, with laboratories, engineering and electronic workshops, living quarters, kitchens and much else. Yet today there is a new development. Telescopes can be operated by remote control, so that the astronomer need not be in the observatory at all – or even in the same continent. For example, it is quite practicable to sit in a control room in Cambridge and operate a telescope thousands of kilometres away in Chile or Hawaii.

New observatories are being planned. One particularly interesting site is the South Pole, where viewing conditions are excellent even though the climate is somewhat daunting. It is hoped that the Polar Observatory will be operational well before the next decade.

▼ ***Domes on Mauna Kea***. *Mauna Kea, in Hawaii, is an extinct volcano over 4000 m (14,000 feet) high. On its summit several large telescopes have been erected, and more are planned. The main advantage is the thinness of the atmosphere, and the fact that most of the atmospheric water vapour lies below. The main disadvantage is that one's lungs take in less than 39 per cent of the normal amount of oxygen, and care must always be taken.*

▲ ***Dome of the Palomar 5.08-m (200-inch) reflector***. *The Hale reflector was brought into action in 1948, and was for many years in a class of its own. Though it is no longer the world's largest, it maintains its position in the forefront of research, and is now used with electronic equipment, so that it is actually far more effective than it was when first completed.*

Great Telescopes

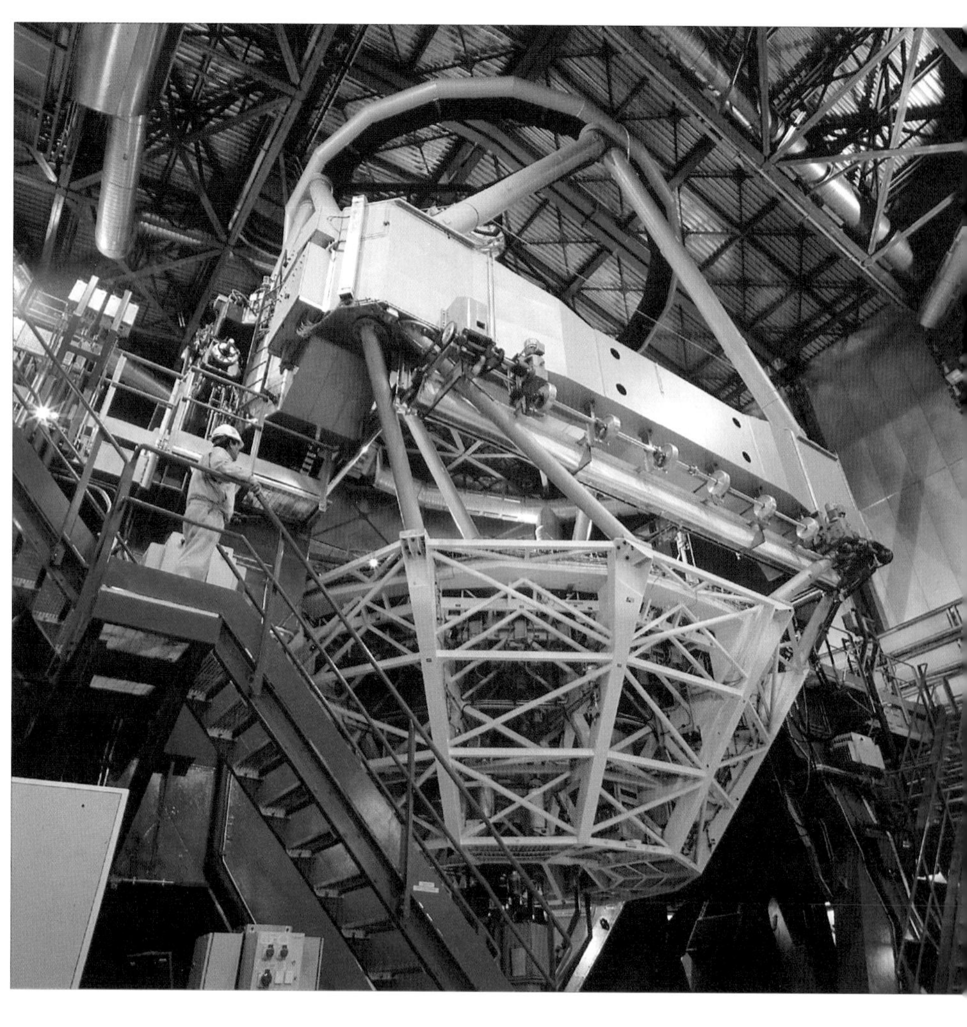

▲ ***The New Technology Telescope (NTT) at La Silla****. The NTT, at the site of the European Southern Observatory, has a mirror 3.5 m (138 inches) in diameter. The telescope is of very advanced design; it moves only in altitude, and the entire observatory rotates. New techniques such as active and adaptive optics have been introduced, and the NTT has proved to be extremely successful. It was completed in 1989.*

For many years the Mount Wilson 2.5-metre (100-inch) reflector was not only the world's largest telescope, but was in a class of its own. It was set up through the untiring energy of George Ellery Hale, an American astronomer who not only planned huge telescopes but also had the happy knack of persuading friendly millionaires to pay for them! Hale had already been responsible for the Yerkes refractor; later he planned the 5-metre (200-inch) Palomar reflector, though he died before the telescope was completed in 1948. The Palomar telescope is still in full operation, and is indeed more effective than it used to be, because it is now used with the latest electronic equipment. What is termed a CCD, or Charge-Coupled Device, is far more sensitive than any photographic plate.

In 1975 the Russians completed an even larger telescope, with a 6-metre (236-inch) mirror, but it has never been a success, and is important mainly because of its mounting, which is of the altazimuth type. With an altazimuth, the telescope can move freely in either direction – up or down (*alt*itude) or east to west (*azimuth*). This means using two driving mechanisms instead of only one, as with an equatorial, but this is easy enough with the latest computers, and in all other respects an altazimuth mounting is far more convenient. All future large telescopes will be mounted in this way.

The New Technology Telescope (NTT), at La Silla in Chile, looks very different from the Palomar reflector. It is short and squat, with a 3.5-metre (138-inch) mirror which is only 24 centimetres (10 inches) thick and weighs 6 tonnes (13,440 pounds). Swinging a large mirror around means distorting it, and with the NTT two systems are used to compensate for this. The first is termed 'active optics', and involves altering the shape of the mirror so that it always retains its perfect curve; this is done by computer-controlled pads behind the mirror. With 'adaptive optics' an extra computer-controlled mirror is inserted in the telescope, in front of a light-sensitive detector. By monitoring the image of a relatively bright star in the field of view, the mirror can be continuously modified to compensate for distortions in the image due to air turbulence.

The VLT or Very Large Telescope, at Cerro Paranal in the northern Atacama Desert of Chile, is operated by the European Southern Observatory. It has four 8.2-metre (323-inch) mirrors working together. The first two were operational by mid-1999, and the other two will be ready very early in the new century.

The Keck Telescope on Mauna Kea has a 9.8-metre (387-inch) mirror which has been made from 36 hexagonal segments, fitted together to form the correct optical curve; the final shape has to be accurate to a limit of one thousandth the width of a human hair. A twin Keck has been built beside it, and when the two telescopes are operating together they could, in theory, be capable of distinguishing a car's headlights separately from a distance of over 25,000 kilometres (over 15,000 miles).

Some telescopes have been constructed to meet special needs. With a Schmidt telescope, the main advantage is a very wide field of view, so that large areas of the sky can be photographed with a single exposure; the United Kingdom Infra-Red Telescope (UKIRT) on Mauna Kea was designed to collect long-wavelength (infra-red) radiations, though in fact it has proved to be so good that it can be used at normal wavelengths as well.

◀ ***VLT Kueyen,*** *the second unit of the VLT (Very Large Telescope). The VLT, at Paranal in Chile, will be much the most powerful telescope ever built. It will have four 8.2-metre (323-inch) mirrors, working together, named Antu (the Sun), Kueyen (Moon), Melipal (Southern Cross) and Yepun (Sirius). Kueyen, shown here, came into operation in 1999, following Antu in 1998. These names come from the Mapuche language of the people of Chile south of Santiago. [Reproduced by kind permission of the European Southern Observatory]*

◀ ***The Nicholas U. Mayall reflector at Kitt Peak, Arizona.*** *Kitt Peak is the national observatory of the United States. Its largest optical telescope is the Mayall reflector, with a 3.81-m (150-inch) mirror; the altitude is 2064 m (6770 feet). The telescope is equatorially mounted – one of the last really great instruments which will be mounted in this manner. The telescope was completed in 1973, and works closely with the very similar 3.5-m (138-inch) reflector at La Silla in Chile.*

THE WORLD'S LARGEST TELESCOPES 1999

Telescopes	Observatory	Aperture m	Aperture in	Lat.	Long.	Elev., m	Completed
REFLECTORS							
Keck I	W. M. Keck Observatory, Mauna Kea, Hawaii, USA	9.82	387	19° 49′ N	155° 28′ W	4150	1992
Keck II	W. M. Keck Observatory, Mauna Kea, Hawaii, USA	9.82	387	19° 49′ N	155° 28′ W	4150	1996
Hobby-Eberly Telescope	Mt Fowlkes, Texas, USA	9.2	362	30° 40′ N	101° 01′ W	2072	1998
Subaru	Mauna Kea, Hawaii, USA	8.3	327	19° 50′ N	155° 28′ W	4100	1998
Antu (first unit of VLT)	Cerro Paranal, Chile	8.2	323	24° 38′ S	70° 24′ W	2635	1998
Kueyen (second unit of VLT)	Cerro Paranal, Chile	8.2	323	24° 38′ S	70° 24′ W	2635	1999
Bolshoi Teleskop Azimutalnyi	Special Astrophysical Observatory, Mt Pastukhov, Russia	6.0	236	43° 39′ N	41° 26′ E	2100	1975
Hale Telescope	Palomar Observatory, Palomar Mtn, California, USA	5.08	200	33° 21′ N	116° 52′ W	1706	1948
Multiple Mirror Telescope (MMT)	Mount Hopkins Observatory, Arizona, USA	*(equiv.)* 4.5	177	31° 41′ N	110° 53′ W	2608	1979
William Herschel Telescope	Obs. del Roque de los Muchachos, La Palma, Canary Is	4.2	165	28° 46′ N	17° 53′ W	2332	1987
Victor Blanco Telescope	Cerro Tololo Interamerican Observatory, Chile	4.001	158	30° 10′ S	70° 49′ W	2215	1976
Anglo-Australian Telescope (AAT)	Anglo-Australian Telescope Siding Spring, Australia	3.893	153	31° 17′ S	149° 04′ E	1149	1975
Nicholas U. Mayall Reflector	Kitt Peak National Observatory, Arizona, USA	3.81	150	31° 58′ N	111° 36′ W	2120	1973
United Kingdom Infra-Red Telescope (UKIRT)	Joint Astronomy Centre, Mauna Kea, Hawaii, USA	3.802	150	19° 50′ N	155° 28′ W	4194	1978
Canada-France-Hawaii Telescope (CFH)	Canada-France-Hawaii Tel. Corp., Mauna Kea, Hawaii, USA	3.58	141	19° 49′ N	155° 28′ W	4200	1979
3.6-m Telescope	European Southern Observatory, La Silla, Chile	3.57	141	29° 16′ S	70° 44′ W	2387	1977
3.5-m Telescope	Calar Alto Observatory, Calar Alto, Spain	3.5	138	37° 13′ N	02° 32′ W	2168	1984
New Technology Telescope (NTT)	European Southern Observatory, La Silla, Chile	3.5	138	29° 16′ S	70° 44′ W	2353	1989
Astrophys. Research Consortium (ARC)	Apache Point, New Mexico, USA	3.5	138	32° 47′ N	105° 49′ W	2800	1993
WIYN	Kitt Peak, Arizona, USA	3.5	138	31° 57′ N	111° 37′ W	2100	1998
Starfire	Kirtland AFB, New Mexico, USA	3.5	138	classified	classified	1900	1998
Galileo	La Palma, Canary Islands	3.5	138	28° 45′ N	17° 53′ W	2370	1998
C. Donald Shane Telescope	Lick Observatory, Mt Hamilton, California, USA	3.05	120	37° 21′ N	121° 38′ W	1290	1959
Nodo (liquid mirror)	New Mexico, USA	3.0	118	32° 59′ N	105° 44′ W	2758	1999
NASA Infra-Red Facility (IRTF)	Mauna Kea Observatory, Mauna Kea, Hawaii, USA	3.0	118	19° 50′ N	155° 28′ W	4208	1979
Harlan Smith Telescope	McDonald Observatory, Mt Locke, Texas, USA	2.72	107	30° 40′ N	104° 01′ W	2075	1969
UBC-Laval Telescope (LMT)	Univ. of Brit. Col. and Laval Univ., Vancouver, Canada	2.7	106	49° 07′ N	122° 35′ W	50	1992
Shajn 2.6-m Reflector	Crimean Astrophys. Observatory, Crimea, Ukraine	2.64	104	44° 44′ N	34° 00′ E	550	1960
Byurakan 2.6-m Reflector	Byurakan Observatory, Mt Aragatz, Armenia	2.64	104	40° 20′ N	44° 18′ E	1500	1976
Nordic Optical Telecscope (NOT)	Obs. del Roque de los Muchachos, La Palma, Canary Is	2.56	101	28° 45′ N	17° 53′ W	2382	1989
Irénée du Pont Telescope	Las Campanas Observatory, Las Campanas, Chile	2.54	100	29° 00′ N	70° 42′ W	2282	1976
Hooker Telescope (100 inch)	Mount Wilson Observatory, California, USA	2.5	100	34° 13′ N	118° 03′ W	1742	1917
Isaac Newton Telescope (INT)	Obs. del Roque de los Muchachos, La Palma, Canary Is	2.5	100	28° 46′ N	17° 53′ W	2336	1984
Hubble Space Telescope (HST)	Space Telescope Science Inst., Baltimore, USA	2.4	94	orbital	orbital		1990
REFRACTORS							
Yerkes 40-inch Telescope	Yerkes Observatory, Williams Bay, Wisconsin, USA	1.01	40	42° 34′ N	88° 33′ W	334	1897
36-inch Refractor	Lick Observatory, Mt Hamilton, California, USA	0.89	35	37° 20′ N	121° 39′ W	1290	1888
33-inch Meudon Refractor	Paris Observatory, Meudon, France	0.83	33	48° 48′ N	02° 14′ E	162	1889
Potsdam Refractor	Potsdam Observatory, Germany	0.8	31	52° 23′ N	13° 04′ E	107	1899
Thaw Refractor	Allegheny Observatory, Pittsburgh, USA	0.76	30	40° 29′ N	80° 01′ W	380	1985
Lunette Bischoffscheim	Nice Observatory, France	0.74	29	43° 43′ N	07° 18′ E	372	1886
SCHMIDT TELESCOPES							
2-m Telescope	Karl Schwarzschild Observatory,Tautenberg, Germany	1.34	53	50° 59′ N	11° 43′ E	331	1950
Oschin 48-inch Telescope	Palomar Observatory, California , USA	1.24	49	33° 21′ S	116° 51′ W	1706	1948
United Kingdom Schmidt Telescope (UKS)	Royal Observatory, Edinburgh, Siding Spring, Australia	1.24	49	31° 16′ S	149° 04′ E	1145	1973
Kiso Schmidt Telescope	Kiso Observatory, Kiso, Japan	1.05	41	35° 48′ N	137° 38′ E	1130	1975
3TA-10 Schmidt Telescope	Byurakan Astrophys. Observatory, Mt Aragatz, Armenia	1.00	39	40° 20′ N	44° 30′ E	1450	1961
Kvistaberg Schmidt Telescope	Uppsala University Observatory, Kvistaberg, Sweden	1.00	39	59° 30′ N	17° 36′ E	33	1963
ESO 1-m Schmidt Telescope	European Southern Observatory, La Silla, Chile	1.00	39	29° 15′ S	70° 44′ W	2318	1972
Venezuela 1-m Schmidt Telescope	Centro F. J. Duarte, Merida, Venezuela	1.00	39	08° 47′ N	70° 52′ W	3610	1978

Invisible Astronomy

▲ ***The Swedish Sub-millimetre Telescope (SEST)****. This telescope is designed specifically for research in the sub-millimetre range of the electromagnetic spectrum. It is sited at La Silla, in the Atacama Desert of Northern Chile; this photograph of it was taken in 1990.*

▼ ***The Lovell Telescope****. This 76-m (250-foot) 'dish' at Jodrell Bank, in Cheshire, UK, was the first really large radio telescope; it has now been named in honour of Professor Sir Bernard Lovell, who master-minded it. It came into use in 1957 – just in time to track Russia's Sputnik 1, though this was not the sort of research for which it was designed! It has been 'upgraded' several times; this photograph was taken in 1993.*

The colour of light depends upon its wavelength – that is to say, the distance between two successive wave-crests. Red light has the longest wavelength and violet the shortest; in between come all the colours of the rainbow – orange, yellow, green and blue. By everyday standards the wavelengths are very short, and we have to introduce less familiar units. One is the Ångström (Å), named in honour of the 19th-century Swedish physicist Anders Ångström; the founder of modern spectroscopy, one Å is equal to one ten-thousand millionth of a metre. The other common unit is the nanometre (nm). This is equal to one thousand millionth of a metre, so that 1 nanometre is equivalent to 10 Ångströms.

Visible light extends from 400 nm or 4000 Å for violet up to 700 nm or 7000 Å for red (these values are only approximate; some people have greater sensitivity than others). If the wavelength is outside these limits, the radiations cannot be seen, though they can be detected in other ways; for example, if you switch on an electric fire you will feel the infra-red, in the form of heat, well before the bars become hot enough to glow. To the long-wave end of the total range of wavelengths, or electromagnetic spectrum, we have infra-red (700 nanometres to 1 millimetre), microwaves (1 millimetre to 0.3 metre) and then radio waves (longer than 0.3 metre). To the short-wave end we have ultra-violet (400 nanometres to 10 nanometres), X-rays (10 nanometres to 0.01 nanometre) and finally the very short gamma rays (below 0.01 nanometre). Note that what are called cosmic rays are not rays at all; they are high-speed sub-atomic particles coming from outer space.

Until recently, astronomers had to depend solely upon visible light, so that they were rather in the position of a pianist trying to play a waltz on a piano which lacks all its notes except for a few in the middle octave. Things are very different now; we can study the whole range of wavelengths, and what may be called 'invisible astronomy' has become of the utmost importance.

Radio telescopes came first. In 1931 Karl Jansky, an American radio engineer of Czech descent, was using a home-made aerial to study radio background 'static' when he found that he was picking up radiations from the Milky Way. After the end of the war Britain took the lead, and Sir Bernard Lovell master-minded the great radio

▶ ***UKIRT****. The United Kingdom Infra-Red Telescope, on the summit of Mauna Kea in Hawaii. It has a 3.8-m (150-inch) mirror. UKIRT proved to be so good that it can also be used for ordinary optical work, which was sheer bonus.*

▼ ***The Arecibo Telescope****. The largest dish radio telescope in the world, it was completed in 1963; the dish is 304.8 m (approximately 1000 feet) in diameter. However, it is not steerable; though its equipment means that it can survey wide areas of the sky.*

telescope at Jodrell Bank in Cheshire; it is a 'dish', 76 metres (250 feet) across, and is now known as the Lovell Telescope.

Just as an optical collects light, so a radio telescope collects and focuses radio waves; the name is somewhat misleading, because a radio telescope is really more in the nature of an aerial. It does not produce an optical-type picture, and one certainly cannot look through it; the usual end product is a trace on a graph. Many people have heard broadcasts of 'radio noise' from the Sun and other celestial bodies, but the actual noise is produced in the equipment itself, and is only one way of studying the radiations.

Other large dishes have been built in recent times; the largest of all, at Arecibo in Puerto Rico, is set in a natural hollow in the ground, so that it cannot be steered in the same way as the Lovell telescope or the 64-metre (210-foot) instrument at Parkes in New South Wales. Not all radio telescopes are the disk type, and some of them look like collections of poles, but all have the same basic function. Radio telescopes can be used in conjunction with each other, and there are elaborate networks, such as MERLIN (Multi-Element Radio Link Interferometer Network) in Britain. Resolution can now be obtained down to 0.001 of a second of arc, which is the apparent diameter of a cricket ball seen from a range of 16,000 kilometres (10,000 miles).

The sub-millimetre range of the electromagnetic spectrum extends from 1 millimetre down to 0.3 of a millimetre. The largest telescope designed for this region is the James Clerk Maxwell Telescope (JCMT) on Mauna Kea, which has a 15-metre (50-foot) segmented metal reflector; sub-millimetre and microwave regions extend down to the infra-red, where we merge with more 'conventional' telescopes; as we have noted, the UKIRT in Hawaii can be used either for infra-red or for visual work. The infra-red detectors have to be kept at a very low temperature, as otherwise the radiations from the sky would be swamped by those from the equipment. High altitude – the summit of Mauna Kea is over 4000 metres (14,000 feet) – is essential, because infra-red radiations are strongly absorbed by water vapour in the air.

Some ultra-violet studies can be carried out from ground level, but virtually all X-rays and most of the gamma rays are blocked by layers in the upper atmosphere, so that we have to depend upon artificial satellites and space probes. This has been possible only during the last few decades, so all these branches of 'invisible astronomy' are very young. But they have added immeasurably to our knowledge of the universe.

▲ ***The Parkes Radio Dish in New South Wales.*** *This is one of the largest 'dishes', with a diameter of 64 m (210 feet). It was completed in 1961, and is one of the oldest of the major radio telescopes, but it has remained in the forefront of research, and is part of the new 'Australia Telescope', which is a combination of several separate radio instruments.*

▼ ***The electromagnetic spectrum*** *extends far beyond what we can see with the human eye. These days, gamma-ray, X-ray and ultra-violet radiation from hotter bodies and infra-red radiation and radio waves from cooler are also studied.*

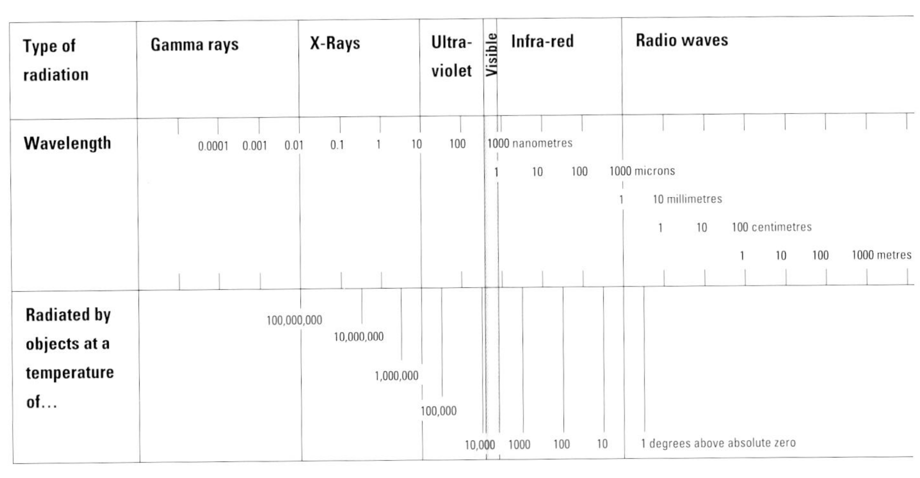

Rockets into Space

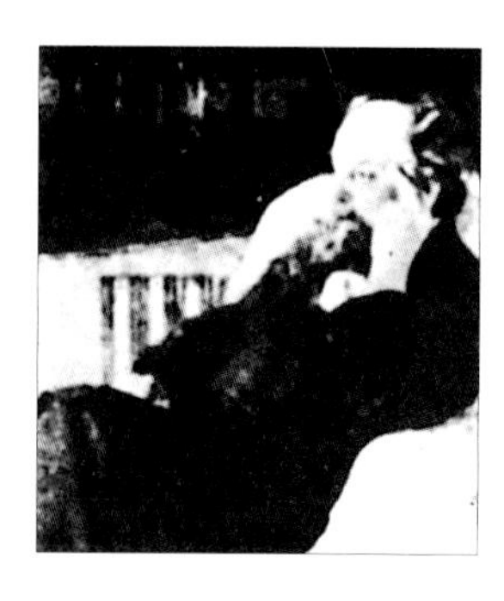

▲ ***Tsiolkovskii***. *Konstantin Eduardovich Tsiolkovskii is regarded as 'the father of space research'; it was his work which laid down the general principles of astronautics.*

▲ ***Goddard***. *Robert Hutchings Goddard, the American rocket engineer, built and flew the first liquid-propellant rocket in 1926. His work was entirely independent of that of Tsiolkovskii.*

▼ ***The V2 weapon***. *The V2 was developed during World War II by a German team, headed by Wernher von Braun.*

The idea of travelling to other worlds is far from new. As long ago as the second century AD a Greek satirist, Lucian of Samosata, wrote a story in which a party of sailors passing through the Strait of Gibraltar were caught up in a vast waterspout and hurled on to the Moon. Even Johannes Kepler wrote 'science fiction'; his hero was taken to the Moon by obliging demons! In 1865 Jules Verne published his classic novel in which the travellers were put inside a projectile and fired moonward from the barrel of a powerful gun. This would be rather uncomfortable for the intrepid crew members, quite apart from the fact that it would be a one-way journey only (though Verne cleverly avoided this difficulty in his book, which is well worth reading even today).

The first truly scientific ideas about spaceflight were due to a Russian, Konstantin Eduardovich Tsiolkovskii, whose first paper appeared in 1902 – in an obscure journal, so that it passed almost unnoticed. Tsiolkovskii knew that ordinary flying machines cannot function in airless space, but rockets can do so, because they depend upon what Isaac Newton called the principle of reaction: every action has an equal and opposite reaction. For example, consider an ordinary firework rocket of the type fired in England on Guy Fawkes' night. It consists of a hollow tube filled with gunpowder. When you 'light the blue touch paper and retire immediately' the powder starts to burn; hot gas is produced, and rushes out of the exhaust, so 'kicking' the tube in the opposite direction. As long as the gas streams out, the rocket will continue to fly.

This is all very well, but – as Tsiolkovskii realized – solid fuels are weak and unreliable. Instead, he planned a liquid-fuel rocket motor. Two liquids (for example, petrol and liquid oxygen) are forced by pumps into a combustion chamber; they react together, producing hot gas which is sent out of the exhaust and makes the rocket fly. Tsiolkovskii also suggested using a compound launcher made up of two separate rockets joined together. Initially the lower stage does all the work; when it has used up its propellant it breaks away, leaving the upper stage to continue the journey by using its own motors. In effect, the upper stage has been given a running start.

Tsiolkovskii was not a practical experimenter, and the first liquid-propellant rocket was not fired until 1926, by the American engineer Robert Hutchings Goddard (who at that time had never even heard about Tsiolkovskii's work). Goddard's rocket was modest enough, moving for

a few tens of metres at a top speed of below 100 kilometres per hour (60 miles per hour), but it was the direct ancestor of the spacecraft of today.

A few years later a German team, including Wernher von Braun, set up a 'rocket-flying field' outside Berlin and began experimenting. They made progress, and the Nazi Government stepped in, transferring the rocket workers to Peenemünde, an island in the Baltic, and ordering them to produce military weapons. The result was the V2, used to bombard England in the last stages of the war (1944–5). Subsequently, von Braun and many other Peenemünde scientists went to America, and were largely responsible for the launching of the first United States artificial satellite, Explorer 1, in 1958. But by then the Russians had already ushered in the Space Age. On 4 October 1957 they sent up the first of all man-made moons, Sputnik 1, which carried little on board apart from a radio transmitter, but which marked the beginning of a new era.

Remarkable progress has been made since 1957. Artificial satellites and space stations have been put into orbit; men have reached the Moon; unmanned probes have been sent past all the planets apart from Pluto, and controlled landings have been made on the surfaces of Mars and Venus. Yet there are still people who question the value of space research. They forget – or choose to ignore – the very real benefits to meteorology, physics, chemistry, medical research and many other branches of science, quite apart from the practical value of modern communications satellites. Moreover, space research is truly international.

Principle of the rocket

The liquid-propellant rocket uses a 'fuel' and an 'oxidant'; these are forced into a combustion chamber, where they react together, burning the fuel. The gas produced is sent out from the exhaust; and as long as gas continues to stream out, so the rocket will continue to fly. It does not depend upon having atmosphere around it, and is at its best in outer space, where there is no air-resistance.

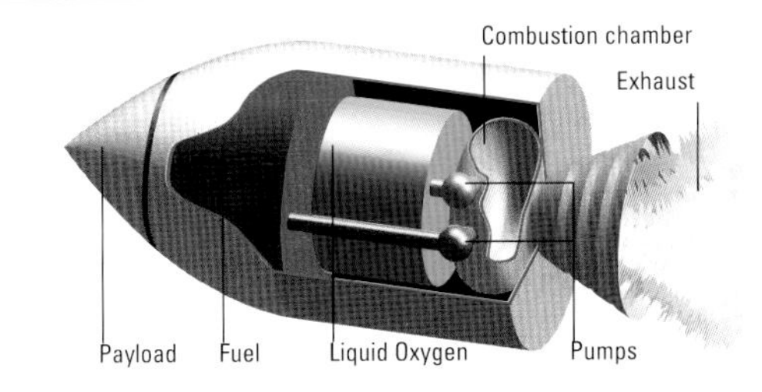

◀ ***Launch of Ulysses.*** *Ulysses, the spacecraft designed to survey the poles of the Sun, was launched from Cape Canaveral on 6 October 1990; the probe itself was made in Europe. The photograph here shows the smoke trail left by the departing spacecraft.*

▶ ***Russian rocket launch 1991.*** *This photograph was taken from Baikonur, the Russian equivalent of Cape Canaveral. It shows a Progress unmanned rocket just before launch; it was sent as a supply vehicle to the orbiting Mir space station.*

Satellites and Space Probes

▲ ***Sputnik 1**. Launched on 4 October 1957, by the Russians; this was the first artificial satellite, and marked the opening of the Space Age. It orbited the Earth until January 1958, when it burned up.*

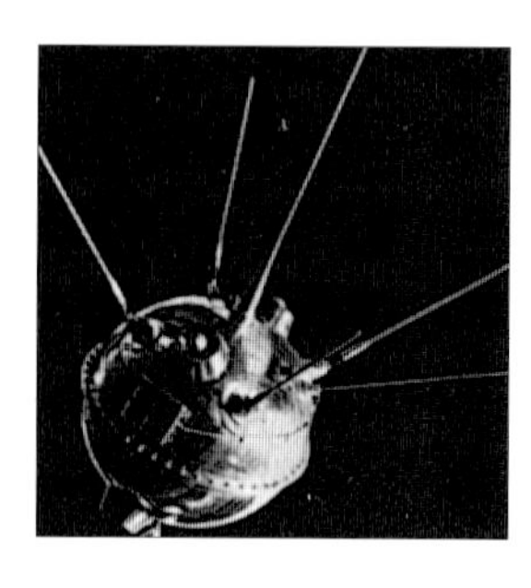

▲ ***Lunik 1 (or Luna 1)**. This was the first space probe to pass by the Moon. It was launched by the Russians on 2 January 1959, and bypassed the Moon at a range of 5955 km (3700 miles) on 4 January.*

▼ ***ROSAT** – the Röntgen satellite. It provided a link between studies of the sky in X-radiation and in EUV (Extreme Ultra-Violet); it carried a German X-ray telescope and also a British wide-field camera.*

If an artificial satellite is to be put into a closed path round the Earth, it must attain 'orbital velocity', which means that it must be launched by a powerful rocket; the main American launching ground is at Cape Canaveral in Florida, while most of the Russian launches have been from Baikonur in what is now Kazakstan. If the satellite remains sufficiently high above the main part of the atmosphere it will be permanent, and will behave in the same way as a natural astronomical body, obeying Kepler's Laws; but if any part of its orbit brings it into the denser air, it will eventually fall back and burn away by friction. This was the fate of the first satellite, Sputnik 1, which decayed during the first week of January 1958. However, many other satellites will never come down – for example Telstar, the first communications vehicle, which was launched in 1962 and is presumably still orbiting, silent and unseen, at an altitude of up to 5000 kilometres (3000 miles).

Communications satellites are invaluable in the modern world, as we all know. Without them, there could be no direct television links between the continents. Purely scientific satellites are of many kinds, and are used for many different programmes; thus the International Ultra-violet Explorer (IUE) has surveyed the entire sky at ultra-violet wavelengths and operated until 1997, while the Infra-Red Astronomical Satellite (IRAS) carried out a full infra-red survey during 1983. There are X-ray satellites, cosmic-ray vehicles and long-wavelength vehicles, but there are also many satellites designed for military purposes – something which true scientists profoundly regret.

To leave the Earth permanently a probe must reach the escape velocity of 11.2 kilometres per second (7 miles per second). Obviously the first target had to be the Moon, because it is so close, and the first successful attempts were made by the Russians in 1959. Lunik 1 bypassed the Moon, Lunik 2 crash-landed there, and Lunik 3 went on a 'round trip' sending back the first pictures of the far side of the Moon which can never be seen from Earth because it is always turned away from us. During the 1960s controlled landings were made by both Russian and American vehicles, and the United States Orbiters circled the Moon, sending back detailed photographs of the entire surface and paving the way for the manned landings in 1969.

Contacting the planets is much more of a problem, because of the increased distances involved and because the planets do not stay conveniently close to us. The first successful interplanetary vehicle was Mariner 2, which bypassed Venus in 1962; three years later Mariner 4 sent

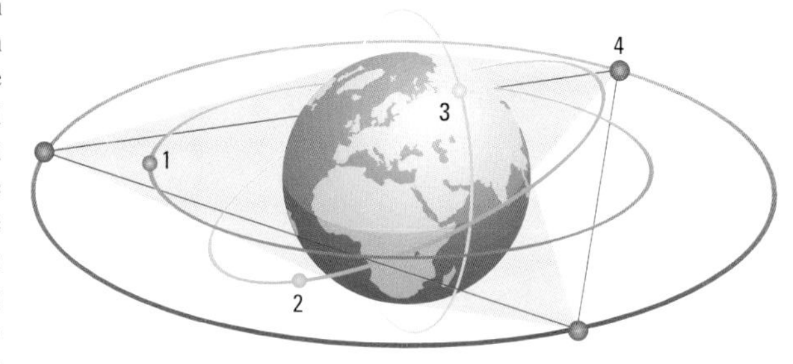

▲ ***Satellites** can orbit the Earth in the plane of the equator (1) or in inclined orbits (2). Polar orbiting satellites (3) require less powerful rockets than those in geostationary orbits (4), which need to be much higher at 36,000 km (22,500 miles) above the Earth.*

back the first close-range photographs of Mars. During the 1970s controlled landings were made on Mars and Venus, and Mariner 10 made the first rendezvous with the inner planet Mercury. Next came the missions to the outer planets, first with Pioneers 10 and 11, and then with the two Voyagers. Pride of place must go to Voyager 2, which was launched in 1977 and bypassed all four giants – Jupiter (1979), Saturn (1981), Uranus (1986) and finally Neptune (1989). This was possible because the planets were strung out in a curve, so that the gravity of one could be used to send Voyager on to a rendezvous with its next target. This situation will not recur for well over a century, so that it came just at the right moment. The Voyagers and the Pioneers will never return; they are leaving the Solar System for ever, and once we lose contact with them we will never know their fate. (In case any alien civilization finds them, they carry pictures and identification tapes, though one has to admit that the chances of their being found do not seem to be very high.) Neither must we forget the 'armada' to Halley's Comet in 1986, when no less than five separate satellites were launched in a co-ordinated scientific effort. The British-built Giotto went right into the comet's head and sent back close-range pictures of the icy nucleus.

Other planetary missions were launched during the 1990s, including several landings on Mars. Finance is always a problem, and several very interesting and important missions have had to be postponed or cancelled, but a great deal has been learned, and we now know more about our neighbour worlds than would have seemed possible in October 1957, when the Space Age began so suddenly.

▼ ***IUE***. *The International Ultra-violet Explorer, launched on 26 January 1978, operated until 1997, though its planned life expectancy was only three years! It has carried out a full survey of the sky at ultra-violet wavelengths, and has actually provided material for more research papers than any other satellite.*

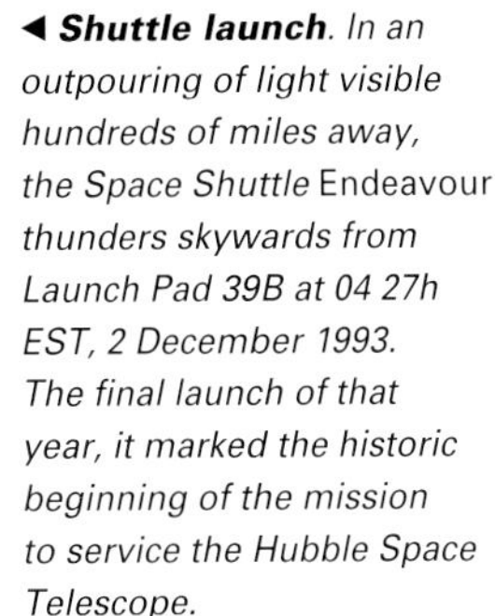

◀ ***Shuttle launch***. *In an outpouring of light visible hundreds of miles away, the Space Shuttle* Endeavour *thunders skywards from Launch Pad 39B at 04 27h EST, 2 December 1993. The final launch of that year, it marked the historic beginning of the mission to service the Hubble Space Telescope.*

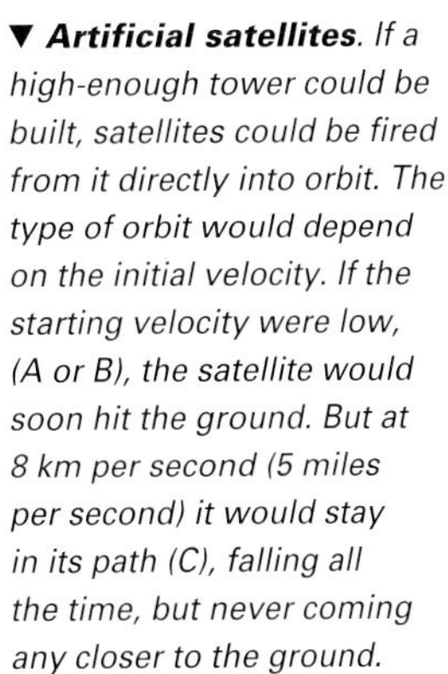

▼ ***Artificial satellites***. *If a high-enough tower could be built, satellites could be fired from it directly into orbit. The type of orbit would depend on the initial velocity. If the starting velocity were low, (A or B), the satellite would soon hit the ground. But at 8 km per second (5 miles per second) it would stay in its path (C), falling all the time, but never coming any closer to the ground.*

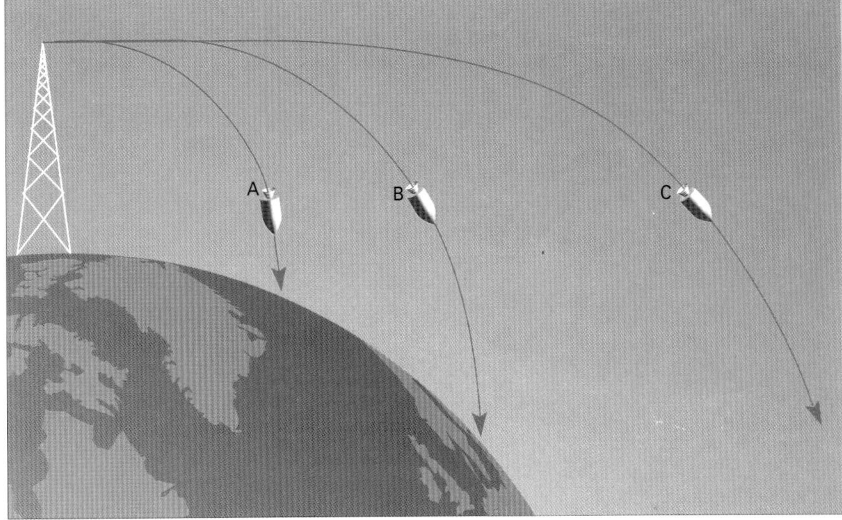

Man in Space

▲ ***Yuri Gagarin**, the first man in space; on 12 April 1961 he completed a full circuit of the Earth in Vostok 1. The maximum altitude was 327 km (203 miles), and tne flight time 1h 48m. Tragically, Gagarin later lost his life in an ordinary aircraft accident.*

▲ ***Valentina Tereshkova**, the first woman in space; she flew in Vostok 6 from 16–18 June 1963. At the same time, Vostok 5 was in orbit piloted by Valery Bykovsky. During her spaceflight she carried out an extensive research programme, and has since been active in the educational and administrative field. I took this photograph of her in 1992.*

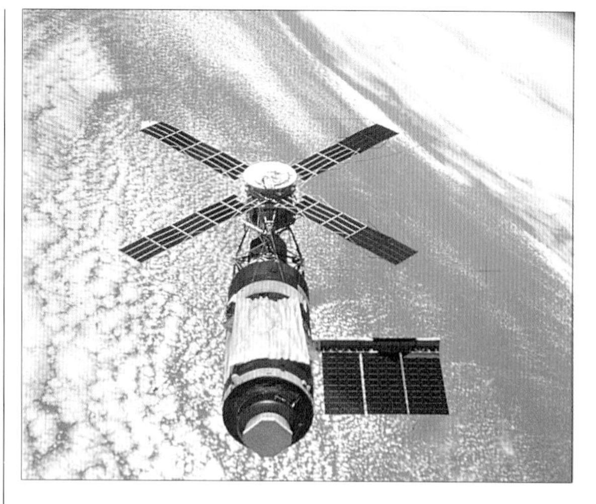

▲ ***Skylab**. The first American space station; during 1973–4 it was manned by three successive crews. It continued in orbit until 11 July 1979, when it broke up in the atmosphere.*

▶ ***Retrieval of a satellite in low Earth orbit**. Shuttle astronaut Joseph P. Allen is holding on to the Palapa B-2 communications satellite, with Dale Gardener (lower right) assisting.*

Manned spaceflight began on 12 April 1961, when Major Yuri Gagarin of the Soviet Air Force was launched in the spacecraft Vostok 1 and made a full circuit of the Earth before landing safely in the pre-arranged position. His total flight time was no more than 1 hour 40 minutes, but it was of immense significance, because it showed that true spaceflight could be achieved.

Up to that time there had been severe misgivings. For example, how would an astronaut react to the condition of weightlessness, or zero gravity? Once in orbit, all sensation of weight vanishes, because the astronaut and the spacecraft are falling in the same direction at the same rate. (Lie a coin on top of a book, and drop both to the floor; during the descent the coin will not press on the book – with reference to the book, it will be weightless.) In fact, zero gravity did not prove to be at all uncomfortable, and neither was there any obvious danger from cosmic radiation or meteoritic particles. The stage was set for further flights, and they were not long delayed. Following a brief sub-orbital 'hop' by Commander Alan Shepard, Colonel John Glenn became the first American to make an orbit of the Earth; the date was 20 February 1962, and the total flight time was almost five hours.

During the 1960s many missions were undertaken; there were space links, space dockings and space walks, the first of which was undertaken by Alexei Leonov in 1965. The climax came in 1969, when Neil Armstrong and Buzz Aldrin made the initial landing on the Moon in the lunar module of Apollo 11. Other missions followed, and the Apollo programme continued until the end of 1972. It was wise to call a halt at that point; there was no provision for rescue in the event of a faulty landing, and the consequences would have been tragic.

There have in fact been eleven deaths during the space missions (plus a ground disaster, when three Apollo astronauts were killed during a test). This is far fewer than the number of people who lost their lives during the early days of aeronautics, but it is a timely reminder that space is a dangerous environment.

Space stations have been set up; during 1973–4 America's Skylab was manned by three successive crews, and since then there have been long-duration Russian stations such as Mir, launched in 1986. With the abrupt collapse of the USSR the whole situation has changed, and there has been serious talk of combining the American and Russian space-station programmes, though whether this will happen remains to be seen.

One vital development has been the Space Shuttle, which is a 'reusable' vehicle. The first flight was made in 1981, and many have been undertaken since. There have been notable triumphs, such as the repair of the Hubble Space Telescope by astronauts in December 1993, but there has also been the *Challenger* disaster of 1986, when seven astronauts died; quite apart from the human tragedy,

▲ ***The first American space walk***. *Major Edward White remained outside Gemini 4 for 21 minutes on 3 June 1965. Tragically White later lost his life in a capsule fire on the ground.*

► ***The rendezvous of Geminis 6 and 7***. *Walter Schirra and Thomas Stafford (Gemini 6) met up with Frank Borman and James Lovell (Gemini 7) on 4 December 1965. It is easy to understand why the open 'jaws' were likened to an angry alligator.*

▲ ***One small step for a man***. *Neil Armstrong stepped from Apollo 11's lunar module* Eagle *into the history books when he became the first man on the Moon in July 1969.*

the accident delayed the advance of manned spaceflight for several years. The situation has been further affected by the Russians' shortage of money.

However, the International Space Station has already been started; the first unit – due to the Russians – was launched in 1998, followed in 1999 by the first American component. This is highly encouraging, and it is clear that all future missions must be on an international basis.

The value of manned space missions has often been questioned, and it has been claimed that they are unnecessary, since all the work needed can be carried out by unpiloted vehicles. Yet if we are to continue with the exploration of space, it does seem that both types of programme are needed.

The idea of a permanent Lunar Base no longer seems far-fetched; it should be established fairly early in the 21st century. Mars is presumably next on the list, and if all goes well the first expedition there should set out within the next few decades; it is quite likely that 'the first man on Mars' has already been born. It is worth remembering that there could have been a meeting between the first airman (Orville Wright) and the first man on the Moon (Neil Armstrong). Their lives overlapped, and I have had the pleasure of knowing both!

The Hubble Space Telescope

▶ ***The HST mirror.*** *The mirror of the Hubble Space Telescope was perfectly made – but to the wrong curve! Human error resulted in an unacceptable amount of spherical aberration. Fortunately, the 1993 repair mission restored the situation.*

One of the most ambitious experiments in the history of science began on 24 April 1990, with the launch of the Hubble Space Telescope (HST) – named in honour of the American astronomer Edwin Hubble, who was the first to prove that the objects once called spiral nebulae are independent star-systems. The HST is a reflector with a 2.4-metre (94-inch) mirror; it is 13 metres (43 feet) long, and weighs 11,000 kilograms (24,200 pounds). It was launched in the Space Shuttle *Discovery*, and put into a near-circular orbit which takes it round the Earth in a period of 94 minutes at a distance of almost 600 kilometres (370 miles).

It is an American project, controlled by NASA, but with strong support from the European Space Agency; the solar panels, which provide the power for the instruments, were made by British Aerospace in Bristol. Five main instruments are carried, of which the most important are probably the Wide Field and Planetary Camera (WFPC) and the Faint Object Camera (FOC). Operating under conditions of perfect seeing, high above the atmosphere, the HST was expected to far outmatch any Earth-based telescope, even though its mirror is so much smaller than that of instruments such as the Keck.

The first images were received on 20 May 1990, and it was at once plain that the results would indeed be superb; the HST can 'see' more than any ground-based instrument could hope to do. Moreover, its range extends from visible light well into the ultra-violet. Yet there was also an unwelcome discovery. The mirror had been wrongly made, and was of an incorrect shape; it was too 'shallow' a curve. The error was tiny – no more than 0.002 of a millimetre – but it was enough to produce what is termed spherical aberration. Images were blurred, and it was said, rather unkindly, that the telescope was short-sighted. Some of the original programmes had to be modified or even abandoned.

Regular servicing missions had been planned during the estimated operating time of fifteen years. The first of these was undertaken in December 1993 by a team of astronauts sent up in the Space Shuttle *Endeavour*. They 'captured' the telescope, brought it into the Shuttle bay, and carried out extensive repairs and maintenance before putting it back into orbit. The WFPC was replaced, and extra optical equipment was introduced to compensate for the error in the main mirror.

A second servicing mission was carried out in February 1997, and a third in December 1999; new instruments were installed, and essential repairs undertaken. The telescope is now even better than was originally planned.

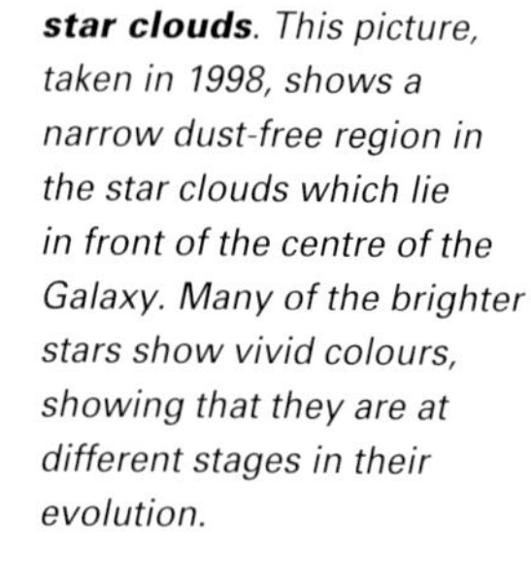

▼ ***Inside the Sagittarius star clouds.*** *This picture, taken in 1998, shows a narrow dust-free region in the star clouds which lie in front of the centre of the Galaxy. Many of the brighter stars show vivid colours, showing that they are at different stages in their evolution.*

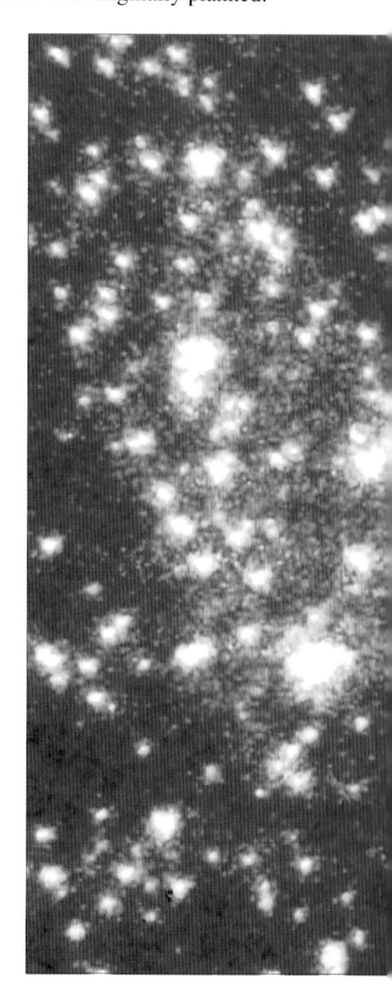

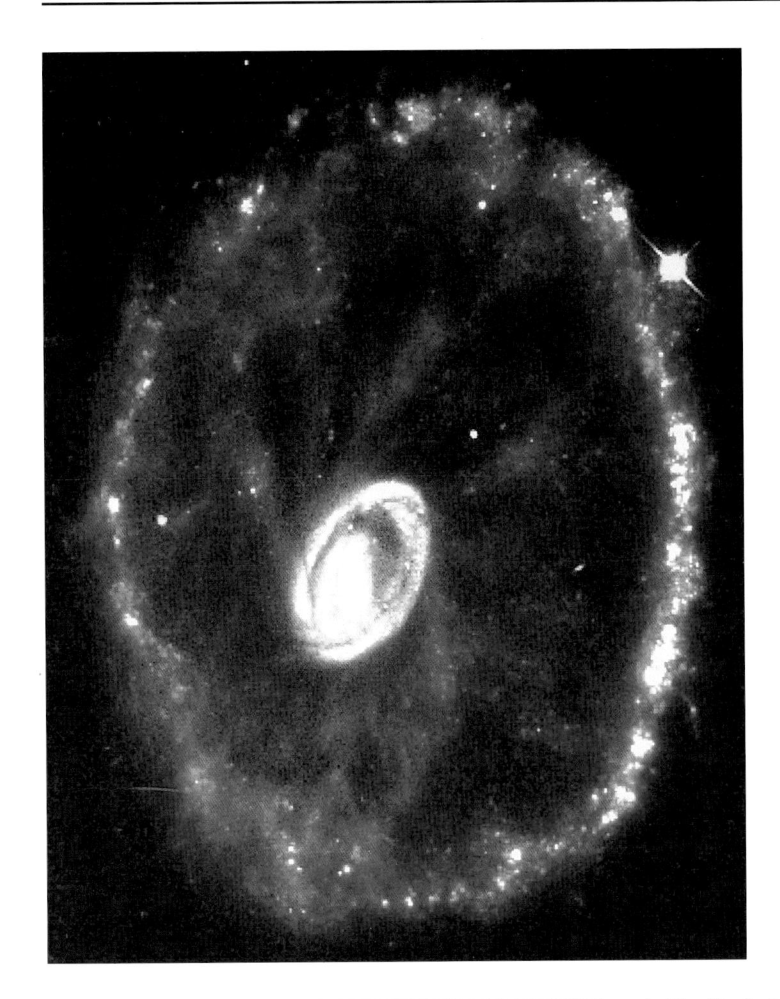

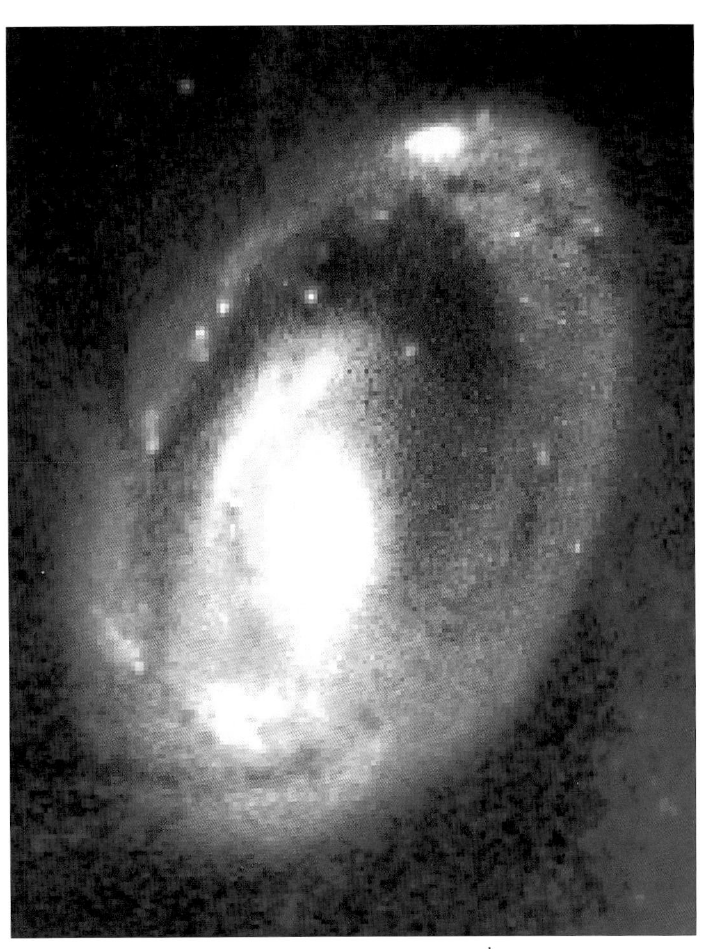

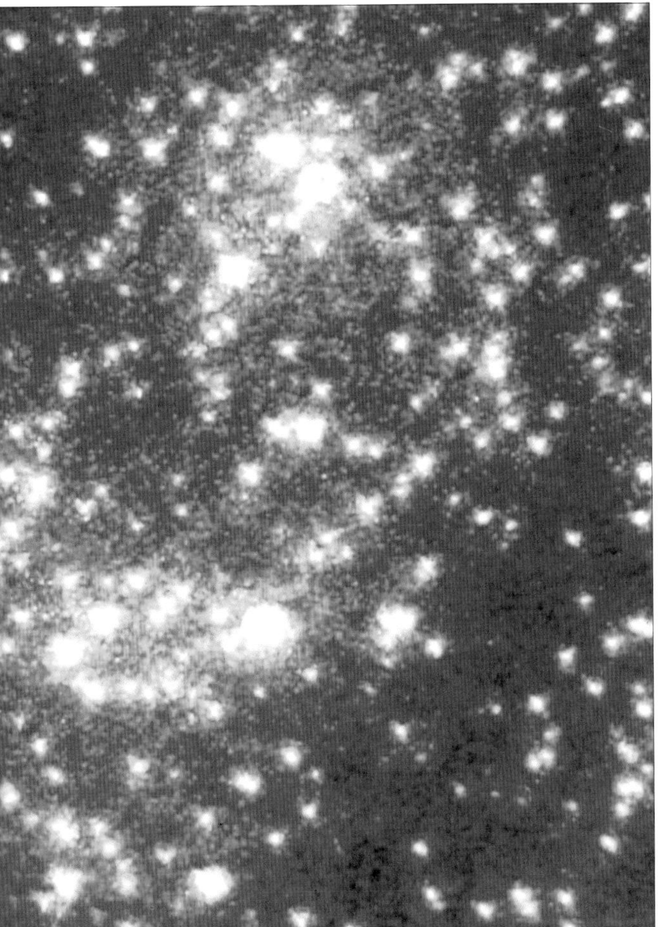

▲ ***The Cartwheel Galaxy**, a system in Sculptor, 500 million light-years away. The nucleus is the bright object in the centre of the left-hand image; the spoke-like structures are wisps of material connecting the nucleus to the outer ring of young stars. The galaxy's unusual configuration was created by an almost head-on collision with a smaller galaxy about 200 million years ago. The right-hand image shows the galaxy's nucleus in close-up.*

◀ ***The Faint Object Camera** has been able to obtain detailed pictures of star fields; here, the core of the globular cluster 47 Tucanae is resolved into individual stars.*

▶ ***The repaired HST, 1994**. During the repair mission, the faulty solar panels were replaced. This picture was taken just after the telescope was released from the Shuttle bay; the new solar panels are in place.*

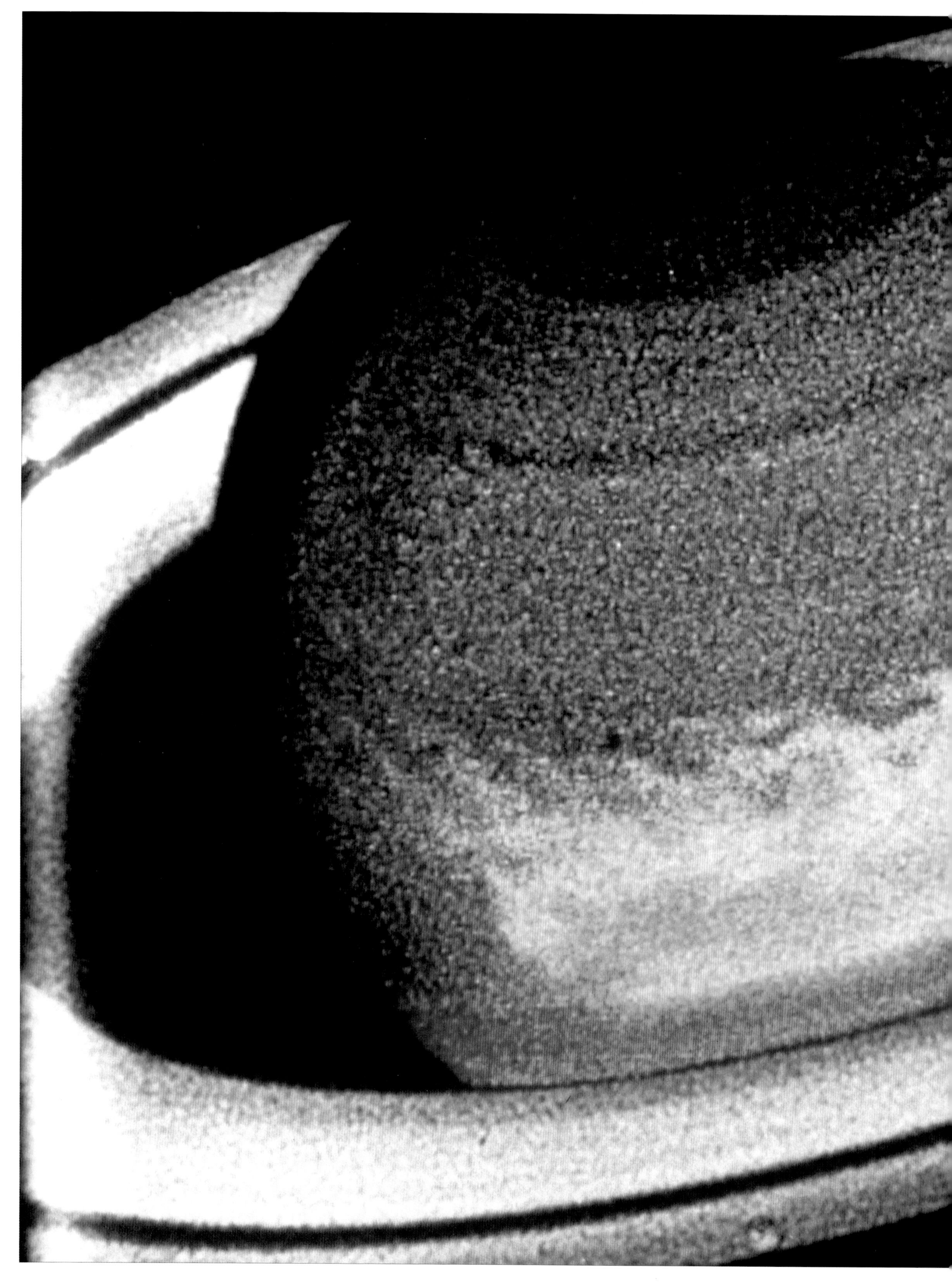

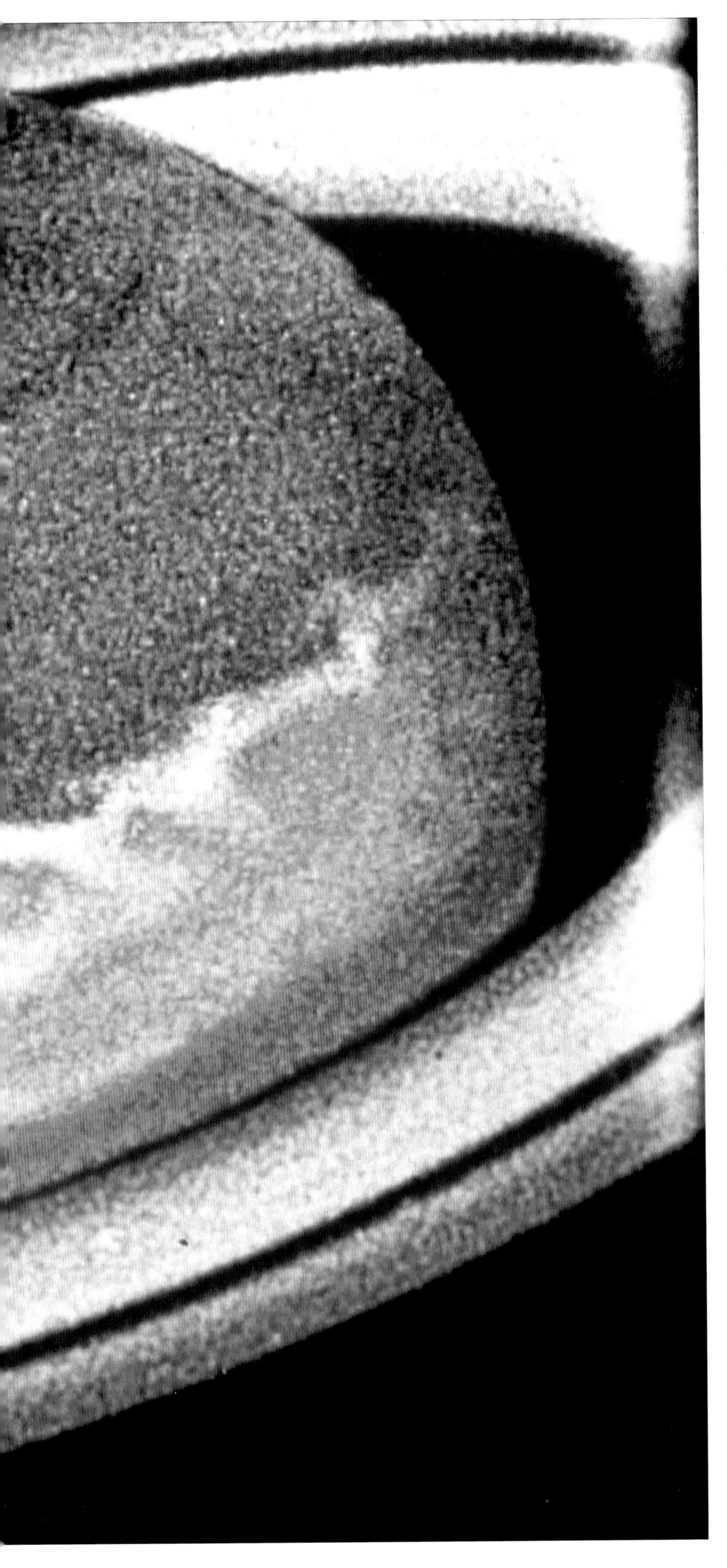

The Solar System

◄ ***The white spot on Saturn***, *taken by the Hubble Space Telescope on 9 November 1990. The spot had been originally seen in September, and by November had spread out to become a white zone. The pictures were taken in blue and in infra-red light, and were then combined to produce the image shown here; the lower parts of the cloud are in blue, and higher clouds in red.*

The Sun's Family

The Solar System is the only part of the universe which we can explore with spacecraft of the kind we can build today. It is made up of one star (the Sun), nine planets (of which the Earth comes third in order of distance), and various lesser bodies, such as the satellites, asteroids, comets and meteoroids.

The Sun is a normal star (astronomers even relegate it to the status of a dwarf), but it is the supreme controller of the Solar System, and all the other members shine by reflected sunlight. It is believed that the planets formed by accretion from a cloud of material which surrounded the youthful Sun; the age of the Earth is known to be about 4.6 thousand million years, and the Solar System itself must be rather older than this.

It is very noticeable that the Solar System is divided into two parts. First there are four small, solid planets: Mercury, Venus, Earth and Mars. Then comes a wide gap, in which move thousands of midget worlds known variously as asteroids, planetoids and minor planets. Beyond we come to the four giants: Jupiter, Saturn, Uranus and Neptune, together with a maverick world, Pluto, which is too small and lightweight to be classed as a bona-fide planet. It seems that the four inner planets lost their original light gases because of the heat of the Sun, so that they are solid and rocky; the giants, which formed in a colder region, were able to retain their lighter gases.

The Earth has one satellite: our familiar Moon, which is much the closest natural body in the sky (excluding occasional wandering asteroids). Of the other planets, Mars has two satellites, Jupiter sixteen, Saturn eighteen, Uranus twenty, Neptune eight and Pluto one, though only four of these (three in Jupiter's system and one in Saturn's) are as large as our Moon.

Comets may be spectacular, but are of very low mass compared with planets. The only solid part is the nucleus, which has been described as a 'dirty ice-ball'; when the comet nears the Sun the ices begin to evaporate, and the comet may produce a gaseous 'head' with a long tail. Bright comets have very eccentric orbits, so that they come back to the inner part of the Solar System only at intervals of many centuries, and we cannot predict them. There are many short-period comets which return regularly, but all these are faint; each time a comet passes relatively close to the Sun it loses a certain amount of material, and the short-period comets have to a great extent wasted away.

As a comet moves along it leaves a 'dusty trail' behind it. When the Earth ploughs through one of these trails it collects dusty particles, which burn away in the upper air and produce the luminous streaks which we call shooting-stars. Larger objects, which may survive the fall to the ground, are termed meteorites; they come from the asteroid belt, and are not associated either with comets or with shooting-star meteors.

There is also a great deal of thinly-spread 'dust', especially in the main plane of the Solar System. Small particles of this kind catch the sunlight, and produce the glows which we call the Zodiacal Light and the Gegenschein.

How far does the Solar System extend? This is not an easy question to answer. There may well be another planet beyond Neptune and Pluto, and it is thought that comets come from a cloud of icy objects orbiting the Sun at a distance of around one to two light-years, but we cannot be sure. The nearest star beyond the Sun is just over four light-years away, so that if we give the limit of the Solar System as being at a distance of two light-years we are probably not very far wrong.

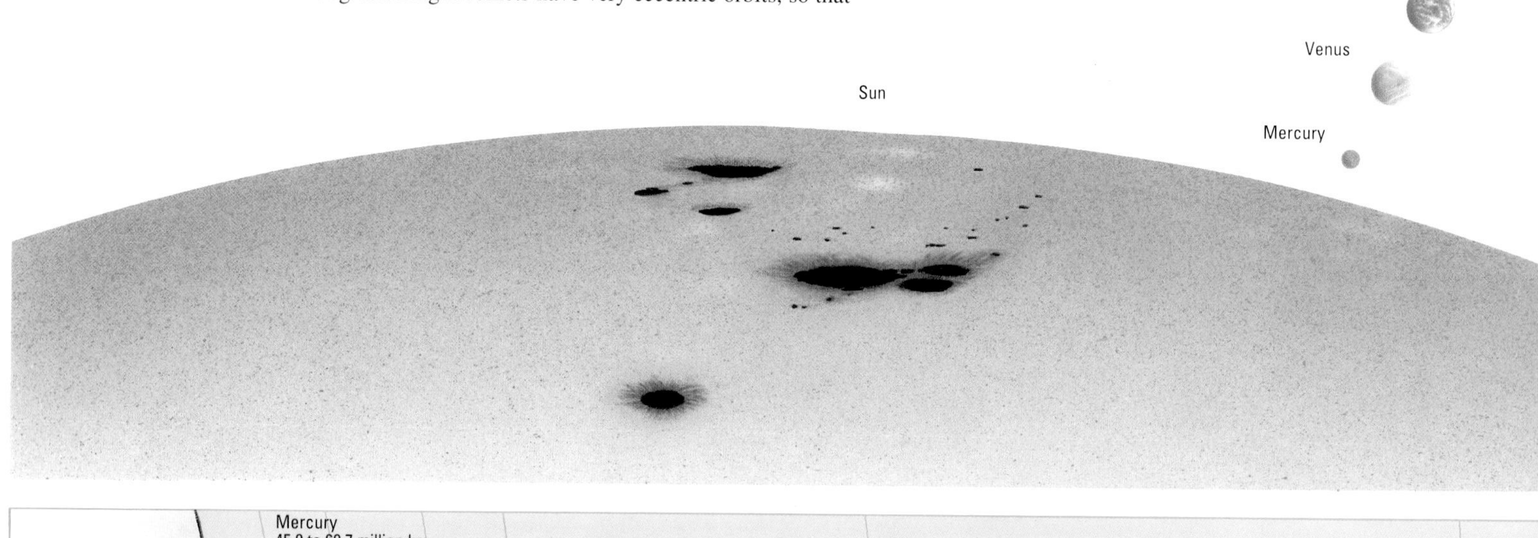

PLANETARY DATA

		Mercury	Venus	Earth	Mars	Jupiter	Saturn	Uranus	Neptune	Pluto
Distance from	**max.**	69.7	109	152	249	816	1507	3004	4537	7375
Sun, millions	**mean**	57.9	108.2	149.6	227.9	778	1427	2870	4497	5900
of km	**min.**	45.9	107.4	147	206.7	741	1347	2735	4456	4425
Orbital period		87.97d	224.7d	365.3d	687.0d	11.86y	29.46y	84.01y	164.8y	247.7y
Synodic period, days		115.9	583.92	—	779.9	398.9	378.1	369.7	367.5	366.7
Rotation period (equatorial)		58.646d	243.16d	23h 56m 04s	24h 37m 23s	9h 55m 30s	10h 13m 59s	17h 14m	16h 7m	6d 9h 17s
Orbital eccentricity		0.206	0.007	0.017	0.093	0.048	0.056	0.047	0.009	0.248
Orbital inclination, °		7.0	3.4	0	1.8	1.3	2.5	0.8	1.8	17.15
Axial inclination, °		2	178	23.4	24.0	3.0	26.4	98	28.8	122.5
Escape velocity, km/s		4.25	10.36	11.18	5.03	60.22	32.26	22.5	23.9	1.18
Mass, Earth =1		0.055	0.815	1	0.11	317.9	95.2	14.6	17.2	0.002
Volume, Earth = 1		0.056	0.86	1	0.15	1319	744	67	57	0.01
Density, water =1		5.44	5.25	5.52	3.94	1.33	0.71	1.27	1.77	2.02
Surface gravity, Earth =1		0.38	0.90	1	0.38	2.64	1.16	1.17	1.2	0.06
Surface temp., °C		+427	+480	+22	−23	−150	−180	−214	−220	−230
Albedo		0.06	0.76	0.36	0.16	0.43	0.61	0.35	0.35	0.4
Diameter, km (equatorial)		4878	12,104	12,756	6794	143,884	120,536	51,118	50,538	2324
Maximum magnitude		−1.9	−4.4	—	−2.8	−2.6	−0.3	+5.6	+7.7	+14

Uranus
2,735 to 3,004 million km

Neptune
4,456 to 4,537 million km

Pluto
4,425 to 7,375 million km

The Earth in the Solar System

▼ ***Planet Earth**, seen from the command module of the lunar spacecraft Apollo 10 in May 1969. The Earth is coming into view as the spacecraft moves out from the far side of the Moon. The lunar horizon is sharp, as there is no atmosphere to cause blurring.*

Why do we live on the Earth? The answer must be: 'Because we are suited to it'. There is no other planet in the Solar System which could support Earth-type life except under very artificial conditions. Our world has the right sort of temperature, the right sort of atmosphere, a plentiful supply of water, and a climate which is to all intents and purposes stable – and has been so for a very long time.

The Earth's path round the Sun does not depart much from the circular form, and the seasons are due to the tilt of the rotational axis, which is 23½ degrees to the perpendicular. We are actually closer to the Sun in December, when it is winter in the northern hemisphere, than in June – but the difference in distance is not really significant, and the greater amount of water south of the equator tends to stabilize the temperature.

The axial inclination varies to some extent, because the Earth is not a perfect sphere; the equatorial diameter is 12,756 kilometres (7927 miles), the polar diameter only 12,714 kilometres (7901 miles) – in fact, the equator bulges out slightly. The Sun and Moon pull on this bulge, and the result is that over a period of 25,800 years the axis sweeps out a cone of angular radius about 23°26′ around the perpendicular to the plane of the Earth's orbit. Because of this effect – termed precession – the positions of the celestial poles change. At the time when the Egyptian Pyramids were built, the north pole star was Thuban in the constellation of Draco; today we have Polaris in Ursa Minor, and in 12,000 years from now the pole star of the northern hemisphere will be the brilliant Vega, in Lyra.

We have found out a great deal about the history of the Earth. Its original atmosphere was stripped away, and was replaced by a secondary atmosphere which leaked out from inside the globe. At first this new atmosphere contained much more carbon dioxide and much less free oxygen than it does now, so that we would have been quite unable to breathe it. Life began in the sea; when plants spread on to the lands, around 430 million years ago, they removed much of the carbon dioxide by the process known as photosynthesis, replacing it with oxygen.

Life was slow to develop, as we know from studies of fossils; we can build up a more or less complete geological record, and it has been found that there were several great 'extinctions', when many life-forms died out. One of these occurred about 65 million years ago, when the dinosaurs became extinct – for reasons which are still not clear, though it has been suggested that the cause was a major climatic change due to the impact of a large asteroid. In any case, man is a newcomer to the terrestrial scene. If we give a time-scale in which the total age of the Earth is represented by one year, the first true men will not appear until 11pm on 31 December.

Throughout Earth history there have been various cold spells or Ice Ages, the last of which ended only 10,000 years ago. In fact, the last Ice Age was not a period of continuous glaciation; there were several cold spells interrupted by warmer periods, or 'interglacials', and it is by no means certain that we are not at the moment simply in the middle of an interglacial. The reasons for the Ice Ages is not definitely known, and may be somewhat complex, but we have to remember that even though the Sun is a steady, well-behaved star its output is not absolutely constant; in historical times there have been marked fluctuations – for example, the so-called 'little ice age' between 1645 and 1715, when the Sun was almost free of spots and Europe, at least, was decidedly colder than it is at the present moment.

Neither can the Earth exist for ever. Eventually the Sun will change; it will swell out to become a giant star, and the Earth will certainly be destroyed. Luckily there is no immediate cause for alarm. The crisis will not be upon us for several thousands of millions of years yet, and it is probably true to say that the main danger to the continued existence of life on Earth comes from ourselves.

Earth's history is divided into different 'eras', which are subdivided into 'periods'. The most recent periods are themselves subdivided into 'epochs'. The main divisions and subdivisions are shown on the table opposite.

◀ ***The Great Pyramid of Cheops***. *Although it is essentially a tomb, the Great Pyramid is astronomically aligned with the position of the north pole of the sky at the time of the ancient Egyptians. Due to precession the position of the pole changes, describing a circle in a period of 25,800 years.*

▶ ***The seasons*** *are due not to the Earth's changing distance from the Sun, but to the fact that the axis of rotation is inclined at 23$\frac{1}{2}$° to the perpendicular to the plane of the Earth's orbit around the Sun. During northern summer, the northern hemisphere is inclined towards the Sun; during southern summer it is the turn of the southern hemisphere. The Sun crosses the celestial equator around 22 March (vernal equinox – the Sun moving from south to north) and 22 September (autumnal equinox – the Sun moving from north to south). The solstices are the times when the Sun is at its furthest from the equator of the sky. The dates of the eqinoxes and solstices are not quite constant, owing to the vagaries of our calendar.*

Spring Northern Hemisphere
Autumn Southern Hemisphere
Winter Northern Hemisphere
Summer Southern Hemisphere
Sun
Summer Northern Hemisphere
Winter Southern Hemisphere
Autumn Northern Hemisphere
Spring Southern Hemisphere

PERIODS IN EARTH'S HISTORY

	Began (million years ago)	Ended (million years ago)	
PRE-CAMBRIAN ERA			
Archæan	3800	2500	Start of life
Proterozoic	2500	590	Life in the seas
PALAEOZOIC ERA			
Cambrian	590	505	Sea life
Ordovician	505	438	First fishes
Silurian	438	408	First land plants
Devonian	408	360	Amphibians
Carboniferous	360	286	First reptiles
Permian	286	248	Spread of reptiles
MESOZOIC ERA			
Triassic	248	213	Reptiles and early mammals
Jurassic	213	144	Age of dinosaurs
Cretaceous	144	65	Dinosaurs, dying out at the end
CENOZOIC ERA			
Tertiary Period			
Palaeocene	65	55	Large mammals
Eocene	55	38	Primates begin
Oligocene	38	25	Development of primates
Miocene	25	5	Modern-type animals
Pliocene	5	2	Ape-men
Quaternary Period			
Pleistocene	2	0.01	Ice Ages. True men
Holocene	0.01	Present	Modern men

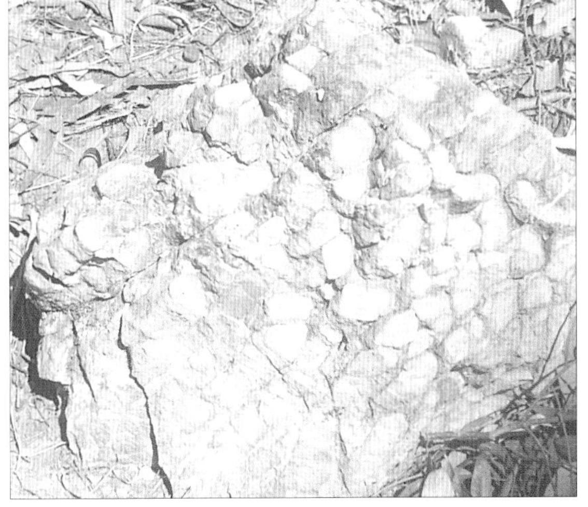

◀ ***Stromatolites****, Australia, 1993. These are made up of calcium carbonate, precipitated or accumulated by blue-green algae. They date back for at least 3,500,000 years, and are among the oldest examples of living organisms.*

The Earth as a Planet

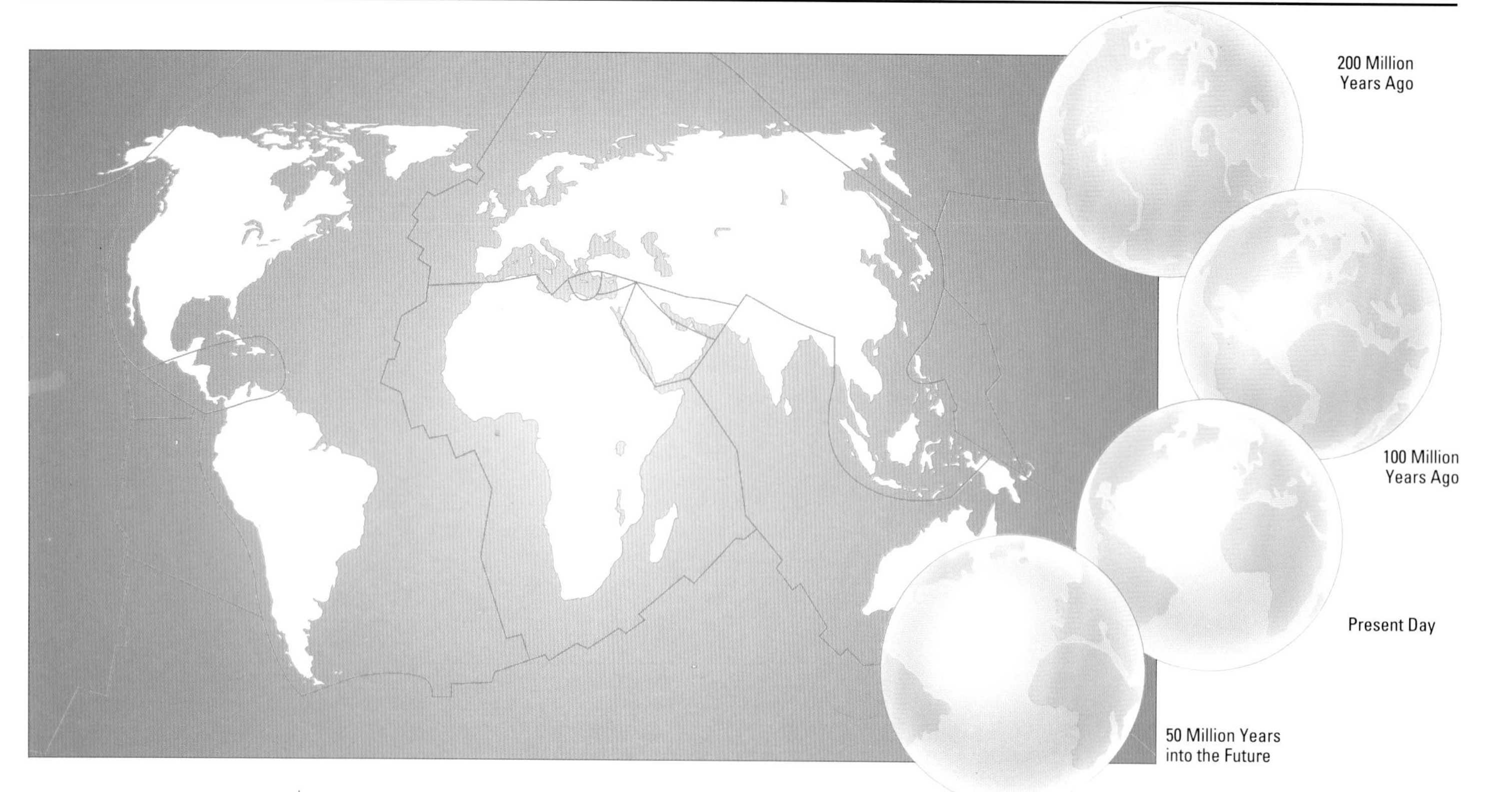

▲ ***The Earth's crust*** *is divided into six large tectonic plates and a number of minor ones. They are separated by mid-ocean ridges, deep-sea trenches, active mountain belts and fault zones. Volcanic eruptions and earthquakes are largely confined to the areas where plates meet. During the geological history of the Earth, these plates have moved around, creating and re-creating continents.*

The Earth's crust, on which we live, does not extend down very far – some 10 kilometres (6 miles) below the oceans and 50 kilometres (30 miles) below the continents. Temperature increases with depth, and at the bottom of the world's deepest mines, those in South Africa, the temperature rises to 55 degrees C. Below the crust we come to the mantle, where the solid rocks behave as though plastic. The mantle extends down to 2900 kilometres (1800 miles), and then we come to the iron-rich liquid core. Inside this is the solid core, which accounts for only 1.7 per cent of the Earth's mass and has been said to 'float' in the liquid. The central temperature is thought to be 4000–5000 degrees C.

A glance at a world map shows that if the continents could be cut out in the manner of a jigsaw puzzle, they would fit neatly together. For example, the bulge on the east coast of South America fits into the hollow of west Africa. This led the Austrian scientist Alfred Wegener to suggest that the continents were once joined together, and have now drifted apart. His ideas were ridiculed for many years, but the concept of 'continental drift' is now well established, and has led on to the relatively young science of plate tectonics.

The Earth's crust and the upper part of the mantle (which we call the lithosphere) is divided into well-marked plates. When plates are moving apart, hot mantle material rises up between them to form new oceanic crust. When plates collide, one plate may be forced beneath another – a process known as subduction – or they may buckle and force up mountain ranges. Regions where the tectonic plates meet are subject to earthquakes and volcanic activity, and it is from earthquake waves that we have drawn much of our knowledge of the Earth's internal constitution.

The point on the Earth's surface vertically above the origin or 'focus' of an earthquake is termed the epicentre. Several types of waves are set up in the globe. First there are the P or primary waves, which are waves of compression and are often termed 'push' waves; there are also S or secondary waves, which are also called 'shake-waves' because they may be likened to the waves set up in a mat when it is shaken by one end. Finally there are the L or long waves, which travel round the Earth's surface and cause most of the damage. The P waves can travel through liquid, but the S waves cannot, and by studying how they are transmitted through the Earth it has been possible to measure the size of the Earth's liquid core.

If earthquakes can be destructive, then so can volcanoes, which have been called 'the Earth's safety valves'. The mantle, below the crust, contains pockets of magma (hot, fluid rock), and above a weak point in the crust the magma may force its way through, building up a volcano. When the magma reaches the surface it solidifies and cools, to become lava. Hawaii provides perhaps the best example of long-continued vulcanism. On the main island there are two massive shield volcanoes, Mauna Kea and Mauna Loa, which are actually loftier than Everest, though they do not rise so high above the surface because, instead of rising above the land, they have their roots deep in the ocean-bed. Because the crust is shifting over the mantle, Mauna Kea has moved away from the 'hot spot' and has become extinct – at least, one hopes so, because one of the world's major observatories has been built upon its summit. Mauna Loa now stands over the 'hot spot', and is very active indeed, though in time it too will be carried away and will cease to erupt.

Other volcanoes, such as Vesuvius in Italy, are cone-shaped. The magma forces its way up through a vent, and if this vent is blocked the pressure may build up until there is a violent explosion – as happened in AD 79, when the Roman cities of Pompeii and Herculaneum were destroyed. There have been many devastating volcanic eruptions, one of the latest being that of Mount Pinotubo in the Philippines, which sent vast quantities of dust and ash into the upper atmosphere.

The Earth is not the only volcanic world in the Solar System. There are constant eruptions upon Io, one of the satellites of Jupiter; there are probably active volcanoes on Venus, and we cannot be certain that all the Martian volcanoes are extinct. However, it does not seem that plate tectonics can operate upon any other planet or satellite, so that in this respect the Earth is unique in our experience.

Volcanoes
Volcanoes form where tectonic plates meet. Pockets of magma force themselves up from the mantle through weak points in the crust. The molten magma may bubble inside the crater or give off clouds of ash and gas. Magma may also find its way to the surface via side vents. A volcano may be inactive for a considerable time, allowing the magma to solidify near the surface. Huge pressure can then build up beneath it, often with devastating results.

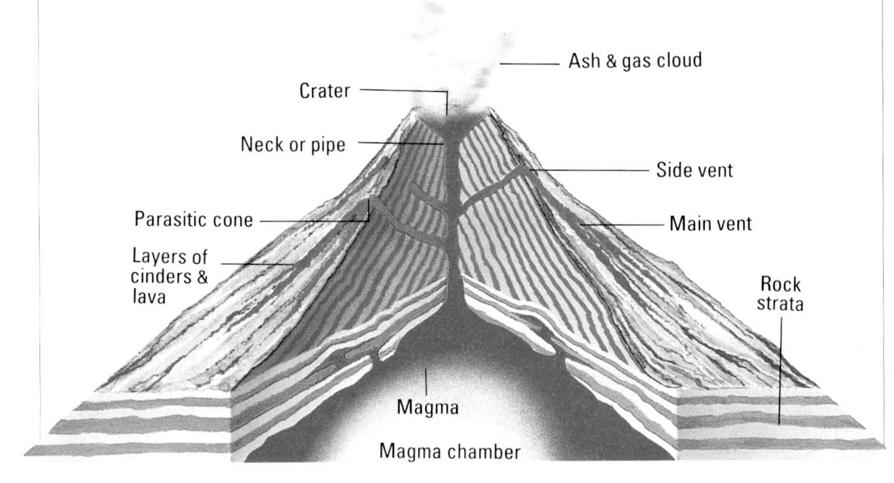

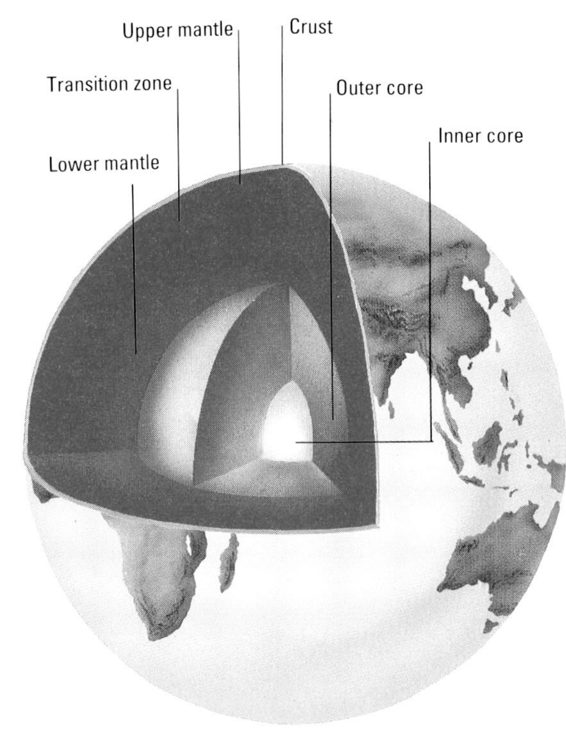

▲ ***Seismic activity*** *has allowed scientists to study the inner structure of the Earth. The crust is only on average 10 km (6 miles) thick beneath the oceans and 50 km (30 miles) thick beneath the land. Below is the 2900-km-thick (1800 miles) mantle of hot, plastic rock. Inside that is an outer liquid core, 2100 km (1300 miles) thick, with a solid core inside it, 2700 km (1700 miles) in diameter.*

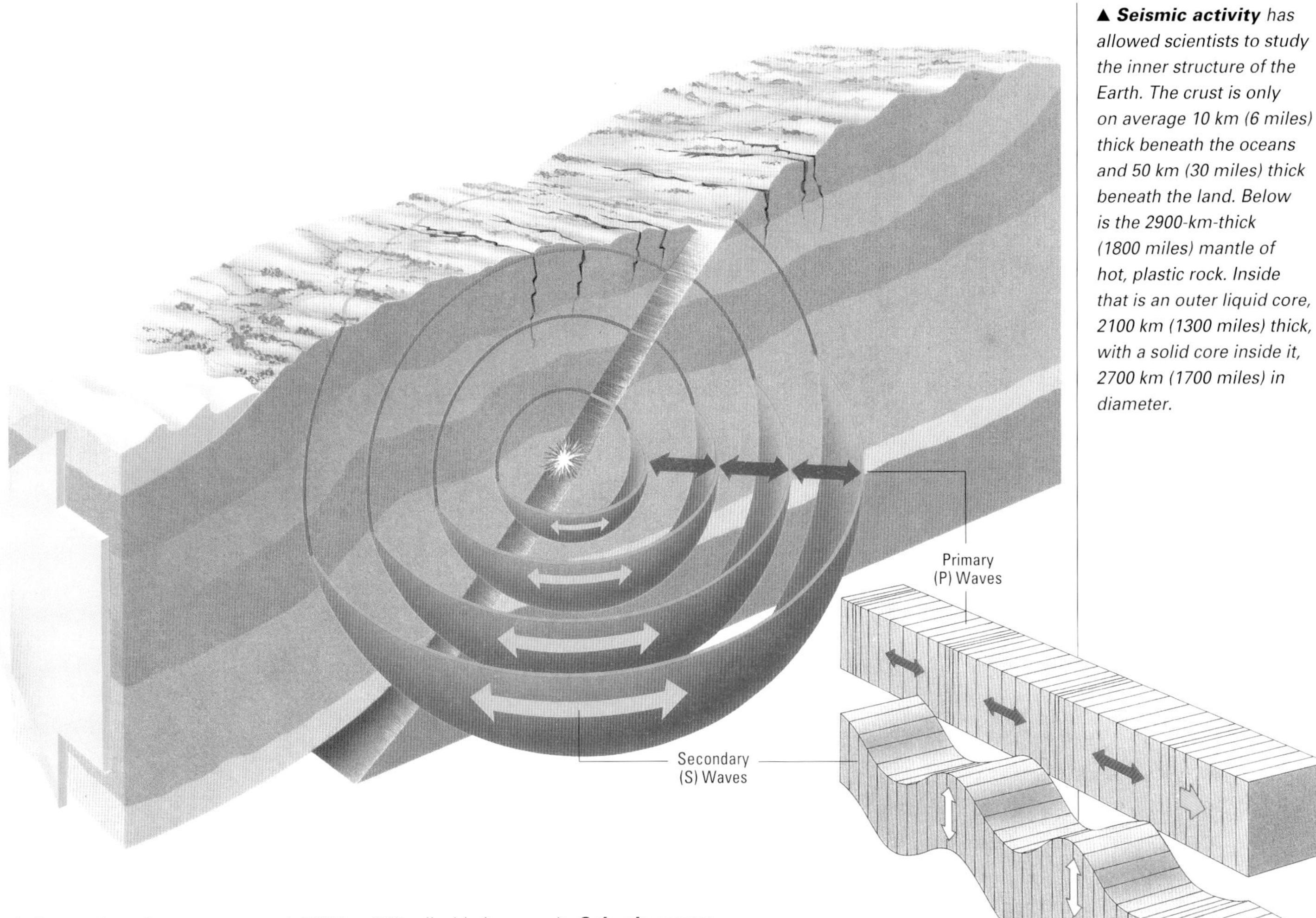

▲ ***An earthquake*** *occurs along a fault line when the crust on either side is being forced to move in different directions. The focus, where the fault gives, can be up to 700 km (450 miles) below the surface. The epicentre is the point on the surface directly above the focus where the damage is usually most severe.*

▶ ***Seismic waves.*** *P waves are compression waves that travel through solid and fluid alike. S waves are transverse waves that only travel through solids.*

The Earth's Atmosphere and Magnetosphere

▶ ***Aurora**: 21 October 1989, as seen from the Belfast area of Northern Ireland (by T. J. C. A. Moseley). This was a very widespread display, and was seen as far south as Sussex. During the 1989–91 period, when the Sun was near the maximum of its cycle of activity, there were several exceptionally brilliant aurorae.*

▼ ***The Earth's magnetosphere** is the region of space in which Earth's magnetic field is dominant. On the sunward side of the Earth, the solar wind compresses the magnetosphere to within eight to ten Earth radii (RE). On the opposite side, interaction with the solar wind draws the field lines out into a magnetotail, extending well beyond the orbit of the Moon. The boundary of the magnetosphere, across which the solar wind cannot easily flow, is the magnetopause; a bow shock is produced in the solar wind preceding the magnetopause by three to four Earth radii.*

As seen from space the Earth is truly magnificent, as we have been told by all the astronauts – particularly those who have observed it from the Moon, although it is quite impossible to see features such as the Great Wall of China, as has often been claimed! The outlines of the seas and continents show up clearly, and there are also clouds in the atmosphere, some of which cover wide areas.

The science of meteorology has benefited greatly from space research methods, because we can now study whole weather systems instead of having to rely upon reports from scattered stations. The atmosphere is made up chiefly of nitrogen (78 per cent) and oxygen (21 per cent), which does not leave much room for anything else; there is some argon, a little carbon dioxide, and traces of gases such as krypton and xenon, together with a variable amount of water vapour.

The atmosphere is divided into layers. The lowest of these, the troposphere, extends upwards for about 8 kilometres (5 miles) out to more than 17 kilometres (over 10 miles) – it is deepest over the equator. It is here that we find clouds and weather. The temperature falls with increasing height, and at the top of the layer has dropped to −44 degrees C; the density is, of course, very low.

Above the troposphere comes the stratosphere, which extends up to about 50 kilometres (30 miles). Surprisingly, the temperature does not continue to fall; indeed it actually rises, reaching +15 degrees C at the top of the layer. This is because of the presence of ozone, the molecule of which is made up of three oxygen atoms instead of the usual two; ozone is warmed by short-wave radiations from the Sun. However, the rise in temperature does not mean increased heat. Scientifically, temperature is defined by the rate at which the atoms and molecules fly around; the greater the speeds, the higher the temperature. In the stratosphere, there are so few molecules that the 'heat' is negligible. It is the 'ozone layer' which shields us from harmful radiations coming from space. Whether it is being damaged by our own activities is a matter for debate, but the situation needs to be watched.

Above the stratosphere comes the ionosphere, which extends from about 50 to 600 kilometres (30 to 370 miles); it is here that some radio waves are reflected back to the ground, making long-range communication possible. In the ionosphere we find the lovely noctilucent clouds, which are quite unlike ordinary clouds, and may possibly be due to water droplets condensing as ice on to meteoritic particles; their average height is around 80 kilometres (50 miles). The ionosphere is often divided into the mesosphere, up to 80 kilometres (50 miles), and the thermosphere, up to 200 kilometres (125 miles). Beyond comes the exosphere, which has no definite boundary, but simply thins out until the density is no more than that of the interplanetary medium. There is also the Earth's geocorona, a halo of hydrogen gas which extends out to about 95,000 kilometres (60,000 miles).

Aurorae, or polar lights – aurora borealis in the northern hemisphere, aurora australis in the southern – are also found in the ionosphere; the usual limits are from 100 to 700 kilometres (60 to 440 miles), though these limits may sometimes be exceeded. Aurorae are seen in various forms: glows, rays, bands, draperies, curtains and 'flaming patches'. They change very rapidly, and can be

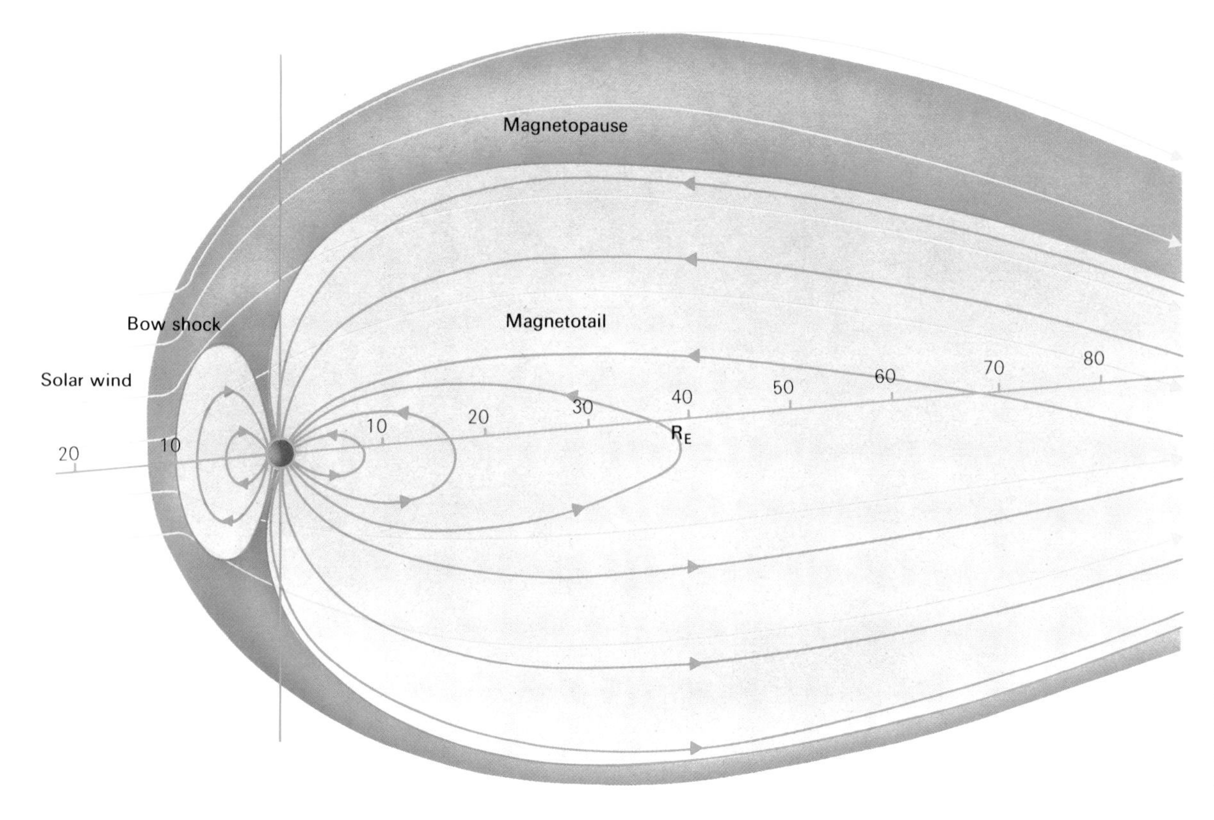

▶ ***The Earth's atmosphere** consists of the troposphere, extending from ground level to a height of between 8 and 17 km (5–10 miles); the stratosphere extends up to around 50 km (30 miles); the mesosphere, between 50 and around 80 km (50 miles); the thermosphere from around 80 up to 200 km (125 miles); beyond this height lies the exosphere.*

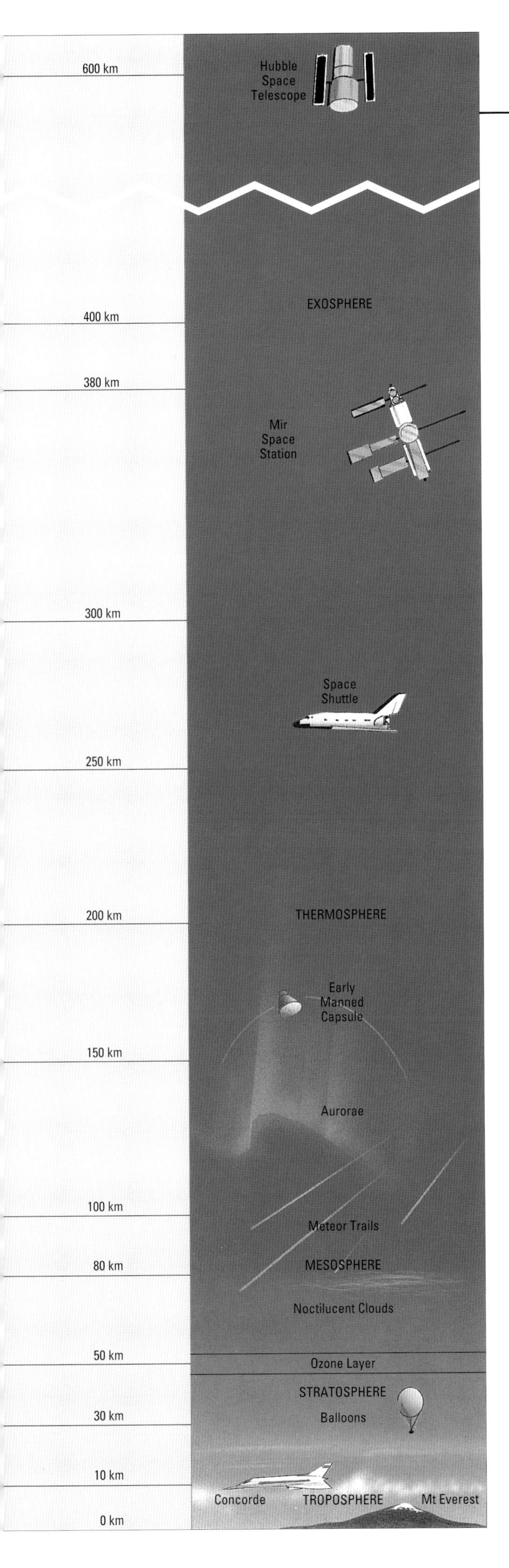

extremely brilliant. They are due to electrified particles from space, mainly originating in the Sun, which collide with atoms and molecules in the upper atmosphere and make them glow. Because the particles are electrically charged, they tend to cascade down towards the magnetic poles, so that aurorae are best seen from high latitudes. They are very common in places such as Alaska, northern Norway, northern Scotland and Antarctica, but are much rarer from lower latitudes such as those of southern England, and from the equator they are hardly ever seen. Auroral activity is more or less permanent around the so-called auroral ovals, which are 'rings' placed asymmetrically round the magnetic poles. When there are violent disturbances in the Sun, producing high-speed particles, the ovals broaden and expand, producing displays further from the main regions. Aurorae have been known for many centuries. The Roman emperor Tiberius, who reigned from AD 14 to 37, once dispatched his fire-engines to the port of Ostia because a brilliant red aurora led him to believe that the whole town was ablaze.

The Earth has a strong magnetic field. The region over which this field is dominant is called the magnetosphere; it is shaped rather like a tear-drop, with its tail pointing away from the Sun. On the sunward side of the Earth, it extends to about 65,000 kilometres (40,000 miles), but on the night side it spreads out much further.

Inside the magnetosphere there are two zones of strong radiation; they were detected by the first successful American satellite, Explorer 1 of February 1958, and are known as the Van Allen zones, in honour of the scientist who designed the equipment. There are two main zones, one with its lower limit at just under 8000 kilometres (5000 miles) and the other reaching out to 37,000 kilometres (23,000 miles). The inner belt, composed chiefly of protons, dips down towards the Earth's surface over the South Atlantic, because the Earth's magnetic field is offset from the axis of rotation, and this 'South Atlantic Anomaly' presents a distinct hazard to sensitive instruments carried in artificial satellites.

It cannot be said that we understand the Earth's magnetic field completely, and there is evidence of periodic reversals, as well as changes in intensity. At least it is certain that the field is due to currents in the iron-rich liquid core. Incidentally, it is worth noting that the Moon and Venus have no detectable magnetic fields, and if Mars has one it is extremely weak. In this respect, the Earth is quite unlike the other inner planets.

▼ ***Noctilucent clouds.*** *These strange, beautiful clouds can often become conspicuous; their origin is uncertain, but they may be due to water droplets condensing on meteoritic particles. This photograph was taken from Alaska in January 1993 (A. Watson).*

The Earth–Moon System

The Moon is officially classed as the Earth's satellite, but in many ways it may be better to regard the Earth–Moon system as a double planet; the mass ratio is 81 to 1, whereas for example Titan, the largest satellite of Saturn, has a mass only 1/4150 that of Saturn itself – even though Titan is considerably larger than our Moon.

We are by no means certain about the origin of the Moon. The attractive old theory according to which it simply broke away from the Earth, leaving the hollow now filled by the Pacific Ocean, has long since been discounted. It may be that the Earth and the Moon were formed together from the solar nebula, but there is increasing support for the idea that the origin of the Moon was due to a collision between the Earth and a large wandering body, so that the cores of the Earth and the impactor merged, and debris from the Earth's mantle, ejected during the collision, formed a temporary ring round the Earth from which the Moon subsequently built up. The Earth's mantle is much less massive than its core, and this theory would explain why the Moon is not so dense as the Earth; moreover, analyses of the lunar rocks show that the Moon and the Earth are of about the same age.

It is often said that 'the Moon goes round the Earth'. In a way this is true. To be strictly accurate the two bodies move together round their common centre of gravity, or barycentre; however, since the barycentre lies deep inside the Earth's globe, the simple statement is good enough for most purposes.

The orbital period is 27.3 days, and everyone is familiar with the phases, or apparent changes of shape, from new to full. When the Moon is in the crescent stage, the 'dark' side may often be seen shining faintly. There is no mystery about this; it is due to light reflected on to the Moon from the Earth, and is therefore known as earthshine. It can be quite conspicuous. Note, incidentally, that because the Earth and the Moon are moving together round the Sun, the synodic period (that is to say, the interval between one New Moon and the next) is not 27.3 days, but 29.5 days.

The Moon's axial rotation period is exactly equal to its orbital period. This is due to tidal friction over the ages. During its early history, the Moon was much closer to the Earth than it is now, and the Earth's rotation period was shorter; even today the 'day' is becoming gradually longer, while the Moon is being driven outwards from the

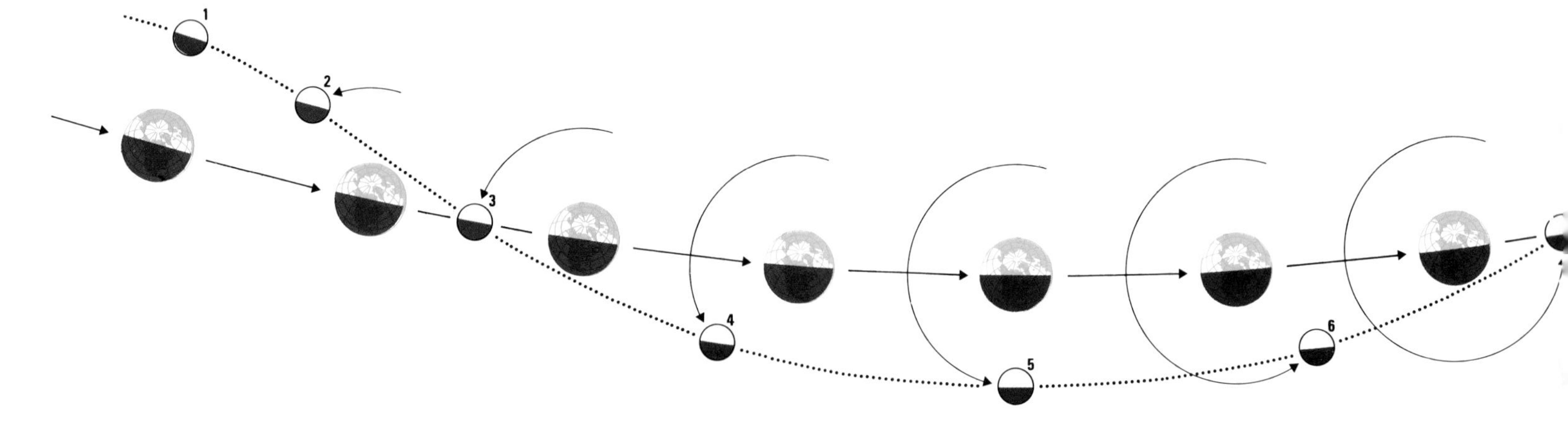

The New Moon *(1 and 9) occurs when the Moon is closest to the Sun. In the Crescent Moon (2), Mare Crisium is prominent between the eastern limb and the terminator. Earthshine is often seen.*

The Half Moon, First Quarter *(3) reveals Mare Serenitatis with the great chain of craters near the central meridian. Since the Sun is still low over the area that can be seen, the features are well defined.*

The Gibbous Moon *(4) reveals the great ray-craters Tycho and Copernicus. Although the craters are well illuminated and readily identifiable, their spectacular rays are not yet as striking as they will soon become.*

The Full Moon *(5). There are no shadows, and the rays from Tycho and Copernicus are so prominent that crater identification becomes difficult. The lunar maria take on a decidedly dark hue against the brilliant rays.*

The Waning Moon *(6). This is not as brilliant as the waxing Gibbous Moon. More of the dark maria which were once thought to be seas are illuminated. They are, in fact, gigantic plains of volcanic lava.*

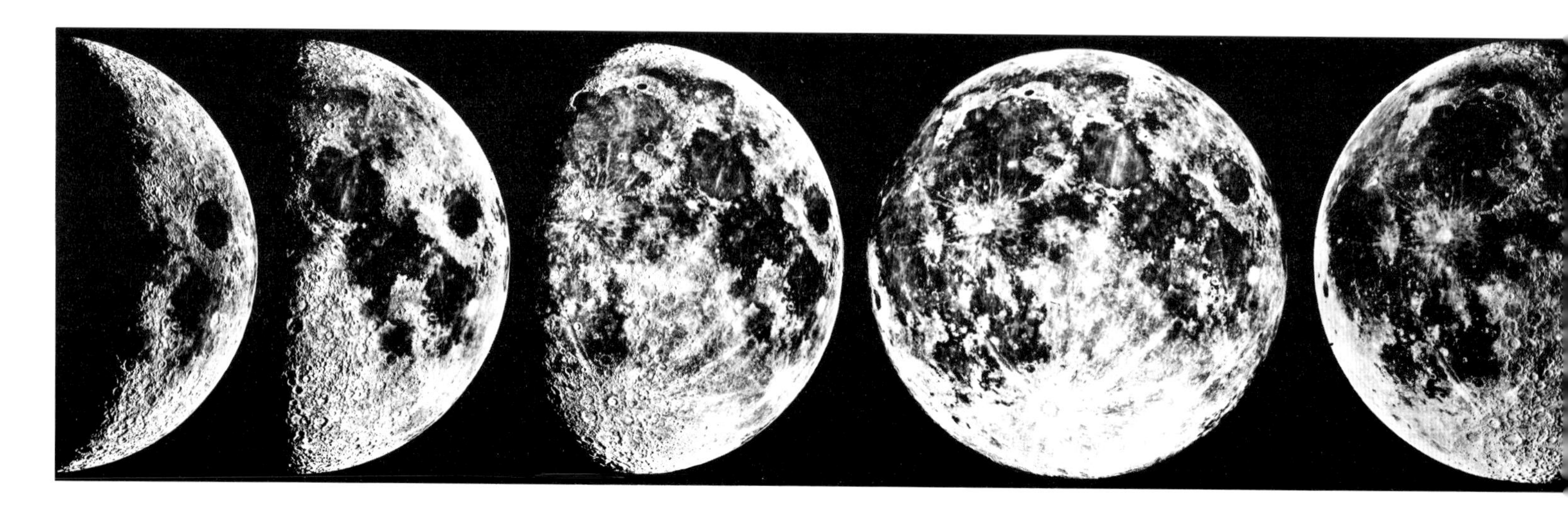

Earth. However, these effects are very slight. The Moon's distance is increasing at a rate of less than 4 centimetres (1.5 inches) per year.

The captured or 'synchronous' rotation means that there is a part of the Moon which is always turned away from us, so that until 1959, when the Russians sent their probe Lunik 3 on a 'round trip', we knew nothing definite about it. In fact it has proved to be basically the same as the region we have always known, though the surface features are arranged in a somewhat different manner.

The Moon's low escape velocity means that it has been unable to hold on to any atmosphere it may once have had. Like the Earth it has a crust, a mantle and a core. There is a loose upper layer, termed the regolith, from 1 to 20 metres (3 to 65 feet) deep; below comes a layer of shattered bedrock about 1 kilometre (0.6 miles) thick, and then a layer of more solid rock going down to about 25 kilometres (15 miles). Next comes the mantle, and finally the core, which is metal-rich and is probably between 1000 and 1500 kilometres (600 to 930 miles) in diameter. The core is hot enough to be molten, though the central temperature is much less than that of the Earth.

LUNAR DATA

Distance from Earth, centre to centre:	
max. (apogee)	406,697 km (252.681 miles)
mean	384,400 km (238,828 miles)
min. (perigee)	356,410 km (221,438 miles)
Orbital period	27.321661 days
Axial rotation period	27.321661 days
Synodic period (interval between successive New Moons)	29d 12h 44m 3s
Mean orbital velocity	3680 km/h (2286 miles/h)
Orbital inclination	5° 9′
Apparent diameter:	max. 33′ 31″
	mean 31′ 6″
	min. 29′ 22″
Density, water = 1	3.34
Mass, Earth = 1	0.012
Volume, Earth = 1	0.020
Escape velocity	2.38 km/s (1.48 miles/s)
Surface gravity, Earth = 1	0.165
Albedo	0.07
Mean magnitude at Full:	−12.7
Diameter	3476.6 km (2160 miles)

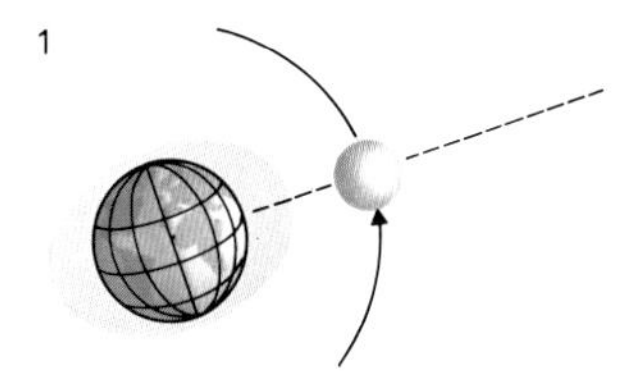

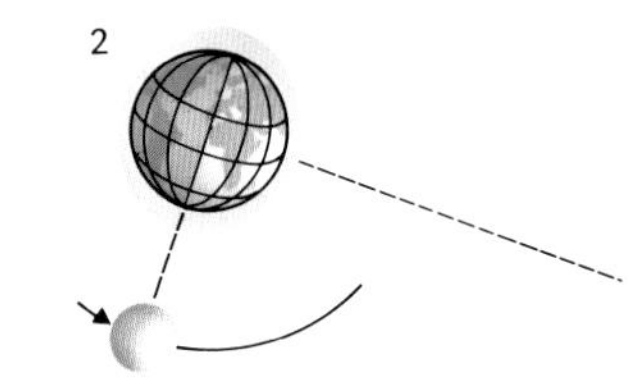

▼ ***The tides*** *are largely raised by the Moon, but the Sun also has an effect. When they act together (1), the tides are strong (spring tides). When they act at right angles (2), the tides are weak (neap tides).*

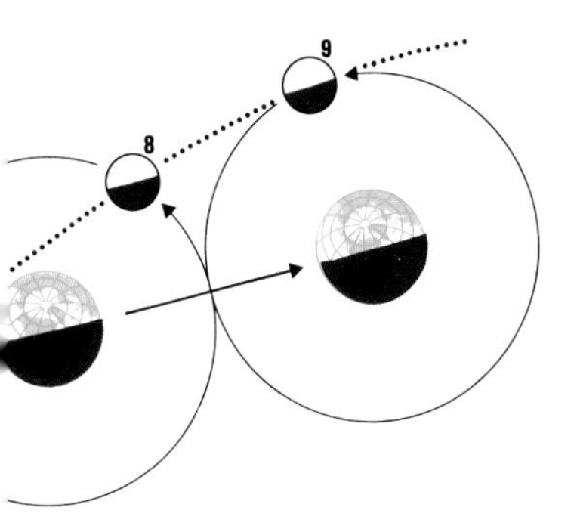

The Half Moon, Last Quarter *(7). The rays are less striking; shadows inside the large craters are increasing. The Old Moon (8) occurs just before the New, seen in the dawn sky. Earth-shine may often be seen.*

Eclipse of the Moon

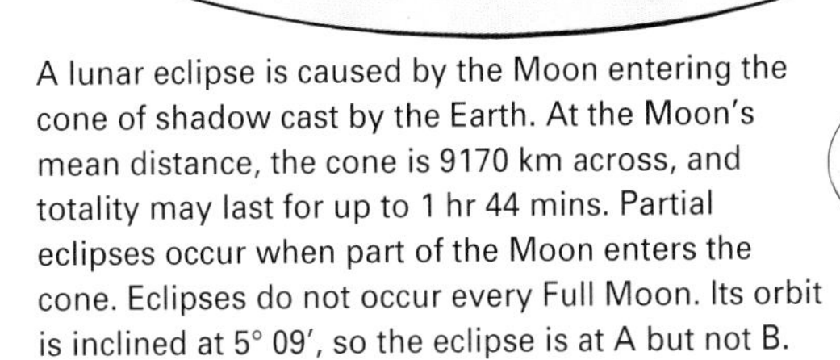

A lunar eclipse is caused by the Moon entering the cone of shadow cast by the Earth. At the Moon's mean distance, the cone is 9170 km across, and totality may last for up to 1 hr 44 mins. Partial eclipses occur when part of the Moon enters the cone. Eclipses do not occur every Full Moon. Its orbit is inclined at 5° 09′, so the eclipse is at A but not B.

The Moon does not generally disappear during eclipses, for some sunlight is refracted on to it by the Earth's atmosphere. Some eclipses are 'bright', with beautiful colours, others are dark, and records exist of the Moon vanishing completely, usually because of dust or volcanic ash in the Earth's upper air. The eclipse of 24–25 June 1964, shown here, was bright.

Features of the Moon

The Moon is much the most spectacular object in the sky to the user of a small telescope. There is an immense amount of detail to be seen, and the appearance changes dramatically from one night to the next because of the changing angle of solar illumination. A crater which is imposing when close to the terminator, or boundary between the daylight and night hemispheres, may be almost impossible to identify near Full Moon, when there are virtually no shadows.

The most obvious features are the wide dark plains known as seas or maria. For centuries now it has been known that there is no water in them (and never has been!), but they retain their romantic names such as the Mare Imbrium (Sea of Showers), Sinus Iridum (Bay of Rainbows) and Oceanus Procellarum (Ocean of Storms). They are of various types. Some, such as the Mare Imbrium, are essentially circular in outline, with mountainous borders; the diameter of the Mare Imbrium is 1300 kilometres (800 miles). Other seas, such as the vast Oceanus Procellarum, are irregular and patchy, so that they give the impression of being lava 'overflows'. There are bays, such as the Sinus Iridum which leads off the Mare Imbrium and is a superb sight when the Sun is rising or setting over it, catching the mountain-tops while the floor is still in shadow and producing the appearance often nicknamed the 'Jewelled Handle'.

▼ ***The Alps**. Part of the Mare Imbrium; the craters to the lower part of the picture are Archimedes (left), Aristillus and Autolycus. The low-walled formation with two interior craterlets is Cassini. The Alpine Valley can be seen cutting through the Alpine range near the top of the picture.*

Most of the major maria form a connected system. There is, however, one exception: the isolated, well-formed Mare Crisium, near the Moon's north-east limb, which is easily visible as a separate object with the naked eye. It appears elongated in a north–south direction, but this is because of the effect of foreshortening; the north–south diameter is 460 kilometres (285 miles), while the east–west diameter is 590 kilometres (370 miles). Maria still closer to the limb are so foreshortened that they can be made out only under favourable conditions.

The whole lunar scene is dominated by the craters, which range from vast enclosures such as Bailly, 293 kilometres (182 miles) in diameter, down to tiny pits. No part of the Moon is free from them; they cluster thickly in the uplands, but are also to be found on the floors of the maria and on the flanks and crests of mountains. They break into each other, sometimes distorting each other so completely the original forms are hard to trace; some have had their walls so reduced by lava flows that they have become 'ghosts', and some craters have had their seaward walls so breached that they have become bays. Fracastorius, at the edge of the Mare Nectaris, is a good example of this.

Riccioli, a Jesuit astronomer who drew a lunar map in 1651, named the main craters after various personalities, usually scientists. His system has been followed up to the present time, though it has been modified and extended, and later astronomers such as Newton have come off second-best. Some unexpected names are found. Julius Caesar has his own crater, though this was for his association with calendar reform rather than his military prowess.

Central peaks, and groups of peaks, are common, and the walls may be massive and terraced. Yet in profile a crater is not in the least like a steep-sided mine-shaft. The walls rise to only a modest height above the outer surface, while the floor is sunken; the central peaks never rise as high as the outer ramparts, so that in theory a lid could be dropped over the crater! Some formations, such as Plato in the region of the Alps and Grimaldi near the western limb, have floors dark enough to make them identifiable under any conditions of illumination; Aristarchus, in the Oceanus Procellarum, is only 37 kilometres (23 miles) across, but has walls and central peak so brilliant that when lit only by earthshine it has sometimes been mistaken for a volcano in eruption. One crater, Wargentin in the south-west limb area, has been filled with lava to its brim, so that it has taken on the form of a plateau. It is almost 90 kilometres (55 miles) across.

The most striking of all the craters are Tycho, in the southern uplands, and Copernicus in the Mare Nubium. Under high light they are seen to be the centres of systems of bright rays, which spread out for hundreds of kilometres. They are surface features, casting no shadows, so that they are well seen only when the Sun is reasonably high over them; near Full Moon they are so prominent that they drown most other features. Interestingly, the Tycho rays do not come from the centre of the crater, but are tangential to the walls. There are many other minor ray-centres, such as Kepler in the Oceanus Procellarum and Anaxagoras in the north polar area.

The main mountain ranges form the borders of the regular maria; thus the Mare Imbrium is bordered by the Alps, Apennines and Carpathians. Isolated peaks and hills abound, and there are also domes, which are low swellings often crowned by summit craterlets. One feature of special interest is the Straight Wall, in the Mare Nubium – which is not straight, and is not a wall! The land to the west drops by about 300 metres (1000 feet), so that the 'wall' is simply a fault in the surface. Before Full Moon its shadow causes it to appear as a black line; after full it reappears as a bright line, with the Sun's rays shining on its inclined face. It is by no means sheer, and the gradient seems to be no more than 40 degrees. In the future it will no doubt become a lunar tourist attraction....

Valleys are found here and there, notably the great gash cutting through the Alps. The so-called 'Rheita Valley' in the south-eastern uplands is really a chain of craters which have merged, and crater-chains are very common on the Moon, sometimes resembling strings of beads. There are also rills – alternatively known as rilles or clefts – which are crack-like collapse features. Some of these, too, prove to be crater-chains either wholly or in part. The most celebrated rills are those of Hyginus and Ariadaeus, in the region of the Mare Vaporum, but there are intricate rill-systems on the floors of some of the large craters, such as Gassendi, near the northern boundary of the Mare Humorum, and Alphonsus, the central member of a chain of great walled plains, near the centre of the Moon's disk of which the flat-floored, 148-kilometre (92-mile) Ptolemaeus is the largest crater.

Many of the maria are crossed by ridges, which are low, snaking elevations of considerable length. Ridges on the seas are often the walls of ghost craters which have been so completely inundated by lava that they are barely recognizable.

It is now agreed that the craters were produced by a violent meteoritic bombardment which began at least 4500 million years ago and ended about 3850 million years ago. There followed widespread vulcanism, with magma pouring out from below and flooding the basins. The lava flows ended rather suddenly about 3200 million years ago, and since then the Moon has shown little activity apart from the formation of an occasional impact crater. It has been claimed that the ray-craters Tycho and Copernicus may be no more than a thousand million years old, though even this is ancient by terrestrial standards. Even the youngest craters are very ancient by terrestrial standards.

There is little activity now; there are occasional localized glows and obscurations, known as Transient Lunar Phenomena (TLP), thought to be due to gaseous release from below the crust, but all in all the Moon today is essentially changeless.

▲ ***Copernicus, Stadius and Eratosthenes.*** *Copernicus is the large crater to the lower left; it is one of the major ray-centres of the Moon. Eratosthenes, smaller but equally well formed, is to the upper right, at the end of the Apennine range. Stadius, to the right of Copernicus, is a 'ghost' crater whose walls have been levelled by the mare lava to such an extent that they are now barely traceable.*

◀ ***Ptolemaeus group.*** *A great chain of walled plains near the centre of the Moon's disk. Ptolemaeus is to the top of the picture, at the centre; below it comes Alphonsus, with a reduced central peak and dark patches on its floor; below Alphonsus is Arzachel, smaller but with higher walls and a central peak. The walled plain Albategnius lies to the left of Ptolemaeus.*

Lunar Landscapes

Lunar photographs taken with even small telescopes can show a surprising amount of detail, and there always appears to be something new to see. It is not hard to compile one's own lunar photographic atlas.

NAMED NEAR-SIDE SEAS (MARIA)

Sinus Aestuum	*The Bay of Heats*
Mare Australe	*The Southern Sea*
Mare Crisium	*The Sea of Crises*
Palus Epidemiarum	*The Marsh of Epidemics*
Mare Foecunditatis	*The Sea of Fertility*
Mare Frigoris	*The Sea of Cold*
Mare Humboldtianum	*Humboldt's Sea*
Mare Humorum	*The Sea of Humours*
Mare Imbrium	*The Sea of Showers*
Sinus Iridum	*The Bay of Rainbows*
Mare Marginis	*The Marginal Sea*
Sinus Medii	*The Central Bay*
Lacus Mortis	*The Lake of Death*
Palus Nebularum	*The Marsh of Mists*
Mare Nectaris	*The Sea of Nectar*
Mare Nubium	*The Sea of Clouds*
Mare Orientale	*The Eastern Sea*
Oceanus Procellarum	*The Ocean of Storms*
Palus Putredinis	*The Marsh of Decay*
Sinus Roris	*The Bay of Dews*
Mare Serenitatis	*The Sea of Serenity*
Mare Smythii	*Smyth's Sea*
Palus Somnii	*The Marsh of Sleep*
Lacus Somniorum	*The Lake of the Dreamers*
Mare Spumans	*The Foaming Sea*
Mare Tranquillitatis	*The Sea of Tranquility*
Mare Undarum	*The Sea of Waves*
Mare Vaporum	*The Sea of Vapours*

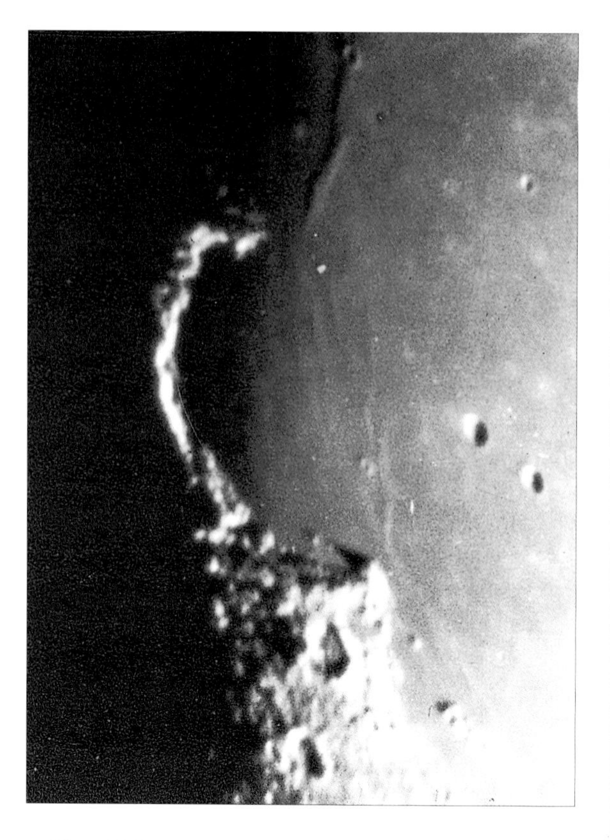

▲ ***Sinus Iridum*** – *the 'Jewelled Handle' appearance. The Sun is rising over the bay, and the rays are catching the mountainous border while the lower-lying land is still in shadow. The two small craterlets to the right of the Bay are Helicon and Le Verrier.*

► ***The Posidonius area.*** *This region is the large-walled plain at the bottom of the picture; its smaller companion is Chacornac. The two large craters at the top are Atlas and Hercules; to their left is Bürg, with the dark plain of the Lacus Mortis and an extensive system of rills.*

◄ ***Lunar rills.*** *Two of the best-known rills are shown here. The Hyginus Rill, near the centre, is actually a crater-chain; the largest feature, Hyginus, is 6 km (4 miles) in diameter. The long Ariadaeus Rill, to the right, is more 'cracklike'; Ariadaeus itself is 15 km (9 miles) in diameter. To the right are the dark-floored Boscovich and Julius Caesar.*

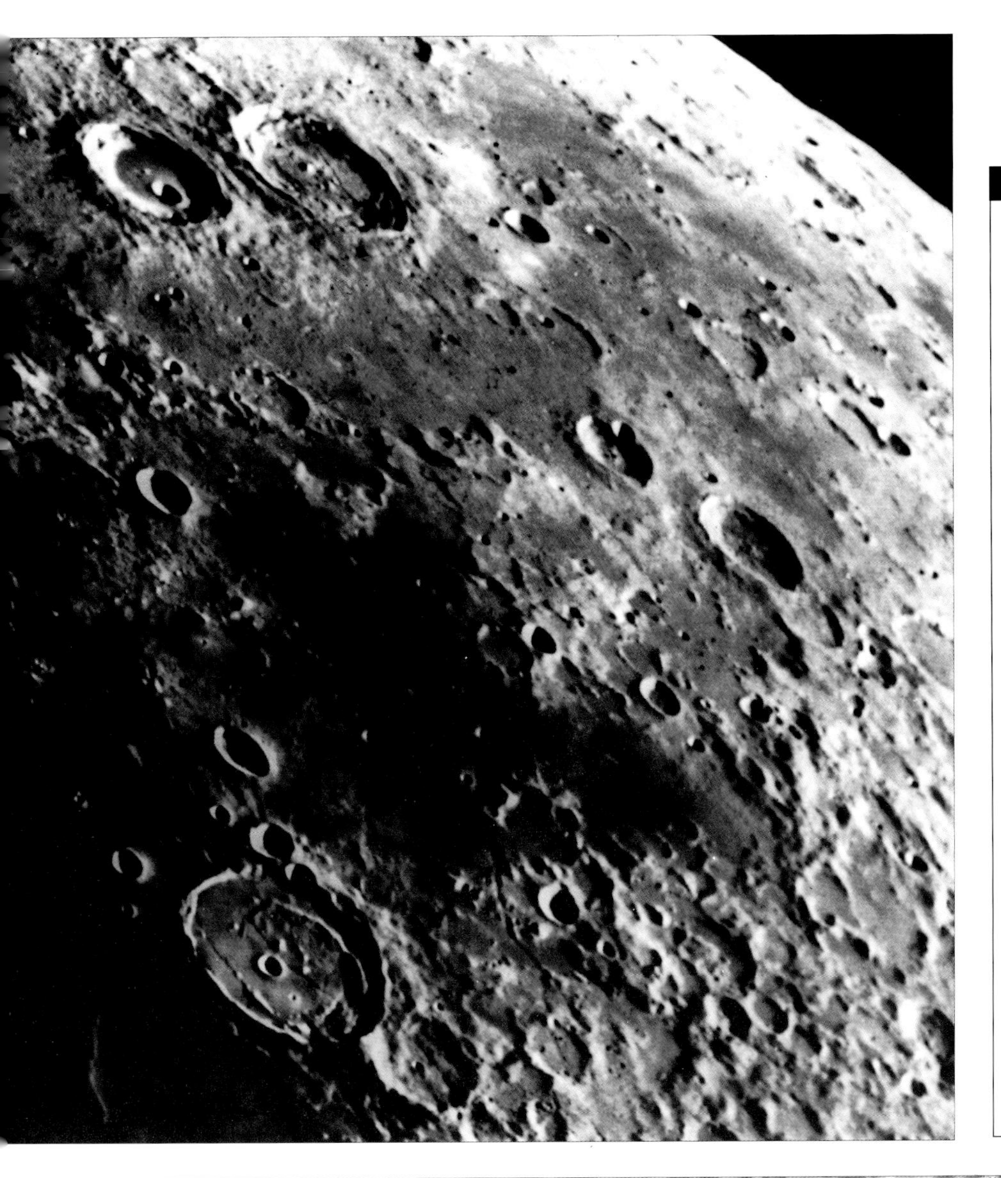

MAIN MOUNTAIN RANGES	
Alps	Northern border of Imbrium
Altai Scarp	South-west of Nectaris, from Piccolomini
Apennines	Bordering Imbrium
Carpathians	Bordering Imbrium to the south
Caucasus	Separating Serenitatis and Nebularum
Cordillera	Limb range, near Grimaldi
Haemus	Southern border of Serenitatis
Harbinger	Clumps of peaks in Imbrium, near Aristarchus
Jura	Bordering Iridum
Percy	NW border of Humorum; not a major range
Pyrenees	Clumps of hills bordering Nectaris to the east
Riphaeans	Short range in Nubium
Rook	Limb range, associated with Orientale
Spitzbergen	Mountain clump in Iridum, north of Archimedes
Straight Range	In Imbrium, near Plato; very regular
Taurus	Mountain clumps east of Serenitatis
Tenerife	Mountain clumps in Imbrium, south of Plato
Ural	Extension of the Riphaeans

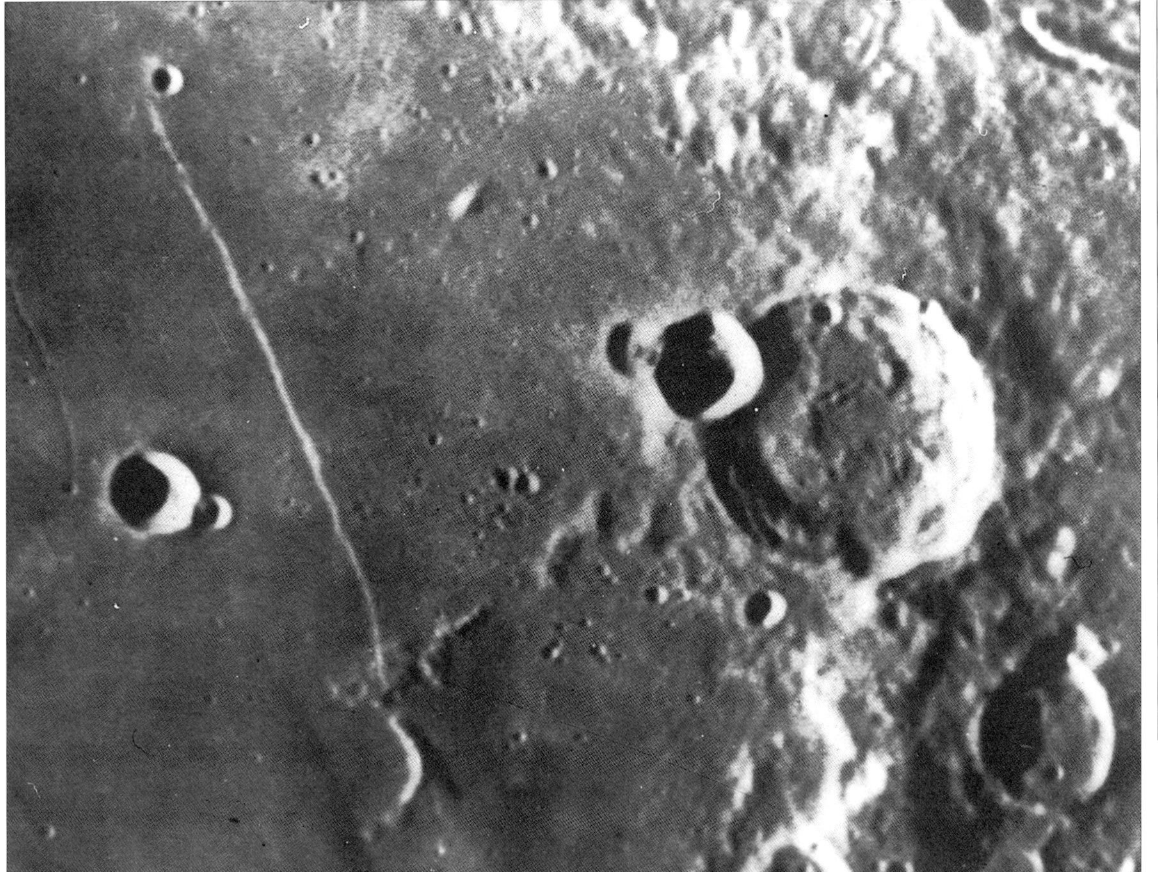

◀ ***Thebit and the straight wall****. Thebit, on the edge of the Mare Nubium, is 60 km (37 miles) in diameter. It is interrupted by a smaller crater, Thebit A, which is in turn broken by the smaller Thebit F – demonstrating the usual arrangement of lunar cratering. To the left of Thebit is the fault misnamed the Straight Wall; to the left of the Wall is the well-formed, 18-km (11-mile) crater Birt, associated with a rill which ends in a craterlet.*

The Far Side of the Moon

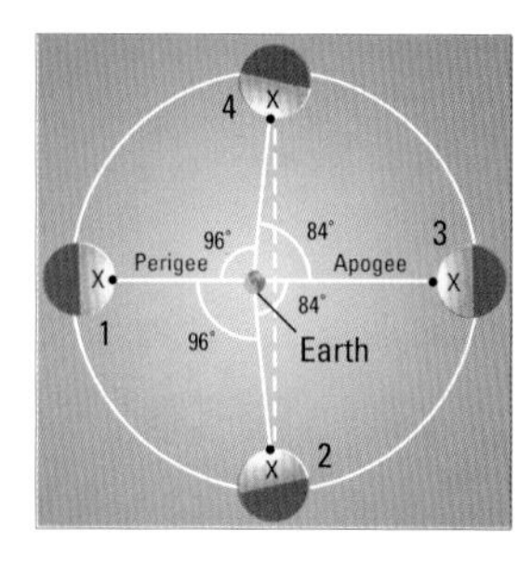

▲ ***Libration in longitude****. X is the centre of the Moon's disk, as seen from Earth. At position 1 the Moon is at perigee. After a quarter of its orbit it has reached position 2; but since it has travelled from perigee it has moved slightly faster than its mean rate, and has covered 96° instead of 90°. As seen from Earth, X lies slightly east of the apparent centre of the disk, and a small portion of the far side has come into view in the west. After a further quarter-month the Moon has reached position 3. It is now at apogee, and X is again central. A further 84° is covered between positions 3 and 4, and X is displaced towards the west, so that an area beyond the mean eastern limb is uncovered. At the end of one orbit the Moon has arrived back at 1, and X is once more central on the Moon's disk as seen from Earth.*

Look at the Moon, even with the naked eye, and you will see the obvious features such as the principal maria. The positions of these features on the disk are always much the same, because of the synchronous rotation. Yet there are slight shifts, due to the effects known as librations. All in all we can examine a grand total of 59 per cent of the lunar surface, and only 41 per cent is permanently averted, though of course we can never see more than 50 per cent.

The most important libration – the libration in longitude – is due to the fact that the Moon's path round the Earth is elliptical rather than circular, and it moves at its fastest when closest to us (perigee). However, the rate of axial rotation does not change, so that the position in orbit and the amount of axial spin become periodically 'out of step'; we can see a little way round alternate mean limbs. There is also a libration in latitude, because the Moon's orbit is inclined by over 5 degrees, and we can see for some distance beyond the northern and southern limbs. Finally there is a diurnal or daily libration, because we are observing from the surface, not the centre of the globe.

All these effects mean that the 'libration regions' are carried in and out of view. They are so foreshortened that it is often difficult to distinguish between a crater and a ridge, and before 1959 our maps of them were very imperfect. About the permanently hidden regions nothing definite was known. It was reasonable to assume that they were basically similar to the familiar areas – though some strange ideas had been put forward from time to time. The last-century Danish astronomer Andreas Hansen once proposed that all the Moon's air and water had been drawn round to the far side, which might well be inhabited! The first pictures of the far side were obtained in October 1959 by the Russian space probe Lunik 3 (also known as Luna 3). It went right round the Moon, taking pictures of the far side and later sending them back by television techniques. The pictures are very blurred and lacking in detail, but they were good enough to show that, as expected, the far side is just as barren and just as crater-scarred as the areas we have always known. Later spacecraft, both manned and automatic, have enabled us to draw up very complete maps of the entire lunar surface.

There is a definite difference between the near and the far sides, no doubt because the Moon's rotation has been synchronous since a fairly early stage in the evolution of the Earth–Moon system; the crust is thickest on the far side. One major sea, the Mare Orientale, lies mainly on the hidden regions; only a small part of it can be seen from Earth, and then only under conditions of favourable libration. The spacecraft pictures have shown it to be a vast, multi-ringed structure which is probably the youngest of all the lunar seas. Otherwise there are no large maria on the far side, and this is the main difference between the two hemispheres.

One very interesting object is Tsiolkovskii, 240 kilometres (150 miles) in diameter. It has a dark floor which gives the impression of being shadowed in many photographs, though the real cause of the darkness is the hue of the floor itself; there is no doubt that we are seeing a lake of solidified lava, from which a central peak rises. In many ways Tsiolkovskii seems to be a sort of link between a crater and a mare. It intrudes into a larger but less regular basin, Fermi, which has the usual light-coloured interior.

Many of the familiar types of features are seen on the far side, and the distribution of the craters is equally non-random; when one formation breaks into another, it is always the smaller crater which is the intruder. Valleys, peaks and rays systems exist. Though the Moon has no overall magnetic field that we can detect, there are regions of localized magnetism here and there; one of these lies near the rather irregular far-side crater Van de Graaff. It has been suggested that the Moon used to have a definite magnetic field which has now died away.

On the original Lunik 3 picture a long, bright feature running for hundreds of kilometres was shown, and was thought to be a mountain range which was promptly named in honour of the Soviet Union. Alas, it was later found that the feature is nothing more than a surface ray, and the Soviet Mountains were tactfully deleted from the maps. However, it was surely right to name the most imposing far-side feature in honour of Konstantin Eduardovich Tsiolkovskii, the great pioneer who was writing about spaceflight almost a hundred years ago.

Moon
at its maximum height above the horizon
Moon
rising
Moon
setting
A
Earth

◀ ***Diurnal libration****. We are observing from the Earth's surface at A, not its centre, so that we can see a little way alternately round the northern and southern limbs.*

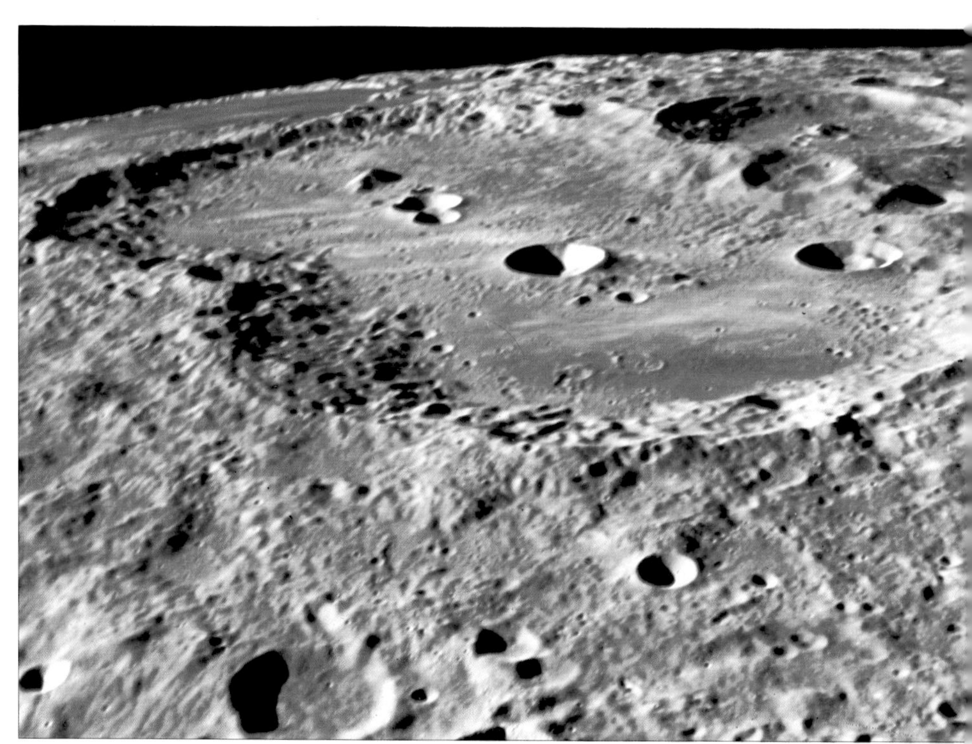

▶ ***Van de Graaff*** *is a large but rather irregular formation, perhaps compound, notable because of the amount of remnant magnetism in and near it. Its floor contains several smaller craters. The wall is broken (top right-hand corner) by Birkeland, a well-formed crater with a prominent central peak.*

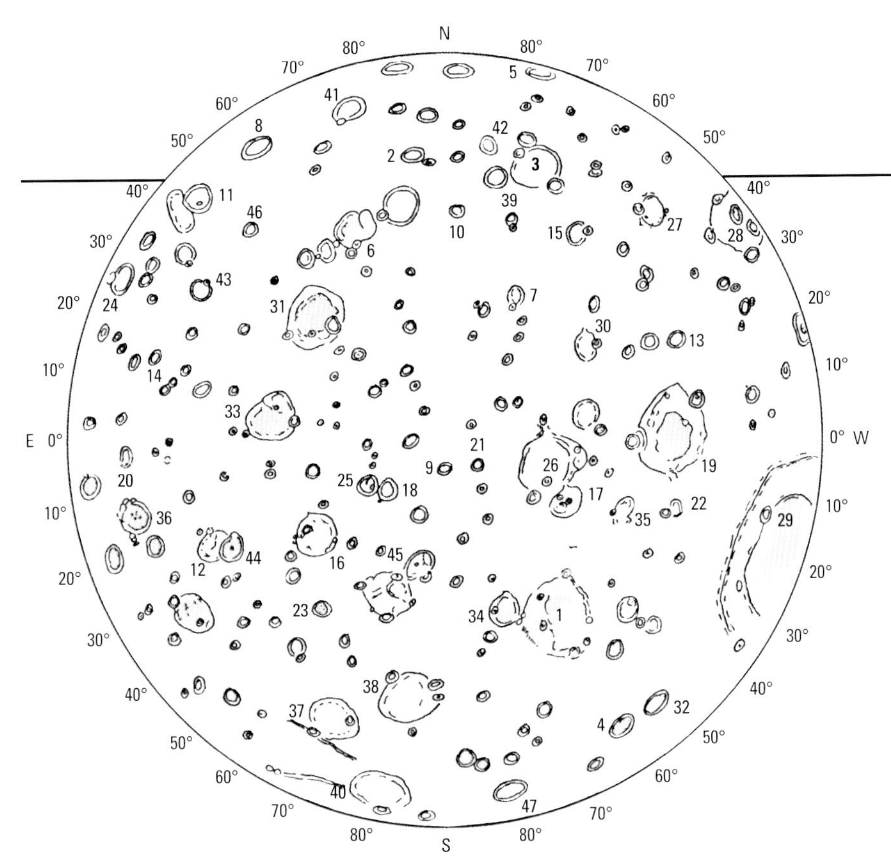

◀ ***The far side** of the Moon – first recorded from the Soviet space probe Lunik 3 in 1959, and now fully mapped.*

▼ ***Tsiolkovskii** is exceptional in many ways. It is 240 km (150 miles) in diameter, with terraced walls and a massive central mountain structure. The darkness is caused by lava; in fact Tsiolkovskii seems to be intermediate in type of lunar feature, falling somewhere between a crater and a mare, or sea.*

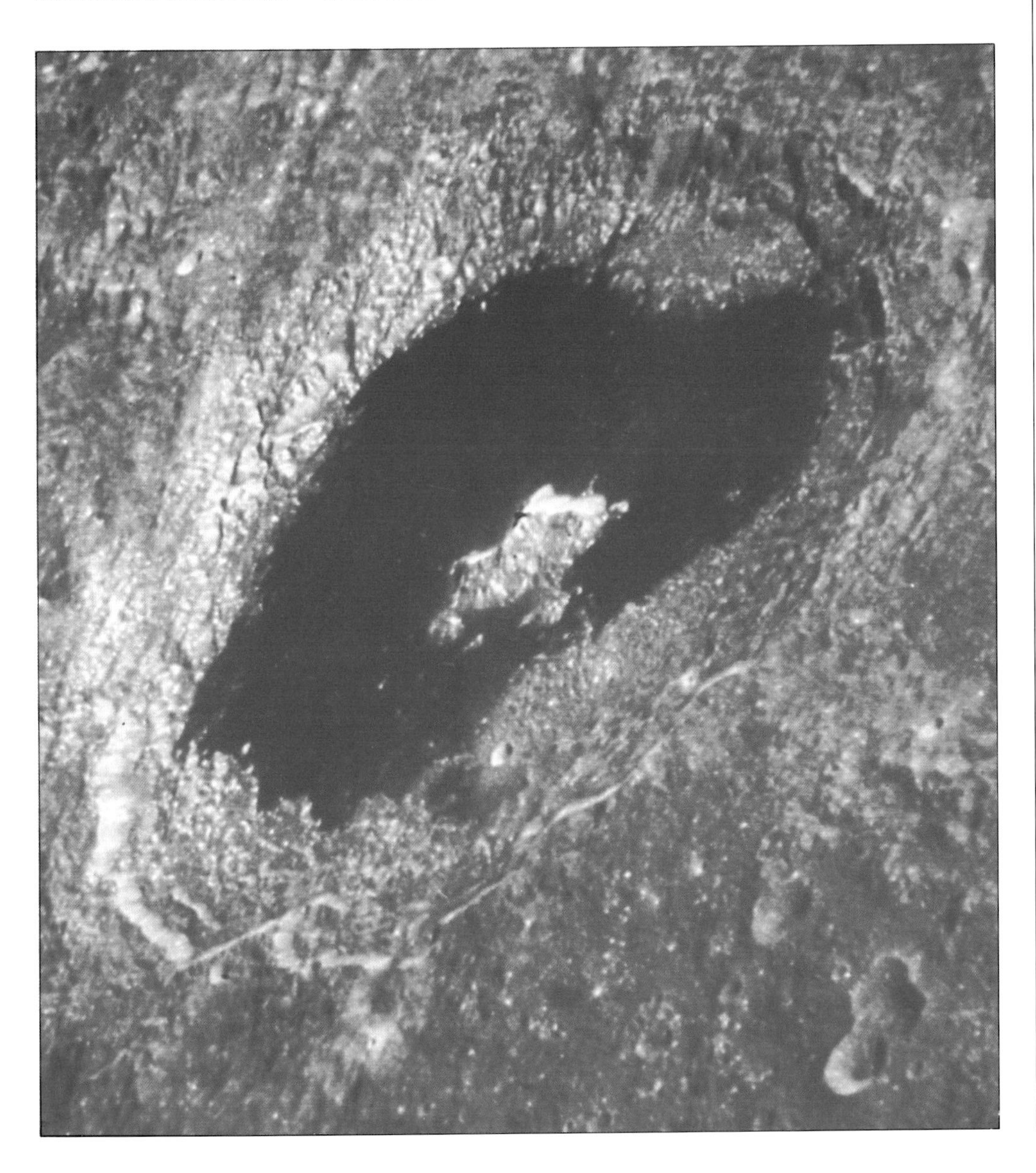

FAR-SIDE FEATURES

	Name	Lat. °	Long. °
1	Apollo	37 S	153 W
2	Avogadro	64 N	165 E
3	Birkhoff	59 N	148 W
4	Boltzmann	55 S	115 W
5	Brianchon	77 N	90 W
6	Campbell	45 N	152 E
7	Cockcroft	30 N	164 W
8	Compton	55 N	104 E
9	Daedalus	6 S	180
10	Dunér	45 N	179 E
11	Fabry	43 N	100 E
12	Fermi	20 S	122 E
13	Fersman	18 N	126 W
14	Fleming	15 N	109 E
15	Fowler	43 N	145 W
16	Gagarin	20 S	150 E
17	Galois	16 S	153 W
18	Heaviside	10 S	167 E
19	Hertzsprung	0	130 W
20	Hirayama	6 S	93 E
21	Icarus	6 S	173 W
22	Joffe	15 S	129 W
23	Jules Verne	36 S	146 E
24	Joliot	26 N	94 E
25	Keeler	10 S	162 E
26	Korolev	5 S	157 W
27	Landau	42 N	119 W
28	Lorentz	34 N	100 W
29	Lowell	13 S	103 W
30	Mach	18 N	149 W
31	Mare Moscoviense	27 N	147 E
32	Mendel	49 S	110 W
33	Mendeléev	6 N	141 E
34	Oppenheimer	35 S	166 W
35	Paschen	14 S	141 W
36	Pasteur	12 S	105 E
37	Planck	57 S	135 E
38	Poincaré	57 S	161 E
39	Rowland	57 N	163 W
40	Schrödinger	75 S	133 E
41	Schwarzschild	71 N	120 E
42	Sommerfeld	65 N	161 W
43	Szilard	34 N	106 E
44	Tsiolkovskii	21 S	129 E
45	Van de Graaff	27 S	172 E
46	H. G. Wells	41 N	122 E
47	Zeeman	75 S	135 W

Missions to the Moon

▼ ***Apollo 15**. The first mission in which a 'Moon Car', or Lunar Roving Vehicle, was taken to the Moon, enabling the astronauts to explore much greater areas. Astronaut Irwin stands by the Rover, with one of the peaks of the Apennines in the background. The electrically-powered Rovers performed faultlessly. The peak is much further from the Rover than may be thought; distances on the Moon are notoriously difficult to estimate.*

The Russians took the lead in exploring the Moon with spacecraft. Their Luniks contacted the Moon in 1959, and they were also the first to make a controlled landing with an automatic probe. Luna 9 came gently down in the Oceanus Procellarum, on 3 February 1966, and finally disposed of a curious theory according to which the lunar seas were coated with deep layers of soft dust. Later, the Russians were also able to send vehicles to the Moon, collect samples of lunar material, and bring them back to Earth. It is now known that they had planned a manned landing there in the late 1960s, but had to abandon the idea when it became painfully clear that their rockets were not sufficiently reliable. By 1970 the 'Race to the Moon' was definitely over.

American progress had been smoother. The Ranger vehicles crash-landed on the Moon, sending back data and pictures before being destroyed; the Surveyors made soft landings, obtaining a tremendous amount of information; and between 1966 and 1968 the five Orbiters went round and round the Moon, providing very detailed and accurate maps of virtually the entire surface. Meanwhile, the Apollo programme was gathering momentum.

By Christmas 1968 the crew of Apollo 8 were able to go round the Moon, paving the way for a landing. Apollo 9 was an Earth-orbiter, used to test the lunar module which would go down on to the Moon's surface. Apollo 10, the final rehearsal, was another lunar orbiter; and then, in July 1969, first Neil Armstrong, then Edwin 'Buzz' Aldrin, stepped out on to the bleak rocks of the Mare Tranquillitatis from the *Eagle,* the lunar module of Apollo 11. Millions of people on Earth watched Armstrong make his immortal 'one small step' on to the surface of the Moon. The gap between our world and another had at last been bridged.

Apollo 11 was a preliminary mission. The two astronauts spent more than two hours outside their module, setting up the first ALSEP (Apollo Lunar Surface Experimental Package), which included various instruments – for example a seismometer, to detect possible 'moon-quakes'; a device for making a final search for any trace of lunar atmosphere, and an instrument designed to collect particles from the solar wind. Once their work was completed (interrupted only briefly by a telephone call from President Nixon) the astronauts went back into the lunar module; subsequently they lifted off, and rejoined Michael Collins, the third member of the expedition, who had remained in lunar orbit. The lower part of the lunar module was used as a launching pad, and was left behind, where it will remain until it is collected and removed to a lunar museum. The return journey to Earth was flawless.

Apollo 12 (November 1969) was also a success; astronauts Conrad and Bean were even able to walk over to an old Surveyor probe, which had been on the Moon ever since 1967, and bring parts of it home. Apollo 13 (April 1970) was a near-disaster; there was an explosion during the outward journey, and the lunar landing had to be abandoned. With Apollo 14 (January 1971) astronauts Shepard and Mitchell took a 'lunar cart' to carry their equipment, and with the three final missions, Apollo 15 (July 1971), 16 (April 1972) and 17 (December 1972) a Lunar Roving Vehicle (LRV) was used, which increased the range of exploration very considerably. One of the Apollo 17 astronauts, Dr Harrison Schmitt, was a professional geologist who had been given training specially for the mission.

The Apollo programme has increased our knowledge of the Moon beyond all recognition – and yet in a way it

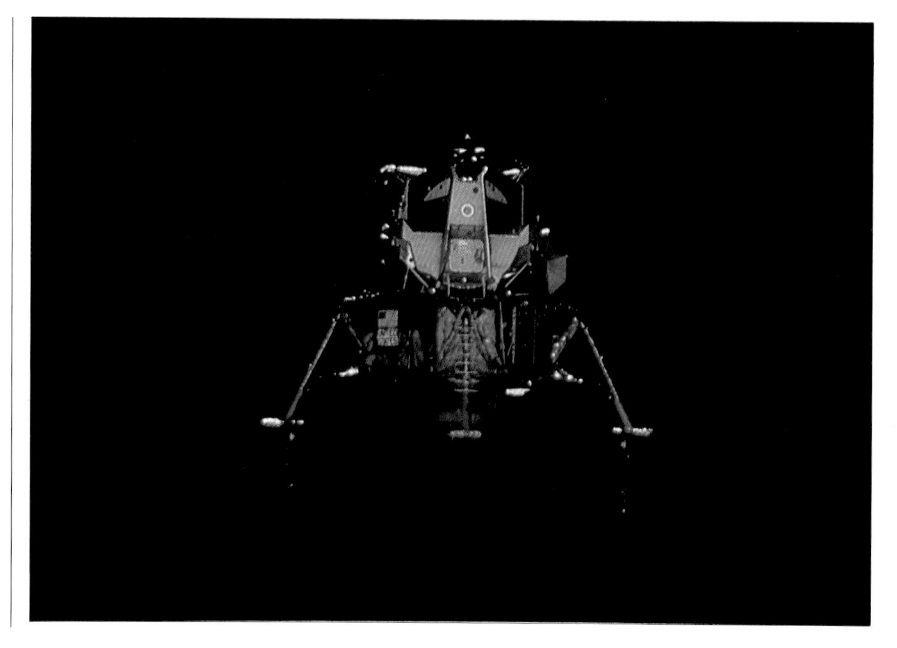

◀ ***Apollo 16**. The landing site for Apollo 16 was in the highlands of Descartes, in one of the rougher parts of the Moon. Again a Rover was taken, and again it performed faultlessly. Astronauts Duke and Young explored a wide area, and set up a number of scientific experiments in the Apollo Lunar Surface Experimental Package (ALSEP). The Lunar Module was designed to make a landing on the Moon and return the astronauts to orbit. The upper section has one ascent engine only, and there can be no second chance. The photograph was taken from the orbiting Command Module.*

was limited; the missions were really in the nature of reconnaissances. The various ALSEPs continued operating for some years, until they were eventually switched off mainly on financial grounds.

No men have been to the Moon since 1972, though there have been a few unmanned missions. In 1993 the Japanese probe Hiten crashed on the Moon near the crater Furnerius, and images were also sent back by the Galileo probe en route to Jupiter. We may hope for the establishment of a Lunar Base and a Lunar Observatory in the foreseeable future. To quote Eugene Cernan, commander of Apollo 17, when I talked with him:

> 'I believe we'll go back. We went to the Moon not initially for scientific purposes, but for national and political ones which was just as well, because it enabled us to get the job done! When there is real motivation, for instance to use the Moon as a base for exploring other worlds in the Solar System, or set up a full-scale scientific base, then we'll go back. There will be others who will follow in our steps.'

▲ ***The scene from Apollo 17****. The Lunar Rover is well shown. The sky is of course jet-black; one is reminded of Buzz Aldrin's description of the Moon as 'magnificent desolation'.*

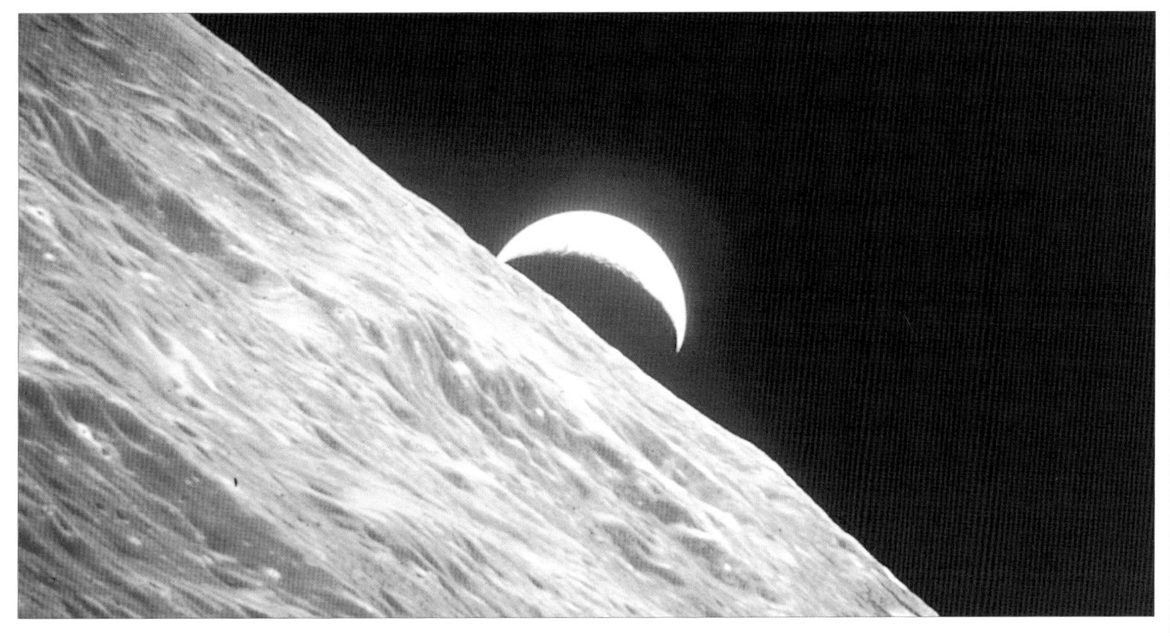

▲ ***Earthrise****. This picture was taken from Apollo 17, the last manned lunar mission. It shows the crescent Earth rising over the limb of the Moon; when the photograph was taken, Apollo 17 was in lunar orbit.*

▶ ***Apollo 17****. During one of the Moon walks, Dr Schmitt, the geologist, suddenly called attention to what seemed to be orange soil inside a small crater, unofficially named Shorty. At first it was thought to indicate recent fumarole activity, but the colour was due to very small, very ancient glassy 'beads'.*

The Clementine Mission

▶ ***Clementine***. *The latest American lunar probe was paid for by the Department of Defense. It weighed 140 kg (300 lb), and carried an array of sensors.*

The 1994 mission to the Moon, Clementine, was named after an old American mining song – because, after the lunar part of its programme, the probe was scheduled to go on to an asteroid, and it has been suggested that in the future it may be possible to carry out mining operations on asteroids.

Clementine was funded partly by NASA and partly by the US Department of Defense. The military authorities were anxious to test instruments and techniques capable of locating hostile ballistic missiles, and the only way to circumvent the strict regulations about this sort of activity was to go to the Moon. Therefore, the Department could test its anti-ballistic missile system and do some useful scientific work as well.

Clementine was launched on 24 January 1994 from the Vandenburg Air Force Base in California, and began its Earth programme. It weighed 140 kilograms (300 lb) and carried an array of advanced sensors. After completing this part of its mission several manoeuvres were carried out, and Clementine entered lunar orbit on 21 February 1995. For 2$^1/_2$ months it orbited the Moon in a highly inclined path, which took it from 415 kilometres (260 miles) to 2940 kilometres (1830 miles) from the Moon; a full research programme was successfully completed.

Clementine surveyed the whole of the Moon. Many gravity measurements were made, and superb images obtained; the inclined orbit meant that the polar regions could be mapped more accurately than ever before. For example, there were detailed views of the vast South Pole–Aitken Basin which is 2250 kilometres (1400 miles) in diameter and 12 kilometres (7 miles) deep. There was also the Mendel-Rydberg Basin, 630 kilometres (390 miles) across, which is less prominent because it lies under a thick blanket of debris from the adjacent Mare Orientale.

It was claimed that Clementine had detected indications of ice inside some of the polar craters, whose floors are always in shadow – but ice did not seem at all probable on a world such as the Moon. Clementine left lunar orbit on 3 May 1995; it had been hoped to rendezvous with a small asteroid, Geographos, but a programming error ruled this out. The next lunar probe, Prospector, was launched on 3 January 1998, and carried out an extensive mapping survey. On 31 July 1999 it was deliberately crashed into a polar crater, in the hope that water might be detected in the debris, but no signs of water were found and the idea of lunar ice has been generally abandoned.

▶ ***Launch of Clementine***. *The probe was launched on 24 January 1994 from the Vandenburg Air Force Base, and put into a circular orbit round the Earth. It departed for lunar orbit on 21 February, and spent two months mapping the Moon – particularly the polar areas, which were less well known than the rest of the surface.*

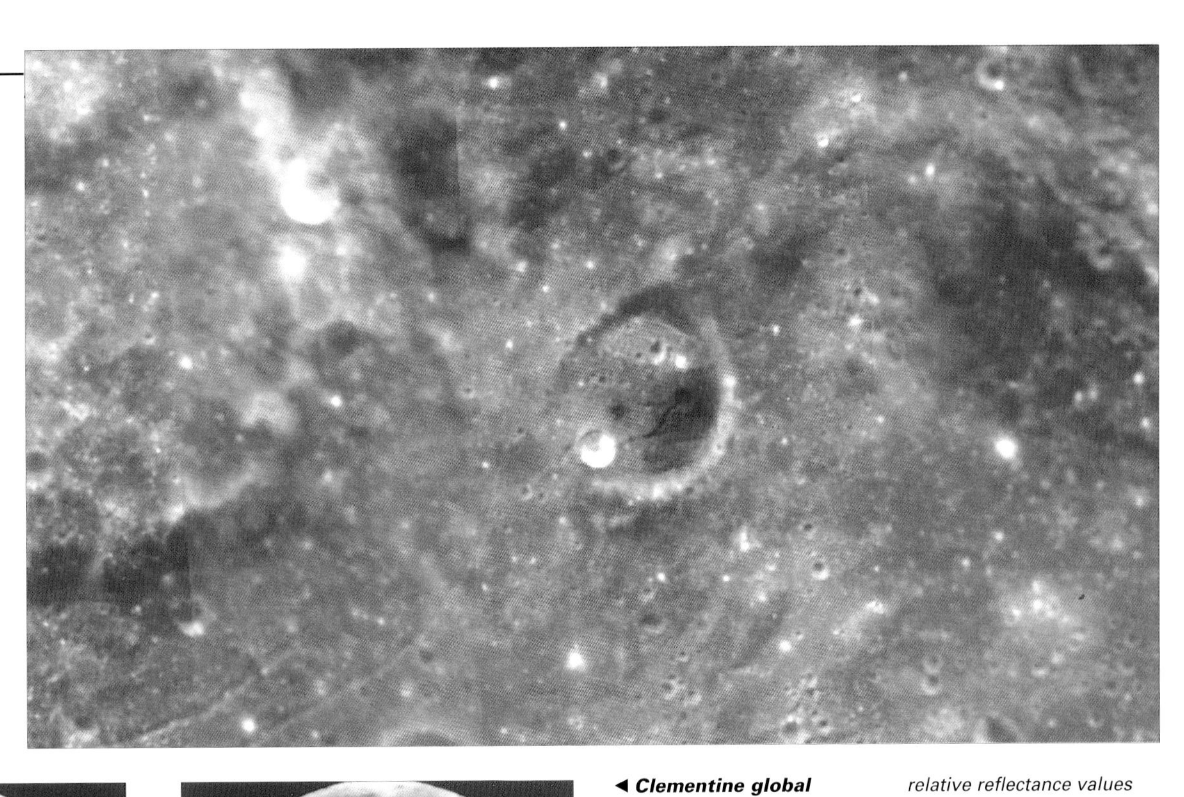

▶ ***Colour mosaic of Clementine images** over a part of the floor of the Apollo Basin, on the Moon's far side. An interesting feature of this area is the volcanic crater at latitude 30°S, longitude $153\frac{1}{2}$°W, which is surrounded by dark pyroclastics (fragmental volcanic ejecta). The Apollo Basin is exceptionally deep, and lies on the floor of the huge South Pole–Aitken Basin. The pyroclastics probably formed from gas-driven eruptions that leave deposits rich in glass droplets frozen from the spray of molten lava. [Image processing by Mark Robinson, USGS.]*

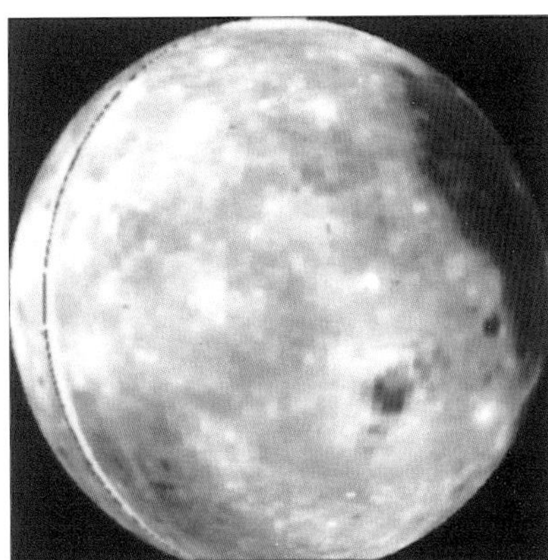

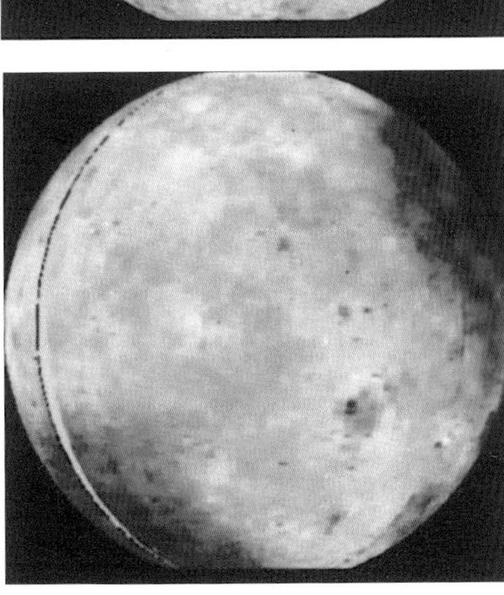

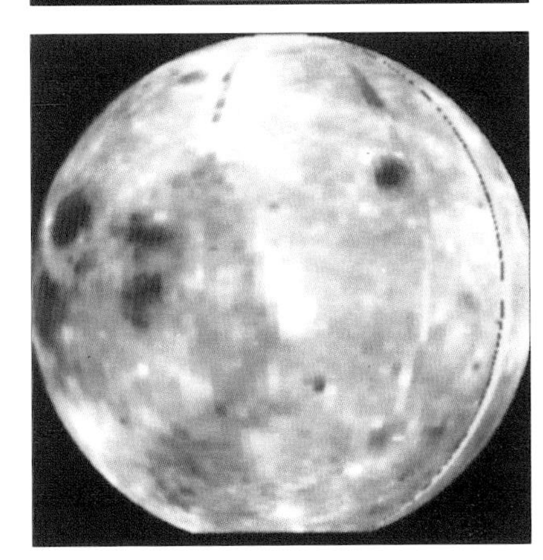

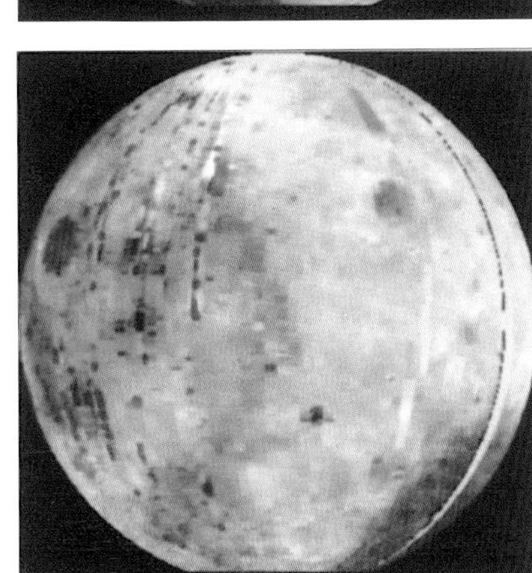

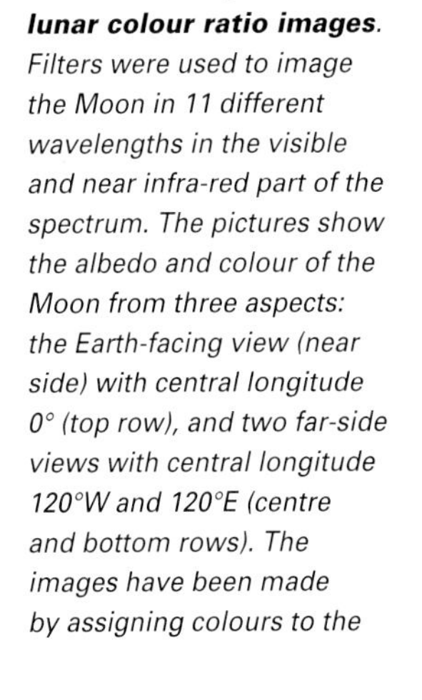

◀ ***Clementine global lunar colour ratio images**. Filters were used to image the Moon in 11 different wavelengths in the visible and near infra-red part of the spectrum. The pictures show the albedo and colour of the Moon from three aspects: the Earth-facing view (near side) with central longitude 0° (top row), and two far-side views with central longitude 120°W and 120°E (centre and bottom rows). The images have been made by assigning colours to the relative reflectance values obtained through various filters, resulting in a map showing compositional variations. The large dark red-grey region on the far side is the huge South Pole–Aitken Basin; the pink area near the centre of the 120°E image may relate to ancient lava flows; the colour picture shows the titanium-rich lavas (deep blue) in the Mare Imbrium, Mare Serenitatis and Oceanus Procellarum (near side).*

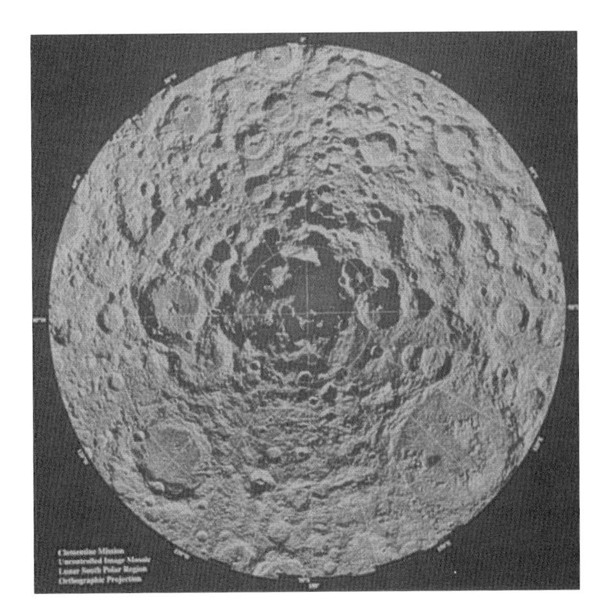

▲ ***The South Polar Region of the Moon** as imaged from Clementine, showing some of the huge craters.*

The Moon: First Quadrant (North-east)

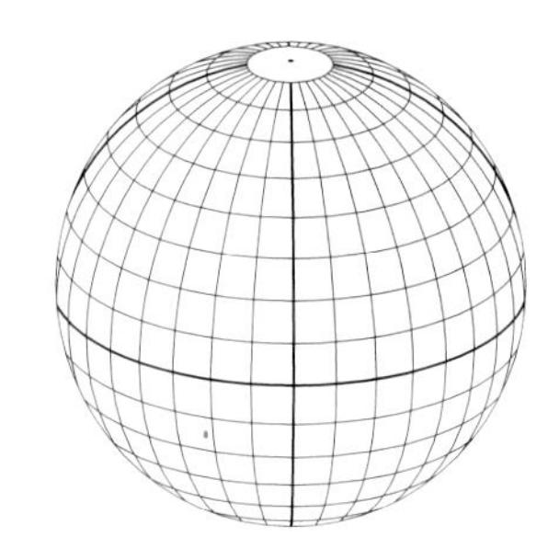

The First Quadrant is occupied largely by sea. The whole of the Mare Serenitatis and Mare Crisium are included, and most of the Mare Tranquillitatis and the darkish Mare Vaporum, with parts of the Mare Frigoris and Mare Foecunditatis. There are also some small seas close to the limb (Smythii, Marginis, Humboldtianum) which are never easy to observe because they are so foreshortened. There are also large walled plains close to the limb, such as Neper and Gauss.

In the south, the Mare Serenitatis is bordered by the Haemus Mountains, which rise to 2400 metres (7800 feet). The Alps run along the southern border of the Mare Frigoris, and here we find the magnificent Alpine Valley, which is 130 kilometres (80 miles) long and is much the finest formation of its type on the Moon; a delicate rill runs along its floor, and there are obscure parallel and transverse valleys. Mont Blanc, in the Alps, rises to 3500 metres (11,500 feet). Part of the Apennine range extends into this quadrant, with lofty peaks such as Mount Bradley and Mount Hadley, both over 4000 metres (13,000 feet) high. There are several major rill systems (Ariadaeus, Hyginus, Triesnecker, Ukert, Bürg) and an area near Arago which is rich in domes. The Apollo 11 astronauts landed in the Mare Tranquillitatis, not far from Maskelyne, and Apollo 17 came down in the area of Littrow and the clumps of hills which are called the Taurus Mountains.

Agrippa *A fine crater with a central peak and terraced walls. It forms a notable pair with its slightly smaller neighbour* ***Godin****.*

Arago *A well-formed crater, with the smaller, bright* ***Manners*** *to the south-east. Close to Arago is a whole collection of domes – some of the finest on the Moon; many of them have summit craterlets.*

Archytas *The most prominent crater on the irregular Mare Frigoris. It has bright walls and a central peak.*

Ariadaeus *A small crater associated with a major rill system. The main rill is almost 250 kilometres (150 miles) long, and has various branches, one of which connects the system with that of* ***Hyginus*** *– which is curved, and is mainly a craterlet-chain. Another complex rill system is associated with* ***Triesnecker*** *and* ***Ukert****. All these features are visible with a small telescope under good conditions.*

Aristillus *This makes up a group together with* ***Archimedes*** *(which is shown on the map of the Second Quadrant) and* ***Autolycus****. All three are very prominent. Under high illumination Autolycus is also seen to be the centre of a minor ray-system.*

Aristoteles *This and* ***Eudoxus*** *form a prominent pair of walled plains. Aristoteles has walls rising to 3300 metres (11,000 feet) above the floor.* ***Atlas*** *and* ***Hercules*** *form another imposing pair. Atlas has complex floor-detail, while inside Hercules there is one very bright crater.*

Bessel *The main formation on the Mare Serenitatis; a well-formed crater close to a long ray which crosses the mare and seems to belong to the Tycho system.*

Bürg *A crater with a concave floor; the very large central peak is crowned by a craterlet (****Rømer*** *is another example of this.) Bürg stands on the edge of a dark plain which is riddled with rills.*

Challis *This and* ***Main*** *form a pair of 'Siamese twins' – a phenomenon also found elsewhere, as with Steinheil and Watt in the Fourth Quadrant.*

Cleomedes *A magnificent enclosure north of Mare Crisium. The wall is interrupted by one very deep crater,* ***Tralles****.*

Dionysius *One of several very brilliant small craterlets in the rough region between Mare Tranquillitatis and Mare Vaporum; others are* ***Cayley****,* ***Whewell*** *and* ***Silberschlag****.*

Endymion *A large enclosure with a darkish floor. It joins the larger but very deformed* ***De la Rue****.*

Gioja *The north polar crater – obviously not easy to examine from Earth. It is well formed, and intrudes into a larger but low-walled formation.*

Julius Caesar *This and* ***Boscovich*** *are low-walled, irregular formations, notable because of their very dark floors.*

Le Monnier *A fine example of a bay, leading off the Mare Serenitatis. Only a few mounds of its seaward wall remain.*

Linné *A famous formation. It was once suspected of having changed from a craterlet into a white spot at some time between 1838 and 1866, but this is certainly untrue. It is a small, bowl-shaped crater standing on a white patch.*

Manilius *A fine crater near Mare Vaporum, with brilliant walls; it is very prominent around the time of Full Moon. So too is* ***Menelaus****, in the Haemus Mountains.*

Picard *This and* ***Peirce*** *are the only prominent craters in the Mare Crisium.*

Plinius *A superb crater 'standing sentinel' on the strait between Mare Serenitatis and Mare Tranquillitatis. It has high, terraced walls; the central structure takes the form of a twin crater.*

Posidonius *A walled plain with low, narrow walls and a floor crowded with detail. It forms a pair with its smaller neighbour* ***Chacornac****.*

Proclus *One of the most brilliant craters on the Moon. It is the centre of an asymmetrical ray system; two rays border the* ***Palus Somnii****, which has a curiously distinctive tone.*

Sabine *This and* ***Ritter*** *make up a pair of almost perfect twins – one of many such pairs on the Moon.*

Struve *A small crater, easy to find as it lies on a dark patch.*

Taruntius *A fine example of a concentric crater. There is a central mountain with a summit pit, and a complete inner ring on the floor. This sort of arrangement is difficult to explain by random impact.*

Thales *A crater near De la Rue, prominent near Full Moon because it is a ray-centre.*

Vitruvius *On the Mare Tranquillitatis, near the peak of Mount Argaeus. It has bright walls, with a darkish floor and a central peak.*

SELECTED CRATERS: FIRST QUADRANT

Crater	Diameter, km	Lat. °N	Long. °E
Agrippa	48	4	11
Apollonius	48	5	61
Arago	29	6	21
Archytas	34	59	5
Ariadaeus	15	5	17
Aristillus	56	34	1
Aristoteles	97	50	18
Atlas	69	47	44
Autolycus	36	31	1
Bessel	19	22	18
Bond, W.C.	160	64	3
Boscovich	43	10	11
Bürg	48	45	28
Cassini	58	40	5
Cauchy	13	10	39
Cayley	13	4	15
Challis	56	78	9
Chacornac	48	30	32
Cleomedes	126	27	55
Condorcet	72	12	70
De la Rue	160	67	56
Democritus	37	62	35
Dionysius	19	3	17
Endymion	117	55	55
Eudoxus	64	44	16
Firmicus	56	7	64
Gärtner	101	60	34
Gauss	136	36	80
Geminus	90	36	57
Gioja	35	North polar	
Godin	43	2	10
Hercules	72	46	39
Hooke	43	41	55
Hyginus	6	8	6
Jansen	26	14	29
Julius Caesar	71	9	15
Le Monnier	55	26	31
Linné	11	28	12
Littrow	35	22	31
Macrobius	68	21	46
Main	48	81	9
Manilius	36	15	9
Manners	16	5	20
Maskelyne	24	2	30
Mason	31	43	30
Menelaus	32	16	16
Messala	128	39	60
Neper	113	7	83
Peirce	19	18	53
Picard	34	15	55
Plana	39	42	28
Plinius	48	15	24
Posidonius	96	32	30
Proclus	29	16	47
Rømer	37	25	37
Ritter	32	2	19
Sabine	31	2	20
Struve	18	43	65
Sulpicius Gallus	13	20	12
Taquet	10	17	19
Taruntius	60	6	48
Thales	39	59	41
Theaetetus	26	37	6
Tralles	48	28	53
Triesnecker	23	4	4
Ukert	23	8	1
Vitruvius	31	18	31

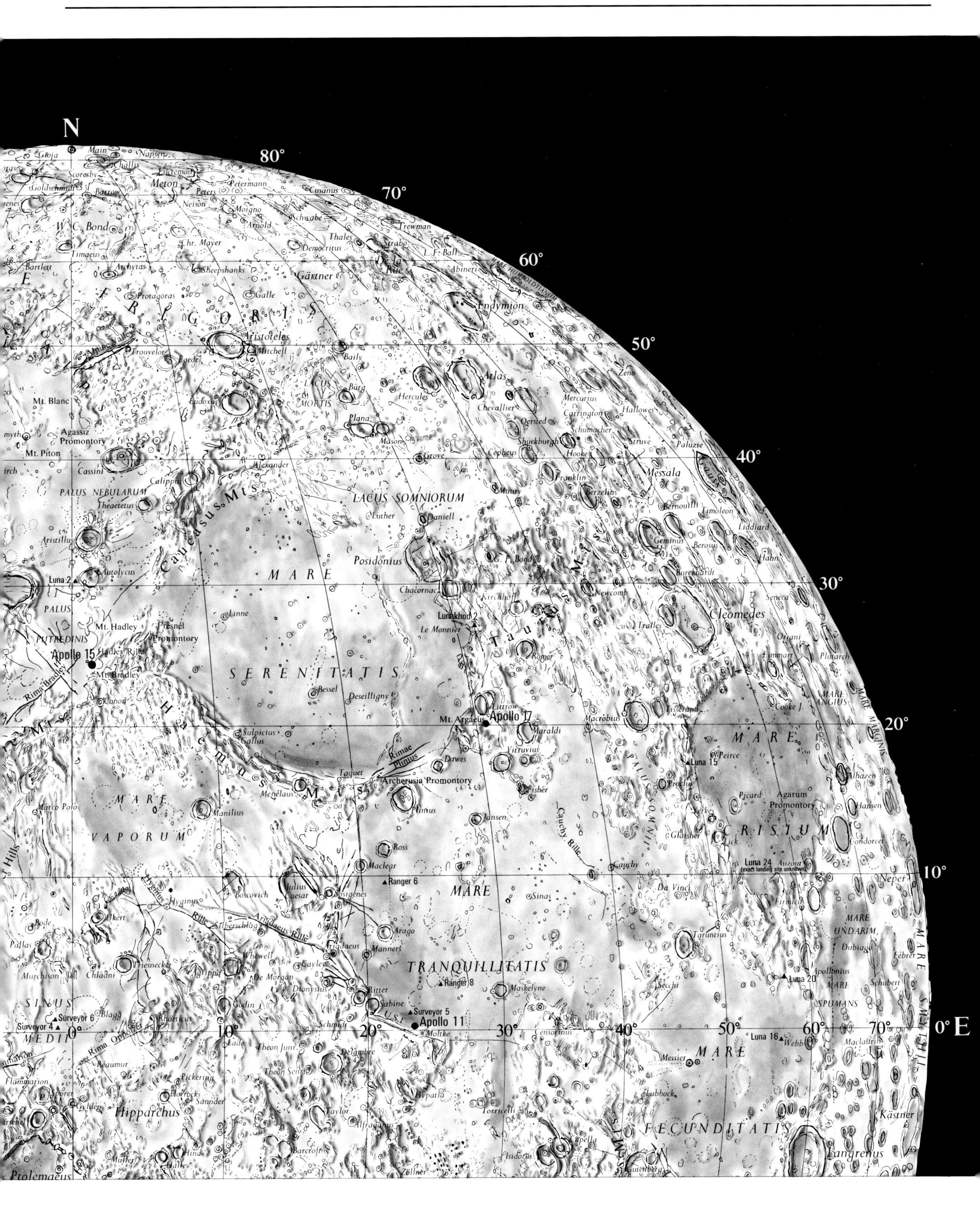
N
80°
70°
60°
50°
40°
30°
20°
10°
0° E
0°
10°
20°
30°
40°
50°
60°
70°
Gioja
Main
Nansen
Challis
Scoresby
Goldschmidt
Barrow
Meton
Peters
Petermann
Cusanus
Neison
Moigno
Schwabe
Trewman
W. C. Bond
Arnold
Thales
Strabo
Chr. Mayer
Democritus
L. F. Ball
Timaeus
De la Rue
Abineri
Bartlett
Archytas
Sheepshanks
Gärtner
Protagoras
Galle
Endymion
Aristoteles
Mitchell
Trouvelot
Egede
Baily
Atlas
Zeno
LACUS MORTIS
Bürg
Hercules
Mercurius
Chevallier
Carrington
Hallowes
Mt. Blanc
Eudoxus
Plana
Oersted
Schumacher
Agassiz Promontory
Mason
Shuckburgh
Struve
Paluzie
Mt. Piton
Grove
Cepheus
Hooke
Cassini
Alexander
Calippus
Franklin
Messala
PALUS NEBULARUM
LACUS SOMNIORUM
Maury
Berzelius
Theaetetus
Caucasus Mts.
Luther
Daniell
Bernoulli
Timoleon
Liddiard
Aristillus
Geminus
Berosus
Posidonius
Hahn
G. P. Bond
Luna 2
Autolycus
MARE
Burckhardt
Chacornac
Kirchhoff
Newcomb
Seneca
PALUS PUTREDINIS
Linne
Lunakhod 2
Le Monnier
Tralles
Cleomedes
Mt. Hadley
Fresnel Promontory
Apollo 15
Hadley Rille
Taurus Mts.
Römer
Plutarch
SERENITATIS
Mt. Bradley
Rima Bradley
Bessel
Deseilligny
Littrow
MARE ANGUIS
MARE MARGINIS
Conon
Haemus Mts.
Mt. Argaeus
Apollo 17
Macrobius
Tisserand
Cooke J.
Sulpicius Gallus
Maraldi
MARE
Peirce
Rimae Plinius
Vitruvius
Luna 15
Dawes
Taquet
Archerusia Promontory
PALUS SOMNII
Alhazen
Proclus
Menelaus
Fisher
Picard
Agarum Promontory
Marco Polo
MARE
VAPORUM
Manilius
Plinius
Hansen
Jansen
Yerkes
Cauchy Rille
Glaisher
Lick
CRISIUM
Condorcet
Ross
Luna 24 (exact landing site unknown)
Auzout
Maclear
Cauchy
Neper
Ranger 6
MARE
Sinas
Da Vinci
Boscovich
Julius Caesar
Sosigenes
Hyginus
Firmicus
Ukert
Silberschlag
Ariadaeus Rille
Arago
MARE UNDARUM
Dubiago
Bode
Manners
Taruntius
Pallas
Whewell
Cayley
TRANQUILLITATIS
Apollonius
Schubert
Murchison
Chladni
Triesnecker
Agrippa
De Morgan
Dionysius
Maskelyne
Ranger 8
Secchi
Luna 20
MARE SPUMANS
SINUS
MEDII
Godin
Ritter
Sabine
Surveyor 5
Apollo 11
Surveyor 6
Blagg
Rhaeticus
Surveyor 4
Schmidt
Moltke
Censorinus
Luna 16
Webb
Maclaurin
MARE SMYTHII
Rima Oppolzer
Lade
Theon Junr.
Delambre
Messier
MARE
Réaumur
Theon Senr.
Flammarion
Pickering
Horrocks
Lubbock
Sanders
Hipparchus
Hypatia
Torricelli
Halley
Taylor
Alfraganus
FECUNDITATIS
Kästner
Langrenus
Müller
Hind
Isidorus
Gutenberg
Zöllner
Ptolemaeus

The Moon: Second Quadrant (North-west)

This is the 'marine quadrant', containing virtually the whole of the Mare Imbrium and most of the Oceanus Procellarum, as well as the Sinus Aestuum, Sinus Roris, a small part of the Sinus Medii and a section of the narrow, irregular Mare Frigoris. The Sinus Iridum, leading off the Mare Imbrium, is perhaps the most beautiful object on the entire Moon when observed at sunrise or sunset, when the solar rays catch the top of the Jura Mountains which border it. There are two prominent capes, Laplace and Heraclides; the seaward wall of the bay has been virtually levelled. It was in this area that Russia's first 'crawler', Lunokhod 1, came down in 1970.

The Apennines make up the most conspicuous mountain range on the Moon; with the lower Carpathians in the south, they make up much of the border of the Mare Imbrium. The Straight Range, in the northern part of the Mare, is made up of a remarkable line of peaks rising to over 1500 metres (5000 feet); the range is curiously regular, and there is nothing else quite like it on the Moon. The Harbinger Mountains, in the Aristarchus area, are made up of irregular clumps of hills. Isolated peaks include Pico and Piton, in the Mare Imbrium. Pico is very conspicuous, and is 2400 metres (7900 feet) high; the area between it and Plato is occupied by a ghost ring which was once called Newton, though the name has now been transferred to a deep formation in the southern uplands and the ghost has been relegated to anonymity.

Anaxagoras *A well-formed crater with high walls and central peak. It is very bright, and is the centre of a major ray-system, so that it is easy to find under all conditions of illumination.*

Archimedes *One of the best-known of all walled plains; regular, with a relatively smooth floor. It forms a splendid trio with Aristillus and Autolycus, which lie in the First Quadrant.*

Aristarchus *The brightest crater on the Moon. Its brilliant walls and central peak make it prominent even when lit only by earthshine; there are strange darkish bands running from the central peak to the walls. Close by is* ***Herodotus****, of similar size but normal brightness. This is the area of the great* ***Schröter Valley****, which begins in a 6-kilometre (4-mile) crater outside Herodotus; broadens to 10 kilometres (6 miles), producing the feature nicknamed the Cobra-Head, and then winds its way across the plain. The total length is 160 kilometres (100 miles), and the maximum depth 1000 metres (3300 feet). It was discovered by the German astronomer Johann Schröter, and is called after him, though Schröter's own crater is a long way away in the area of Sinus Medii and Sinus Aestuum. Many TLP have been recorded in this area.*

Beer *This and* ***Feuillé*** *are nearly identical twins – one of the most obvious craterlet-pairs on the Moon.*

Birmingham *Named not after the city, but after an Irish astronomer. It is low-walled and broken, and one of several formations of similar type in the far north; others are* ***Babbage****,* ***South*** *and* ***John Herschel****.*

Carlini *One of a number of small, bright-walled craterlets in the Mare Imbrium. Others are* ***Caroline Herschel****,* ***Diophantus****,* ***De l'Isle*** *and* ***Gruithuisen****.*

Copernicus *The 'Monarch of the Moon', with high, terraced walls and a complex central mountain group. Its ray-system is second only to that of Tycho, so that at or near Full Moon it dominates the entire area.*

Einstein *A great formation in the limb region, beyond the low, double and very reduced* ***Otto Struve****. Einstein has a large central crater. It is visible only under conditions of favourable libration – as when I discovered it in 1945, using my 30-centimetre (12-inch) reflector.*

Eratosthenes *A magnificent crater, with massive walls and a high central peak; it marks one end of the Apennines, and is very like Copernicus apart from the fact that it lacks a comparable ray-system. East of it is* ***Stadius****, a typical ghost ring; it has a diameter of 70 kilometres (44 miles), but its walls have been so reduced that they are barely traceable. Probably the walls can be nowhere more than about 10 metres (33 feet) high.*

Hevel *One of the great chain which includes Grimaldi and Riccioli (in the Third Quadrant) and* ***Cavalerius****. Hevel has a convex floor and a low central peak; a system of rills lies on the floor. West of Hevel is* ***Sven Hedin****, visible only under conditions of extreme libration; it is 98 kilometres (61 miles) in diameter, with irregular, broken walls.*

Kepler *A bright crater, and the centre of a major ray-system. Its southern neighbour,* ***Encke****, is of about the same size, but is much less bright and has no comparable ray-system.*

Le Verrier *This and* ***Helicon*** *make up a prominent crater-pair in the Mare Imbrium, near Sinus Iridum.*

Lichtenberg *A small crater which glows against the dark mare surface. Unusual coloration effects have been reported here.*

Plato *A large walled plain with fairly low walls, and an iron-grey floor which makes it readily identifiable under any conditions of illumination. There are a few craterlets on the floor, some of which can be 'missed' when they ought logically to be visible. Plato is perfectly circular, though as seen from Earth it is foreshortened into an oval.*

Pythagoras *Were it further on the disk, Pythagoras would be truly magnificent, with its high, terraced walls and massive central peak. Further along the limb, to the south, is the smaller, similar but still very imposing* ***Xenophanes****.*

Timocharis *A well-marked formation with a central crater (a peculiarity which it shares with* ***Lambert****). Timocharis is the centre of a rather obscure system of rays.*

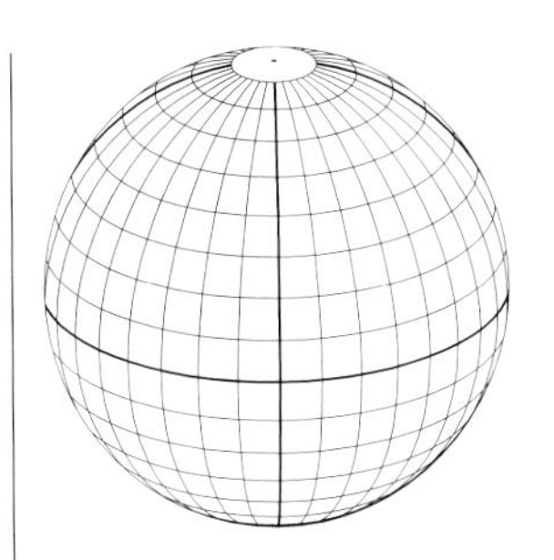

SELECTED CRATERS: SECOND QUADRANT

Crater	Diameter, km	Lat. °N	Long. °W	Crater	Diameter, km	Lat. °N	Long. °W
Anaxagoras	52	75	10	Hevel	122	2	67
Anaximander	87	66	48	Horrebow	32	59	41
Anaximenes	72	75	45	Hortensius	16	6	28
Archimedes	75	30	4	Kepler	35	8	38
Aristarchus	37	24	48	Kirch	11	39	6
Beer	11	27	9	Krafft	51	17	72
Bessarion	15	37	10	Kunowsky	31	3	32
Bianchini	40	49	34	Lambert	29	26	21
Birmingham	106	64	10	Lansberg	42	0	26
Bode	18	7	2	La Voisier	71	36	70
Briggs	38	26	69	Le Verrier	25	40	20
Cardanus	52	13	73	Lichtenberg	19	32	68
Carlini	8	34	24	Marius	42	12	51
Cleostratus	70	60	74	Mayer, Tobias	35	16	29
Condamine	48	53	28	Milichius	13	10	30
Copernicus	97	10	20	Oenopides	68	57	65
De l'Isle	22	30	35	Olbers	64	7	78
Diophantus	18	28	34	Otto Struve	160	25	75
Encke	32	5	37	Pallas	47	5	2
Epigenes	52	73	4	Philolaus	74	75	33
Einstein	160	18	86	Piazzi Smyth	10	42	3
Eratosthenes	61	15	11	Plato	97	51	9
Euler	25	23	29	Pythagoras	113	65	65
Feuillé	13	27	10	Pytheas	19	21	20
Gay-Lussac	24	14	21	Reinhold	48	3	23
Gambart	26	1	15	Repsold	140	50	70
Gérard	87	44	75	Schiaparelli	29	23	59
Goldschmidt	109	75	0	Schröter	32	3	7
Gruithuisen	16	33	40	Seleucus	45	21	66
Harding	23	43	70	Sümmering	27	0	7
Harpalus	52	53	43	South	98	57	50
Helicon	29	40	23	Timaeus	34	63	1
Herodotus	37	23	50	Timocharis	35	17	13
Herschel, Caroline	13	34	31	Ulugh Beigh	70	29	85
Herschel, John	145	62	41	Xenophanes	108	57	77

The Moon: Third Quadrant (South-west)

Highlands occupy a large part of the Third Quadrant, though part of the huge Mare Nubium is included together with the whole of the Mare Humorum. There are some high mountains on the limb, and a very small part of the Mare Orientale can be seen under really favourable libration; otherwise the main mountains are those of the small but prominent Riphaean range, on the Mare Nubium. Of course the most prominent crater is Tycho, whose rays dominate the entire surface around the time of Full Moon. This quadrant also includes two of the most prominent chains of walled plains, those of Ptolemaeus and Walter; the dark-floored Grimaldi and Riccioli; the celebrated plateau Wargentin, and the inappropriately-named Straight Wall. The most important rill systems are those of Sirsalis, Ramsden, Hippalus and Mersenius.

Bailly *One of the largest walled plains on the Moon, but unfortunately very foreshortened. It has complex floor detail, and has been described as 'a field of ruins'.*

Billy *This and **Crüger** are well-formed, and notable because of their very dark floors, which make them easily identifiable.*

Bullialdus *A particularly fine crater, with massive walls and central peak. It is not unlike Copernicus in structure, though it is not a ray-centre.*

Capuanus *A well-formed crater, with a darkish floor upon which there is a whole collection of domes.*

Clavius *A vast walled plain, with walls rising to over 4000 metres (13,000 feet). The north-western walls are broken by a large crater, **Porter**, and there is a chain of craters arranged in an arc across the floor. Near the terminator, Clavius can be seen with the naked eye.*

Euclides *A small crater near the Riphæan Mountains, easy to find because it is surrounded by a bright nimbus.*

Fra Mauro *One of a group of low-walled, reduced formations on the Mare Nubium (the others are **Bonpland**, **Parry** and **Guericke**). Apollo 14 landed near here.*

Gassendi *A grand crater on the north border of the Mare Humorum. The wall has been reduced in places, and is broken in the north by a large crater. There is a rill-system on the floor, and TLP have been seen here. North of Gassendi is a large bay, **Letronne**.*

Grimaldi *The darkest formation on the Moon. The walls are discontinuous, but contain peaks rising to 2500 metres (8000 feet). Adjoining it is **Riccioli**, which is less regular but has one patch on its floor almost as dark as any part of Grimaldi.*

Hippalus *A fine bay in the Mare Humorum, associated with a system of rills. Like another similar bay, **Doppelmayer**, Hippalus has the remnant of a central-peak.*

Kies *A low-walled crater on the Mare Nubium, with a flooded floor. Near it lies a large dome with a summit craterlet.*

Maginus *A very large formation with irregular walls; other large walled plains of the same type in the area are **Longomontanus** and **Wilhelm I**. Maginus is curiously obscure around the time of Full Moon.*

Mercator *This and **Campanus** form a notable pair. They are alike in form and shape, but Mercator has the darker floor.*

Mersenius *A prominent walled plain closely west of the Mare Humorum, associated with a fine system of rills.*

Moretus *A very deep formation in the southern uplands, with a particularly fine central peak.*

Newton *One of the deepest formations on the Moon, but never well seen because it is so close to the limb.*

Pitatus *This has been described as a 'lagoon' on the coast of the Mare Nubium. It has a dark floor and a low central peak. A pass connects it with the neighbour **Hesiodus**, which is associated with a long rill extending south-westwards.*

Ptolemaeus *The largest member of the most imposing line of walled plains on the Moon. Ptolemaeus has a flattish floor with one large crater, **Ammonius**; **Alphonsus** has a central peak and a system of rills on its floor; **Arzachel** is smaller, but with higher walls and a more developed central peak. Several TLP have been seen in Alphonsus. Nearby is **Alpetragius**, with regular walls and a central peak crowned by a craterlet.*

Purbach *One of a line of three major walled plains on the edge of the Mare Nubium. The other members are **Walter**, which has fairly regular walls, and **Regiomontanus**, which gives the impression of having been squashed between Walter to the south and Purbach to the north.*

Scheiner *This and **Blancanus** are two large, important walled plains close to Clavius.*

Schickard *One of the major walled plains on the Moon. The walls are rather low and irregular; the floor contains some darkish patches as well as various hills and craterlets.*

Schiller *A compound formation, produced by the fusion of two old rings.*

Sirsalis *One of 'Siamese twins' with its neighbour Watt. It is associated with a long and very prominent rill.*

Thebit *Near the Straight Wall, the crater is broken by Thebit A, which is in turn broken by Thebit F.*

Tycho *The great ray-crater. Its bright walls make it prominent even in low illumination. Near Full Moon it is clear the rays come tangentially from the walls rather than from the centre.*

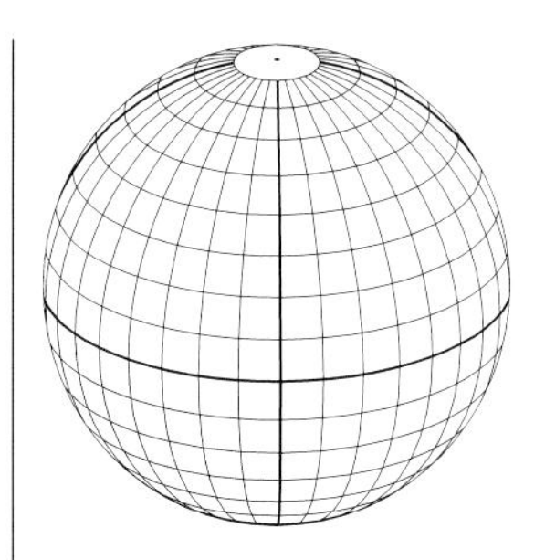

SELECTED CRATERS: THIRD QUADRANT

Crater	Diameter, km	Lat. °S	Long. °W	Crater	Diameter, km	Lat. °S	Long. °W
Agatharchides	48	20	31	Lalande	24	4	8
Alpetragius	43	16	4	Lassell	23	16	8
Alphonsus	129	13	3	Legentil	140	73	80
Arzachel	97	18	2	Letronne	113	10	43
Bailly	294	66	65	Lexell	63	36	4
Bayer	52	51	35	Lohrmann	45	1	67
Bettinus	66	63	45	Longomontanus	145	50	21
Billy	42	14	50	Maginus	177	50	6
Birt	18	22	9	Mercator	38	29	26
Blancanus	92	64	21	Mersenius	72	21	49
Bonpland	58	8	17	Moretus	105	70	8
Bullialdus	50	21	22	Mösting	26	1	6
Bouvard	127	37	87	Nasireddin	48	41	0
Byrgius	64	25	65	Newton	113	78	20
Cabaeus	140	85	20	Nicollet	15	22	12
Campanus	38	28	28	Orontius	84	40	4
Capuanus	56	34	26	Parry	42	8	16
Casatus	104	75	35	Phocylides	97	54	58
Clavius	232	56	14	Piazzi	90	36	68
Crüger	48	17	67	Pictet	48	43	7
Cysatus	47	66	7	Pitatus	86	30	14
Damoiseau	35	5	61	Ptolemaeus	148	14	3
Darwin	130	20	69	Purbach	120	25	2
Davy	32	12	8	Regiomantanus	129 × 105	28	0
Deslandres	186	32	61	Ricciolli	160	3	75
Doppelmayer	68	28	41	Rocca	97	15	72
Euclides	12	7	29	Saussure	50	43	4
Flammarion	72	3	4	Scheiner	113	60	28
Flamsteed	19	5	44	Schickard	202	44	54
Fra Mauro	81	6	17	Schiller	180 × 97	52	39
Gassendi	89	18	40	Segner	74	59	48
Gauricus	64	34	12	Short	70	76	5
Grimaldi	193	6	68	Sirsalis	32	13	60
Gruemberger	87	68	10	Thebit	60	22	4
Guericke	53	12	14	Tycho	84	43	11
Hainzel	97	41	34	Vieta	52	29	57
Hansteen	36	44	83	Vitello	38	30	38
Heinsius	72	32	18	Walter	129	33	1
Hell	31	32	8	Wargentin	89	50	60
Herigonius	16	13	14	Watt	71	50	51
Herschel	45	6	2	Weigel	55	58	39
Hesiodus	45	29	16	Wichmann	13	8	38
Hippalus	61	25	30	Wilhelm I	97	43	20
Inghirami	97	48	70	Wilson	74	69	33
Kies	42	26	23	Wurzelbauer	80	34	16
Kircher	74	67	45	Zucchius	63	61	50
Klaproth	119	70	26	Zupus	26	17	52
Lagrange	165	33	72				

The Moon: Fourth Quadrant (South-east)

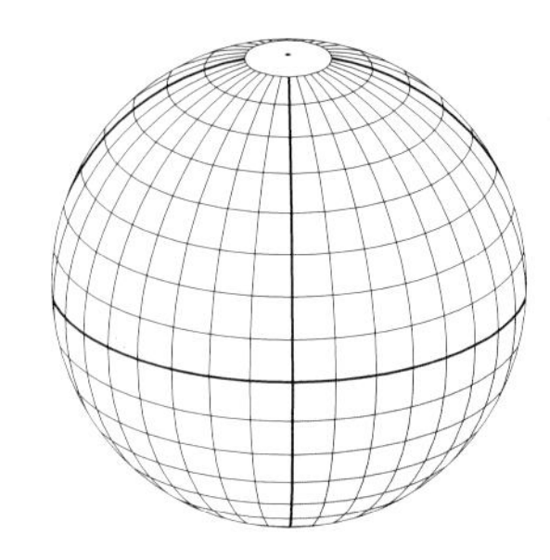

The Fourth Quadrant is made up mainly of highland, though it does contain the Mare Nectaris, part of the Mare Foecunditatis and the irregular limb-sea Mare Australe. There are some large ruined enclosures such as Janssen and Hipparchus, and three imposing formations in a group – Theophilus, Cyrillus and Catharina. We also find four members of the great Eastern Chain: Furnerius, Petavius, Vendelinus and Langrenus. There are two crater-valleys, those of Rheita and Reichenbach, plus the fascinating little Messier, which was once (wrongly) suspected of recent structural change. The feature once called the Altai Mountain range is now known as the Altai Scarp, which is certainly a better name for it; it is concentric with the border of the Mare Nectaris, and runs north-west from the prominent crater Piccolomini.

Alfraganus *A small, very bright crater; minor ray-centre.*
Aliacensis *This crater and its neighbour* ***Werner*** *are very regular. There are several rather similar crater-pairs in the Fourth Quadrant; others are* ***Abenezra-Azophi*** *and* ***Almanon-Abulfeda****.*
Capella *Crater cut by a valley, with a particularly large central peak with summit pit. Adjoins* ***Isidorus****, of similar size.*
Fracastorius *A great bay opening out of the Mare Nectaris. Its seaward wall has been virtually destroyed. Between it and Theophilus is a smaller bay,* ***Beaumont****.*
Goclenius*. A fairly regular crater, making up a group with less perfect* ***Gutenberg*** *and deformed* ***Magelhaens****.*
Hipparchus *A very large enclosure not far from Ptolemaeus. It is very broken, but under low light is still impressive. It adjoins* ***Albategnius****, which is rather better preserved and has a low central peak.*
Humbolt, Wilhelm *A huge formation, too foreshortened to be well seen – though the space probe pictures show that it has considerable floor detail, including a system of rills. It adjoins the rather smaller formation of* ***Phillips****, which is of similar type.*
Janssen *A vast enclosure, but in a very poor state of repair. Its walls are broken in the north by* ***Fabricius*** *and in the south by the bright-walled* ***Lockyer****.*
Langrenus *One of the great Eastern Chain. It has high, terraced walls, rising to over 3000 metres (10,000 feet), and a bright twin-peaked central elevation. Near full moon, Langrenus appears as a bright patch.*
Mädler *A prominent though irregular crater on the Mare Nectaris. It is crossed by a ridge.*
Messier *This and its twin, Messier A (formerly known as W. J. Pickering) lie on the Mare Foecunditatis. They show remarkable changes in appearance over a lunation, though there has certainly been no real change in historic times. The unique 'comet' ray extends to the west.*
Metius *A well-formed walled plain near Janssen.*
Oken *A crater along the limb from the Mare Australe, easy to identify because of its darkish floor.*
Petavius *A magnificent crater – one of the finest on the Moon. Its walls rise to over 3500 metres (11,500 feet) in places; the slightly convex floor contains a complex central mountain group, and there is a prominent rill running from the centre to the south-west wall. Oddly enough, Petavius is none too easy to identify at full moon. Immediately outside it is* ***Palitzsch****, once described as a 'gorge'. In fact, it is a crater-chain – several major rings which have coalesced.*
Piccolomini *The prominent, high-walled crater at the arc of the Altai Scarp.*
Rheita *A deep crater with sharp walls. Associated with it is the so-called 'Valley' over 180 kilometres (110 miles) long and in places up to 25 kilometres (15 miles) broad; it is not a true valley, but is made up of craterlets. Not far away is the* ***Reichenbach*** *valley, which is of similar type but is not so conspicuous or so well-formed.*
Steinheil *This and its neighbour* ***Watt*** *make up a pair of 'Siamese twins', not unlike Scheiner and Blancanus in the Third Quadrant.*
Stöfler *A grand enclosure, with an iron-grey floor which makes it easy to find. Part of the rampart has been destroyed by the intrusion of* ***Faraday****.*
Theon Senior *and* ***Theon Junior****. Very bright craterlets near the regular, conspicuous* ***Delambre****. In many ways they resemble Alfraganus.*
Theophilus *One of the most superb features of the Moon, and in every way the equal of Copernicus except that it is not a ray-centre. It is very deep, with peaks rising to 4400 metres (14,400 feet) above the floor. There is a magnificent central mountain group. It adjoins* ***Cyrillus****, which is less regular and in turn adjoins very rough-floored* ***Catharina****.*
Vendelinus*. A member of the Eastern Chain, but less regular than Langrenus or Petavius, and presumably older. It has no central peak, and in places the walls are broken.*
Vlacq *A deep, well-formed crater with a central peak; it is a member of a rather complex group, of which other members are* ***Hommel*** *and* ***Hagecius****.*
Webb *A crater very near the lunar equator, with a darkish floor and a central hill; centre of system of short, faint rays.*

SELECTED CRATERS: FOURTH QUADRANT

Crater	Diameter, km	Lat. °S	Long. °E	Crater	Diameter, km	Lat. °S	Long. °E
Abenezra	43	21	12	La Péyrouse	72	10	78
Abulfeda	64	14	14	Legendre	74	29	70
Airy	35	18	6	Legentil	140	73	80
Alfraganus	19	6	19	Licetus	74	47	6
Aliacensis	84	31	5	Lilius	52	54	6
Apianus	63	27	8	Lindenau	56	32	25
Azophi	43	22	13	Lockyer	48	46	37
Barocius	80	45	17	Maclaurin	45	2	68
Beaumont	48	18	29	Mädler	32	11	30
Blanchinus	53	25	3	Magelhaens	40	12	44
Boguslawsky	97	75	45	Manzinus	90	68	25
Bohnenberger	35	16	40	Marinus	48	50	75
Boussingault	78	70	50	Messier	13	2	48
Brisbane	47	50	65	Metius	81	40	44
Buch	48	39	18	Mutus	81	63	30
Büsching	58	38	20	Neander	48	31	40
Capella	48	8	36	Nearch	61	58	39
Catharina	89	18	24	Oken	80	44	78
Cyrillus	97	13	24	Palitzsch	97 × 32	28	64
Delambre	52	2	18	Parrot	64	15	3
Demonax	121	85	35	Petavius	170	25	61
Donati	35	21	5	Phillips	120	26	78
Fabricius	89	43	42	Piccolomini	80	30	32
Faraday	64	42	18	Pitiscus	80	51	31
Faye	35	21	4	Playfair	43	23	9
Fermat	40	23	20	Pons	32	25	22
Fernelius	64	38	5	Pontécoulant	97	69	65
Fracastorius	97	21	33	Rabbi Levi	80	35	24
Furnerius	129	36	60	Réaumur	45	2	1
Goclenius	52	10	45	Reichenbach	48	30	48
Gutenberg	72	8	41	Rheita	68	37	47
Hagecius	81	60	46	Riccius	80	37	26
Halley	35	8	6	Rosse	16	18	35
Hekataeus	180	23	84	Sacrobosco	84	24	17
Helmholtz	97	72	78	Steinheil	70	50	48
Hind	26	8	7	Stevinus	70	33	54
Hipparchus	145	6	5	Stöfler	145	41	6
Albategnius	129	12	4	Tacitus	40	16	19
Hommel	121	54	33	Theon Junior	16	2	16
Horrocks	29	4	6	Theon Senior	17	1	15
Humboldt, Wilhelm	193	27	81	Theophilus	101	12	26
				Torricelli	19	5	29
Isidorus	48	8	33	Vendelinus	165	16	62
Janssen	170	46	40	Vlacq	90	53	39
Kant	30	11	20	Watt	72	50	51
Lacaille	53	24	1	Webb	26	1	60
Langrenus	137	9	61	Werner	66	28	3

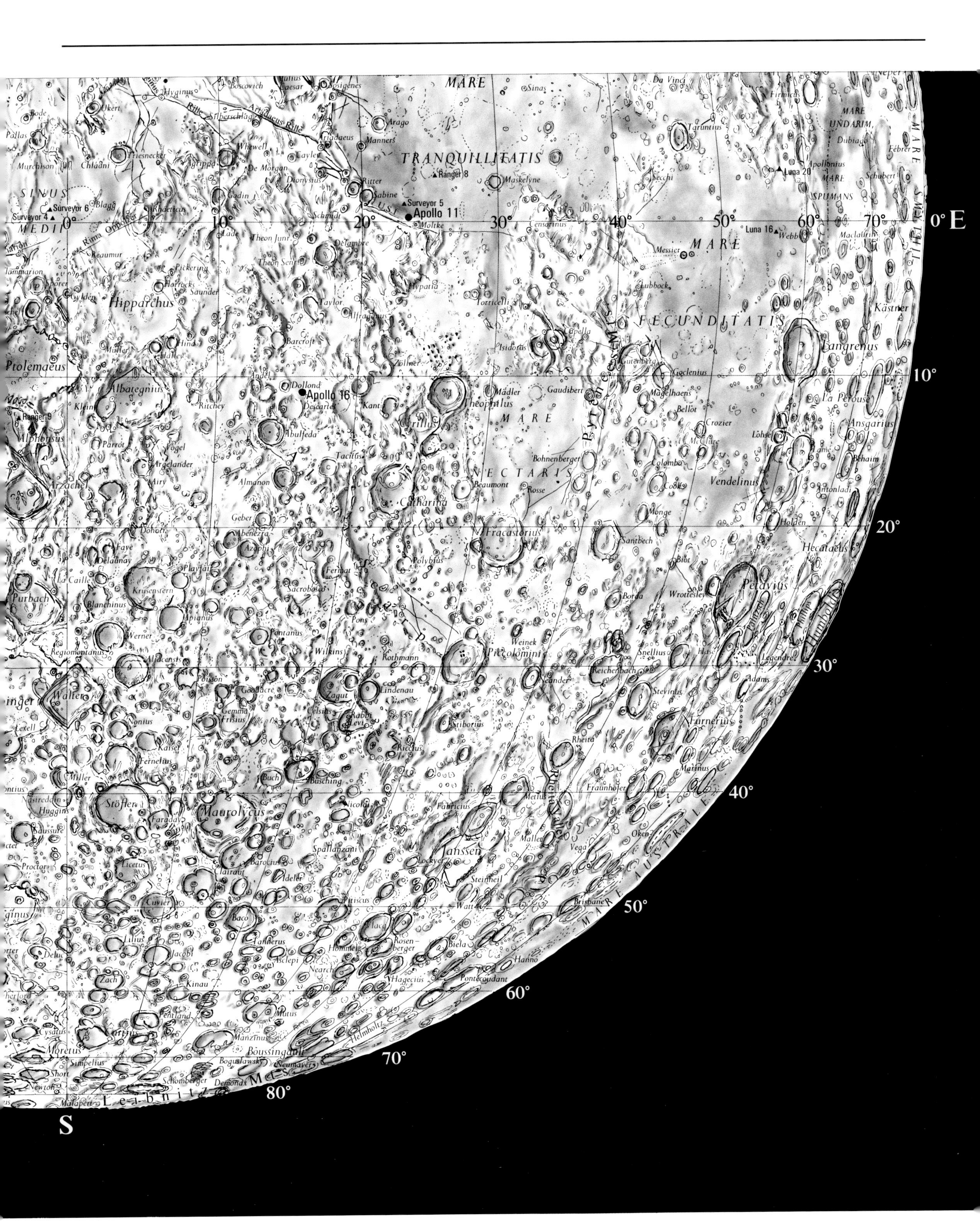
MARE
TRANQUILLITATIS
MARE
FECUNDITATIS
MARE
NECTARIS
MARE UNDARUM
MARE SPUMANS
MARE AUSTRALE
SINUS MEDII
Apollo 11
Apollo 16
Surveyor 5
Surveyor 6
Surveyor 4
Ranger 8
Ranger 9
Luna 20
Luna 16
Hipparchus
Ptolemaeus
Albategnius
Alphonsus
Arzachel
Catharina
Cyrillus
Theophilus
Fracastorius
Langrenus
Vendelinus
Petavius
Humboldt
Piccolomini
Walter
Stöfler
Maurolycus
Janssen
Furnerius
Boussingault
Leibnitz
0° E
10°
20°
30°
40°
50°
60°
70°
80°
S

Movements of the Planets

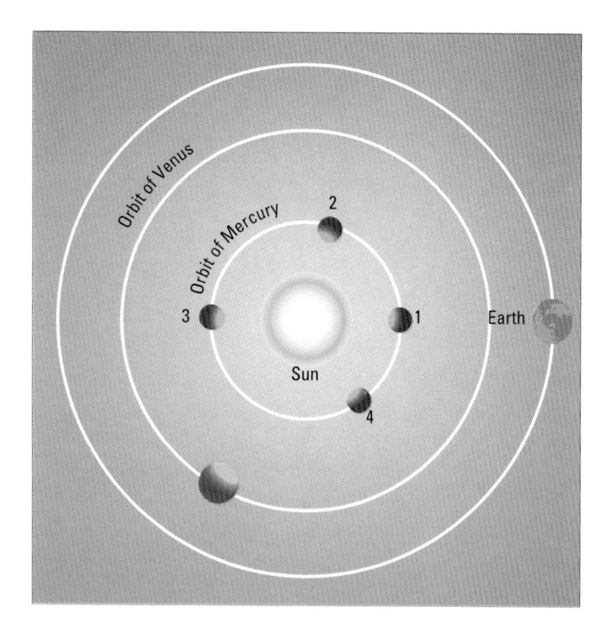

◀ ***Phases of Mercury**. (1) New. (2) Dichotomy (half-phase). (3) Full. (4) Dichotomy. For the sake of clarity, I have not taken the Earth's movement round the Sun into account in this diagram.*

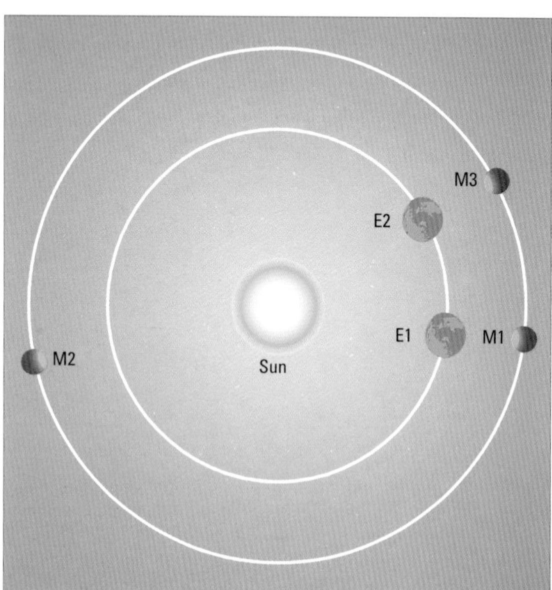

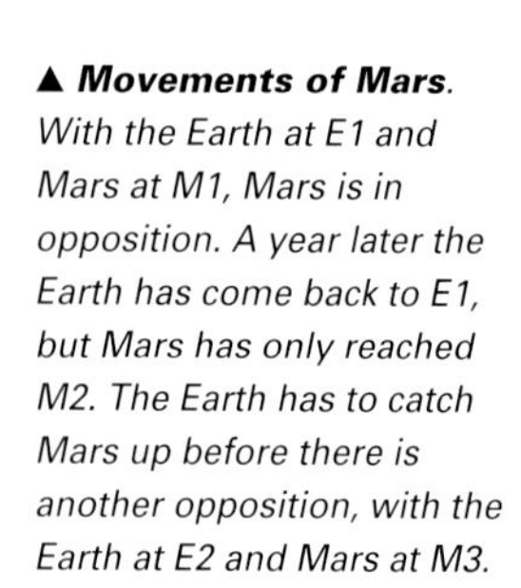

▲ ***Movements of Mars**. With the Earth at E1 and Mars at M1, Mars is in opposition. A year later the Earth has come back to E1, but Mars has only reached M2. The Earth has to catch Mars up before there is another opposition, with the Earth at E2 and Mars at M3.*

▶ ***Retrograde motion of Mars**. As the Earth catches Mars up and passes it, the movement will seem to be retrograde, so that between 3 and 5 Mars will appear to go backwards in the sky – east to west, against the stars, instead of west to east.*

The word 'planet' really means 'wanderer', and the planets were first identified in ancient times by their movements against the starry background. Because their orbits are not greatly inclined to that of the Earth – less than 4° for all the planets apart from Mercury and the exceptional Pluto – they seem to keep to a well-defined band around the sky, termed the Zodiac. There are twelve official Zodiacal constellations, though a thirteenth, Ophiuchus (the Serpent-bearer) does cross the zone for some distance.

The 'inferior' planets, Mercury and Venus, are closer to the Sun than we are, and have their own way of behaving. They seem to stay in the same general area of the sky as the Sun, which makes them awkward to observe – particularly in the case of Mercury, where the greatest elongation from the Sun can never be as much as 30 degrees. They show phases similar to those of the Moon, from new to full, but there are marked differences. At new phase, the dark side of the planet is turned towards us, and we cannot see it at all unless the alignment is perfect, when the planet will appear in transit as a dark disk crossing the face of the Sun. This does not happen very often; Venus was last in transit in 1882, and will not be so again until 8 June 2004. Transits of Mercury are less uncommon; the last was on 6 November 1993; followed by that of 15 November 1999. (*En passant*, there can surely be nobody now living who can remember seeing a transit of Venus!)

When an inferior planet is full, it is on the far side of the Sun, and is to all intents and purposes out of view. At other times the phase may be crescent, half (dichotomy), or gibbous (between half and full). At new, the planet is at inferior conjunction; when full, it is at superior conjunction. These movements mean that the inferior planets are best seen either in the west after sunset, or in the east before sunrise. They never remain above the horizon throughout a night.

The superior planets, the orbits of which lie beyond that of the Earth in the Solar System, can reach superior conjunction – though for obvious reasons they can never pass through inferior conjunction. When seen at right angles to the Sun, they are said to be at quadrature. When near quadrature Mars can show an appreciable phase – down to 85 per cent – so that when viewed through a telescope its shape resembles that of the Moon a day or two from full. The giant planets are so far away that their phases are inappreciable.

When the Sun, the Earth and a planet are lined up, with the Earth in the mid position, the planet is at opposi-

◀ ***Venus near the Moon***. *Venus almost occulted by the Moon on 5 October 1980, photographed through a 30-cm (12-inch) reflector.*

tion; it is exactly opposite to the Sun in the sky, and is best placed for observation. The interval between one opposition and the next is known as the synodic period.

The movements of Mars are shown in the next two diagrams. It is clear that oppositions do not occur every year; the Earth has to 'catch Mars up', and the mean synodic period is 780 days. Oppositions of Mars occur in 1999, 2001 and 2003, but not in 1998 or 2000. As the Earth 'passes' Mars, there is a period when the planet will move against the stars in an east-to-west or retrograde direction. The giant planets are so much further away, and move so much more slowly, that they come to opposition every year. Jupiter's synodic period is 399 days, but that of Neptune is only 367.5 days, so that it comes to opposition less than two days later every year.

There should be no trouble in identifying Venus and Jupiter, because they are always so brilliant; Mercury is unlikely to be seen unless deliberately looked for, while Uranus is on the fringe of naked-eye visibility, and Neptune and Pluto are much fainter. Mars at its best can actually outshine all the planets apart from Venus, but when at its faintest it is little brighter than the Pole Star, though its strong red colour will usually betray it. Saturn is brighter than most of the stars, and because it takes almost 30 years to complete one journey round the Zodiac it can be found without difficulty once initially identified.

Planets can pass behind the Moon, and be occulted. The planets themselves may occult stars, and these events are interesting to watch, but they do not happen very often, and an occultation of a bright star by a planet is very rare.

Mercury

Mercury, the innermost planet, is never easy to study from Earth. It is small, with a diameter of only 4878 kilometres (3030 miles); it always stays in the same region of the sky as the Sun, and it never comes much within 80 million kilometres (50 million miles) of us. Moreover, when it is at its nearest it is new, and cannot be seen at all except during the rare transits.

Mercury has a low escape velocity, and it has always been clear that it can have little in the way of atmosphere. The orbital period is 88 days. It was once assumed that this was also the length of the axial rotation period, in which case Mercury would always keep the same face turned towards the Sun, just as the Moon does with respect to the Earth; there would be an area of permanent day, a region of everlasting night, and a narrow 'twilight zone' in between, over which the Sun would bob up and down over the horizon – because the orbit of Mercury is decidedly eccentric, and there would be marked libration effects. However, this has been shown to be wrong. The real rotation period is 58.6 days, or two-thirds of a Mercurian year, and this leads to a very curious calendar indeed. To an observer on the planet's surface, the interval between sunrise and sunset would be 88 Earth-days.

The orbital eccentricity makes matters even stranger, because the heat received at perihelion is 2½ times greater than at aphelion. At a 'hot pole', where the Sun is overhead at perihelion, the temperature rises to +127 degrees C, but at night a thermometer would register −183 degrees C. Mercury has an extremely uncomfortable climate.

To an observer situated at a hot pole, the Sun will rise when Mercury is at aphelion, and the solar disk will be at its smallest. As the Sun nears the zenith, it will grow in size, but for a while the orbital angular velocity will be greater than the constant spin angular velocity; our observer will see the Sun pass the zenith, stop, and move backwards in the sky for eight Earth-days before resuming its original direction of motion. There are two hot poles, one or the other of which will always receive the full blast of solar radiation when Mercury is at perihelion. An observer 90 degrees away will have a different experience; the Sun will rise at perihelion, so that after first coming into view it will sink again before starting its climb to the zenith. At sunset it will disappear, and then rise again briefly before finally departing, not to rise again for another 88 Earth-days.

Mercury has a globe which is denser than that of any other planet apart from the Earth. There seems to be an iron-rich core about 3600 kilometres (2250 miles) in diameter (larger than the whole of the Moon), containing about 80 per cent of the total mass; by weight Mercury is 70 per cent iron and only 30 per cent rocky material. The core is presumably molten, and above it comes a 600-kilometre (370-mile) mantle and crust composed of silicates.

Most of our detailed knowledge of Mercury has been obtained from one probe, Mariner 10. It was launched on 3 November 1973, and after by-passing the Moon made rendezvous with Venus on 5 February 1974. The gravity field of Venus was used to send Mariner in towards an encounter with Mercury, and altogether there were three active passes before contact was lost: on 29 March and 21 September 1974, and 16 March 1975, by which time the equipment was starting to fail. The last messages were received on 24 March 1975, though no doubt Mariner is still orbiting the Sun and still making periodical approaches to Mercury.

As expected, the atmosphere proved to be almost non-existent. The ground pressure is about 1/10,000,000,000 of a millibar, and the main constituent is helium, presumably drawn from the solar wind. A magnetic field was detected, with a surface value about one per cent of the Earth's field; there are two magnetic poles of opposite polarity, inclined by 11 degrees to the rotational axis. The polarity of the field is the same as ours; that is to say, a compass needle would point north. The field is just strong enough to deflect the solar wind away from the planet's surface.

It has to be admitted that Mercury is not a rewarding telescopic object, and little will be seen apart from the characteristic phase. Any form of life there seems to be totally out of the question.

PLANETARY DATA – MERCURY

Sidereal period	87.969 days
Rotation period	58.6461 days
Mean orbital velocity	47.87 km/s (29.76 miles/s)
Orbital inclination	7° 00′ 15″.5
Orbital eccentricity	0.206
Apparent diameter	max. 12″.9, min. 4″.5
Reciprocal mass, Sun = 1	6,000,000
Density, water = 1	5.5
Mass, Earth = 1	0.055
Volume, Earth = 1	0.056
Escape velocity	4.3 km/s (2.7 miles/s)
Surface gravity, Earth = 1	0.38
Mean surface temperature	350°C (day); −170°C (night)
Oblateness	Negligible
Albedo	0.06
Maximum magnitude	−1.9
Diameter	4878 km (3030 miles)

▼ ***Mariner 10**. So far, this is the only spacecraft to have by-passed Mercury; it was also the first to use the gravity-assist technique. It has provided us with our only good maps of the surface, and has shown that the Earth-based maps (even Antoniadi's) were very inaccurate. Even so, it was able to image less than half the surface, so that our knowledge of the topography of Mercury is still very incomplete.*

◀ ***Mercury from Mariner 10****. Six hours after its closest approach, Mariner 10 took this series of 18 images of Mercury's surface, which have been combined to make a photomosaic. Note that, in general, the arrangement of the craters follows the lunar pattern; small craters break into larger ones, not vice versa. There are also ray centres. The north pole is at the top.*

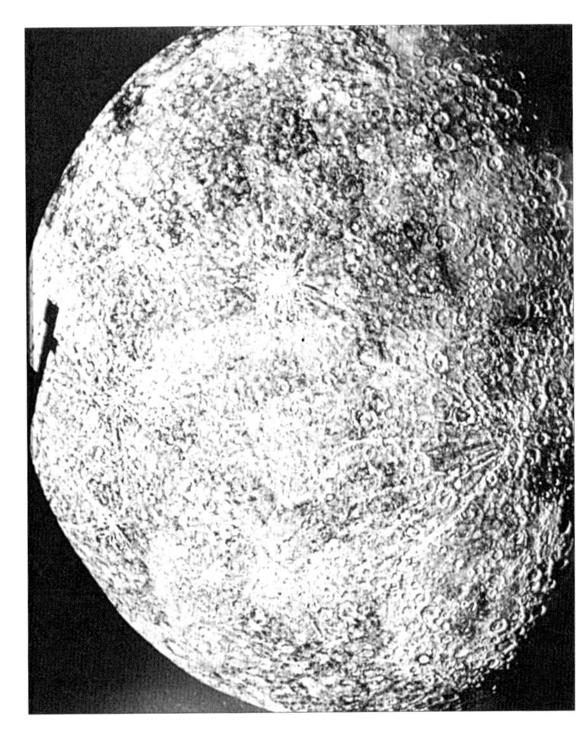

▲ ***The cratered surface of Mercury****. This is also a mosaic produced from Mariner 10 images; the curved black streak to the far left indicates a small portion of the surface which was not covered. The most obvious difference between Mercury and the Moon is that Mercury lacks broad plains similar to the lunar maria.*

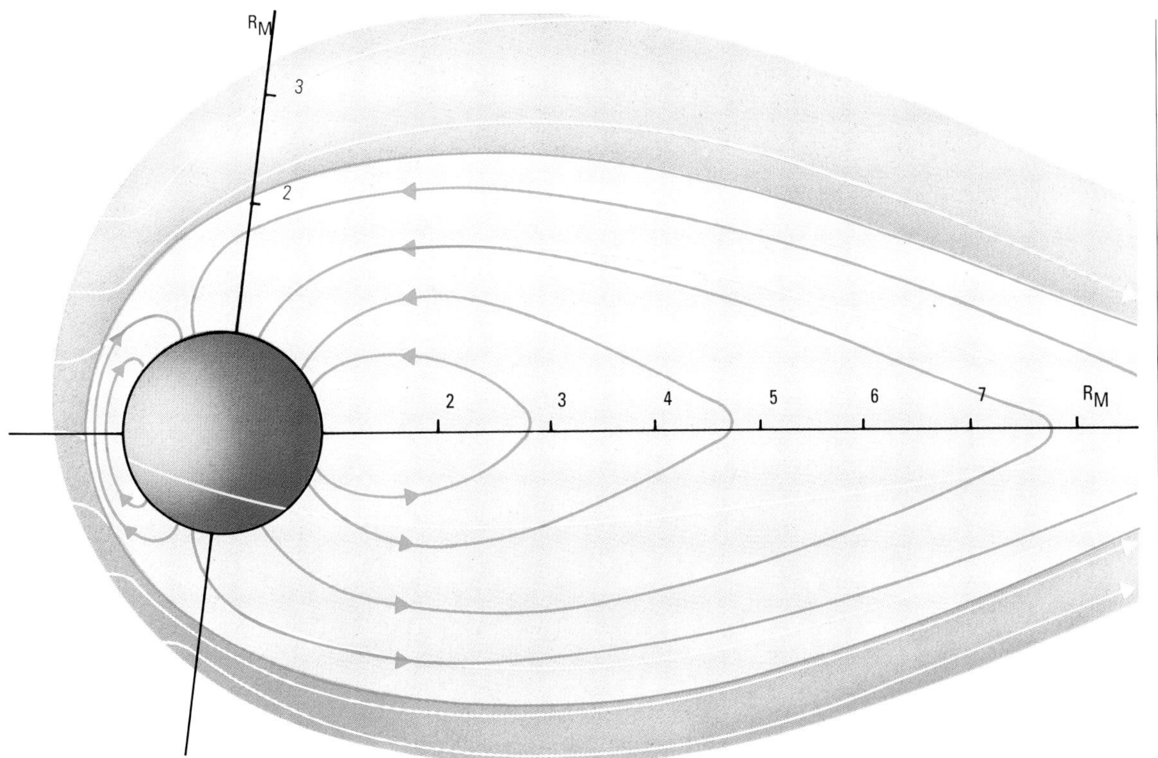

◀ ***The magnetosphere of Mercury****. The discovery of a Mercurian magnetic field was something of a surprise. No radiation belts can form, but there is a definite interaction between the Mercurian field and the solar wind; there is a well-defined bow-shock. The presence of a magnetic field is certainly due to the comparatively large iron-rich core of the planet.*

Features of Mercury

▲ ***Brahms*** *is a large crater north of the Caloris Basin. It has a central peak complex, terraced walls, and displays ejecta deposits.*

Mercury is above all a cratered world. The formations range from small pits up to colossal structures larger than anything comparable on the Moon. (Beethoven, Mercury's largest crater, is well over 600 kilometres [370 miles] across.) Small craters below 20 kilometres (12 miles) in diameter are, in general, bowl-shaped; larger craters have flatter floors, often with terraced walls and central peaks. As with the Moon, the distribution is non-random. There are lines, chains and groups, and where one formation breaks into another it is virtually always the smaller crater which is the intruder. Between the heavily cratered areas are what are termed intercrater plains, with few large structures but many craterlets in the 5 to 10 kilometres (3 to 6 miles) range; these are not found on the Moon or Mars. Neither are the lobate scarps, cliffs from 20 to 500 kilometres (12 to 300 miles) long and up to 3 kilometres (almost 2 miles) high; they seem to be thrust faults, cutting through features and displacing older ones.

Of the basins, much the most imposing is Caloris, which has some points in common with the lunar Mare Imbrium. It is 1500 kilometres (over 900 miles) in diameter and surrounded by mountains rising to between 2000 and 3000 metres (6600 and 10,000 feet) above the floor.

Antipodal to the Caloris Basin is the 'hilly and lineated terrain', often called 'weird terrain'. It covers 360,000 square kilometres (139,000 square miles), and consists of hills, depressions and valleys which have destroyed older features. Evidently the formation of this terrain is linked with the origin of Caloris – however that may have come about.

Presumably the ages of the surface features are much the same as those of the Moon. It has been estimated that the Caloris Basin is about 4000 million years old, and that extensive vulcanism ended about 3900 million years ago. Certainly there can be virtually no activity there now. It has been suggested, on the basis of radar observations, that there may be ice inside some of the polar craters, whose floors are always shadowed and are therefore intensely cold, but the idea of ice on a world such as Mercury does not seem very plausible.

It is a pity that our coverage of Mercury is so incomplete; for example, only half the Caloris Basin was in sunlight during the three active passes of Mariner 10. For more detailed information we must await the results from a new spacecraft, and no more missions to Mercury have been funded as yet.

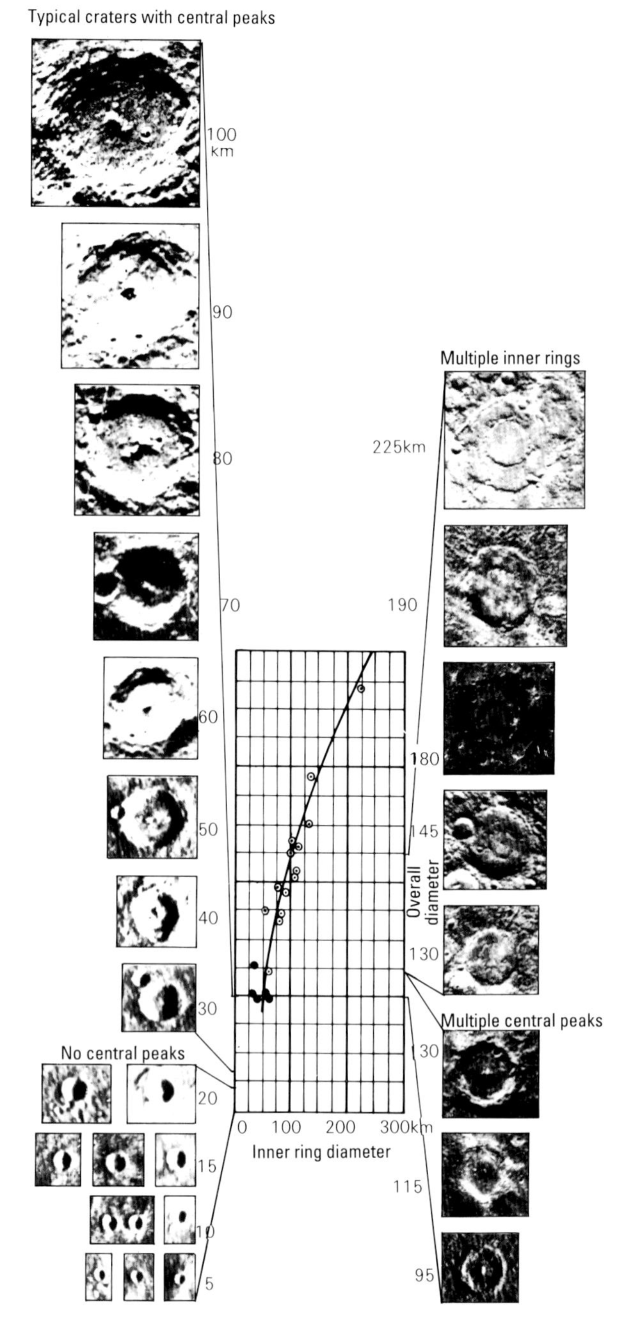

▲ ***Crater morphology*** *is similar to the Moon's. Craters are generally circular, have ejecta rim deposits, fields of secondary craters, terraced inner walls and central peaks, or even concentric inner rings. The smallest craters are bowl-shaped; with increasing size there may be a central peak, then inner terracing of the walls. Still larger craters have more frequent central peaks, and in the very largest structures complete or partial concentric inner rings may develop. The change from one type to the next occurs at lower diameters on Mercury than on the Moon.*

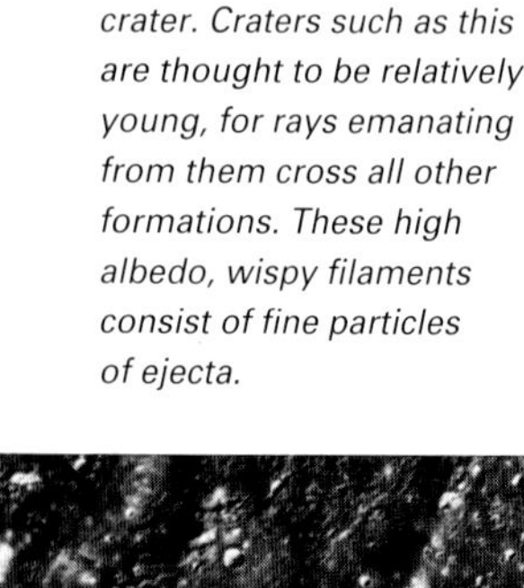

▼ ***Degas*** *is a bright ray-crater. Craters such as this are thought to be relatively young, for rays emanating from them cross all other formations. These high albedo, wispy filaments consist of fine particles of ejecta.*

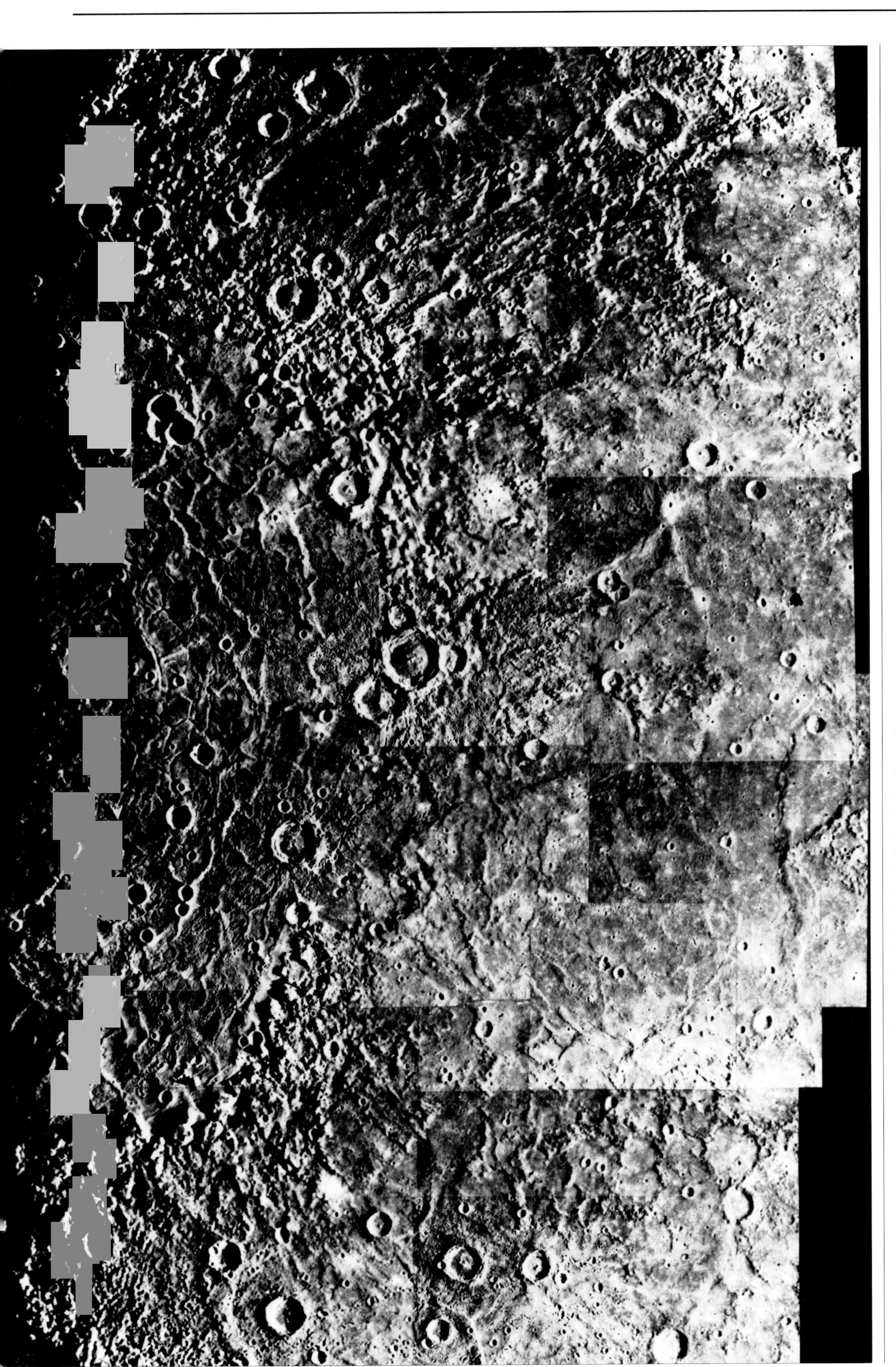

◀ ***The Caloris Basin**.*
This composite mosaic was compiled from Mariner 10 images which have different resolution limits, so that the degree of detail is not the same everywhere. The floor of the basin, with its central fractures and its outer region of sinuous ridges, is well shown. There are many small craters on the outer eastern floor; the large crater in the extreme north-east corner of the image is Van Eyck, 235 km (156 miles) in diameter. The Basin is 1500 km (over 900 miles) in diameter, and is bounded by a ring of smooth mountain blocks rising 1 to 2 km (0.6 to 1.2 miles) above the surrounding surface. Unfortunately only part of it was recorded from Mariner 10; at each encounter the same regions were available. About 80 per cent of 10–20 km (6–12 mile) Mercurian craters are terraced; on the Moon only 12 per cent of craters in the same class are.

Map of Mercury

The first serious attempts at mapping Mercury were made between 1881 and 1889 by the Italian astronomer G.V. Schiaparelli, who used 22-centimetre (8½-inch) and 49-centimetre (19-inch) refractors. Schiaparelli observed in broad daylight, when both Mercury and the Sun were high in the sky. He believed the rotation period to be synchronous, so that the same regions were always in sunlight, and he recorded various bright and dark features.

A more detailed map was published in 1934 by E. M. Antoniadi, who used the 83-centimetre (33-inch) refractor at the Meudon Observatory, near Paris. He too believed in a synchronous rotation, and also thought (wrongly) that the Mercurian atmosphere was dense enough to support clouds. He drew various features, and named them; thus a large dark patch was called the Solitudo Hermae Trismegisti (the Wilderness of Hermes the Thrice Greatest). However, when the Mariner 10 results were received, it was found that the earlier maps were so inaccurate that their nomenclature had to be abandoned.

Mariner mapped less than half the total surface; at each active pass the same regions were sunlit, but there is no reason to believe that the remaining areas are basically different. Superficially the surface looks very like the Moon; it is coated with a layer of porous silicate 'dust' forming a regolith which probably extends downwards for a few metres or a few tens of metres (10 to 100 feet). There are craters, which have been named after people; plains (planitia) named from the names of Mercury in different languages; mountains (montes); valleys (valles) named after radar installations; scarps (dorsa) named after famous ships of exploration and discovery; and ridges (rupes) named after astronomers who have paid particular attention to Mercury. Some craters have ray systems, notably Kuiper, which is named after the Dutch astronomer who played such a major role in the early days of planetary exploration by spacecraft.

The south pole of Mercury lies in the crater Chao Meng Fu. It has been agreed that the 20th meridian passes through the centre of the 1.5-kilometre (1-mile) crater Hun Kal, 0.58 degrees south of the Mercurian equator. The name Hun Kal is taken from the word for the numeral 20 in the language of the Maya, who used a base-20 number system.

It has been suggested that there may be ice inside some of the polar craters, whose floors are always in shadow. If ice exists there, it is presumably of cometary origin.

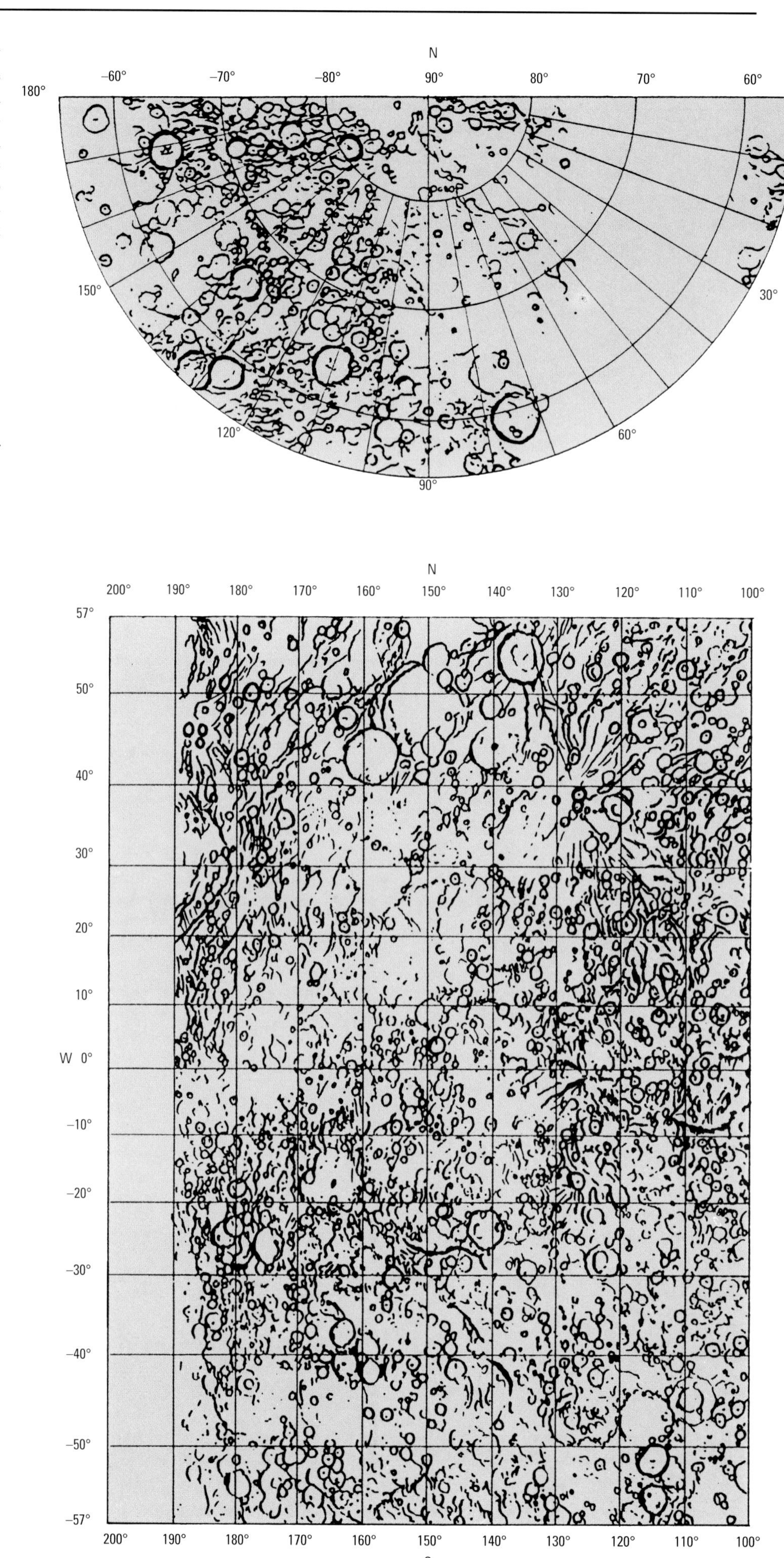

▶ ***Maps of Mercury**, prepared by Paul Doherty using data from Mariner 10.*

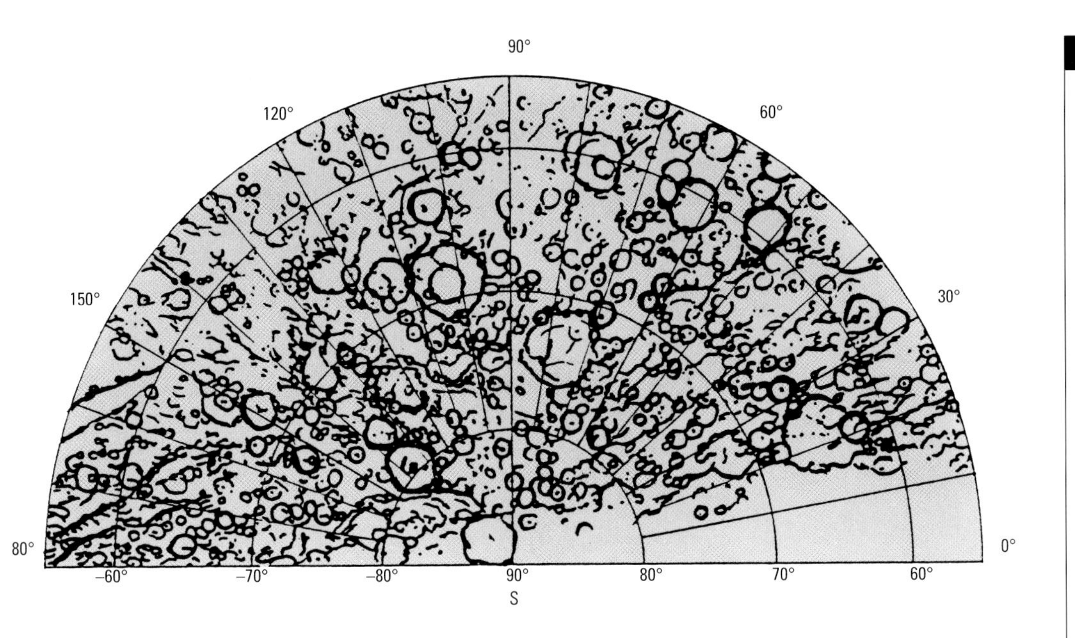

N
100° 90° 80° 70° 60° 50° 40° 30° 20° 10° 0°
57° 50° 40° 30° 20° 10° -0° E -10° -20° -30° -40° -50° -57°
100° 90° 80° 70° 60° 50° 40° 30° 20° 10° 0°
S

MAIN FEATURES ON MERCURY

Craters	Lat. °	Long. °	Diameter, km	Diameter, miles
Ahmad Baba	58.5 N	127	115	71
Andal	47 S	38.5	90	56
Aristoxenes	82 N	11	65	40
Bach	69 S	103	225	139
Beethoven	20 S	124	625	388
Boccaccio	80.5 S	30	135	84
Botticelli	64 N	110	120	75
Chao Meng Fu	87.5 S	132	150	93
Chong Chol	47 N	116	120	72
Chopin	64.5 S	124	100	62
Coleridge	54.5 S	66.5	110	68
Copley*	37.5 S	85.5	30	19
Goethe	79.5 N	44	340	211
Hitomaro	16 S	16	105	65
Homer	1 S	36.5	320	199
Hun Kal	0.5 S	20	1.5	0.9
Khansa	58.5 S	52	100	62
Kuiper*	11 S	31.5	60	37
Lermontov	15.5 S	48.5	160	99
Mena*	0.5 N	125	20	12
Michelangelo	44.5 S	110	200	124
Monteverdi	64 N	77	130	80
Murasaki	12 S	31	125	78
Nampeyp	39.5 S	50.5	40	25
Petrarch	30 S	26.5	160	99
Pushkin	65 S	24	200	124
Rabelais	59.5 S	62.5	130	81
Raphael	19.5 S	76.5	350	218
Renoir	18 S	52	220	142
Rubens	59.5 S	73.5	180	93
Shakespeare	40.5 N	151	350	218
Sholem Aleichem	51 N	86.5	190	118
Snorri*	8.5 S	83.5	20	12
Stravinsky	50.5 N	73	170	106
Strindborg	54 N	136	165	103
Tansen*	4.5 S	72	25	16
Tolstoj	15 N	165	400	250
Turgenev	66 N	135	110	68
Valmiki	23.5 S	141.5	220	137
Van Eyck	43.5 N	159	235	146
Verdi	64.5 N	165	150	93
Vivaldi	14.5 N	86	210	130
Vyasa	48.5 N	80	275	170
Wagner	67.5 S	114	135	84
Wang Meng	9.5 N	104	120	75
Wren	24.5 N	36	215	134

OTHER FEATURES

Montes	Lat. °	Long. °
Caloris	22–40 N	180
Planitiae		
Borealis	70 N	80
Budh	18 N	148
Caloris	30 N	195
Odin	25 N	171
Sobkou	40 N	130
Suisei	62 N	150
Tir	3 N	177
Dorsa		
Antoniadi	28 N	30
Schiaparelli	24 N	164
Rupes		
Adventure	64 S	63
Discovery	53 S	38
Heemskerck	25 N	125
Pourquoi-Pas	58 S	156
Santa Maria	6 N	20
Vostok	38 S	19
Valles		
Arecibo	27 N	29
Goldstone	15 S	32
Haystack	5 N	46.5
Simeiz	12.5 S	65

(* = ray centre)

Venus

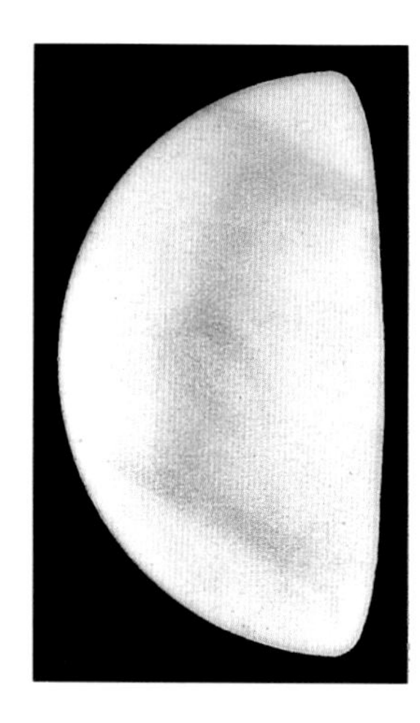

▲ ***Venus** drawn with my 31-cm (12½-inch) reflector. All that could be seen were very vague, cloudy shadings which are necessarily rather exaggerated in the sketch, together with slightly brighter areas near the cusps. The terminator appeared essentially smooth. The internal structure of Venus may not be too unlike that of the Earth, but with a thicker crust and an iron-rich core which is smaller both relatively and absolutely. There is no detectable magnetic field and, like Mercury, Venus has no satellite.*

Venus, the second planet in order of distance from the Sun, is as different from Mercury as it could possibly be. It is far brighter than any other star or planet, and can cast strong shadows; very keen-sighted people can see the phase with the naked eye during the crescent stage, and binoculars show it easily. Yet telescopically Venus is a disappointment. Little can be seen, and generally the disk appears blank. We are looking not at a solid surface, but at the top of a layer of cloud which never clears. Before the Space Age, we knew very little about Venus as a world.

We knew the size and mass; Venus is only very slightly inferior to the Earth, so that the two are near-twins. The orbital period is 224.7 days, and the path round the Sun is almost circular. Estimates of the rotation period ranged from less than 24 hours up to many months, but the favoured value was about a month. The vague shadings sometimes visible on the disk were much too indefinite to give any reliable results. There was also the Ashen Light, or dim visibility of the 'night' side, when Venus was in the crescent phase. It seemed to be real, but few people agreed with the 19th-century astronomer Franz von Paula Gruithuisen that it might be due to illuminations on the planet's surface lit by the local inhabitants to celebrate the accession of a new emperor!

It was suggested that Venus might be in the condition of the Earth during the Coal Forest period, with swamps and luxuriant vegetation of the fern and horse-tail variety; as recently as the early 1960s many astronomers were confident that the surface was mainly covered with water, though it was also thought possible that the surface temperature was high enough to turn Venus into a raging dust-desert. Certainly it had been established that the upper part of the atmosphere, at least, was made up mainly of carbon dioxide, which tends to shut in the Sun's heat.

The first positive information came in December 1962, when the American spacecraft Mariner 2 passed by Venus at a range of less than 35,000 kilometres (21,800 miles) and sent back data which at once disposed of the attractive 'ocean' theory. In 1970 the Russians managed to make a controlled landing with Venera 7, which transmitted for 23 minutes before being put out of action, and on 21 October 1975 another Russian probe, Venera 9, sent back the first picture direct from the surface. It showed a forbidding, rock-strewn landscape, and although the rocks are grey they appear orange by reflection from the clouds above. The atmospheric pressure was found to be around 90 times that of the Earth's air at sea level, and the temperature is over 480°C.

Radar measurements have shown that the rotation period is 243.2 days – longer than Venus's 'year'; moreover, the planet rotates from east to west, in a sense opposite to that of the Earth. If it were possible to see the Sun from the surface of Venus, it would rise in the west and set in the east 118 Earth-days later, so that in its way the calendar of Venus is every bit as strange as that of Mercury. The reason for this retrograde rotation is not known. According to one theory, Venus was hit by a massive body early in its evolution and literally knocked over. This does not sound very plausible, but it is not easy to think of anything better.

It has been found that the top of the atmosphere lies around 400 kilometres (250 miles) above the surface, and that the upper clouds have a rotation period of only 4 days. The upper clouds lie at an altitude of 70 kilometres (44 miles), and there are several definite cloud-layers, though below 30 kilometres (19 miles) the atmosphere is relatively clear and calm. The atmosphere's main constituent is indeed carbon dioxide, accounting for over 96 per cent of the whole; most of the rest is nitrogen. The clouds are rich in sulphuric acid; at some levels there must be sulphuric acid 'rain' which evaporates before reaching ground level.

PLANETARY DATA – VENUS

Sidereal period	224.701 days
Rotation period	243.16 days
Mean orbital velocity	35.02 km/s (21.76 miles/s)
Orbital inclination	3°23′ 39″.8
Orbital eccentricity	0.007
Apparent diameter	max. 65″.2 min. 9″.5 mean 37″.3
Reciprocal mass, Sun = 1	408,520
Density, water = 1	5.25
Mass, Earth = 1	0.815
Volume, Earth = 1	0.86
Escape velocity	10.36 km/s (6.43 miles/s)
Surface gravity, Earth = 1	0.903
Mean surface temperature	cloud-tops 33°C surface +480°C
Oblateness	0
Albedo	0.76
Maximum magnitude	−4.4
Diameter	12,104 km (7523 miles)

Earth

◄ ***Five photographs of Venus** taken on the same scale. As the phase shrinks, the apparent diameter increases. Near inferior conjunction the bright limb can be followed all round the disk – an effect of the planet's atmosphere.*

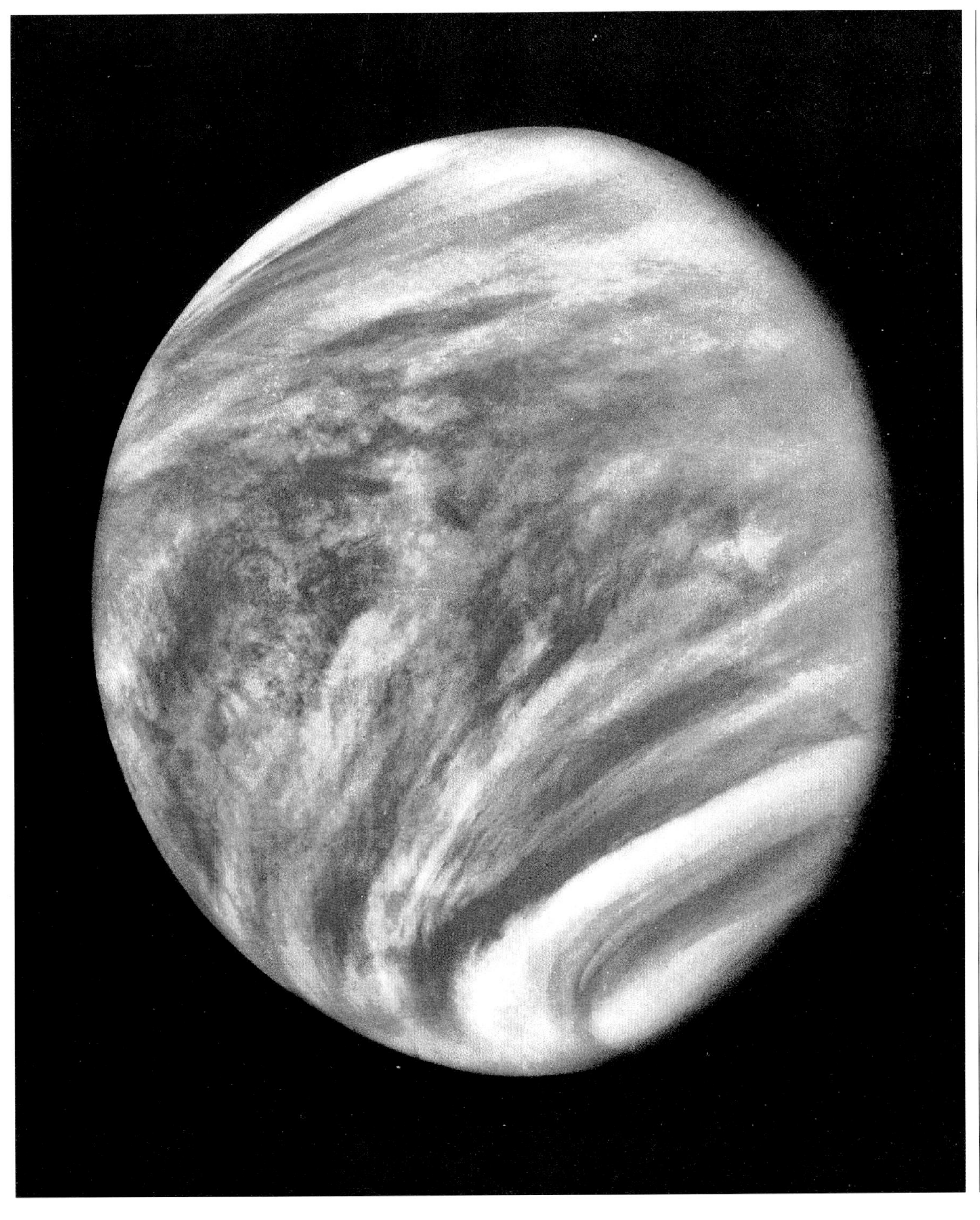

◀ ***Venus from Mariner 10***. *This picture was taken on 6 February 1974, one day after Mariner 10 flew past Venus* en route *for Mercury. The images were taken in ultra-violet light; the blueness is 'false colour', not the actual hue of the planet. The photograph was made by first computer-enhancing several television frames, and then forming a mosaic and retouching them. Note the difference in appearance near the planet's poles, which correspond to the cusps as seen from Earth.*

▼ ***Venera 13*** *on the surface of Venus in March 1982. Part of the spacecraft is shown in this picture; the temperature was measured at 457°C (855°F) and the pressure at 89 atmospheres. The rock was reddish-brown, and the sky brilliant orange.*

Mapping Venus

▲ ***Topographic globes of Venus.*** *Pioneer Venus 2 visited Venus in 1978. The mission involved an entry probe and a 'bus' which dispatched several small landers which sent back data during their descent. The map was compiled as a false-colour representation with blue indicating low levels and yellow and red higher areas. Ishtar and Aphrodite stand out very clearly. It has been suggested that in the future it may be possible to 'seed' the atmosphere, breaking up the carbon dioxide and sulphuric acid and releasing free oxygen.*

Because we can never see the surface of Venus, the only way to map it is by radar. It has been found that Venus is a world of plains, highlands and lowlands; a huge rolling plain covers 65 per cent of the surface, with lowlands accounting for 27 per cent and highlands for only 8 per cent. The higher regions tend to be rougher than the lowlands, and this means that in radar they are brighter (in a radar image, brightness means roughness).

There are two main upland areas, Ishtar Terra and Aphrodite Terra. Ishtar, in the northern hemisphere, is 2900 kilometres (1800 miles) in diameter; the western part, Lakshmi Planum, is a high, smooth, lava-covered plateau. At its eastern end are the Maxwell Mountains, the highest peaks on Venus, which rise to 11 kilometres (nearly 7 miles) above the mean radius and 8.2 kilometres (5 miles) above the adjoining plateau. Aphrodite straddles the equator; it measures 9700 × 3200 kilometres (6000 × 2000 miles), and is made up of several volcanic massifs, separated by fractures. Diana Chasma, the deepest point on Venus, adjoins Aphrodite.

(*En passant*, it has been decreed that all names of features in Venus must be female. The only exception is that of the Maxwell Mountains. The Scottish mathematician James Clerk Maxwell had been placed on Venus before the official edict was passed!)

A smaller highland area, Beta Regio, includes the shield volcano, Rhea Mons and the rifted mountain Theia Mons. Beta, which is cut by a huge rift valley rather like the Earth's East African Rift, is of great interest. It is likely that Rhea is still active, and there can be no doubt that the whole surface of Venus is dominated by vulcanism. Venus's thick crust will not slide over the mantle in the same way as that of the Earth, so that plate tectonics do no apply; when a volcano forms over a hot spot it will remain there for a very long period. Lava flows are found over the whole of the surface.

Craters are plentiful, some of them irregular in shape while others are basically circular. The largest, Mead, has a diameter of 280 kilometres (175 miles), though small craters are less common than on Mercury, Mars or the Moon.

There are circular lowland areas, such as Atalanta Planitia, east of Ishtar; there are systems of faults, and there are regions now called tesserae – high, rugged tracts extending for thousands of square kilometres and characterized by intersecting ridges and grooves. Tesserae used to be called 'parquet terrain', but although the term was graphic it was abandoned as being insufficiently scientific.

Venus has been contacted by fly-by probes, radar-carrying orbiters and soft-landers; in 1985 the two Russian probes *en route* for Halley's Comet even dispatched two balloons into the upper atmosphere of the planet, so that information could be sent back from various levels as the balloons drifted around. The latest probe, Magellan, has confirmed and extended the earlier findings that Venus is overwhelmingly hostile.

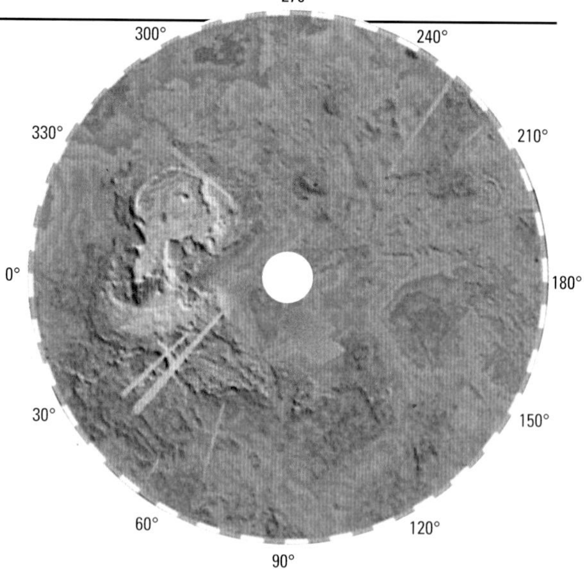

The topography of Venus. *This map was obtained by the Magellan radar altimeter during its 24 months of systematic mapping of the surface of Venus. Colour is used to code elevation (see the colour bar), and simulated shading to emphasize relief. Red corresponds to the highest, blue to the lowest elevations. The image below shows the portion of the planet*

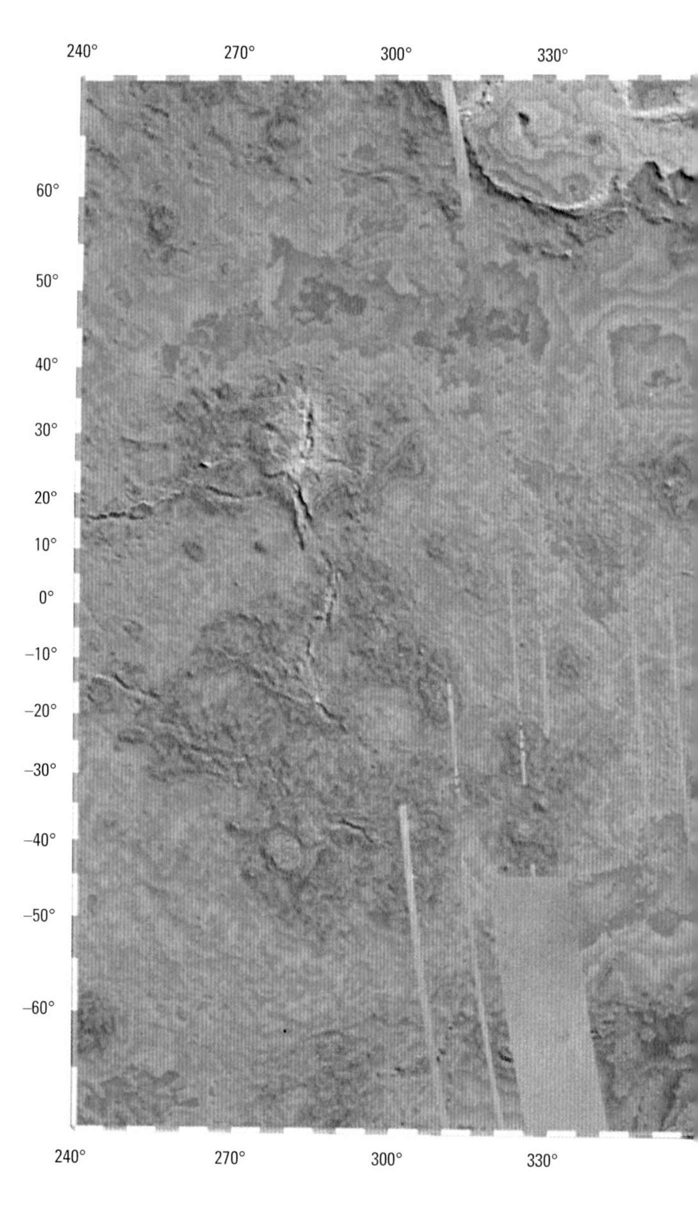

between 69°N and 69°S latitude in Mercator projection; above are the two polar regions covering latitudes above 44° in stereographic projection. Height accuracy is better than 50 m (165 feet); and the horizontal resolution of detail on the surface is about 10 km (6 miles) near the equator, rising to as much as 25 km (15.5 miles) at higher latitudes.

FEATURES ON VENUS – SELECTED LIST

	Feature	Lat.°	Long.°
TERRAE	Aphrodite	40 S–5 N	140–000
	Ishtar	52–75 N	080–305
REGIONES	Alpha	29–32 S	000
	Asteria	18–30 N	228–270
	Beta	20–38 N	292–272
	Metis	72 N	245–255
	Phoebe	10–20 N	275–300
	Tellus	35 N	080
	Thetis	02–15 S	118–140
PLANITIA	Atalanta	54 N	162
	Lakshmi Planum	60 N	330
	Lavinia	45 S	350
	Leda	45 N	065
	Niobe	138 N–10 S	132–185
	Sedna	40 N	335
CHASMA	Artemis	30–42 S	121–145
	Devana	00	289
	Diana	15 S	150
	Heng-O	00–10 N	350–000
	Juno	32 S	102–120
CRATERS	Colette	65 N	322
	Lise Meitner	55 S	322
	Pavlova	14 N	040
	Sacajewa	63 N	335
	Sappho	13 N	027
VOLCANO	Rhea Mons	31 N	285
MOUNTAIN	Theia Mons	29 N	285

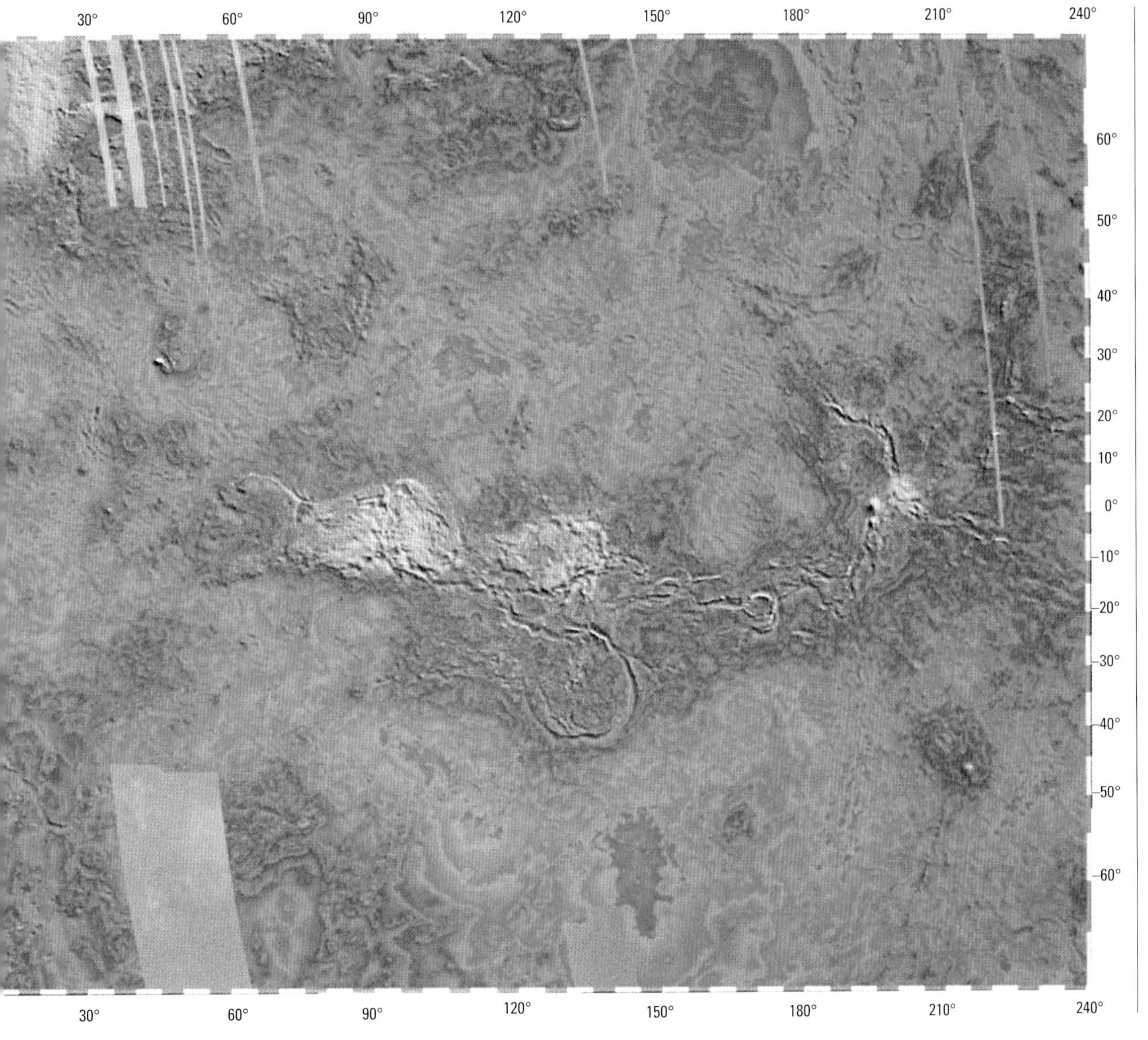

The Magellan Mission

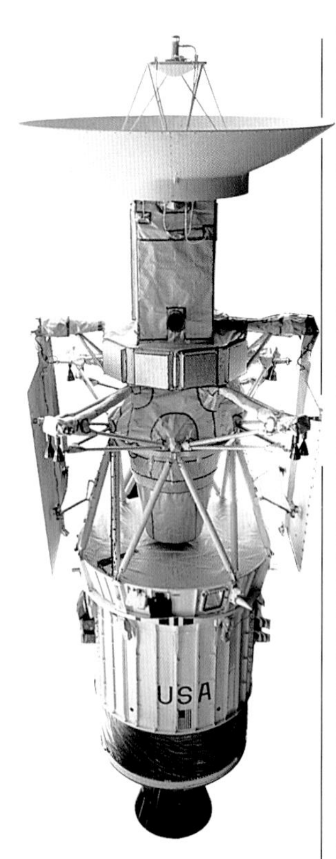

▲ ***Magellan**. Released from the cargo bay of the space Shuttle* Atlantis *on 4 May 1989, the spacecraft reached Venus on 10 August 1990. After 37 silent minutes while it swung around the back of the planet, it emerged in a perfect orbit. Its radar mapping programme was completed by September 1992, and it began a cycle of gravity mapping. Magellan has set a record for the amount of data gathered by a space mission.*

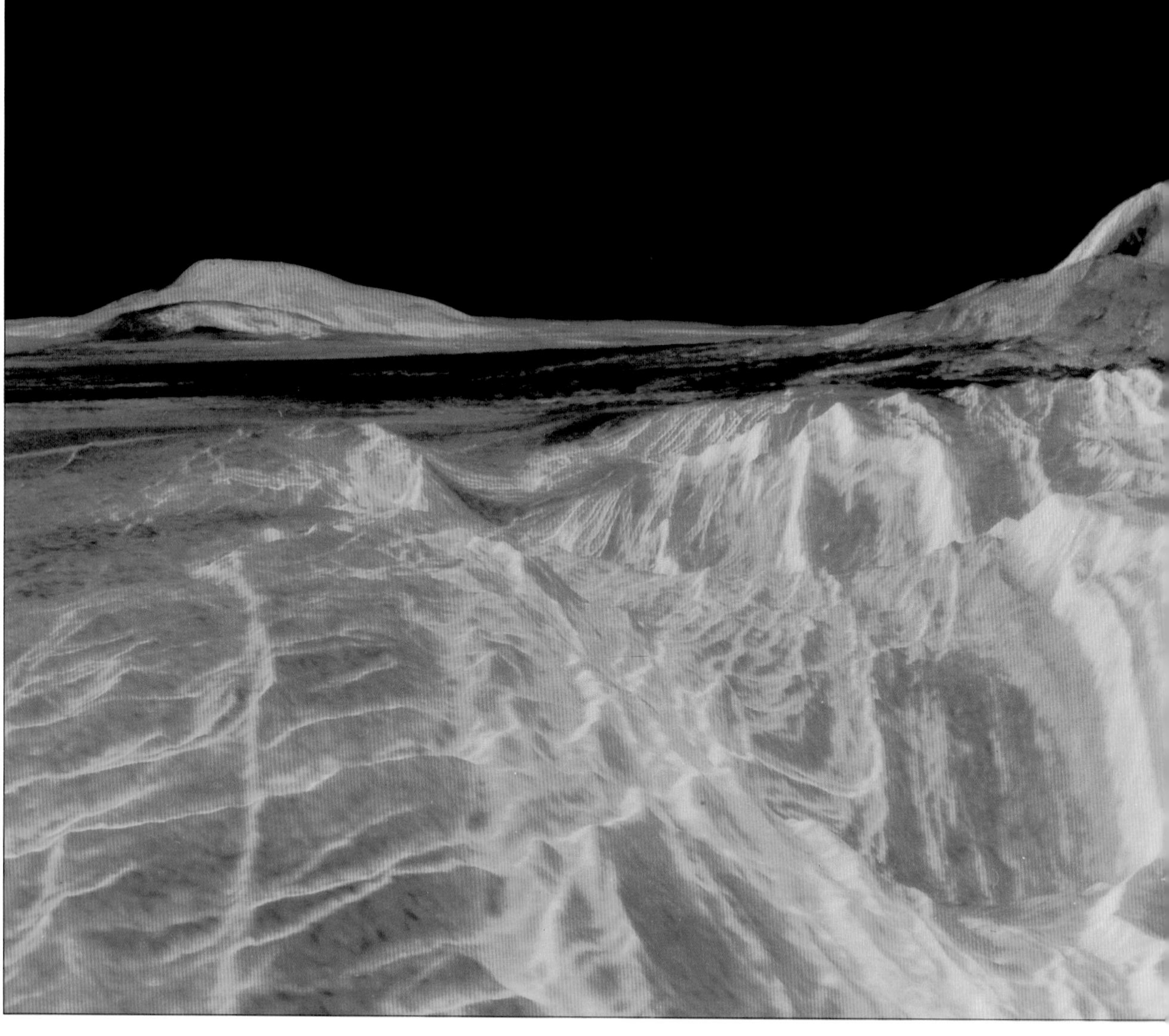

The various American and Russian spacecraft to Venus, launched between 1961 and 1984, had provided a great deal of information about the planet, but there was still need for better radar coverage. This was the purpose of the Magellan probe, launched from the Shuttle *Atlantis* on 4 May 1989. It was hoped that the resolution would be far better than anything achieved by the earlier missions, and so it proved. Radar mapping began in September 1990, and by 1993 over 98 per cent of the planet's surface had been covered. Cycle 4 ended on 24 May of that year. One cycle is 243 Earth-days, during which Venus rotates completely beneath the spacecraft's orbital plane. When Magellan first went into orbit round Venus, the period was 3.2 hours, and the minimum distance was 289 kilometres (180 miles), though in September 1992 this was reduced to 184 kilometres (115 miles).

Magellan could resolve features down to 120 metres (400 feet). The main dish, 3.7 metres (12 feet) across, sent down a pulse at an oblique angle to the spacecraft, striking the surface below much as a beam of sunlight will do on Earth. The surface rocks modify the pulse before it is reflected back to the antenna; rough areas are radar-bright, smooth areas are radar-dark. A smaller antenna sends down a vertical pulse, and the time-lapse between transmission and return gives the altitude of the surface below to an accuracy of 10 metres. Magellan has shown fine details on the volcanic surface. There are for example multiple lava flows, with varying radar reflectivity indicating rocky and smoother areas. There are flows which have clearly been due to very liquid lava, and even show river-like meandering. The features known as tesserae are high, rugged tracts extending for several thousands of kilometres; one of these is Alpha Regio, shown on the facing page. Magellan showed many 'coronae', caused by plumes of hot material rising from below the surface. Arachnoids, so far found only on Venus, are so named because of their superficial resemblance to spiders' webs; they are circular to ovoid in shape, with concentric rings and intricate outward-extending features. They are similar in form to the coronae – circular volcanic structures surrounded by ridges, grooves and radial lines. There are strange-looking objects which have been nicknamed 'pancakes'; these too are of volcanic origin. There are strong indications of explosive vulcanism here and there. The crater Cleopatra on the eastern slopes of the Maxwell Montes is about 100 kilometres (62 miles) in diameter.

The scale and colours of the images shown here are products of the computer processing; for example, the vertical scale in the image of Gula Mons seen at top right in the main picture has been deliberately increased to accentuate its features. The colours are not as they would be seen by an observer on the planet – assuming that he could get there. For example, the bright patches representing lava flows would not appear so to the naked eye.

Maxwell Montes. The Maxwell Mountains are seen as a large bright patch below centre in this image. They are the highest mountains on the planet, with peaks extending more than 7 km (4.5 miles) above the surface.

Atalanta Planitia. This vast plain can be seen to the right of the fault lines which radiate from near the central region.

◀ ***The northern hemisphere of Venus****.* *This false colour projection of the surface of Venus was created from data gathered during the first cycle of Magellan's radar mapping observations. The Magellan data was supplemented with earlier Pioneer Venus data, and the general colour hue comes from the Russian Venera lander's images taken whilst on the surface of the planet in 1972.*

Lakshmi Planum. The Lakshmi Plateau lies just to the left of the Maxwell Mountains. It stands 2.5–4 km (1.5–2.5 miles) above the surface and is covered by lava.

North Pole. The north pole of Venus lies at the very centre of the image. Longitude zero is to the right. There is major faulting above the pole.

▲ ***Eistla Regio****. This false-colour perspective of the western part of the Eistla Regio region of Venus depicts the view looking north-west from a point 700 km (440 miles) from the crater of Gula Mons, the mountain seen at top right which stands 3 km (2 miles) above the surrounding plain. The foreground is dominated by a large rift valley.*

▶ ***Alpha Regio****. This mosaic of radar images shows seven dome-like hills averaging 25 km (16 miles) in diameter and 750 m (2400 feet) in height. They may have been formed by successive lava eruptions.*

Movements of Magellan

The Magellan space probe was put into an orbit around Venus, and mapped the surface in a series of 20-kilometre (12.5-mile) swathes. When this programme was completed, the probe was put into a more elliptical orbit. When the whole of the surface had been mapped, in 1993, Magellan entered a circular orbit to undertake gravitational studies of the planet.

Mars

Mars, the first planet beyond the orbit of the Earth, has always been of special interest, because until relatively recently it was thought that life might exist there. Less than a century ago, there was even a prize (the Guzman Prize) offered in France to be given to the first man to establish contact with beings on another world – Mars being specifically excluded as being too easy!

Mars is considerably smaller and less dense than the Earth, and in size it is intermediate between the Earth and the Moon. The escape velocity of 5 kilometres per second (3.1 miles per second) is high enough to hold down a thin atmosphere, but even before the Space Age it had become clear that the atmosphere is not dense enough to support advanced Earth-type life; neither could oceans exist on the surface. The axial tilt is much the same as ours, so that the seasons are of similar type even though they are much longer. The orbital period is 687 days. The axial rotation period, easily measured from observations of the surface markings, is 24 hours 37 minutes 22.6 seconds, so that a Martian 'year' contains 668 Martian days or 'sols'.

The orbit of Mars is decidedly eccentric. The distance from the Sun ranges between 249 million and 207 million kilometres (between 155 million and 129 million miles), and this has a definite effect upon Martian climate. As with Earth, perihelion occurs during southern summer, so that on Mars the southern summers are shorter and warmer than those of the north, while the winters are longer and colder.

At its nearest to us, Mars may come within 59 million kilometres (36 million miles) of the Earth, closer than any other planet apart from Venus. Small telescopes will then show considerable surface detail. First there are the polar ice-caps, which vary with the seasons; at its greatest extent the southern cap may extend down to latitude 50°, though at minimum it becomes very small. Because of the more extreme climate in the southern hemisphere, the variations in the size of the cap are greater than those in the north.

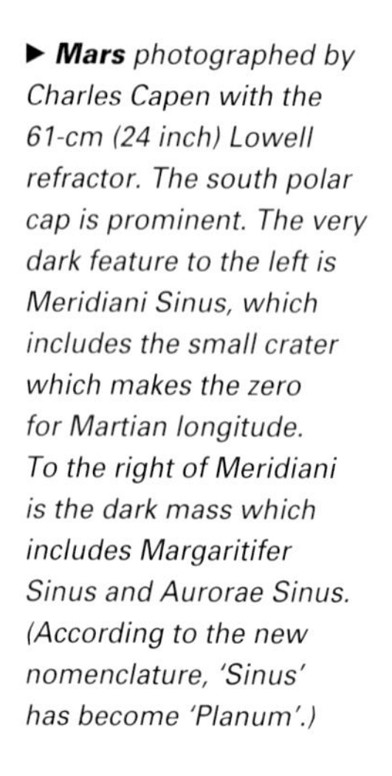

▶ ***Mars*** *photographed by Charles Capen with the 61-cm (24 inch) Lowell refractor. The south polar cap is prominent. The very dark feature to the left is Meridiani Sinus, which includes the small crater which makes the zero for Martian longitude. To the right of Meridiani is the dark mass which includes Margaritifer Sinus and Aurorae Sinus. (According to the new nomenclature, 'Sinus' has become 'Planum'.)*

The dark areas are permanent, though minor variations occur; as long ago as 1659 the most conspicuous dark feature, the rather V-shaped patch now known as the Syrtis Major, was recorded by the Dutch astronomer Christiaan Huygens. Originally it was assumed that the dark areas were seas, while the ochre tracts which cover the rest of the planet represented dry land. When it was found that the atmospheric pressure is too low for liquid water, it was believed that the dark areas were old sea-beds filled with vegetation. This view was generally accepted up to the time of the first fly-by made by Mariner 4, in 1965.

There are various bright areas, of which the most prominent is Hellas, in the southern part of the planet. At times it is so bright that it has been mistaken for an extra polar cap, and it was once thought to be a snow-covered plateau, though it is now known to be a deep basin.

In general the Martian atmosphere is transparent, but clouds can be seen it, and there are occasional dust-storms which may spread over most of the planet, hiding the surface features completely. What apparently happens is that if the windspeed exceeds 50 to 100 metres per second (160 to 320 feet per second), tiny grains of surface material are whipped up and given a 'skipping' motion, known technically as saltation. When they strike the surface they force still smaller grains into the atmosphere, where they remain suspended for weeks. Widespread dust-storms are commonest when Mars is near perihelion, and the surface winds are at their strongest.

The first reasonably reliable maps of Mars date back to the 1860s. The various features were named, mainly after astronomers; the old maps show Mädler Land, Lassell Land, Beer Continent and so on. (The latter name honoured Wilhelm Beer, a German pioneer of lunar and planetary observation.) Then, in 1877, G. V. Schiaparelli produced a more detailed map and renamed the features, so that, for example, the most prominent dark marking on Mars, the V-shaped feature drawn by Huygens so long ago, and formerly known as the Kaiser Sea, was renamed Syrtis Major. It is Schiaparelli's nomenclature, modified and extended, which we use today.

Schiaparelli also drew strange, artificial-looking lines across the ochre deserts, which he called *canali* or channels; inevitably this was translated as 'canals' and the suggestion was made that the features might be artificial waterways. This view was championed by Percival Lowell, who built the great observatory at Flagstaff in

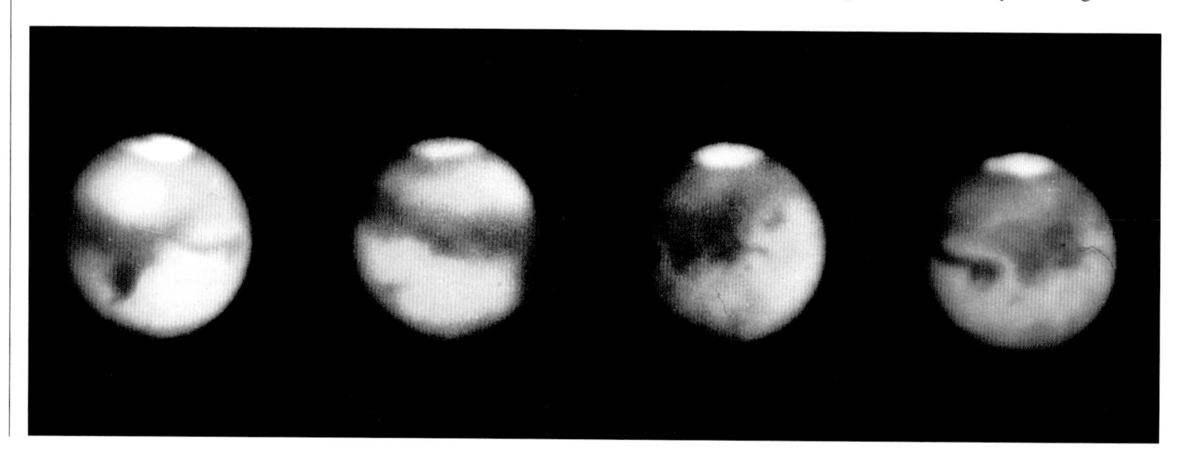

▶ ***The rotation of Mars****. A series of images taken by Charles Capen with the Lowell refractor. The V-shaped Syrtis Major appears on the left-hand picture; the extreme right-hand picture shows the Meridiani Sinus. The south polar cap is much in evidence. The main markings shown on these pictures can be seen with moderate-sized telescopes when Mars is well placed.*

Arizona mainly to study Mars, and equipped it with an excellent 61-centimetre (24-inch) refractor. Lowell believed that the canals represented a planet-wide irrigation system, built by the local inhabitants to pump water from the ice-caps at the poles through to the equator. Disappointingly, it has now been proved that the canals do not exist; they were merely tricks of the eye, and Lowell's Martians have been banished to the realm of science fiction.

The best pre-Space Age maps of Mars were those drawn up by E. M. Antoniadi in the 1920s and early 1930s. The telescope used was the Meudon 83-centimetre (33-inch) refractor, and Antoniadi's charts proved to be amazingly accurate, but the real 'breakthrough' came with Mariner 4 in 1965.

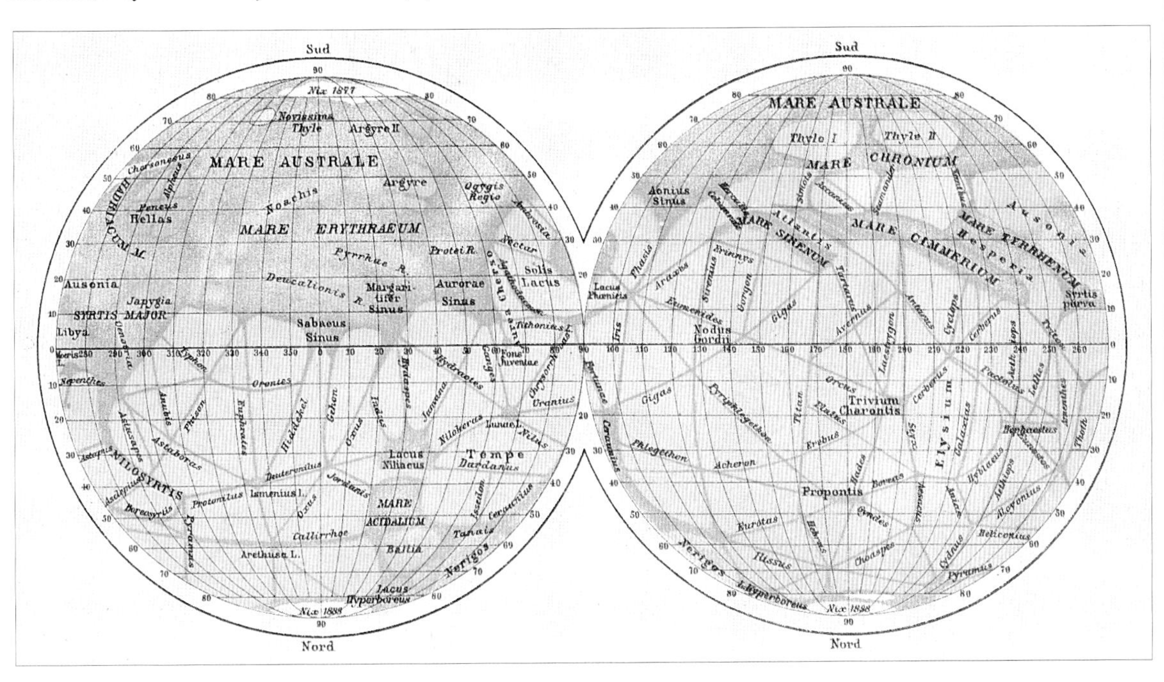

◀ ***Schiaparelli's charts of Mars**, compiled from observations made between 1877 and 1888. The main dark features are clearly shown – but so too are the canals, which are now known to be non-existent! Schiaparelli's map uses the nomenclature which he introduced in 1877, and which is still followed today.*

◀ ***The Lowell 61-cm (24-inch) refractor** at Flagstaff, used by Percival Lowell to draw the Martian 'canals'.*

▼ ***Sketch of Mars** made by Patrick Moore on 23 February 1981, using a magnification of ×815 on the Lowell telescope. The main features are clearly shown – but no canals!*

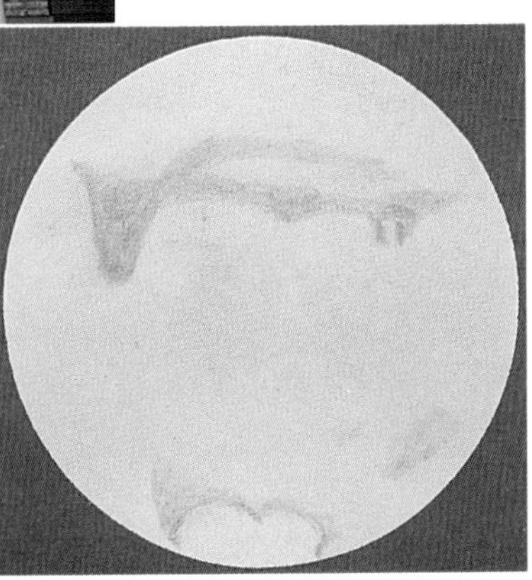

PLANETARY DATA – MARS

Sidereal period	686.980 days
Rotation period	24h 37m 22s.6
Mean orbital velocity	24.1 km/s (15 miles/s)
Orbital inclination	1° 50′ 59″.4
Orbital eccentricity	0.093
Apparent diameter	max. 25″.7, min. 3″.5
Reciprocal mass, Sun = 1	3,098,700
Density, water = 1	3.94
Mass, Earth = 1	0.107
Volume, Earth = 1	0.150
Escape velocity	5.03 km/s (3.1 miles/s)
Surface gravity, Earth = 1	0.380
Mean surface temperature	−23°C
Oblateness	0.009
Albedo	0.16
Maximum magnitude	−2.8
Diameter (equatorial)	6794 km (4222 miles)

Missions to Mars

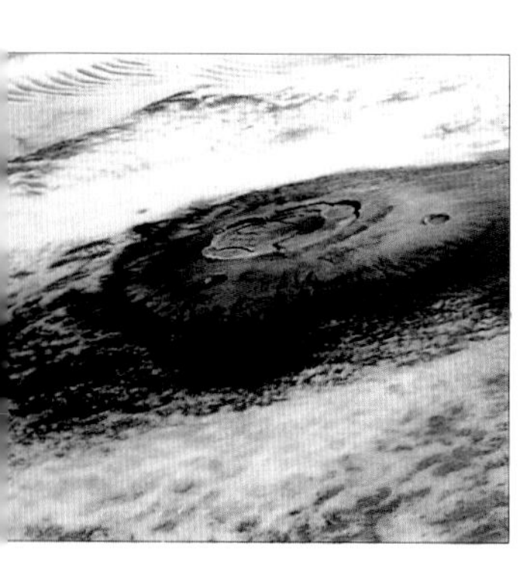

▲ ***Olympus Mons**, the highest volcano on Mars; it is 25 km (15 miles) above the outer surface, and has a base measuring 600 km (375 miles). It is crowned by an 85-km (53-mile) caldera. It is a huge shield volcano, far larger and more massive than terrestrial shield volcanoes such as Mauna Kea and Mauna Loa.*

Mariner 4 bypassed Mars on 14 July 1965, at a range of 9789 kilometres (6080 miles). It carried cameras, and the images showed that Mars is a cratered world rather than having a smooth, gently undulating surface, as had been generally believed. Other probes followed, mainly American; even now the Russians have had very little luck with their Martian spacecraft. Then, on 13 November 1971, Mariner 9 entered a closed orbit, and for the next eleven months sent back splendid pictures of much of the surface. The minimum distance from Mars was 1640 kilometres (1020 miles), over 7000 images were received, and contact was not lost until October 1972.

Mariner 9 changed many of our ideas about Mars. First the atmosphere turned out to be much thinner than anticipated. It had been estimated that the ground pressure should be about 87 millibars, equivalent to the pressure in the Earth's air at rather less than twice the height of Everest, and that the main constituent was likely to be nitrogen; in fact the pressure is below 10 millibars everywhere – so that it corresponds to what we regard as a reasonably good laboratory vacuum – and most of the atmosphere is made of carbon dioxide, with only small amounts of nitrogen and other gases.

It was found that, though the polar caps are made of ice, they are not the same as the caps on Earth, and neither are they identical with each other. In each case there is a seasonal upper cap made of carbon dioxide ice, below which is a permanent residual cap; the residual cap of the northern cap is composed of water ice, while that of the southern cap is a mixture of water ice and carbon dioxide ice. During the southern winter, which is colder than that of the northern, carbon dioxide condenses out of the atmosphere on to the polar cap, and there is a temporary fall in atmospheric pressure.

The pictures of the surface were dramatic; naturally, it had not been realized that all the earlier probes had surveyed the least interesting areas of the planet. For the first time we could examine the giant volcanoes, such as the majestic Olympus Mons, which has an altitude of 25 kilometres (15 miles) – three times the height of Everest – with a 600-kilometre (375-mile) base and an 85-kilometre (50-mile) caldera at its summit. It was found that there are two marked bulges in the Martian crust, those of Tharsis and Elysium, and it is here that most of the volcanoes lie, though there are minor volcanoes elsewhere. Tharsis is the major feature; along it lie the volcanoes of Ascraeus Mons, Arsia Mons and Pavonis Mons, with Olympus Mons not far away. All these had been seen by Earth-based observers, but there had been no way of finding out just what they were; Olympus Mons had been known as Nix Olympia, the Olympic Snow, as it sometimes shows up telescopically as a white patch. North of Tharsis lies Alba Patera, only a few kilometres high but more than 2400 kilometres (1500 miles) across. The Elysium bulge is smaller than that of Tharsis, and the volcanoes are lower.

The two hemispheres of Mars are not alike. In general the southern part of the planet is the higher, more heavily cratered and more ancient, though it does contain two deep and well formed basins, Hellas and Argyre. The northern hemisphere is lower, younger and less cratered, though it does contain part of the Tharsis bulge.

The Mariner Valley (Valles Marineris), just south of the equator, has a total length of 4500 kilometres (2800 miles), with a maximum width of 600 kilometres (375 miles); the deepest part of the floor is 7 kilometres (4 miles) below the rim. There are complex systems, such as Noctis Labyrinthus (once taken for a lake, and named Noctis Lacus), with canyons from 10 to 20 kilometres (6 to 12 miles) wide making up the pattern which has led to the nickname of the Chandelier. There are features which can hardly be anything other than old riverbeds, so that in the past Mars must have had a warmer climate and a denser atmosphere than it does now; and of course there are the craters, which are everywhere and some of which are more than 400 kilometres (250 miles) across. Here and there we find 'islands', and there is strong evidence of past flash-flooding.

Whether there is any active vulcanism going on now is a matter for debate. There is a crust, probably between 15 and 20 kilometres (9 to 12 miles) deep, which overlies a mantle. A magnetic field has been detected but is very weak.

◀ ***Mars in winter**, photographed by Viking Lander 2. In the shadows of rocks and boulders a white condensate can be seen, either water ice or frozen carbon dioxide (dry ice), or a combination of the two precipitated on to the ground as snow or frost. Or it could have come to the surface from below, by cryopumping. Several small trenches can be seen in the centre foreground, dug by the lander's soil-sampler arm as it gathered material for Viking soil experiments. The soil scoop's cover lies to the right of the trenches, where it was dropped when the lander's mission began. Most rocks in the scene are about 50 cm (19–20 inches) across.*

◄ ***Mars from Viking 2.*** *This photo of Mars was taken by Viking 2 from 419,000 km (260,000 miles) away as the spacecraft approached the planet on 5 August 1976. Viking 2 was preceded to Mars by Viking 1. In this view of the crescent-lit planet, contrast and colour ratios have been enhanced to improve visibility of subtle surface topography and colour variations. Water-cloud plumes extend north-west from the western flank of Ascraeus Mons, northernmost of the three great volcanoes that line Tharsis Ridge. The middle volcano, Pavonis Mons, is just visible on the dawn terminator, below and west of Ascraeus Mons. Valles Marineris, the great system of rift canyons, extends from the centre of the picture at the terminator downwards to the east. Including the huge complex at its west and named Noctis Labyrinthus, Valles Marineris stretches nearly 4800 km (3000 miles). The bright basin near the bottom of the photo is the Argyre Basin, one of the largest basins on Mars. The ancient crater lies near the south pole (not visible in this photo) and is brightened by the icy frosts and fogs that lie in the bottom of the basin and are characteristic of the near-polar regions of Mars when each pole is experiencing its winter season.*

Satellites of Mars

Mars has two satellites, Phobos and Deimos, both discovered in 1877 by Asaph Hall, using the large refractor at Washington Observatory; previous searches, by William Herschel and Heinrich D'Arrest, had been unsuccessful. Both are very small, and are not easy telescopic objects, because they are so close to Mars. It is interesting to recall that Jonathan Swift, in his Voyage to Laputa (one of *Gulliver's Travels*), had described how astronomers on the curious flying island had discovered two Martian satellites, one of which revolved round the planet in a time less than that of the planet's axial rotation – as Phobos actually does. However, Swift's reasoning was not strictly scientific. If the Earth had one moon and Jupiter four, how could Mars possibly manage with less than two?

Phobos moves at less than 6000 kilometres (3700 miles) above the planet's surface, and the orbital period is only 7 hours 39 minutes, so that to an observer on Mars Phobos would rise in the east and set in the east $4\frac{1}{2}$ hours later, during which time it would go though more than half its cycle of phases from new to full; the interval between successive risings would be no more than 11 hours. Yet Phobos would be of little use as a source of illumination at night. From Mars it would have an apparent diameter of less than half that of our Moon seen from Earth, and would give little more light than Venus does to us; an observer at a latitude of more than 69° would never see it at all, and for long periods it would be eclipsed by Mars' shadow. It would transit the Sun 1300 times in every Martian year, taking 19 seconds to cross the solar disk.

Phobos is a dark, irregularly-shaped body, with a longest diameter of 27 kilometres (17 miles). Its surface is coated with a 'dusty' regolith, and spacecraft images show that it is cratered; the largest crater, 10 kilometres (6 miles) across, is named Stickney in honour of Asaph Hall's wife (this was her maiden name, and it was she who urged her husband to continue hunting for satellites when he was on the verge of giving up). Other craters, one of which is named after Hall, are around 5 kilometres (3 miles) across; there are also ridges, hills, and strange parallel grooves inclined to the equator at 30°. These grooves are from 100 to 200 metres (330 to 660 feet) wide, and 10 to 20 metres (33 to 66 feet) deep. It has been calcu-

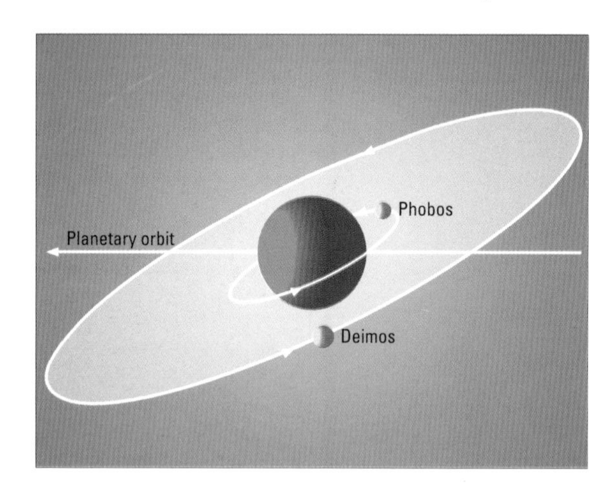

▼ ***The satellite orbits**. Both Phobos and Deimos are very close to Mars; Phobos moves in a circular orbit 9270 km (5800 miles) from the centre of Mars – closer to its parent planet than any other satellite. Deimos orbits at 23,400 km (14,500 miles).*

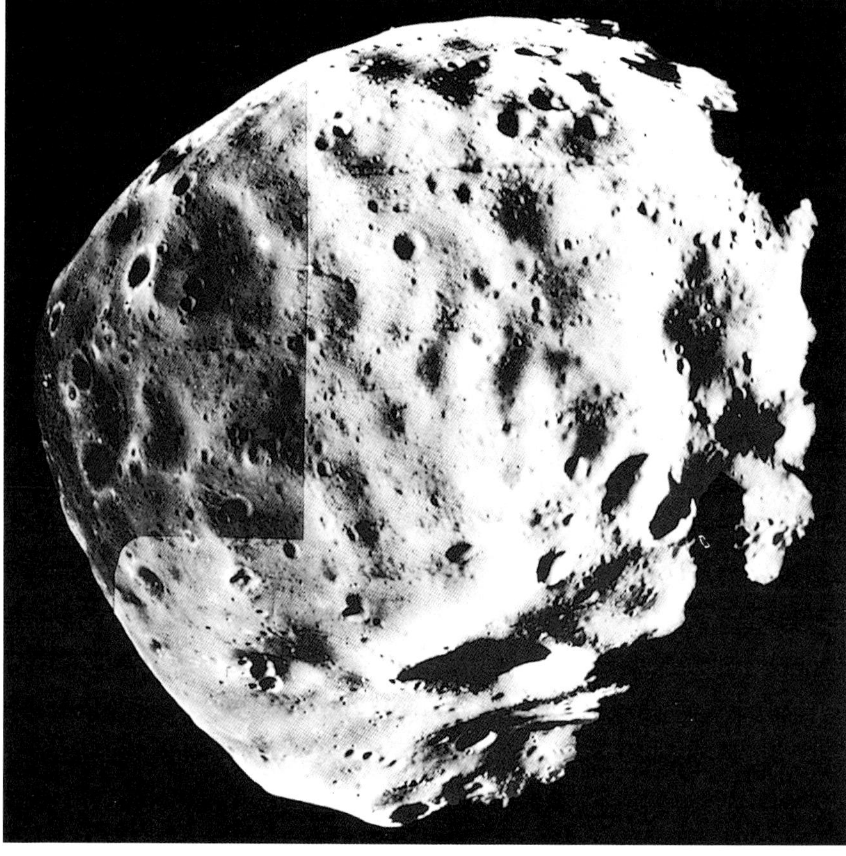

▶ ***Phobos**. The larger satellite of Mars; like Deimos, it is a dark grey colour with a reflectivity of around 5 per cent. Its density indicates that it is made of carbonaceous chondritic material similar to that of some asteroids.*

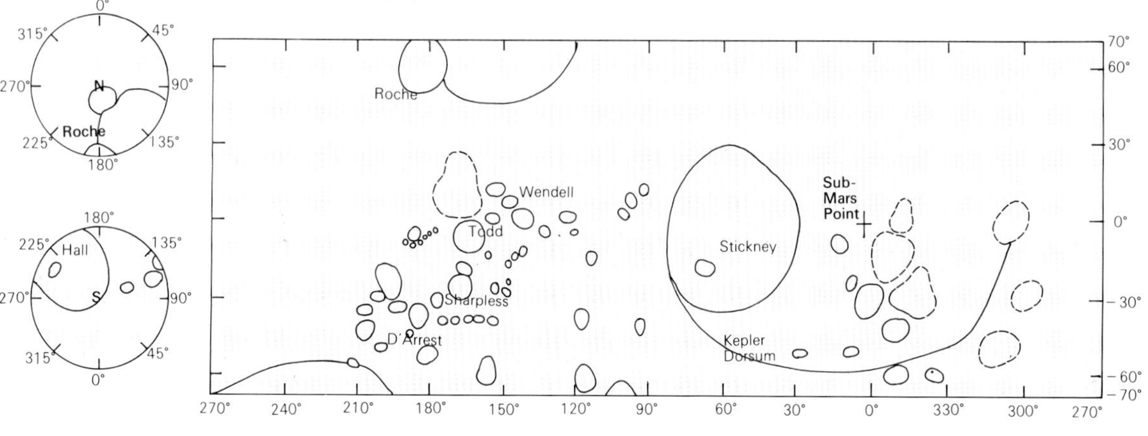

◀ ***Map of Phobos**. Stickney is much the largest crater; it lies close to the sub-Mars point. From Stickney there extends a well-marked ridge, Kepler Dorsum. Of the other main craters, Roche lies in the north and Hall in the south.*

lated that Phobos is spiralling slowly downwards at the rate of about 18 metres (60 feet) per century, in which case it may crash on to Mars in about 40 million years from now.

In July 1988 the Soviet Union launched two probes towards Mars, the main aim being to land on Phobos and examine its surface (in fact, the gravitational pull of the tiny satellite is so slight that an encounter would be more in the nature of a docking operation). Unfortunately, both missions failed. Phobos 1 was lost during the outward journey because of a faulty command sent out by the controllers. Phobos 2 was scheduled to touch down on Phobos, 'hook' on to the surface and then use an ingenious mechanism to hop around the satellite, but contact was lost before Phobos was reached, though some images were obtained. The Russians' ill-fortune with Mars continued, though the failure of America's sophisticated Mars Observer probe in 1993 was even more of a loss. New probes to Mars were launched in 1996 and 1997.

Deimos is even smaller than Phobos, with a longest diameter of no more than 15 kilometres (9 miles). Its regolith is deeper, so that the surface is more subdued; craters and pits are seen. The apparent diameter of Deimos as seen from Mars would be only about twice the maximum apparent diameter of Venus as seen from Earth, and with the naked eye the phases would be none too easy to see. Deimos would remain above the Martian horizon for two and a half 'sols' consecutively; it would transit the Sun about 130 times a Martian year, each passage taking 1 minute 48 seconds, and of course Deimos too would often be eclipsed by the shadow of Mars. It would be invisible to a Martian observer at a latitude higher than 82° North or South. The orbit, unlike that of Phobos, seems to be stable.

The Martian satellites are quite unlike our massive Moon, and it is very probably that they are ex-asteroids which were captured by Mars long ago. This idea is supported by the fact that the first two asteroids surveyed from close range by a spacecraft (Gaspra and Ida) seem to be very similar to Phobos, and are of much the same size. Photographs show similar irregular shapes and cratered surfaces. All in all, Phobos and Deimos are interesting little bodies – and one day they will no doubt be pressed into service as natural space stations.

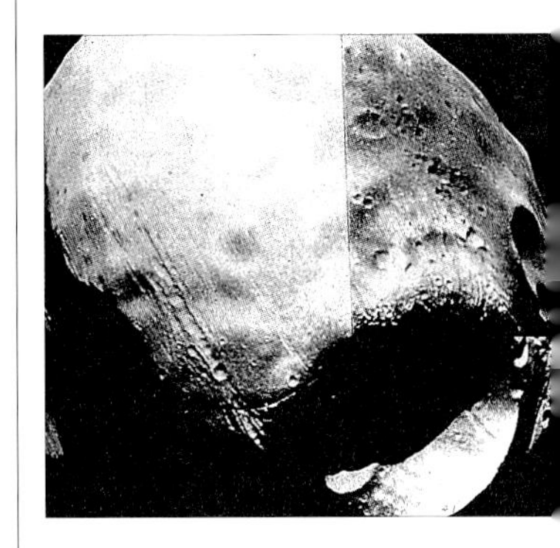

▲ ***Stickney**. The largest crater on Phobos, it is 10 km (6 miles) across. Next to it, half hidden by the shadow from the Kepler Dorsum, is the crater Hall.*

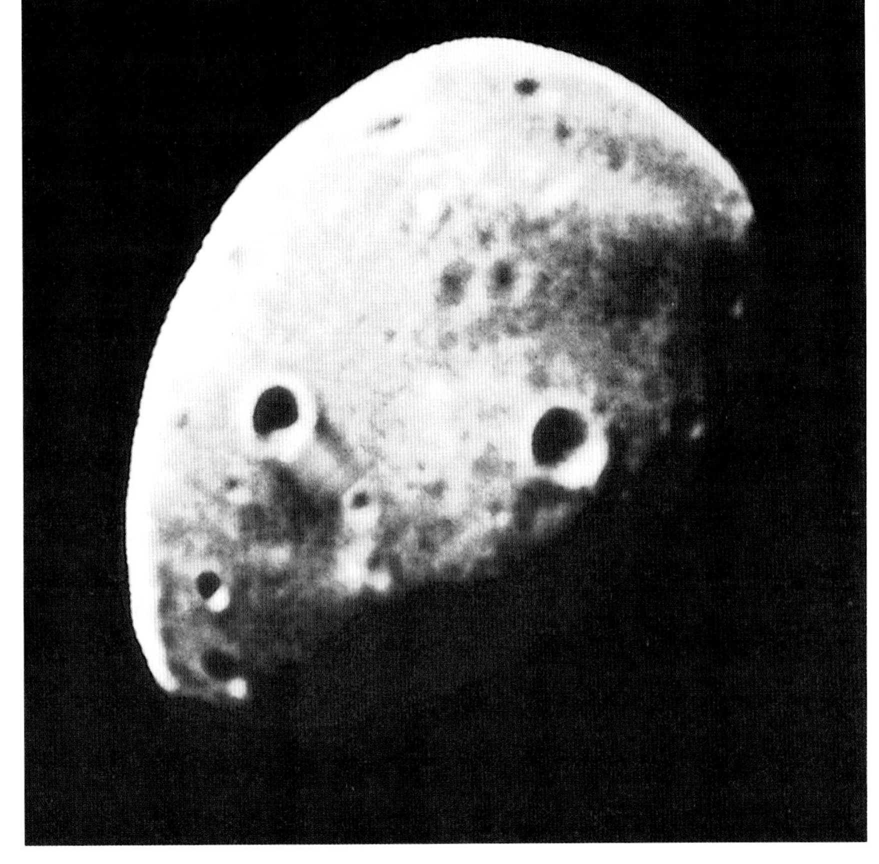

► ***Deimos**. The outer of the two satellites, Deimos is smaller and more irregular in shape. Neither of the satellites is massive enough to become spherical; both have synchronous rotations, always keeping the same face to their parent planet.*

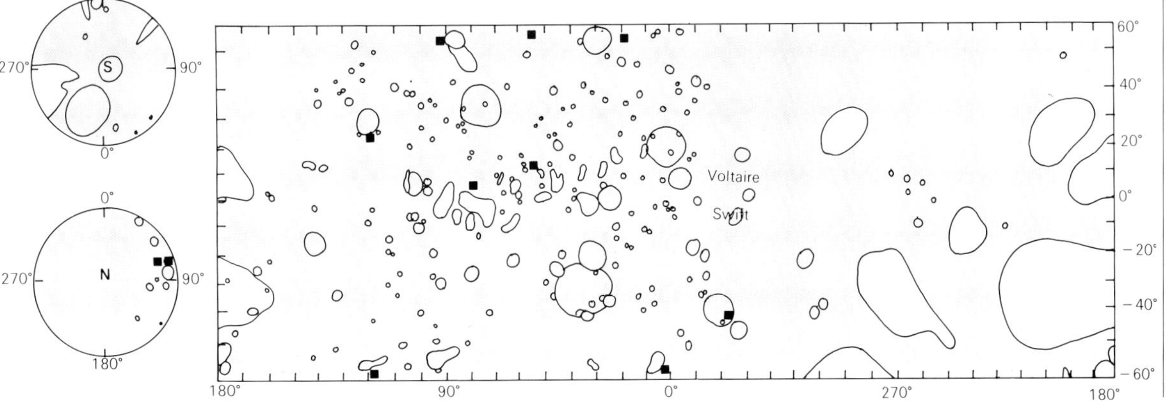

◄ ***Map of Deimos**. This satellite shows fewer well- marked features than Phobos; the sub-Mars point lies some way south of Swift, one of the two named craters..*

Map of Mars

The map of Mars given here shows many features which can be seen with an adequate telescope, such as a 30-centimetre (12-inch) reflector, under good conditions, though others, such as craters, are out of range. The map is drawn with south at the top, as this is the normal telescopic view. The most prominent feature is the triangular Syrtis Major, once thought to be an old sea-bed filled with vegetation but now known to be a plateau. A band of dark markings (the 'Great Diaphragm') runs round the planet rather south of the equator, and there are some features of special interest, such as the Solis Planum, which shows variations in shape and intensity. There are two large basins, Hellas and Argyre, which can sometimes be very bright, Hellas particularly so; this is the deepest basin on Mars, and at the bottom the atmospheric pressure is 8.9 millibars, though this is still not high enough to allow liquid water to exist. In the north the main feature is the wedge-shaped Acidalia Planitia. Observers sometimes continue to use the pre-Space Age names, so that Solis Planum is 'Solis Lacus' and Acidalia Planitia is 'Mare Acidalium'.

SELECTED FEATURES OF MARS

	Lat.°	Long.°	Diameter, km
CATENA			
Coprates	14 S–16 S	067–058	505
Ganges	02 S–03 S	071–067	233
Tithonia	06 S–05 S	087–080	400
CHAOS			
Aromatum	01 S	044	–
Aureum	02 S–07 S	030–024	365
Margaritifer	07 S–13 S	017–025	430
CHASMA			
Australe	80 S–89 S	284–257	501
Candor	04 S–08 S	078–070	400
Capri	15 S–03 S	053–031	1275
Coprates	10 S–16 S	069–053	975
Gangis	06 S–09 S	055–043	575
Tithonium	03 S–07 S	092–077	880
CRATERS			
Antoniadi	22 N	299	380
Barabashov	47 N	069	130
Becquerel	22 N	008	675
Cassini	24 N	328	440
Copernicus	50 S	169	280
Flaugergues	17 S	341	230
Herschel	14 S	230	320
Huygens	14 S	304	495
Kepler	47 S	219	238
Lowell	52 S	081	200
Lyot	50 N	331	220
Newton	40 S	158	280
Ptolemaeus	46 S	158	160
Proctor	48 S	330	160
Schiaparelli	03 S	343	500
Schroter	02 S	304	310
Prouvelot	16 N	013	150

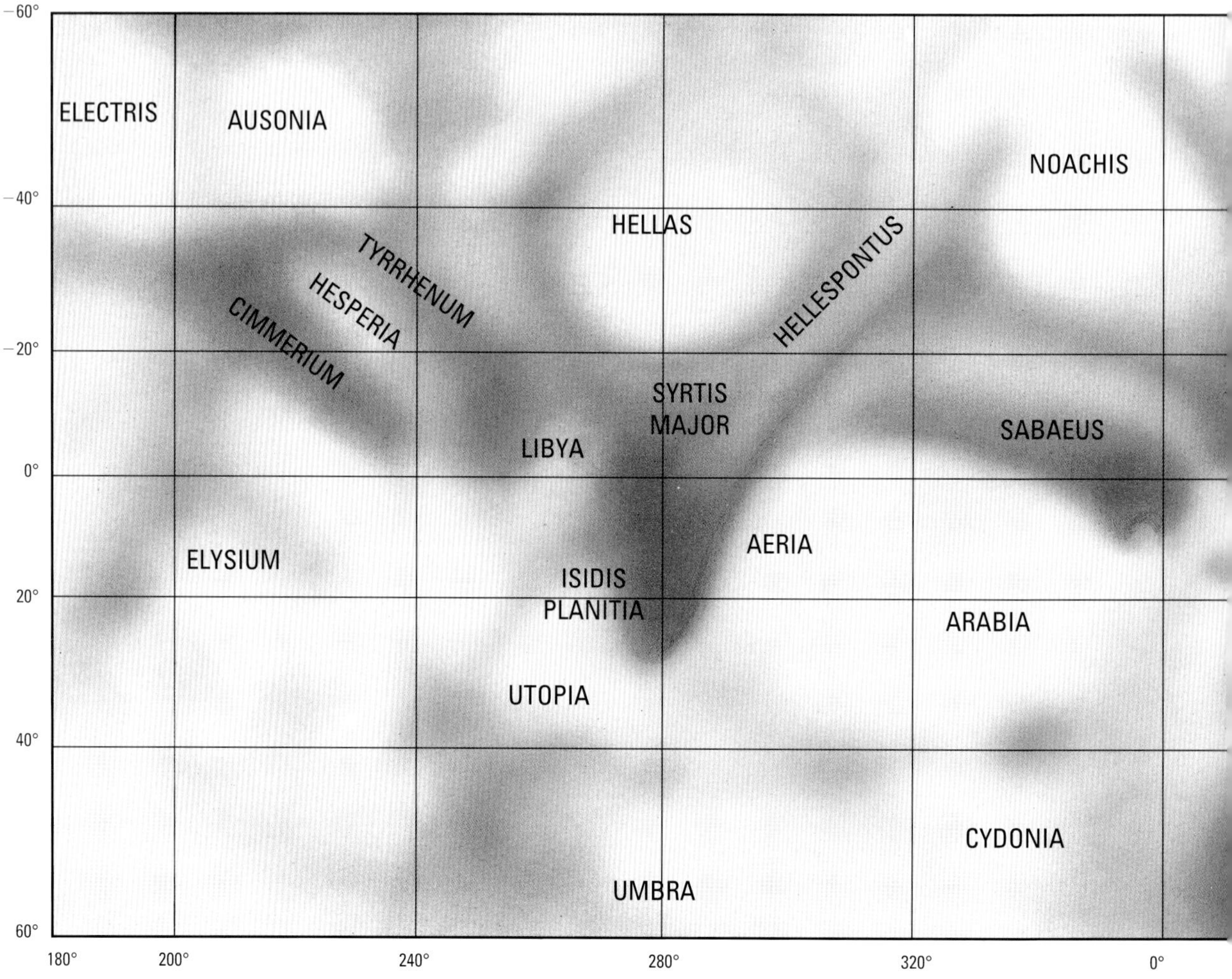

▼ ***Map of Mars*** *prepared by Paul Doherty from observations made by the author using his 39-cm (15-inch) Newtonian reflector.*

	Lat.°	Long.°	Diameter, km
MONS			
Arsia	09 S	121	500
Ascraeus	12 S	104	370
Elysium	25 N	213	180
Olympus	18 N	133	540
Pavonis	01 N	113	340
MONTES			
Charitum	50S–59S	060–027	1279
Hellespontes	35S–50S	319–310	854
Libya	15N–02S	253–282	2015
Nereidum	50S–38S	060–030	1626
Phlegra	30N–46N	195	919
Tharsis	12S–16N	125–101	2175
PLANITIA			
Acidalia	55 N–14 S	060–000	2615
Amazonia	00–40 N	168–140	2416
Arcadia	55 N–40 S	195–110	3052
Argyre	45 S–36 S	043–051	741
Chryse	19 S–30 N	051–037	840
Elysium	10 S–30 S	180–260	5312
Hellas	60 S–30 S	313–272	1955
Isidis	04 S–20 S	279–255	800
Syrtis Major	20 N–01 S	298–293	1262
Utopia	35 N–50 N	310–195	3276
PLANUM			
Aurorae	09 S–15 S	053–043	565
Hesperia	10 S–35 S	258–242	2125
Lunae	05 N–23 N	075–060	1050
Ophir	06 S–12 S	063–054	550
Sinai	09 S–20 S	097–070	1495
Solis	20 S–30 S	098–088	1000
Syria	10 S–20 S	112–097	900
TERRAE			
Arabia	00-43 N	024–280	5625

	Lat.°	Long.°	Diameter, km
Cimmeria	45 S	210	–
Margaritifer	02 N–27 S	012–045	1924
THOLUS			
Albor	19 N	210	115
Ceraunius	24 N	097	135
Hecates	32 N	210	152
Uranius	26 N	098	65
Meridiani	05 N	000	–
Noachis	15 S–83 S	040–300	1025
Promethei	30 S–65 S	240–300	2967
Sabaae	01 S	325	–
Sirenum	50 S	150	–
Tempe	24 N–54 N	050–093	1628
Tyrrhena	10 S	280	–
Xanthe	19 N–13 S	015–065	2797
VASTITAS			
Borealis	Circumpolar	9999	
LABYRINTHUS			
Noctis	04 S–14 S	110–095	1025
VALLIS			
Auqakuh	30 N–27 N	300–297	195
Huo Hsing	34 N–28 N	299–292	662
Kasei	27 N–18 N	075–056	1090
Ma' adim	28 S–16 S	184–181	955
Mangala	04 S–09 S	150–152	272
Marineris	01 N–18 S	024–113	5272
Tiu	03 N–14 N	030–035	680
MONTES			
Charitum	50 S–59 S	060–027	1279
Hellespontes	35 S–50 S	319–310	854
Libya	15 N–02 S	253–282	2015
Nereidum	50 S–38 S	060–030	1626
Phlegra	30 N–46 N	195	919
Tharsis	12 S–16 N	125–101	2175

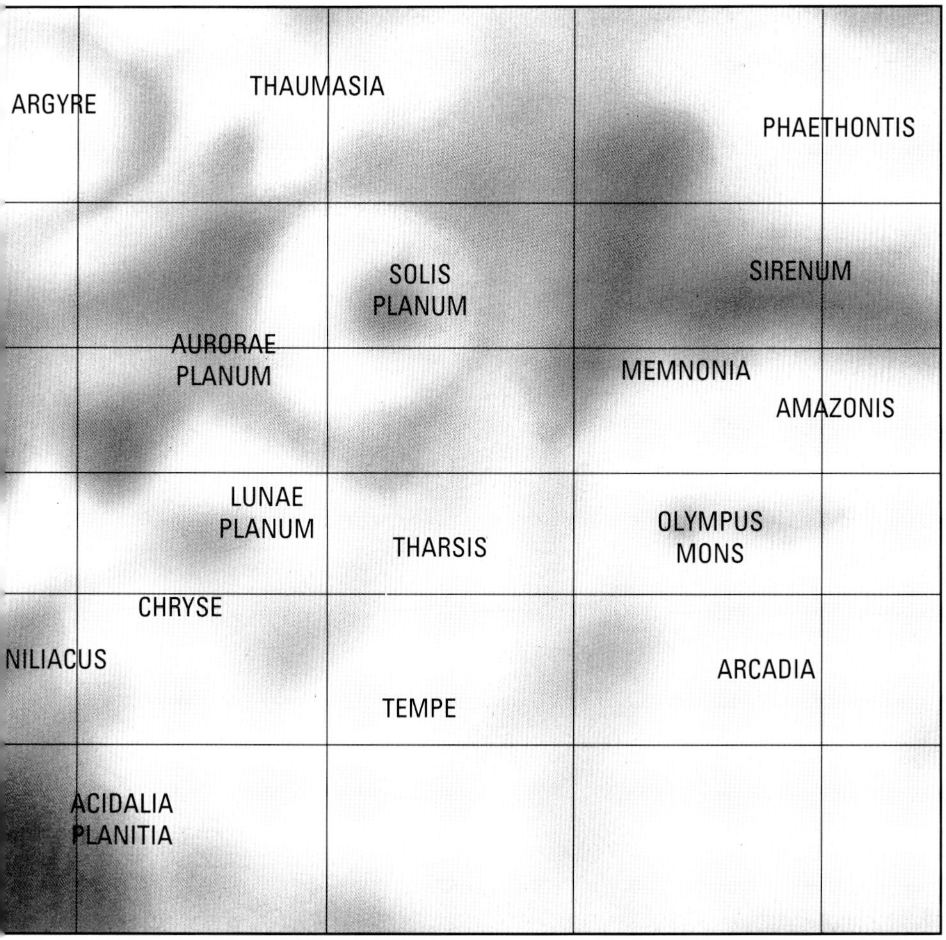

MAIN TYPES OF FEATURES

Catena – line or chain of craters, e.g. Tithonia Catena.
Chaos – areas of broken terrain, e.g. Aromatum Chaos.
Chasma – very large linear chain, e.g. Capri Chasma.
Colles – hills, e.g. Deuteronilus Colles.
Dorsum – ridge, e.g. Solis Dorsum.
Fossa – ditch: long, shallow, narrow depression, e.g. Claritas Fossae.
Labyrinthus – canyon complex. Noctis Labyrinthus is the only really major example.
Mensa – small plateau or table-land, e.g. Nilosyrtis Mensae.
Mons – mountain or volcano, e.g. Olympus Mons.
Patera – saucer-like volcanic structure, e.g. Alba Patera.
Planitia – smooth, low-lying plain, e.g. Hellas Planitia.
Planum – plateau; smooth high area, e.g. Hesperia Planum.
Rupes – cliff, e.g. Ogygis Rupes.
Terrae – lands, names often given to classical albedo features, e.g. Sirenum Terra, formerly known as Mare Sirenum.
Tholus – domed hill, e.g. Uranius Tholus.
Vallis – valley, e.g. Vallis Marineris.
Vastitas – extensive plain. Vastitas Borealis is the main example.

Hubble Views of Mars

Two Hubble Space Telescope images of Mars, taken about a month apart on 18 September and 15 October 1996, reveal a state-sized dust storm churning near the edge of the Martian north polar cap. The polar storm is probably a consequence of large temperature differences between the polar ice and the dark regions to the south, which are heated by the springtime sun. The increased sunlight also causes the dry ice in the polar caps to sublime and shrink.

Mars is famous for large, planet-wide dust storms. Smaller storms resembling the one seen here were observed in other regions by Viking orbiters in the late 1970s. However, this is the first time that such an event has been caught near the receding north polar cap. The Hubble images provide valuable new insights into the behaviour of localized dust storms on Mars, which are typically below the resolution of ground-based telescopes. This kind of advanced planetary 'weather report' has been invaluable in aiding preparation for the landing on Mars of NASA's Pathfinder spacecraft, which took place on 4 July 1997, and the arrival of the Mars Global Surveyor orbiter in September 1997. Fortunately, these images show no evidence for large-scale dust storm activity, which plagued a previous Mars mission in the early 1970s.

However, please note that in order to help compare locations and sizes of features, the map projections (shown to the right of each disk) are centred on the geographic north pole. Maps are oriented with 0 degrees longitude at the top and show meridians every 45 degrees of longitude (longitude increases clockwise); latitude circles are also shown for 40, 60 and 80 degrees north latitude. The colour images were assembled from separate exposures taken with the Wide Field Planetary Camera 2 (WFPC-2).

As with all Hubble images of the planets, these have north at the top – whereas the older maps were always oriented with south uppermost. [Credit: Phil James (University of Toledo), Steve Lee (University of Colorado) and NASA; David Crisp and the WFPC-2 Science Team (Jet Propulsion Laboratory/California Institute of Technology)]

► ***18 September 1996**: the salmon-coloured notch in the white north polar cap is a 1000-km (600-mile) long storm – nearly the width of Texas. The bright dust can also be seen over the dark surface surrounding the cap, where it is caught up in the Martian jet stream and blown easterly. The white clouds at lower latitudes are mostly associated with major Martian volcanoes such as Olympus Mons. This image was taken when Mars was more than 300 million km (186 million miles) from Earth, and the planet was smaller in angular size than Jupiter's Great Red Spot!*

September 18, 1996

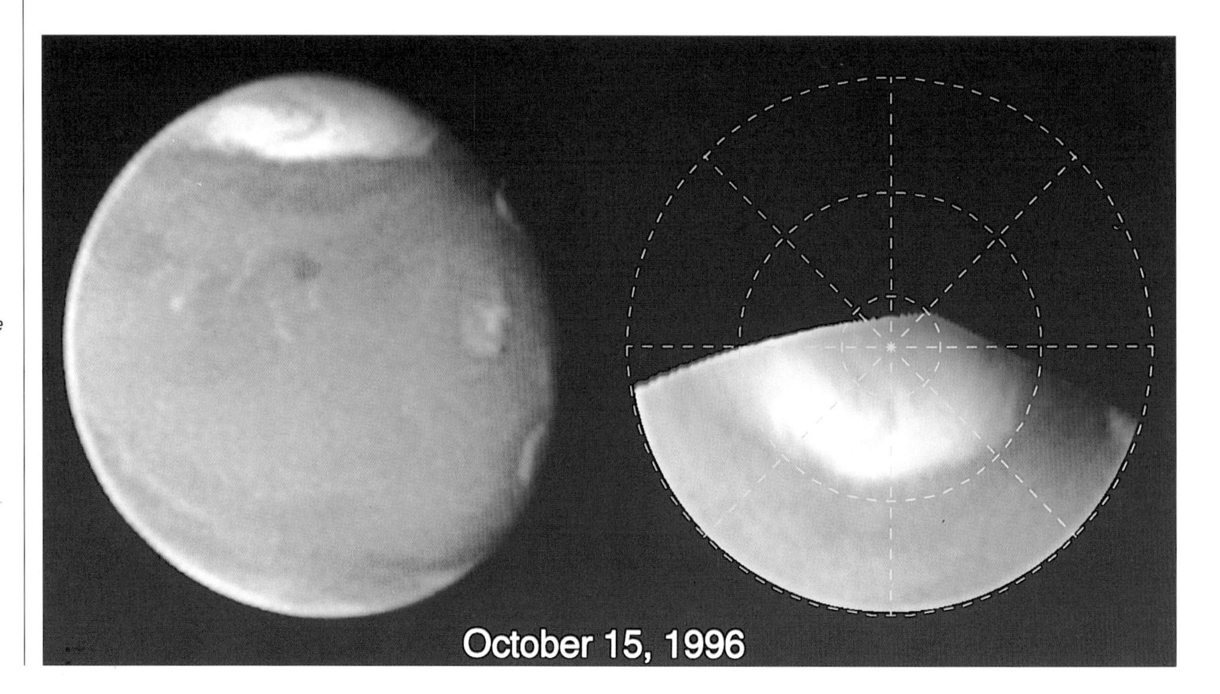

► ***15 October 1996**: though the storm has dissipated by October, a distinctive dust-coloured, comma-shaped feature can be seen curving across the ice cap. The shape is similar to cold fronts on Earth, which are associated with low-pressure systems. Nothing quite like this feature has been seen before, either in ground-based or spacecraft observation. The snow-line marking the edge of the cap receded northwards by approximately 200 km (120 miles), while the distance to Mars narrowed to 275 million km (170 million miles).*

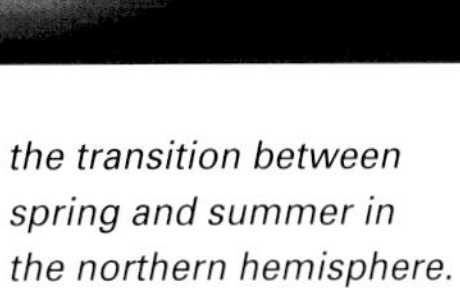

▲ ***Images taken with the WFPC-2*** *on 10 March 1997. The pictures show the transition between spring and summer in the northern hemisphere. The annual north polar carbon dioxide ice cap is rapidly sublimating, revealing the much smaller permanent water ice cap.*

◄ ***The sharpest view of Mars*** *ever taken from Earth was obtained by the recently refurbished Hubble Space Telescope. This stunning portrait was taken with the Wide Field Planetary Camera 2 (WFPC-2) on 10 March 1997, just before Mars opposition, when the red planet made one of its closest passes to the Earth (about 100 million km or 60 million miles). At this distance, a single picture element (pixel) in the Planetary Camera spans 22 km (13 miles) on the Martian surface.*

The Search for Life on Mars

The most successful Mars probes to date have been the two Vikings, each of which consisted of an orbiter and a lander. Viking 1 was launched on 20 August 1975, and in the following June was put into a closed orbit round Mars. It continued the mapping programme of Mariner 9, and some of the images were very detailed; one showed a rock on which light and shadow effects gave an uncanny look of a human face – a fact which was not overlooked by scientific eccentrics. On 20 July 1976, the lander was separated from the orbiter, and was brought gently down partly by parachute braking (useful even in the thin Martian atmosphere) and partly by rocket braking. The landing site was Chryse, (the Golden Plain) at latitude 22 degrees North, longitude 47.5 degrees West. The first images showed a red, rock-strewn landscape under a pink sky; temperatures were very low, reaching a maximum of −31 degrees C near noon and a minimum of −86 degrees C just after dawn.

The main task of Viking 1 was to search for life. Material was scooped up, drawn into the spacecraft and analysed chemically for traces of organic substances. The results were sent back to Earth and were at first decidedly puzzling, but it has to be admitted that no positive signs of life were found. The results from Viking 2, which landed in the more northerly plain of Utopia on 3 September 1976, were similar. Windspeeds were measured, and the first analyses of the surface material carried out; the main constituent is silica (over 40 per cent).

If there is any life on Mars today it must be very lowly indeed. Whether the situation was different in the past is not certain because of the evidence of old riverbeds; at some periods Mars may have been less unfriendly than it is now, and it is at least possible that life appeared, dying out when conditions deteriorated. We will know for

▼ ***First colour picture from Viking 1**. The lander of Viking 1 came down in the 'golden plain' of Chryse, at latitude 22.4°N, longitude 47.5°W. The picture shows a red, rock-strewn landscape; the atmospheric pressure at the time was approximately 7 millibars. The first analysis of the surface material was made from Viking: 1–44 per cent silica, 5.5 per cent alumina, 18 per cent iron, 0.9 per cent titanium and 0.3 per cent potassium.*

▲ ***The search for life**. Viking 1 drew in material from the red 'desert', analysed it, and sent back the results. It had been expected that signs of organic activity would be detected, but this proved not to be the case. It is too early to say definitely that Mars is totally lifeless, but this does seem to be indicated by the available evidence. Both the Viking landers have long since ceased to operate, so that for a final decision we must await the results from a new spacecraft. There are already plans to send a probe to Mars, collect material, and return it to Earth for analysis.*

certain only when we can examine Martian material in our laboratories, and this should be possible before long; an automatic probe should be able to land there and return to Earth with specimens for analysis.

It has been claimed that some meteorites found in Antarctica have come from Mars, blasted away from the Red Planet by a giant impact, and that they contain traces of past primitive organisms. This is an interesting possibility, but the evidence is far from conclusive. We do not know for certain that the meteorites are of Martian origin, or that the features contained in them are indeed indicative of past life.

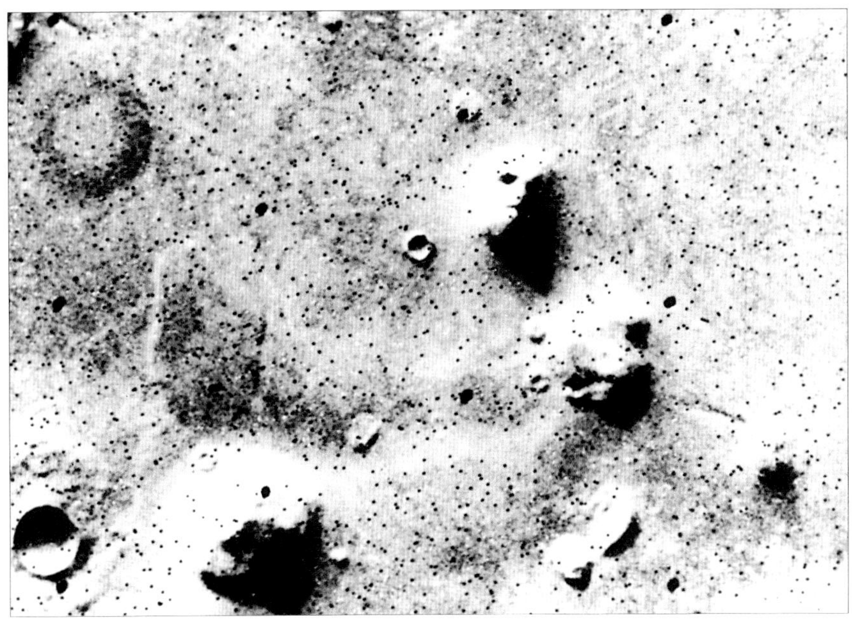

▲ ***The 'face' on Mars.*** *This picture was taken by the orbiting section of Viking 1. The arrangement of light and shadow on a surface rock gives an almost uncanny impression of a human face – and this was naturally seized upon by scientific eccentrics! It is, of course, a purely natural feature.*

◀ ***Landing site of Viking 2.*** *Viking 2 came down in the plain of Utopia; latitude 48°N, longitude 226°W. The site was essentially similar to that of Viking 1, and Viking 2 confirmed the earlier findings. For example, the atmosphere proved to be 95.3 per cent carbon dioxide, 2.7 per cent nitrogen, and 1.6 per cent argon, with smaller quantities of other gases.*

The Pathfinder Mission

On 2 December 1996 a new probe was launched towards Mars: Pathfinder, which carried a small 'rover', Sojourner. This time there was to be no gentle, controlled landing. Pathfinder was encased in tough airbags, and was designed to land at high speed, bouncing several times before coming to rest. It would enter the Martian atmosphere at a speed of 26,700 kilometres (16,600 miles) per hour; the touchdown would undoubtedly be violent, so that everything depended upon the airbags. Once Pathfinder had settled down and assumed an upright position, its 'petals' would open, so that Sojourner could crawl down a ramp on to the Martian surface. It was an ambitious project by any standards, and Sojourner itself was also unusual; it was about the size of a household microwave, but it was in fact a highly sophisticated probe, capable of carrying out on-the-spot analyses of the Martian rocks.

Data began to come through even during the descent. The Martian atmosphere was both clearer and colder than it had been during the Viking landings of 1976; at a height of 80 kilometres (50 miles) above the ground the temperature was −160 degrees C. But everything went according to plan, and on 4 July 1997 – America's Independence Day – Pathfinder landed safely on the old flood plain at the end of Ares Vallis. After a journey of over 480 million kilometres (300 million miles), Pathfinder came down within 20 kilometres (12 miles) of the planned impact point. It was then 190 million kilometres (120 million miles) from Earth; the ground temperature reached a maximum noon value of −13 degrees C, though it plummeted to well below −75 degrees C during the night.

Almost at once, the main station used its camera to transmit a panorama of the entire scene; it seemed more interesting than Chryse or Utopia had been. The site, at the end of Ares Vallis, had been carefully chosen. The valley had once been a raging torrent of water, bringing down rocks of all kinds on to the flood plain, and this did indeed prove to be the case. Next, Sojourner emerged and began its work (there had been a minor delay because one of the airbags had not deflated completely after landing, and had to be manoeuvred away from the ramp). The rocks around were given distinctive nicknames, such as Barnacle Bill, Yogi and Soufflé (the main station itself had already been named in honour of Carl Sagan, the American planetary astronomer who had died not long before the mission). Sojourner's track marks showed that the Martian 'soil' might be compared with the very fine-grained silt found in places such as Nebraska, USA; the grains were less than 50 microns in diameter, which is finer than talcum powder.

Sedimentary rocks were identified, and this was not surprising, since water had once covered the region. There was, however, one major surprise. Though the rocks were essentially basaltic, andesite was also found. This occurs on Earth at the edges of tectonic plates, for instance round the border of the Pacific, but there is no evidence that plate tectonics also applied to Mars, so that the origin of the andesite remains a puzzle.

Pathfinder was not designed to search for traces of life; this will be the task of later missions, and even as Sojourner was crawling around in Ares Vallis, moving at a maximum rate of half a kilometre (one-third of a mile) per hour, a new spacecraft, Mars Global Surveyor, was on its

► ***Pathfinder on Mars.*** *Now renamed the Sagan Memorial Station in honour of Carl Sagan. Beside it is the Sojourner rover, which moved around and analysed the Martian rocks.*

◀ ***Ares Vallis, from Pathfinder/Sojourner.*** *Ares Vallis is an old flood plain; rocks of several different types are shown here.*

◀ ***Crater Galle****, on the edge of the Argyre basin. Certainly it does recall a happy human face!*

way. It entered orbit round Mars in September 1997, and began sending back images of amazingly good quality. One crater, Galle – 230 kilometres (140 miles) across, on the edge of Argyre – gave the impression of a happy smiling face!

However, both of the latest US probes have failed. Mars Climate Orbiter, launched in 1998, was lost in September 1999 because of a programming error. Mars Polar Lander, sent up in January 1999, presumably landed on schedule in December 1999, but no signals were received from it after arrival.

As yet we still do not know whether there is any trace of life there. However, we ought to find out in the near future, when Martian samples will be brought back for study in our laboratories. But whether or not life exists, Mars remains a planet which is, to us, of surpassing interest.

Asteroids

Beyond the orbit of Mars lies the main belt of asteroids of minor planets. Only one (Ceres) is as much as 900 kilometres (560 miles) in diameter, and only one (Vesta) is ever visible with the naked eye; most of the members of the swarm are very small indeed, and there are fewer than 20 main-belt asteroids which are as much as 250 kilometres (150 miles) across.

Ceres, the largest member of the swarm, was discovered on 1 January 1801 – the first day of the new century – by G. Piazzi at the Palermo Observatory. He was not looking for anything of the sort; he was compiling a new star catalogue when he came across a star-like object which moved appreciably from night to night. This was somewhat ironical in view of the fact that a planet-hunt had been organized by a team of astronomers who called themselves the 'Celestial Police'. A mathematical relationship linking the distances of the known planets from the Sun had led to the belief that there ought to be an extra planet between the paths of Mars and Jupiter, and the 'Police' had started work before Piazzi's fortuitous discovery. They did locate three more asteroids – Pallas, Juno and Vesta – between 1801 and 1808, but the next discovery, that of Astraea, was delayed until 1845, long after the 'Police' had disbanded. Since 1847 no year has passed without new discoveries, and the current total of asteroids whose paths have been properly worked out is considerably more than 5000. Some small bodies have been found, lost and subsequently rediscovered; thus 878 Mildred, originally identified in 1916, 'went missing' until its rediscovery in 1990.

The asteroids are not all alike. The largest members of the swarm are fairly regular in shape, though No. 2, Pallas, is triaxial, measuring 580 × 530 × 470 kilometres (360 × 330 × 290 miles), and smaller asteroids are certainly quite irregular in outline; collisions must have been – and still are – relatively frequent. Neither are the compositions the same; some asteroids are carbonaceous, others siliceous, and others metal-rich. No. 3, Vesta, has a surface covered with igneous rock; 16 Psyche is iron-rich; 246 Asporina and 446 Æternitas seem to be almost pure olivine, while in 1990 it was found that there are indications of organic compounds on the surfaces of a few asteroids, including the unusually remote 279 Thule. Some asteroids are fairly reflective, while others, such as 95 Arethusa, are blacker than a blackboard. Obviously, no surface details can be seen from Earth, and almost all our information has been obtained spectroscopically.

No asteroid has an escape velocity high enough to retain atmosphere. The three largest members (Ceres, Pallas and Vesta) account for 55 per cent of the total mass of the main-belt bodies. Two asteroids, 951 Gaspra and 243 Ida, have been surveyed from close range by the Galileo spacecraft, which passed through the main zone during its journey to Jupiter; in 1997 another asteroid, Mathilde, was imaged by the NEAR spacecraft on its way to rendezvous with the asteroid Eros in December 1998.

Asteroids appear so small that to record surface detail in them is far from easy. However, in 1994 Vesta was imaged by the Hubble Space Telescope; the asteroid's apparent diameter was a mere 0.42 of an arc second. There are bright and dark features; part of the surface seems to be covered with quenched lava flows, while another part indicates molten rock which cooled and solidified underground, to be later exposed by impacts on the surface.

Most main-belt asteroids have reasonably circular orbits, though some are highly inclined by 34 degrees in the case of Pallas, for example. They tend to group in 'families', with definite regions which are less populated. This is due to the powerful gravitational pull of Jupiter, and it seems certain that it was Jupiter's disruptive influence which prevented a larger planet from forming.

▼ ***The Celestial Police**. This is an old picture of the observatory at Lilienthal, owned by Johann Hieronymus Schröter. It was here that the 'Celestial Police' met to work out the way in which to search for the missing planet moving between the orbits of Mars and Jupiter. Schröter's main telescope was a 48-cm (19-inch) reflector, but he also used telescopes made by William Herschel.*

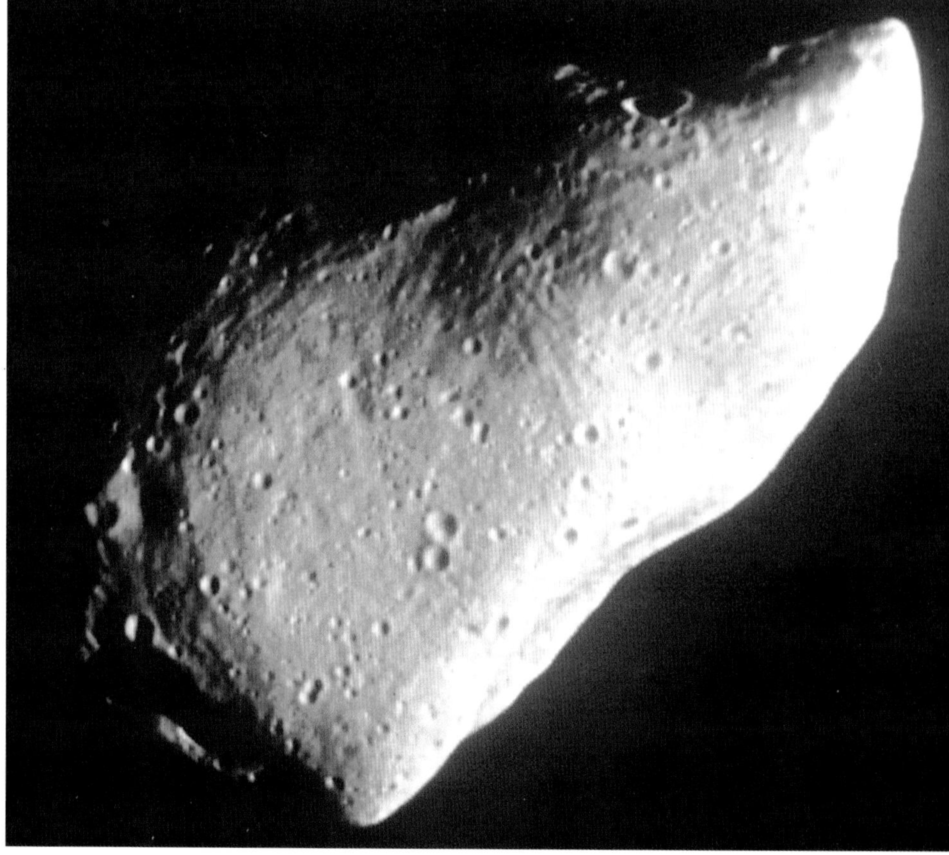

▶ ***Gaspra**. This was the first close-range picture of a main-belt asteroid, obtained by the Galileo probe on 13 November 1991 from a range of 16,000 km (less than 10,000 miles). Gaspra (asteroid 951) proved to be wedge-shaped, with a darkish, crater-scarred surface. Gaspra is irregular in shape; it is 16 km (10 miles) long by 12 km (7.5 miles); and the smallest features recorded are only 55 metres (180 feet) across.*

SOME MAIN-BELT ASTEROIDS							
Name	**Distance from Sun, astronomical units**		**Period, years**	**Type**	**Diameter, km**	**Mag.**	**Rotation period**
	min.	max.			max.		hour
1 Ceres	2.55	2.77	4.60	C	940	7.4	9.08
2 Pallas	2.12	2.77	4.62	CU	580	8.0	7.81
3 Juno	1.98	2.87	4.36	S	288	8.7	7.21
4 Vesta	2.15	2.37	3.63	V	576	6.5	5.34
5 Astraea	2.08	2.57	4.13	S	120	9.8	16.81
10 Hygeia	2.76	3.13	5.54	C	430	10.2	17.50
16 Psyche	2.53	2.92	5.00	M	248	9.9	4.20
44 Nysa	2.06	2.42	3.77	S	84	10.2	5.75
72 Feronia	1.99	2.67	3.41	U	96	12.0	8.1
132 Aethra	1.61	2.61	4.22	SU	38	11.9	?
279 Thule	4.22	4.27	8.23	D	130	15.4	?
288 Glauke	2.18	2.76	4.58	S	30	13.2	1500
704 Interamnia	2.61	3.06	5.36	E	338	11.0	8.7
243 Ida	2.73	2.86	4.84	S	52	14.6	5.0

ASTEROID TYPES		
Designation	**Type**	**Example**
C	Carbonaceous; spectra resemble carbonaceous chondrites	1 Ceres
S	Silicaceous; generally reddish; spectra resemble chondrites	5 Astraea
M	Metallic; perhaps metal-rich cores of former larger bodies which have been broken up by collision	16 Psyche
E	Enstatite; rare, resemble some forms of chondrites in which enstatite ($MgSiO_3$) is a major constituent	434 Hungaria
D	Reddish. surface rich in clays	336 Lacadiera
A	Almost pure olivine	446 Aeternitas
P	Peculiar spectra; not too unlike type M	87 Sylvia
Q	Close-approach asteroids; resemble chondrites	4581 Asclepius
V	Igneous rock surfaces; Vesta is the only large example	4 Vesta
U	Unclassifiable	72 Feronia

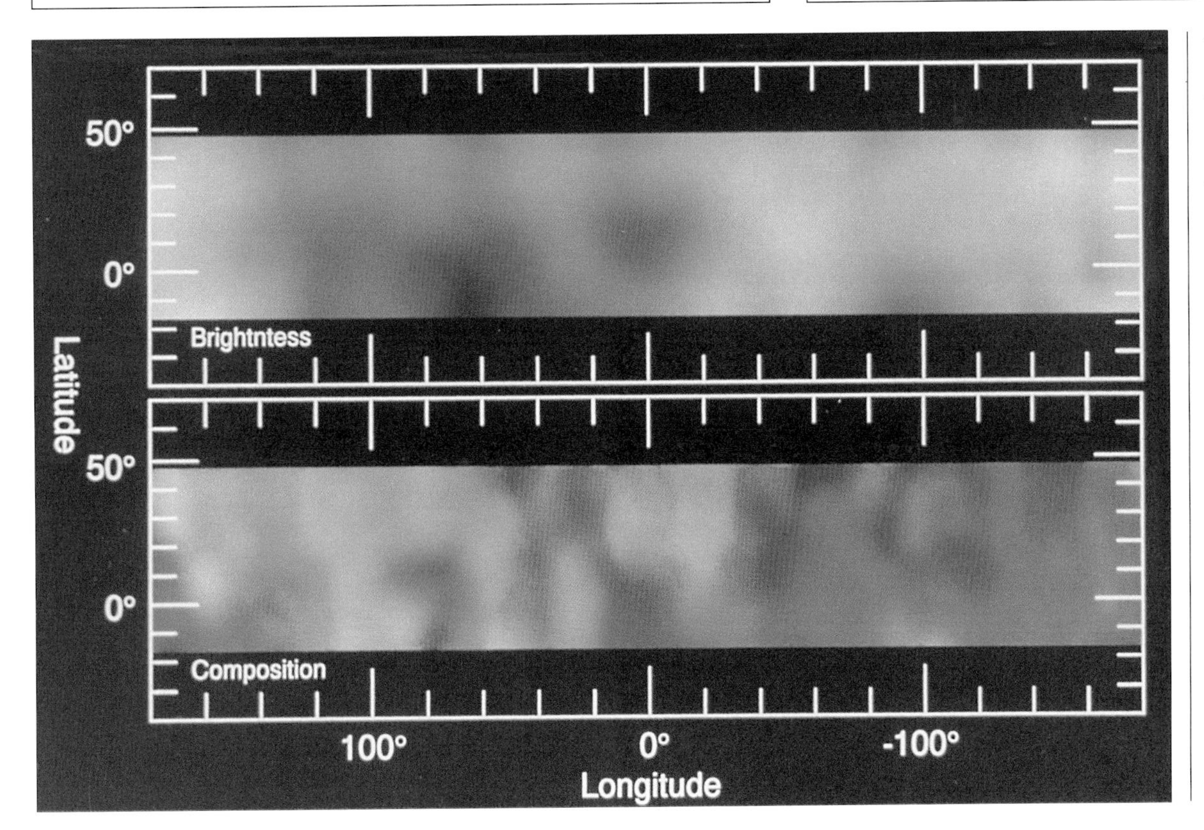

◀ ***Preliminary map of Vesta**, from the Hubble Space Telescope in 1994 (blue light). The dark circular feature (Olbers) may be a 200-km (125-mile) crater. The lower panel is a false-colour composite of visible and near-infra-red images, indicating that different minerals dominate the hemisphere of Vesta on the left and right. [Courtesy Ben Zellner and NASA]*

▼ ***Vesta**, from the Hubble Space Telescope, in November–December 1994. The images in each row were taken in red light, 9 minutes apart; the axial spin is clearly demonstrated. [Courtesy Ben Zellner and Alex Sorrs, HSTI]*

▲ ***Asteroid 243 Ida** with its satellite, Dactyl, imaged on 28 August 1993 by the Galileo spacecraft. Ida is a member of the Koronis group of asteroids, moving around the Sun at a mean distance of 430 million km (267 million miles) in a period of 4.84 years. It measures 56 × 24 × 21 km (35 × 15 × 13 miles); Dactyl has a diameter of about 1 km (0.6 mile). The separation is about 100 km (60 miles). Both bodies were probably formed at the same time when a larger object broke up during a collision.*

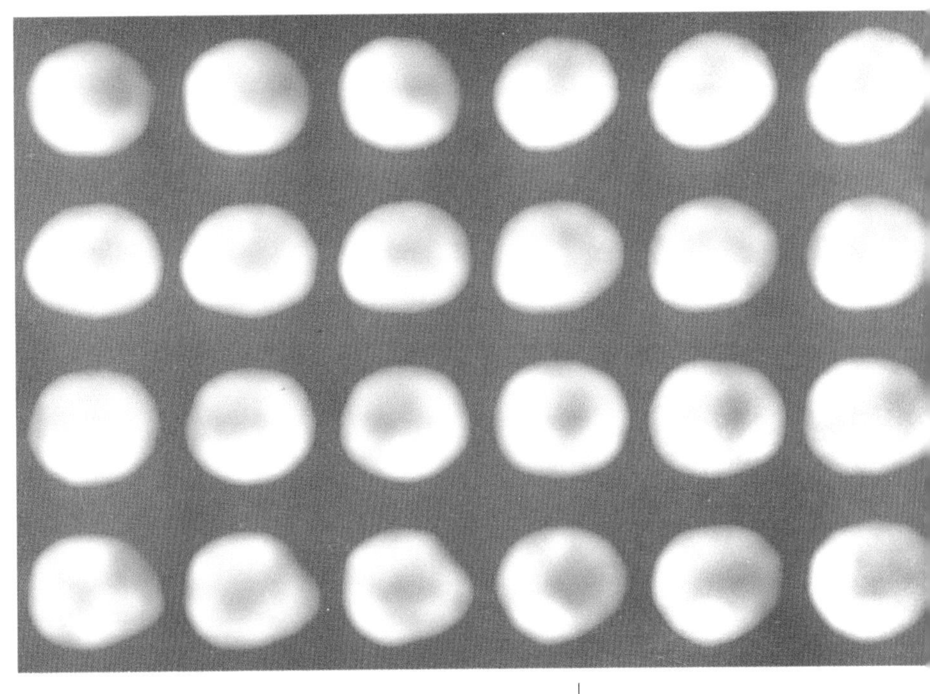

Exceptional Asteroids

Not all asteroids are confined to the main swarm. For example there are the Trojans, which move in the same orbit as Jupiter, occupying what are termed Lagrangian points. In 1772, the French mathematician Joseph Lagrange drew attention to the 'problem of the three bodies', which applies when a massive planet and a small asteroid move round the Sun in the same plane, with virtually circular orbits and in equal periods; if they are 60 degrees apart, they will always remain 60 degrees apart. Therefore, the Trojans are in no danger of being swallowed up, even though they do oscillate to some extent round the actual Lagrangian points. By asteroidal standards they are large; 588 Achilles, the first-known member of the group, is 116 kilometres (72 miles) in diameter and 624 Hektor as much as 232 kilometres (145 miles), though it has been suggested that Hektor may be double. Because of their remoteness, the Trojans are very faint; many dozens are now known. There is also one Martian Trojan, yet to be named, and probably others exist.

▼ ***Asteroid 4015****. This was discovered by Eleanor Helin in 1979. It was subsequently found to be identical with Comet Wilson–Harrington, 1949. The top picture shows it in 1949, with a cometary tail; the bottom in its 1979 guise, when in appearance it was purely asteroidal. This seems to support the suggestion that at least some close-approach asteroids are extinct comets.*

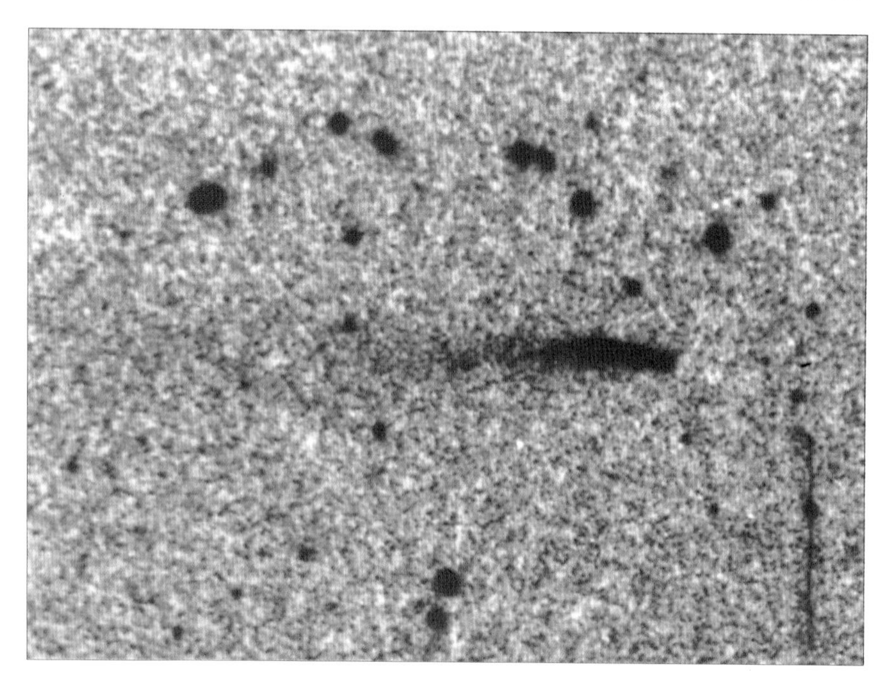

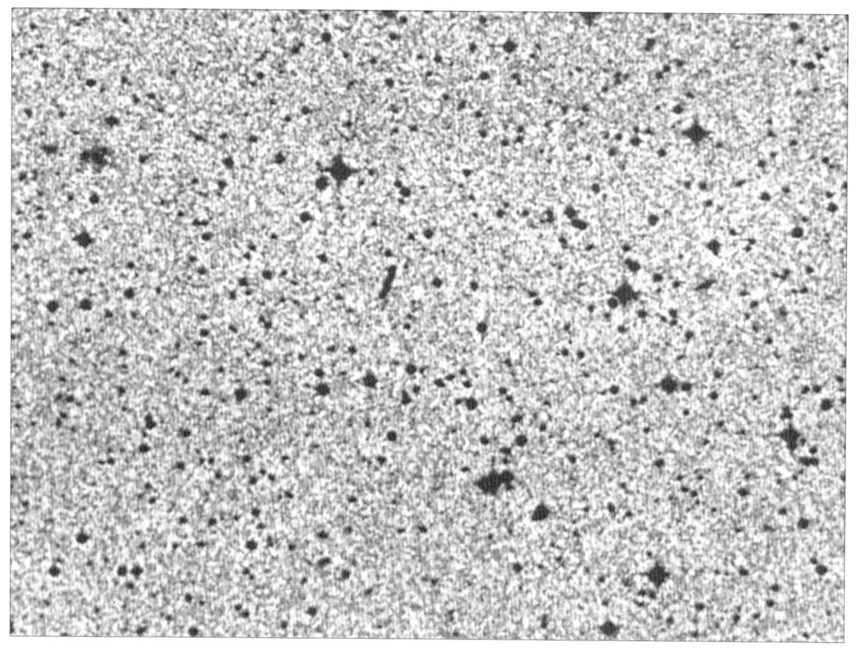

Some small asteroids have very eccentric orbits. Thus 944 Hidalgo has a path which takes it between 300 million and 870 million kilometres (190 million and 550 million miles) from the Sun, so that at aphelion (furthest point) it is more remote than Jupiter. Hidalgo's orbit is in fact like that of a comet, and this may be significant.

Chiron, discovered by C. Kowal in 1977, has a period of 50 years, and spends most of its time between the orbits of Saturn and Uranus. The diameter is at least 150 kilometres (over 90 miles), and when near perihelion it develops what looks like a cometary-type coma, though Chiron is much too large to be regarded as a comet. It has been given an asteroidal number, 2060, although it is quite unlike an ordinary main-belt asteroid.

The first asteroid known to come well within the orbit of Mars was 433 Eros, discovered in 1898; its closest approach to Earth is 23 million kilometres (14 million miles). It is sausage-shaped, with a longest diameter of 22 kilometres (14 miles). Since then many 'close-approach' asteroids have been found, and have been divided into three groups. Amor asteroids have orbits which cross that of Mars, but not that of the Earth; Apollo asteroids do cross the Earth's orbit and have mean distances from the Sun greater than one astronomical unit, while Aten asteroids have paths which lie mainly inside that of the Earth, so that their periods are less than one year.

One member of the Aten group, 2340 Hathor, is only 500 metres (1600 feet) across. At present the holder of the 'approach record' is 1994 XM_1 (yet to be named), which brushed past us on 9 December 1994 at a mere 112,000 kilometres (70,000 miles). Though it is no more than 7 to 12 metres (23 to 40 feet) across, it would have caused tremendous global devastation if it had collided with the Earth.

All the close-approach asteroids are midgets. One, 4179 Toutatis, was contacted by radar and found to be a contact binary, with components 4 kilometres (2.5 miles) and 2.5 kilometres (1.5 miles) across touching each other and moving round their common centre of gravity in 10.5 days.

New close-approach asteroids are being found regularly, and it seems that they are much commoner than used to be thought, so that occasional impacts cannot be ruled out. There have been serious suggestions that we should be keeping a close watch out for them, so that if we see one homing in on us we can try to divert it by sending up a nuclear missile.

Two asteroids, Icarus and Phaethon, have paths which take them inside the orbit of Mercury, so that at perihelion they must be heated to at least 500 degrees C; at aphelion they recede into the main belt, so that their climates must be decidedly uncomfortable.

3200 Phaethon moves in much the same orbit as the Geminid meteor stream, and it may well be a 'dead' comet which has lost all its volatiles. It is also true that Asteroid 4015, discovered in 1979, has been identified with an object observed in 1949 and classed as a comet (Wilson–Harrington). There are similarities, and there may be very close links indeed between meteoroids, comets and close-approach asteroids. Small though they may be, the asteroids have many points of interest. We no longer consider them as the 'vermin of the skies'.

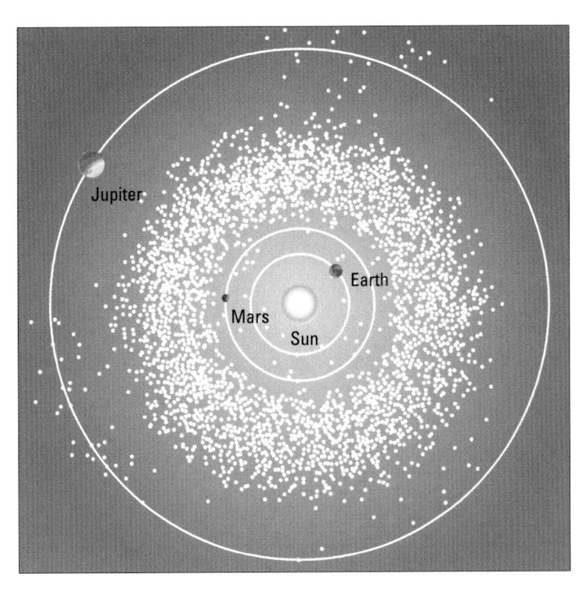

▶ ***Asteroid positions**. This diagram shows the positions of known asteroids in 1990. The orbits of the Earth, Mars and Jupiter are shown. It is clear that most of the asteroids lie in the main belt, between the orbits of Mars and Jupiter; some depart from the main swarm, but the 'close-approach' asteroids are all very small, while the Trojans move in the same orbit as Jupiter.*

▼ ***Asteroid 4179 Toutatis** – a composite of five exposures, obtained with the 3.5-m ESO New Technology Telescope on 21 December 1992. At this time, Toutatis was about 13 million km (8 million miles) from Earth. The unusually rapid motion of Toutatis, caused by its small distance from the Earth, is well illustrated on the photo. The first exposure (at the arrow) was obtained at 8:05 UT and lasted 120 seconds; the next (in the direction towards the upper right) lasted 30 seconds at 8:10 UT; a five-second exposure was then made at 8:15 UT, and the last two (near the right edge of the photo) at 8:59 UT and 9:01 UT also lasted five seconds each. During this one-hour interval, Toutatis moved a distance of more than 3 arc minutes in the sky.*

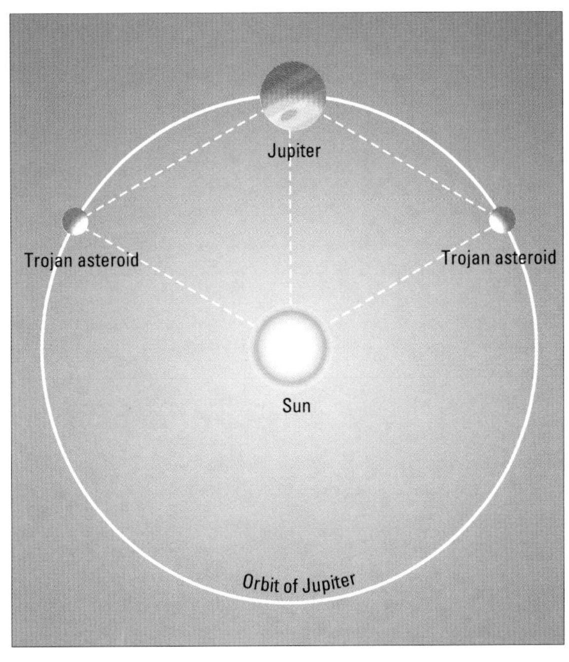

▲ ***The Trojan asteroids**. The Trojans move in the same orbit as Jupiter, but keep prudently either 60 degrees ahead of or 60 degrees behind the Giant Planet, so that they are in no danger of collision – though naturally they oscillate to some extent round their mean points.*

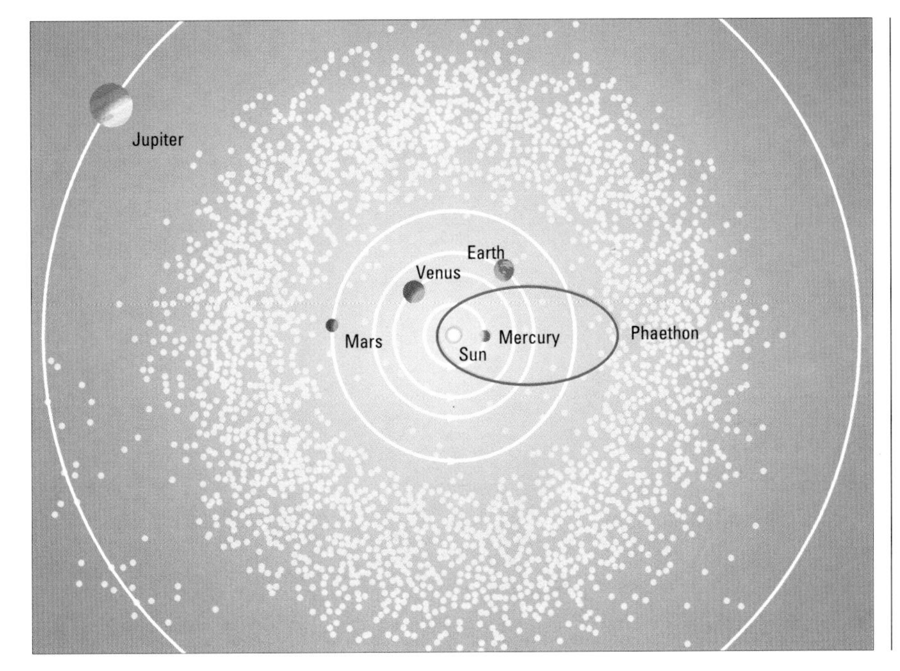

◀ ***Orbit of Phaethon**. 3200 Phaethon, discovered in 1983, is about 5 km (3 miles) in diameter. Its orbit carried it within that of Mercury; the distance from the Sun ranges between 21 million km (13 million miles) and 390 million km (242 million miles). The orbital period is 1.43 years, and the rotation period 4 hours. Phaethon may well be the 'parent' of the Geminid meteor stream. The only other asteroid known to cross Mercury's orbit is 1566 Icarus.*

Jupiter

Jupiter, first of the giant planets, lies well beyond the main asteroid zone. It is the senior member of the Sun's family; indeed, it has been said that the Solar System is made up of the Sun, Jupiter and various minor bodies. Though it has only 1/1047 of the mass of the Sun, it is more massive than all the other planets combined. Despite its distance, it shines more brightly than any other planet apart from Venus and, very occasionally, Mars.

A casual look at Jupiter through a telescope is enough to show that it is quite unlike the Earth or Mars. Its surface is made up of gas; it is yellow, and is crossed by dark streaks which are always called cloud belts. The disk is obviously flattened, because of the rapid rotation. Jupiter's 'year' is almost 12 times ours, but the 'day' amounts to less than ten hours, and this makes the equator bulge out; the polar diameter is over 10,000 kilometres (over 6200 miles) shorter than the diameter measured through the equator. With Earth, the difference is a mere 42 kilometres (26 miles). Jupiter is almost 'upright'; the axial tilt is only just over 3 degrees to the perpendicular.

Until less than a century ago it was believed that the giant planets were miniature suns, warming their satellite systems. In fact the outer clouds are very cold indeed. According to the latest theoretical models, Jupiter has a silicate central core about 15 times as massive as the Earth, and this is admittedly hot; the temperature is rather degrees uncertain, but 30,000 degrees C may be reasonably near the truth. Around the core there is a thick shell of liquid hydrogen, so compressed that it takes on the characteristics of a metal. Further away from the centre there is a shell of liquid molecular hydrogen, and above this comes the gaseous atmosphere, which is of the order of 1000 kilometres (over 600 miles) deep, and is made up of well over 80 per cent hydrogen; most of the rest is helium, with traces of other elements. Spectroscopic analysis shows evidence of uninviting hydrogen compounds such as ammonia and methane.

It is no surprise to find that Jupiter consists mainly of hydrogen, which is, after all, much the most abundant element in the universe. In its make-up Jupiter is not very unlike the Sun, but it would be misleading to describe it as a 'failed star'. For stellar nuclear reactions to be triggered, the temperature must reach 10 million degrees C.

It has been found that Jupiter sends out 1.7 times as much energy as it would do if it depended entirely upon what it receives from the Sun. This is probably because it has not had time to lose all the heat built up during its formation, between four and five thousand million years ago – though it has also been suggested that the excess may be gravitational energy, produced because Jupiter is slowly contracting at a rate of less than a millimetre per year.

The Jovian atmosphere is in constant turmoil. It seems that there are several cloud layers, of which one, at a considerable depth, may be made up of water droplets – with a giant planet it is not easy to define just where the 'atmosphere' ends and the real body of the planet begins! Higher up there are cloud layers of ice crystals, ammonia crystals and ammonium hydrosulphide crystals.

Jupiter is a powerful source of radio waves; this was discovered in 1955 by American researchers (it must be admitted that the discovery was accidental). The main emissions are concentrated in wavelengths of tens of metres (decametric) and tenths of metres (decimetric), and from their variations it seems that the rotation period of the Jovian core is 9 hours 55.5 minutes. It was also found, very unexpectedly, that the decametric radiation is affected by the position in orbit of Io, Jupiter's innermost large satellite – for reasons which did not become clear until the space missions of the 1970s showed that Io is a violently volcanic world.

◄ ***Conjunction of Venus and Jupiter**, June 1991. The two planets are seen close together low in the sky; the bright red glow is an inconvenient light from a neighbouring house! The picture was taken from Selsey, in Sussex. Planetary conjunctions are not uncommon, but the actual occultation of one planet by another is a very rare event.*

▼ ***Three views of Jupiter:** photographs taken by Charles Capen (Lowell Observatory, Arizona). The effects of the planet's rotation are very evident, and a great amount of detail is shown. At this time the surface was very active.*

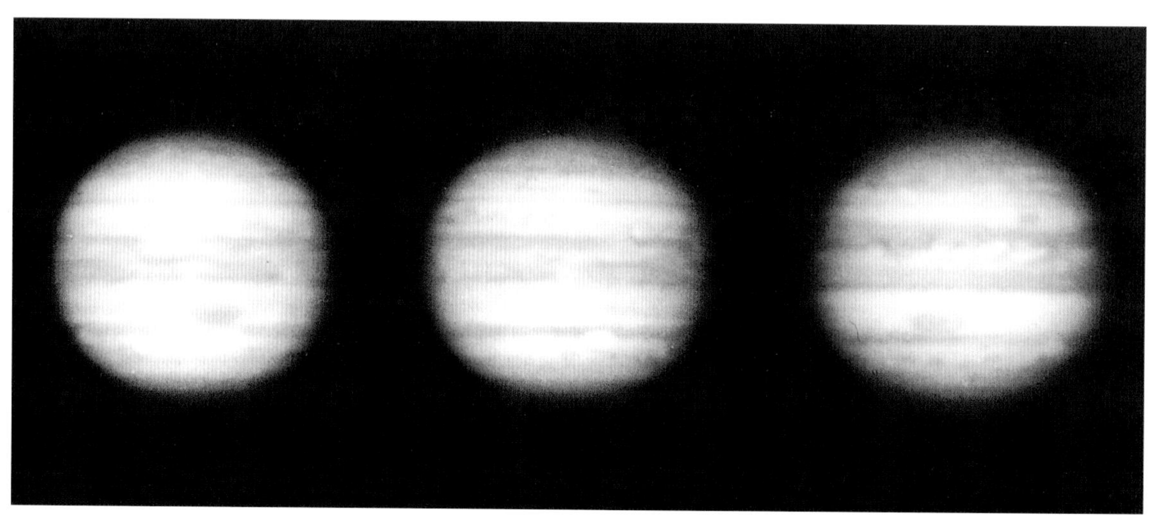

◀ ***Jupiter***, *as imaged from the Hubble Space Telescope on 28 May 1991. At this time the Great Red Spot was very much in evidence; to the left and below the Spot there is one of the celebrated white ovals. This photograph was generated in a computer by combining three separate exposures made in red, green and blue light.*

▼ ***The south-east quadrant of Jupiter****: 11 March 1991, seen from Hubble with the original Wide Field and Planetary Camera (since replaced). The picture shows a striking oval dark ring on the left, with the Great Red Spot to the right about to be carried out of view by virtue of Jupiter's rotation. The picture has about the same resolution (0.15 of a second of arc) as Voyager pictures five days before encounter in 1979. The image is the result of combining photographs taken in blue, green and red light over a period of six minutes. North is up.*

PLANETARY DATA – JUPITER

Sidereal period	4332.59 days
Rotation period (equatorial)	9h 55m 21s
Mean orbital velocity	13.06 km/s (81 miles/s)
Orbital inclination	1° 18′ 15″.8
Orbital eccentricity	0.048
Apparent diameter	max. 50″.1, min. 30″.4
Reciprocal mass, Sun = 1	1047.4
Density, water = 1	1.33
Mass, Earth = 1	317.89
Volume, Earth = 1	1318.7
Escape velocity	60.22 km/s (37.42 miles/s)
Surface gravity, Earth = 1	2.64
Mean surface temperature	−150°C
Oblateness	0.06
Albedo	0.43
Maximum magnitude	−2.6
Diameter (equatorial)	143,884 km (89,424 miles)
Diameter (polar)	133,700 km (83,100 miles)

The Changing Face of Jupiter

▲ ***North Polar Region***
Lat. +90° to +55° approx. Usually dusky in appearance and variable in extent. The whole region is often featureless. The North Polar Current has a mean period of 9hrs 55mins 42secs.

North North North Temperate Belt
Mean Lat. +45°
An ephemeral feature often indistinguishable from the NPR.

North North Temperate Zone *Mean Lat. +41°*
Often hard to distinguish from the overall polar duskiness.

North North Temperate Belt *Mean Lat. +37°*
Occasionally prominent, sometimes fading altogether, as in 1924.

North Temperate Zone
Mean Lat. +33°
Very variable, both in width and brightness.

North Temperate Belt
Mean Lat. +31° to +24°
Usually visible, with a maximum extent of about 8° latitude. Dark spots at southern edge of the North Temperate Belt are not uncommon.

North Tropical Zone
Mean Lat. +24° to +20°
At times very bright. The North Tropical Current, which overlaps the North Equatorial Belt, has a period of 9hrs 55mins 20secs.

North Equatorial Belt
Mean Lat. +20° to +7°
The most prominent of all the Jovian belts. This region is extremely active and has a large amount of detail.

Equatorial Zone
Mean Lat. +7° to −7°
Covering about one-eighth of the entire surface of Jupiter, the EZ exhibits much visible detail.

Equatorial Band
Mean Lat. −0.4°
At times the EZ appears divided into two components by a narrow belt, the EB, at or near to the equator of Jupiter.

South Equatorial Belt
Mean Lat. −7° to −21°
The most variable belt. It is often broader than the NEB and is generally divided into two components by an intermediate zone. The southern component contains the Red Spot Hollow (RSH).

South Tropical Zone
Mean Lat. −21° to −26°
Contains the famous Great Red Spot. The STrZ was the site of the long-lived South Tropical Disturbance.

Great Red Spot
Mean Lat. −22°
Although there are other spots visible on Jupiter's surface, both red and white, the Great Red Spot is much the most prominent. It rotates in an anticlockwise direction.

South Temperate Belt
Mean Lat. −26° to −34°
Very variable in width and intensity; at times it appears double.

South Temperate Zones
Mean Lat. −38°
Often wide; may be extremely bright. Spots are common.

South South Temperate Belt *Mean Lat. −44°*
Variable, with occasional small white spots.

South South South Temperate Belt
Mean Lat. −56°

South Polar Region
Lat. −58° to −90° approx. Like the NPR, very variable in extent.

Jupiter is a favourite target for users of small or moderate telescopes. The main features are the belts and the bright zones; there are also spots, wisps and festoons, with the Great Red Spot often very much in evidence.

Jupiter's rapid spin means that the markings are carried from one side of the disk to the other in less than five hours, and the shifts are noticeable even after a few minutes' observation. Jupiter has differential rotation – that is to say, it does not spin in the way that a rigid body would do. There is a strong equatorial current between the two main belts, known as System I; the main rotation period is 9 hours 50 minutes 30 seconds, while over the rest of the planet (System II) it is 9 hours 55 minutes 41 seconds. However, various discrete features have rotation periods of their own, and drift around in longitude, though the latitudes do not change appreciably.

Generally there are two main belts, one to either side of the equator. The North Equatorial Belt (NEB) is almost always very prominent, and shows considerable detail, but the South Equatorial Belt (SEB) is much more variable, and has been known to become so obscure that it almost vanishes, as happened for a while in 1993. The other belts also show variations in breadth and intensity. Very obvious colours can often be seen on the disk, due to peculiarities of Jovian chemistry.

The most famous of all the features is the Great Red Spot, which has been seen, on and off (more on than off) ever since the first telescopic observations of Jupiter were made, during the 17th century. It is oval, and at its maximum extent it may be 40,000 kilometres (25,000 miles) long by 14,000 kilometres (8700 miles) wide, so that its surface area is then greater than that of the Earth. At times it may be almost brick-red, though at other times the colour fades and the Spot may even disappear completely for a few months or a few years. It forms a hollow in the southern edge of the South Equatorial Belt, and this hollow can sometimes be seen even when the Spot itself cannot. Though its latitude is to all intents and purposes constant at 22° south, the longitude drift over the past century has amounted to 1200 degrees. Between 1901 and 1940 there was also a feature known as the South Tropical Disturbance, which lay in the same latitude as the Spot and took the form of a shaded area between white patches. The rotation period of the South Tropical Disturbance was shorter than that of the Red Spot, so that periodically the Spot was caught up and passed, producing most interesting interactions.

The Disturbance has vanished, and there is no reason to suppose that it will return, but the Red Spot is still with us, though it may be rather smaller than it used to be, and it may not be permanent. For many years it was assumed to be a solid or semi-solid body floating in Jupiter's outer gas, but the space missions have shown that it is a whirling storm – a phenomenon of Jovian 'weather'. It rotates anticlockwise, with a period of 12 days at its edge and 9 days nearer its centre; the centre itself is 8 kilometres (5 miles) above the surrounding clouds, and it is here that material rises, spiralling outwards towards the edge. The cause of the colour is not definitely known, but it may be due to phosphorus, produced by the action of sunlight upon phosphine sent up from the planet's interior. At any rate, the Spot is decidedly colder than the adjacent regions.

Many other spots are seen, some of which are bright, white and well-defined, but generally these features do not last for long. The Great Red Spot itself may be so long-lived simply because of its exceptional size.

Amateur observers have carried out important studies of Jupiter. In particular, they make estimates of the rotation periods of the various features. The procedure is to time the moment when the feature crosses the central meridian of the planet. The central meridian is easy to locate, because of the polar flattening of the globe, and the timings can be made with remarkable accuracy. The longitude of the feature can then be found by using tables given in yearly astronomical almanacs. When Jupiter is suitably placed in the sky, a whole rotation can be covered during a single night's observing.

Jupiter is certainly one of the Solar System's most intriguing worlds. There is always plenty to see, and no-one knows just what will happen next!

▼ ***Voyager 1*** *view of Jupiter at a range of 32 million km (20 million miles).*

Missions to Jupiter

Several spacecraft have now passed by Jupiter. First there were Pioneer 10 (December 1973) and Pioneer 11 (December 1974), which carried out preliminary surveys; Pioneer 11 was subsequently sent on to a brief encounter with Saturn. Next came the much more sophisticated Voyagers, No. 1 (March 1979) and No. 2 (July of the same year). Both these then went on to carry out detailed studies of Saturn and its satellite system; Voyager 2 went on to encounter Uranus and Neptune as well. All these four early probes are now on their way out of the Solar System permanently. In February 1992 the Ulysses solar polar probe passed close to Jupiter, mainly to use the strong gravitational pull of the giant planet to send Ulysses soaring far out of the plane of the ecliptic, but observations of Jupiter were also made – the opportunity was too good to be missed. The latest mission, Galileo, was dispatched in October 1990, though it followed a somewhat circuitous route and did not reach its target until 1996.

The most important results have come from the Voyagers. Particular attention was paid to the magnetosphere, which is very extensive. It is not spherical, but has a long 'magnetotail' stretching away from the direction of the Sun and extends out to well over 700 million kilometres (over 400 million miles), so that at times it may even engulf the planet Saturn. There are zones of radiation ten thousand times stronger than the Van Allen zones of the Earth, so that any astronaut foolish enough to venture into them would quickly die from radiation poisoning. Indeed, the unexpectedly high level of radiation almost crippled the equipment in Pioneer 10, the first probe to pass by the planet, and subsequent spacecraft were aimed so as to pass quickly over the equatorial region, where the danger is at its worst. The Voyagers were carefully constructed so as to tolerate twice the anticipated dose of radiation; only minor effects were noted when Voyager 1 approached Jupiter to within 350,000 kilometres (220,000 miles), but Voyager 2, passing at the greater distance of 650,000 kilometres (400,000 miles), found the level to be three times stronger. Evidently the zones are very variable. The magnetic field itself is extremely complicated, and is reversed relative to that of the Earth, so that a compass needle would point south; the magnetic axis is inclined to the rotational axis by 10 degrees.

An obscure ring was discovered, made up of three components which are now known as Halo, Main and Gossamer. The ring system reaches up to 50,000 kilometres (31,000 miles) above the cloud tops, and is so faint that it would have been well nigh impossible to detect from Earth. It is quite unlike the glorious icy ring-system of Saturn.

Superb pictures of the planet's surface were obtained, showing the turbulent, vividly-coloured clouds and spots. Aurorae and lightning flashes were recorded on the night side, and observations of all kinds were made. It was seen that marked changes had occurred between the Pioneer and Voyager passes, and even in the interval between the Voyager 1 and 2 encounters, the shape of the Red Spot was different, for example. Jupiter is a turbulent place, and the ancients were indeed justified in naming it in honour of the King of the Gods.

▶ ***Voyager views of Jupiter***. *Top left shows the North Temperate Zone from Voyager 1, taken on 2 March 1979. The pale orange line across the lower half of the image marks the North Temperate Current, in which windspeeds reach 120 m/s (390 feet/s). Top right shows the Equatorial Zone from Voyager 2, taken on 28 June 1979; the colours have been deliberately enhanced to bring out more detail. Bottom left is the Great Red Spot from Voyager 2 on 3 July 1979 – long subject to observation, the Spot forms a hollow in the adjoining belt, and though it periodically disappears for a while, it always returns. Bottom right is a view of the Spot showing more structure, from Voyager 1 on 4 May 1979.*

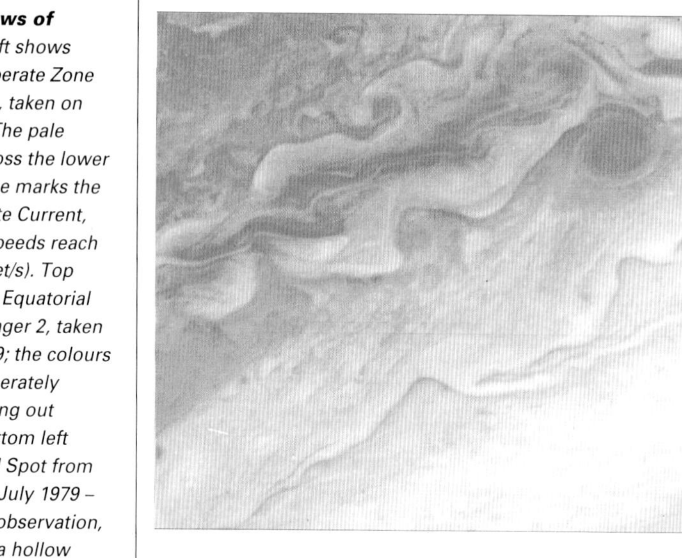

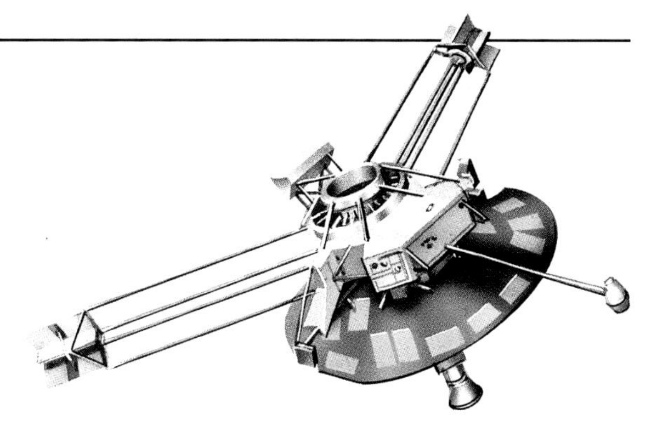

▲ ***Jupiter with satellites*** *seen from Voyager. Europa is to the right, Io is in transit across Jupiter's disk, and Callisto is just visible at bottom left.*

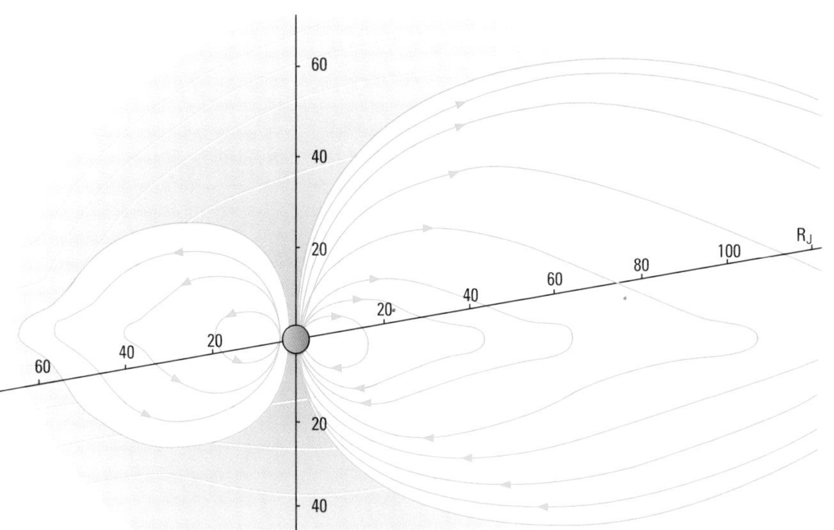

▶ ***The magnetosphere of Jupiter****. Solar wind particles approaching from the left collide with the structurally complex magnetosphere. Inside the bow shock lies the magnetopause. The whole magnetically active region is enveloped by the 'magnetosheath'.*

▲ ***Spacecraft to Jupiter****. Pioneer 10, top, launched 2 March 1972, passed Jupiter on 3 December 1973 at 132,000 km (82,500 miles). Voyager 2, above, launched 20 August 1977, made passes of Jupiter, Saturn, Uranus and Neptune.*

▼ ***Jupiter's ring****. The only images of Jupiter's ring are those from the Voyager probes. This picture was made by Voyager 2 on 10 July 1979; the Voyager was then 2° below the plane of the ring at a distance of 1.55 million km (970,000 miles).*

Impacts on Jupiter

▼ Comet collision – *a Hubble Space Telescope image showing numerous Comet Shoemaker–Levy 9 impact sites on Jupiter shortly after the collision of the last cometary fragment (north is to the top). The impact sites appear as dark 'smudges' lined up across the middle of Jupiter's southern hemisphere and are easily mistaken for 'holes' in the giant planet's atmosphere. In reality, they are the chemical debris 'cooked' in the tremendous fireballs that exploded in Jupiter's atmosphere as each fragment impacted the planet. This material was then ejected high above the bright multicoloured cloud tops where it was caught in the winds of the upper atmosphere and eventually dispersed around the planet.*

There have been two recent impacts on Jupiter – one natural, one man-made. Each has provided us with a great deal of new information.

In March 1993 three American comet-hunters, Eugene and Carolyn Shoemaker and David Levy, discovered what they described as a 'squashed comet'; it was their ninth discovery, so the comet became known as Shoemaker–Levy 9 (SL9). It was unlike anything previously seen. It was orbiting Jupiter, and had been doing so for at least 20 years; calculation showed that on 7 July 1992 it had skimmed over the Jovian cloud tops at a mere 21,000 km (13,000 miles) and the nucleus had been torn apart, so that it had been transformed into a sort of string-of-pearls arrangement. Over 20 fragments were identified, and were conveniently lettered from A to W.

It was soon found that the comet was on a collision course, and that the fragments would hit Jupiter in July 1994. The first fragment, A, impacted on 16 July, just on the side of Jupiter turned away from Earth, but the planet's quick spin soon brought the impact site into view. The other fragments followed during the next several days, and produced dramatic effects; there were huge scars, visible with a very small telescope. The cometary fragments were only a few kilometres in diameter, but were travelling at tremendous speed. Spectacular pictures were obtained from ground observatories and from the Hubble Space Telescope; the effects of the impacts were detectable for months.

The Galileo probe to Jupiter was launched from the Shuttle in October 1989, and after a somewhat roundabout journey reached Jupiter in December 1994. It was made up of two parts: an entry vehicle, and an orbiter, which separated from each other well before arrival. The entry probe was scheduled to plunge into the Jovian clouds, and transmit data until it was destroyed; the orbiter would orbit Jupiter for several years, sending back images of the planet and its satellites.

The high-gain antenna, a particularly important part of the communications link, failed to unfurl; some data were lost, but much of the planned programme could be carried out. The entry probe plunged into the clouds on schedule, and continued to transmit data for 75 minutes, by which time it had penetrated to a depth of 160 kilometres (100 miles).

Some of the results were unexpected. For instance, it had been thought that the strong Jovian winds would be confined to the outer clouds, and would slacken with increasing depth, but this did not happen; by the time contact was lost, the winds were just as strong as they had been at the surface, indicating that the driving force was not the Sun, but heat radiating from the interior. There was much less lightning activity than had been expected, and, most surprising of all, the Jovian atmosphere was dry; the amount of water was very low, so that presumably the water shown after the Shoemaker–Levy impact came not from Jupiter, but from the dying comet.

Later analysis showed that the Galileo entry probe had plunged into the clouds in an unusually 'dry' area of the planet – an equivalent of a Jovian desert, so that the lack of water was not typical. One mystery at least was solved, but it cannot be claimed that we have as yet anything like a complete knowledge of the interior of Jupiter. We can hardly hope for another cometary impact, so that presumably we must wait for another deliberate entry into Jupiter's cloud layer.

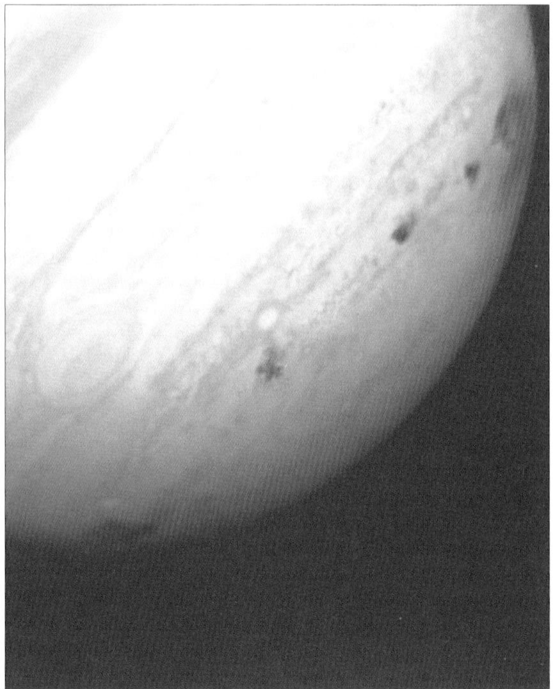

▲ Impact of the largest fragment of Comet Shoemaker–Levy 9, *as viewed from the Hubble Space Telescope.*

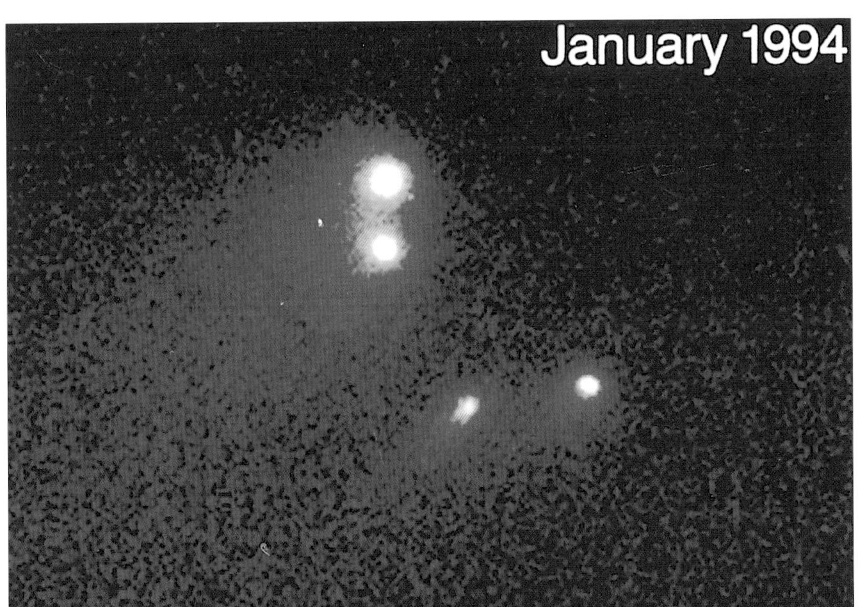

◀ ***Comet Shoemaker–Levy 9**. This is a composite HST image taken in visible light showing the temporal evolution of the brightest region of the comet. In this false-colour representation, different shades of red are used to display different intensities of light:*
***1 July 1993** – data taken prior to the HST servicing mission. The separation of the two brightest fragments is only 0.3″, so ground-based telescopes could not resolve this pair. The other two fragments just to the right of the closely-spaced pair are only barely detectable due to HST's spherical aberration.*
***24 January 1994** – the first HST observation after the successful servicing mission. The two brightest fragments are now about 1″ apart, and the two fainter fragments are much more clearly seen. The light near the faintest fragment is not as concentrated as the light from the others and is elongated in the direction of the comet's tail.*
***30 March 1994** – the latest HST observation shows that the faintest fragment has become a barely discernible 'puff'. Also, the second faintest fragment has clearly split into two distinct fragments by March.*

▼ ***The disruption of Comet Shoemaker–Levy 9**. The comet was broken up following its close approach to Jupiter in July 1992.*

▲ ***Model of the Galileo probe**, at the Jet Propulsion Laboratory, California.*

Satellites of Jupiter

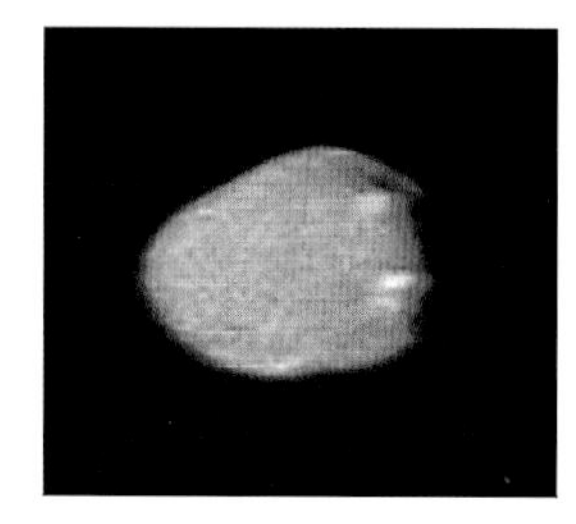

▲ ***Amalthea**, Jupiter's fifth satellite, as seen on 4 March 1979 from Voyager 1 at a range of 425,000 km (255,000 miles). The effective resolution is about 8 km (5 miles). The bright patches, now named Ida and Lyctos, are about 15 km in diameter, and are presumably mountains. Amalthea's red colour may be due, at least in part, to contamination from Io.*

Jupiter has an extensive satellite family. Four are large, and are bright enough to be seen with any small telescope; even powerful binoculars will show them under good conditions. They were observed in January 1610 by Galileo, using the first astronomical telescope, and are therefore known collectively as the Galileans, though they may have been seen slightly earlier by Simon Marius. It was Marius who gave them their names: Io, Europa, Ganymede and Callisto. Perhaps for this reason, the names were not widely used before the onset of the Space Age.

Ganymede and Callisto are much larger than our Moon, and Ganymede is actually larger than the planet Mercury, though less massive; it may have been recorded with the naked eye by the Chinese astronomer Gan De as long ago as 364 BC. Io is slightly larger than the Moon, and Europa only slightly smaller.

Callisto, outermost of the Galileans, is the faintest of the four. There is an icy, cratered crust which may go down to a depth of several hundred kilometres (two or three hundred miles), below which may come a mantle of water or soft ice surrounding a silicate core. There are no signs of past tectonic activity, and certainly Callisto seems totally inert. Ganymede is rather denser, and shows more traces of past activity than in the case of Callisto; in 1996 the Galileo probe detected a weak but appreciable magnetic field, indicating the presence of a metallic core.

Europa too has an icy surface, but there are almost no craters, and the main features are unlike anything found elsewhere; Europa has been likened to a cracked eggshell, and is essentially smooth. According to one theory, the crust lies above an ocean of liquid water, though it is also possible that the mantle is composed of 'slushy' ice lying over the core.

Io is a remarkable world. Its surface is sulphur-coated, and during the Voyager 1 pass several active volcanoes were seen, one of which, Pele, sent a plume up to a height of 280 kilometres (175 miles). By the time of the Voyager 2 pass Pele had ceased to erupt, but several of the other volcanoes were more active than before, and there is of course no reason to suppose that Pele is extinct. The volcanoes can now be monitored by the Hubble Space Telescope, and it is clear that eruptions are going on all the time.

According to one theory, Io's crust may be a 'sea' of sulphur and sulphur dioxide about 4 kilometres (2.5 miles) deep, with only the uppermost kilometre solid. Heat escapes from the interior in the form of lava, erupting below the sulphur ocean, and the result is a violent outrush of a mixture of sulphur, sulphur dioxide gas, and sulphur dioxide 'snow'. Some of the volcanic vents may be as hot as 500 degrees C, though the general surface is at a temperature of below −150 degrees C.

Jupiter and its satellite Io are connected by a powerful electrical flux tube (which is why Io has a marked effect

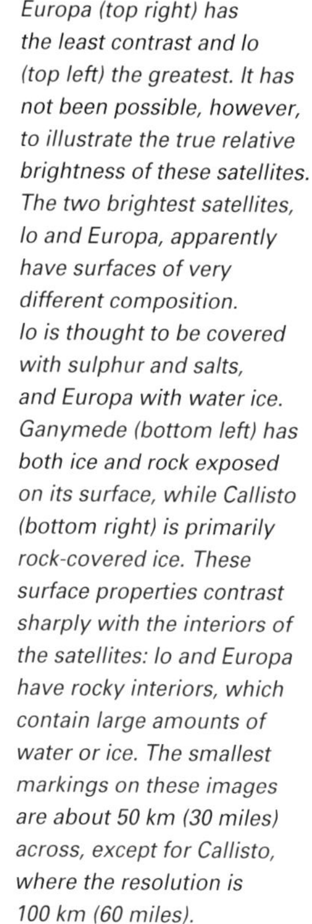

***The four large Galilean satellites** of Jupiter photographed by Voyager 1 between 1 and 3 March, 1979. Here the satellites are shown at their correct relative sizes. The picture processing preserves relative contrasts on the satellites; thus, it is apparent that Europa (top right) has the least contrast and Io (top left) the greatest. It has not been possible, however, to illustrate the true relative brightness of these satellites. The two brightest satellites, Io and Europa, apparently have surfaces of very different composition. Io is thought to be covered with sulphur and salts, and Europa with water ice. Ganymede (bottom left) has both ice and rock exposed on its surface, while Callisto (bottom right) is primarily rock-covered ice. These surface properties contrast sharply with the interiors of the satellites: Io and Europa have rocky interiors, which contain large amounts of water or ice. The smallest markings on these images are about 50 km (30 miles) across, except for Callisto, where the resolution is 100 km (60 miles).*

upon the radio emissions from Jupiter itself), and material from the Ionian volcanoes produces a torus round Jupiter centred on Io's orbit. When the Ulysses space probe swung round Jupiter, on 9 February 1992, there was apprehension about what might happen during the transit of Io's torus, though in the event Ulysses emerged unscathed.

Why is Io so active? It seems that the interior is churned and heated by gravitational flexing by Jupiter and the other Galileans; the orbit of Io is somewhat eccentric, so that the tidal stresses vary. All the same, it is strange that Io should be so active while Europa, which less than twice the distance of Io from Jupiter, is so inert. Incidentally, Io lies in the midst of Jupiter's radiation zones, so that it may qualify as the most lethal world in the Solar System.

The movements of the Galileans can be followed from night to night. They may pass into Jupiter's shadow, and be eclipsed; they may be occulted by the planet; they and their shadows may pass in transit across Jupiter's disk, so that they can be seen as they track slowly along – the shifts in position become evident after only a few minutes. During transits, Io and Europa are usually hard to find except when near the limb, but the less reflective Ganymede and Callisto show up as grey spots. The shadows are always jet-black.

The fifth satellite, Amalthea, was discovered by E. E. Barnard in 1892, with the aid of the great Lick refractor. Amalthea was imaged by Voyager 1, and found to be irregular in shape; the surface is reddish, so that it may have been coloured by contamination from Io. There are two bowl-shaped craters, two bright features which seem to be mountains, and a medley of ridges and troughs. No close-range images were obtained of the other small inner satellites, Metis, Adrastea and Thebe.

The eight outer satellites are very small; only Himalia is as much as 100 kilometres (60 miles) in diameter. They are divided into two groups. The members of the inner group (Leda, Himalia, Lysithea and Elara) have direct motion, but those in the outer group (Ananke, Carme, Pasiphaë and Sinope) have retrograde motion, which adds force to the suggestion that all these small bodies are captured asteroids.

In fact, their orbits are so strongly influenced by the Sun that they are not even approximately circular, and no two cycles are alike. Obviously all Jovian satellites, apart from the Galileans, are beyond the range of amateur-owned telescopes.

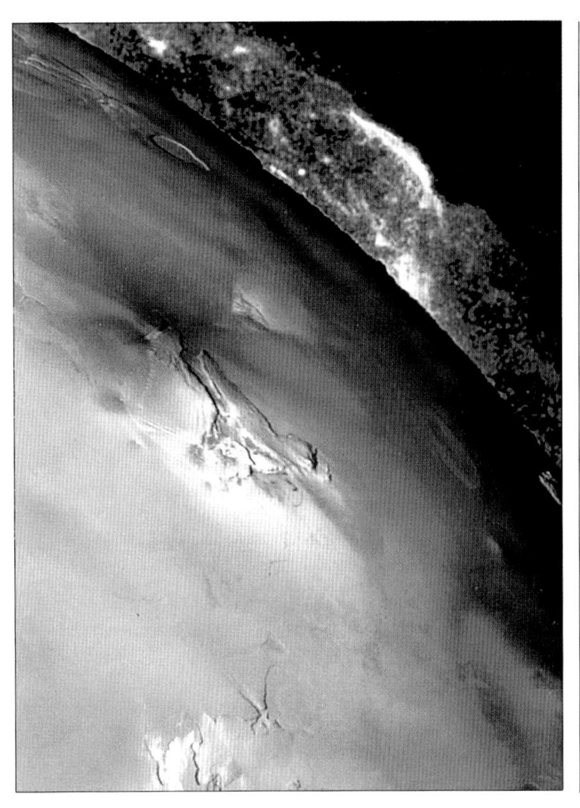

◀ ***The volcanoes of Io*** *from Voyager. Activity, seen along the horizon, seems to be constant; the crust is certainly unstable, and material is sent up from the volcanoes to a height of hundreds of kilometres above the surface.*

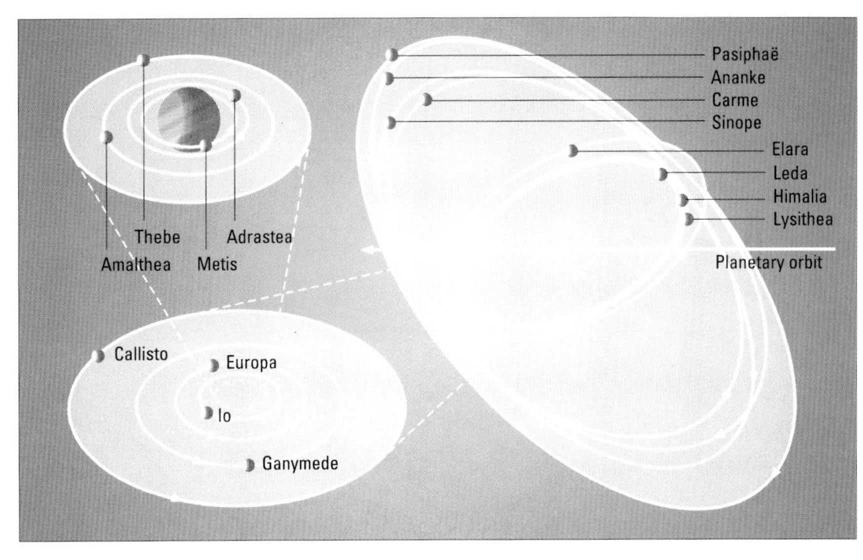

▼ ***Satellites of Jupiter****. The four large satellites, the Galileans, have almost circular orbits, and their inclinations to the Jovian equator are low. This also applies to the small inner satellites. The outer satellites fall into two groups; Leda, Himalia, Lysithea and Elara have prograde motion, while Ananke, Carme, Pasiphaë and Sinope have retrograde motion. The orbits of these outer satellites are so affected by solar gravitation that they are not even approximately circular.*

SATELLITES OF JUPITER

Name	Distance from Jupiter, km	Orbital period, days	Orbital Incl., °	Orbital ecc.	Diameter, km	Density, water = 1	Escape vel., km/s	Mean opp., mag.
Metis	127,900	0.290	0	0	40	3?	0.02?	17.4
Adrastea	128,980	0.298	0	0	26 × 20 × 16	3?	0.01?	18.9
Amalthea	181,300	0.498	0.45	0.003	262 × 146 × 143	3?	0.16?	14.1
Thebe	221,900	0.675	0.9	0.013	110 × 90	3?	0.8?	15.5
Io	421,600	1.769	0.04		3660 × 3637 × 3631	3.55	2.56	5.0
Europa	670,900	3.551	0.47	0.009	3130	3.04	2.10	5.3
Ganymede	1,070,00	7.155	0.21	0.002	5268	1.93	2.78	4.6
Callisto	1,880,000	16.689	0.51	0.007	4806	1.81	2.43	5.6
Leda	11,094,000	238.7	26.1	0.148	10	3?	0.1?	20.2
Himalia	11,480,000	250.6	27.6	0.158	170	3?	0.1?	14.8
Lysithea	11,720,000	259.2	29.0	0.107	24	3?	0.01?	18.4
Elara	11,737,000	259.7	24.8	0.207	80	3?	0.05?	16.7
Ananke	21,200,000	631*	147	0.17	20	3?	0.01?	18.9
Carme	22,600,000	692*	164	0.21	30	3?	0.02	18.0
Pasiphaë	23,500,000	735*	145	0.38	36	3?	0.02	17.7
Sinope	23,700,000	758*	153	0.28	28	3?	0.01?	18.3

(* = retrograde)

The Galilean Satellites – from Galileo

▶ ***Changes on Io (right and bottom right images)****: the right-hand picture was taken from Voyager 1 on 4 March 1979, from 862,000 km (500,000 miles). The bottom-right picture was taken from Galileo on 7 September 1996, from 487,000 km (302,000 miles); the image is centred on the face of Io which is always turned away from Jupiter. The active volcano Prometheus appears near the right centre of the disk. The black and bright red materials correspond to the most recent volcanic deposits, probably no more than a few years old.*

The orbiting section of the Galileo probe has sent back superb images of Jupiter's Galilean satellites. Ganymede and Callisto are icy and cratered; the discovery of a magnetic field on Ganymede was a great surprise. However, the most spectacular images are those of Io and Europa.

The Ionian scene has changed markedly since the Voyager passes, while the existence of an underground ocean inside Europa is a real possibility. Some surface details on the satellites can be seen with the Hubble Space Telescope, and the events on Io can be monitored, but of course the clarity of the Galileo pictures is unrivalled.

Io is the most volcanically active world in the Solar System, and eruptions are going on all the time. It is strange that there is so great a difference between Io and the inert, ice-coated Europa. Io is also the only satellite to show activity apart from Triton, in Neptune's system, with its ice geysers.

▶ ***Three views of Io****: Galileo, June 1996. Some areas are truly red, while others are yellow or light greenish. The major red areas are associated with very recent pyroclastics erupted in the form of volcanic plumes. The most prominent red oval surrounds the volcano Pele (far right); an intense red spot lies near the active plume Marduk, east of Pele. Loki and Amirani are inactive, whereas Volund is active.*

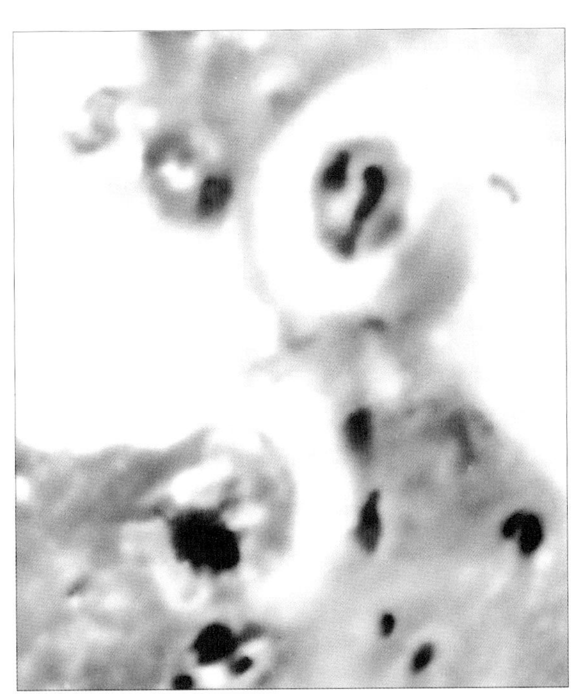

▶ ***Changes on Io in close-up****: left-hand view from Voyager 1 (1979); right-hand view from Galileo (1996). Prometheus is the bright ring, upper right; the dark feature at the lower left is Culann Patera. The Galileo image shows a new dark lava flow emerging from the vent of Prometheus, and the plume is now 75 km (47 miles) west of its 1979 position. The Voyager range was 862,000 km (500,000 miles); the Galileo range 487,000 km (302,000 miles).*

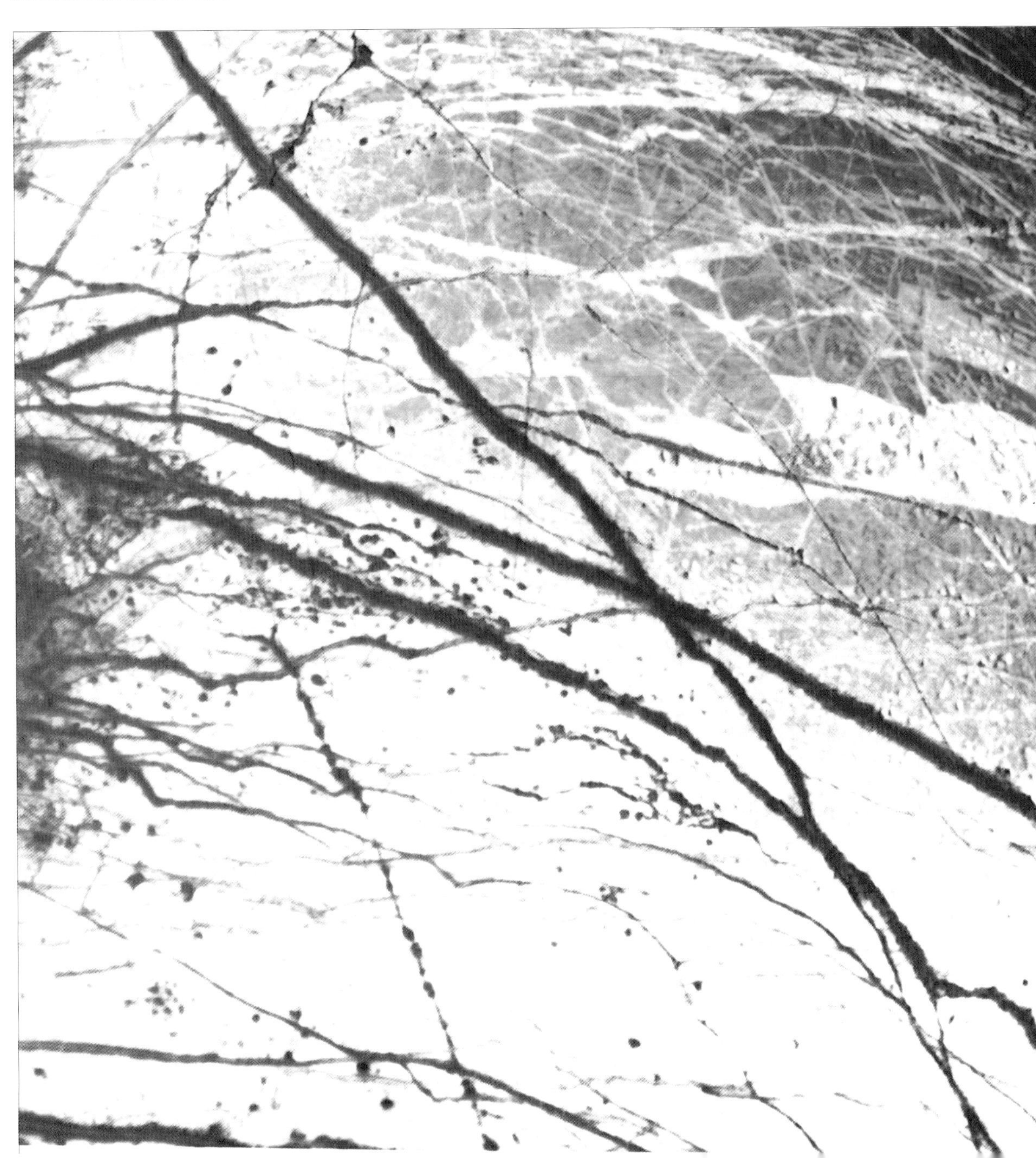

▶ ***Europa, from Galileo****. False colour has been used to enhance the visibility of certain features in this composite of three images of the Minos Linea region on Europa. Triple bands, linear and mottled terrains appear in brown and reddish hues, indicating the presence of contaminants in the ice. The icy plains, shown here in bluish hues, subdivide into units with different albedos at infra-red wavelengths probably because of differences in the grain size of the ice. The composite was produced by Galileo imaging team scientists at the University of Arizona using images with effective wavelenths at 989, 757 and 559 nm. The spatial resolution in the individual images ranges from 1.6 to 3.3 km (1 to 2 miles) per pixel. The area covered, centred at 45°N 221°W, is about 1260 km (about 780 miles) across.*

It has been suggested that beneath Europa's icy crust there may be an ocean of liquid water, perhaps even containing life. However, there is no proof that an ocean exists, and the discovery of life there would indeed be surprising.

Maps of Jupiter's Satellites

The four large satellites are all quite unlike each other. Each has its own special features, and it has been said there is no such thing as an uninteresting Galilean.

Io. *The surface is dominated by volcanoes, notably the heart-shaped Pele and the very active Loki and Prometheus. The constant activity means that the surface must be subject to marked changes even over short periods; these variations may be within the range of the Hubble Space Telescope, and useful infra-red observations can be carried out from the Earth's surface.*

Europa. *A map-maker's nightmare. The main features are the dark, often irregular patches known as maculae, and the complex linea, which are straight or curved, dark or bright elongated markings.*

Io ▶

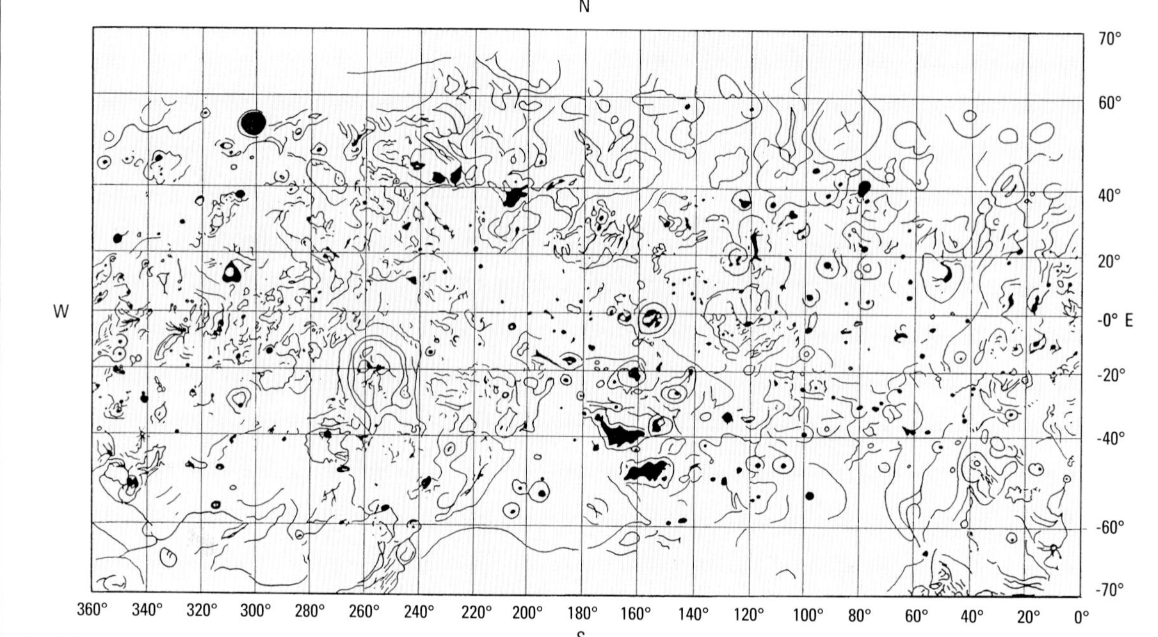

IO

		Lat. °	Long. °W
VOLCANOES	Amirani	27 N	119
	Loki	19 N	30
	Marduk	28 S	210
	Masubi	45 S	053
	Maui	19 N	122
	Pele	19 S	257
	Promethens	03 S	153
	Surt	46 N	336
	Volund	22 N	177
REGIONES	Bactria	45 S	125
	Colchis	10 N	170
	Lerna	65 S	300
	Tarsus	30 S	055
PATERAE	Atar	30 N	279
	Daedalus	19 N	175
	Heno	57 S	312
	Ülgen	41 S	288

EUROPA

		Lat. °	Long. °W
MACULAE	Thera	45 S	178
	Thrace	44 S	169
	Tyre	34 N	144
LINEA	Adonis	38–60 S	112–122
	Belus	14–26 N	170–226
	Minos	45–31 N	199–150
CRATER	Cilix	01 N	182

Europa ▶

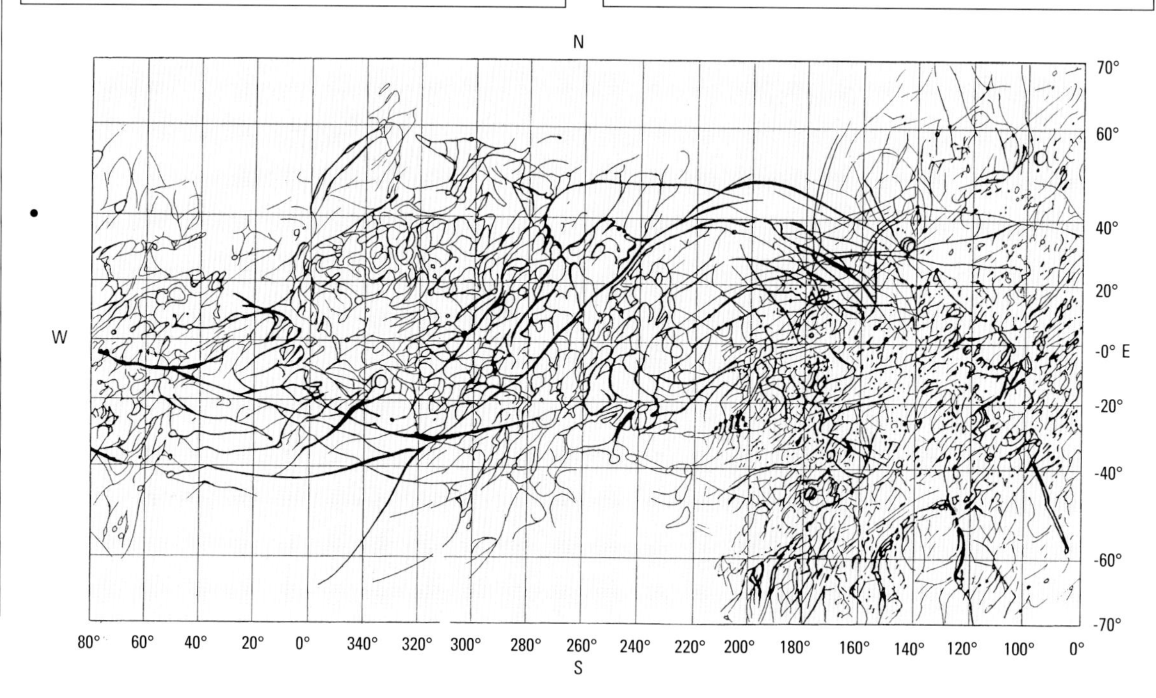

***Ganymede**. The most prominent surface features of this satellite are the dark areas, of which the largest, Galileo Regio, is 4000 kilometres (2500 miles) across, nearly equal to the continental United States. There are brighter, younger regions with 'sulci', i.e. grooves or furrows, with ridges rising to a kilometre or two (half a mile to a mile). There are many craters, some of which are ray-centres.*

***Callisto**. The most prominent surface features on Callisto are the two huge ringed basins. The largest of these, Valhalla, is 600 kilometres (375 miles) across, and is surrounded by concentric rings, one of which has a diameter of more than 3000 kilometres (1900 miles). The other basin, Asgard, is very similar, though much smaller. The surface of Callisto is perhaps the most heavily cratered in the Solar System.*

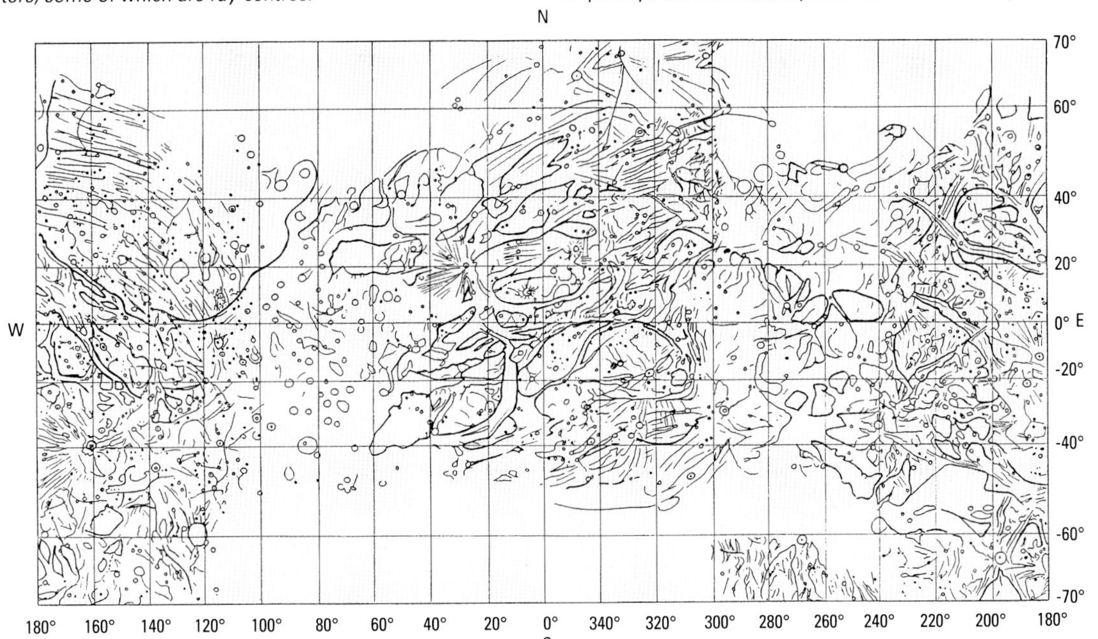

◀ *Ganymede*

GANYMEDE

		Lat. °	Long. °W
REGIONES	Bannard	22 N	010
	Galileo	35 N	145
	Marius	10 S	200
	Nicholson	20 S	000
	Perrine	40 N	030
SULCI	Dardanus	20 S	013
	Aquarius	50 N	010
	Nun	50 N	320
	Tiamat	03 S	210
CRATERS	Achelous	66 N	004
	Eshmun	22 S	187
	Gilgamesh	58 S	124
	Isis	64 S	197
	Nut	61 S	268
	Osiris	39 S	161
	Sebek	65 N	348
	Tros	20 N	28

CALLISTO

		Lat. °	Long. °W
RINGED BASINS	Asgard	30 N	140
	Valhalla	10 N	055
CRATERS	Adlinda	58 S	020
	Alfr	09 S	222
	Bran	25 S	207
	Grimr	43 N	214
	Igaluk	05 N	315
	Lodurr	52 S	270
	Rigr	69 N	240
	Tyn	68 N	229

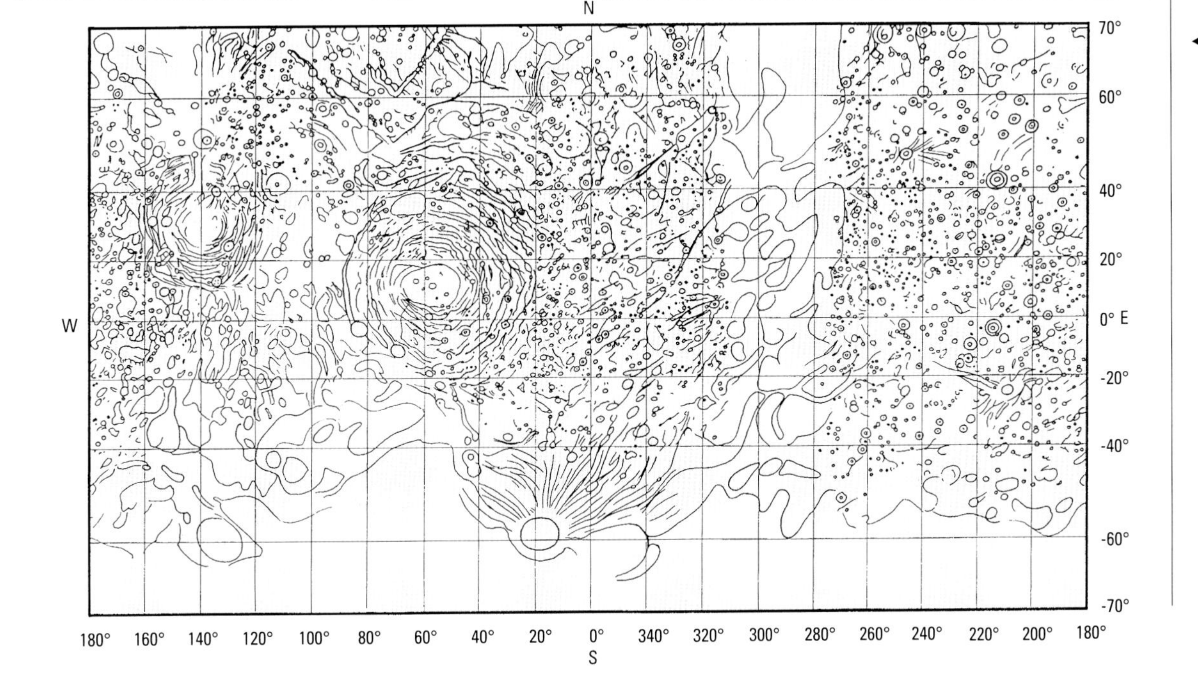

◀ *Callisto*

Saturn

▶ ***North Polar Region***
Lat. +90° to +55° approx. The northernmost part of the disk. Its colour is variable: sometimes bright, sometimes dusky.
North Temperate Zone
Lat. +70° to +40° approx. Generally fairly bright, but from Earth few details can be seen.
North Temperate Belt
Lat. +40°
One of the more active belts on the disk, and usually easy to see telescopically except when covered by the rings.
North Tropical Zone
Lat. +40° to +20°
A generally fairly bright zone between the two dark belts.
North Equatorial Belt
Lat. +20°
A prominent belt, always easy to see and generally fairly dark. Activity within it can sometimes be observed from the Earth.
Equatorial Zone
Lat. +20° to −20°
The brightest part of the planet. Details can be observed in it, and there are occasional white spots. The most prominent example of a white spot in the 20th century was in 1933.
South Equatorial Belt
Lat. −20°
A dark belt, usually about the same intensity as the corresponding belt in the northern hemisphere.
South Tropical Zone
Lat. −20° to −40°
A generally bright zone. Little detail to be seen telescopically.
South Temperate Belt
Lat. −40°
Generally visible when not covered by the rings.
South Temperate Zone
Lat. −40° to −70°
A brightish zone, with little or no visible detail as seen from Earth.
South Polar Region
Lat. −70° to −90° approx. The southernmost part of the disk. Like the north polar region, somewhat variable in its depth of shading.

Saturn, second of the giant planets, is almost twice as remote as Jupiter, and has an orbit period of over 29 years, so that it is a slow mover across the sky – it was natural for the ancients to name it in honour of the God of Time. It can become brighter than any star apart from Sirius and Canopus, and in size and mass it is inferior only to Jupiter.

Telescopically, Saturn may lay claim to being the most beautiful object in the entire sky. It has a yellowish, obviously flattened disk crossed by belts which are much less obvious than those of Jupiter. Around the planet is the system of rings, which can be seen well with even a small telescope except when the system lies edgewise on to us (as in 1995). There are three main rings, two bright and one semi-transparent; others have been detected by the space probes which have flown past Saturn from Earth: Pioneer 11 in 1979, Voyager 1 in 1980 and Voyager 2 in 1981.

Much of our detailed information about Saturn has been drawn from these space missions, but it was already known that in make-up the globe is not unlike that of Jupiter, even though there are important differences in detail – partly because of Saturn's lower mass and smaller size, and partly because of its much greater distance from the Sun. The polar flattening is due to the rapid rotation. The period at the equator is 10 hours 14 minutes, but the polar rotation is considerably longer. Visually, the periods are much less easy to determine than with those of Jupiter because of the lack of well-defined surface markings.

The gaseous surface is made up chiefly of hydrogen, together with helium and smaller quantities of other gases. Below the clouds comes liquid hydrogen, at first molecular and then, below a depth of 30,000 kilometres (19,000 miles), metallic. The rocky core is not a great deal larger than the Earth, though it is much more massive; the central temperature has been given as 15,000 degrees C, though with considerable uncertainty.

One interesting point is that the overall density of the globe of Saturn is less than that of water – it has even been said that if the planet could be dropped into a vast ocean, it would float! Though the mass is 95 times that of the Earth, the surface gravity is only 1.16 times greater. All the same, Saturn has a very powerful gravitational pull, and has a strong perturbing effect upon wandering bodies such as comets.

Saturn, like Jupiter, sends out more energy than it would do if it relied entirely upon what it receives from the Sun, but the cause may be different. Saturn has had ample time to lose all the heat it must have acquired during its formation stage, and there are suggestions that the excess radiation may be gravitational, produced as droplets of helium sink gradually downwards through the lighter hydrogen. This would also explain why Saturn's uppermost clouds contain a lower percentage of helium than in the case of Jupiter.

Saturn emits a radio pulse with a period of 10 hours 39.4 minutes, which is presumably the rotation period of the inner core. The magnetosphere is somewhat variable in extent, but stretches out to approximately the distance of Titan, the largest of Saturn's satellites. Radiation zones exist, though they are much weaker than those of Jupiter.

The magnetic field itself is 1000 times stronger than that of the Earth, and the magnetic axis is almost coincident with the axis of rotation, though the centre of the field is displaced northwards along the axis by about 2400 kilometres (1500 miles) and the field is stronger at the north pole than at the south.

Our view of Saturn today is very different from that of the astronomer R. A. Proctor in 1882, who wrote:

> 'Over a region hundreds of thousands of square miles in extent, the flowing surface of the planet must be torn by sub-planetary forces. Vast masses of intensely hot vapour must be poured forth from beneath, and rising to enormous heights, must either sweep away the enwrapping mantle of cloud which had concealed the disturbed surface, or must itself form into a mass of cloud, recognizable because of its enormous extent.... [Yet] If over a thousand different regions, each as large as Yorkshire, the whole surface were to change from a condition of rest to such activity as corresponds with the tormented surface of seething metal, and vast clouds formed over all such regions so as to hide the actual glow of the surface, our most powerful telescopes would fail to show the slightest trace of change.'

PLANETARY DATA – SATURN

Sidereal period	10,759.20 days
Rotation period (equatorial)	10h 13m 59s
Mean orbital velocity	9.6 km/s (6.0 miles/s)
Orbital inclination	2° 29′ 21″.6
Orbital eccentricity	0.056
Apparent diameter	max. 20.9″, min. 15″.0
Reciprocal mass, Sun = 1	3498.5
Density, water = 1	0.71
Mass, Earth = 1	95.17
Volume, Earth = 1	744
Escape velocity	32.26 km/s (20.05 miles/s)
Surface gravity, Earth = 1	1.16
Mean surface temperature	−180°C
Oblateness	0.1
Albedo	0.61
Maximum magnitude	−0.3
Diameter (equatorial)	120,536 km (74,914 miles)
Diameter (polar)	108,728 km (67,575 miles)

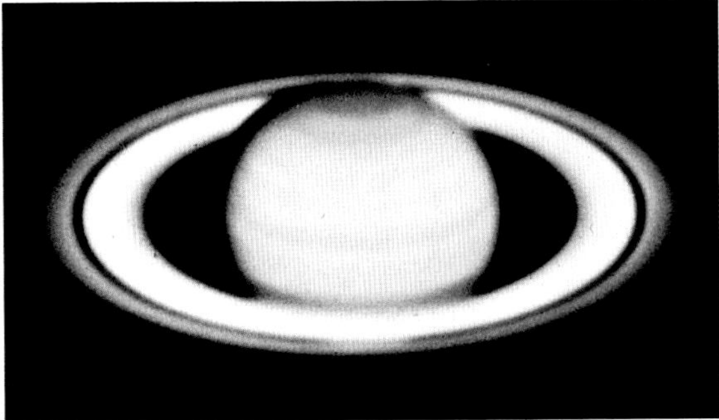

▲ ***Saturn**, photographed by Charles Capen with the 24-inch (61-cm) Lowell refractor. The ring system was then wide open. The Cassini Division in the ring system is well shown; there is not a great amount of detail on the disk – Saturn's surface is much less active than that of Jupiter.*

▶ ***Saturn from space:** 26 August 1990, imaged by the first Wide Field and Planetary Camera from the Hubble Space Telescope. At this moment Saturn was 1390 million km (860 million miles) from Earth. The colour in the image was reconstructed by combining three different pictures, taken in blue, green and red light (4390, 5470 and 7180 Ångströms). The north pole was inclined towards the Earth; the divisions in the ring are clear – including the Encke Division near the outer edge of Ring A.*

Rings of Saturn

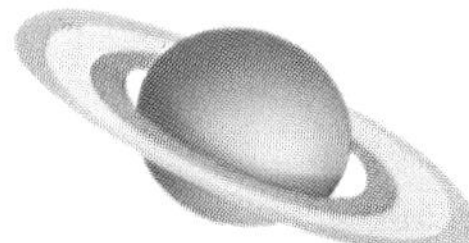
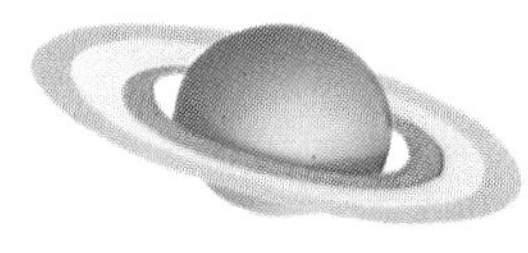
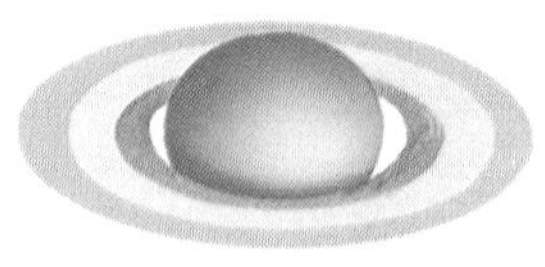

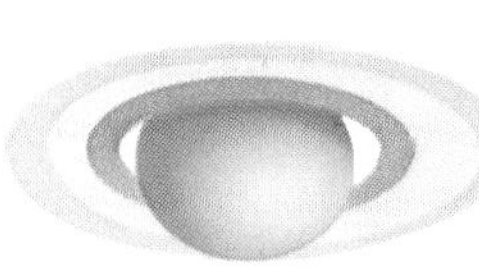
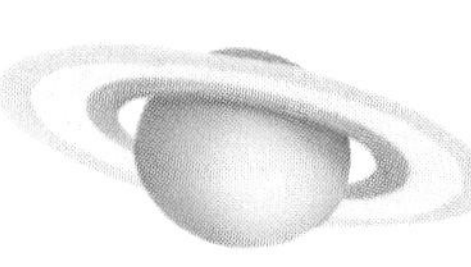
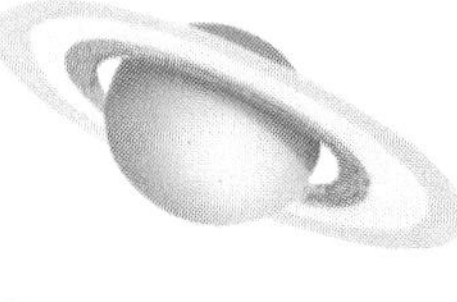
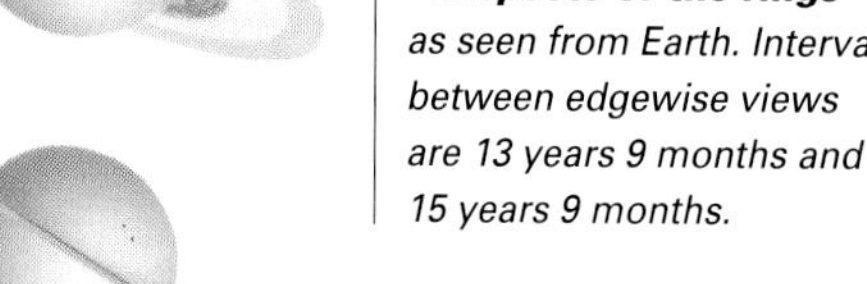

The ring system of Saturn is unique, and quite unlike the dark, obscure rings of Jupiter, Uranus and Neptune. Saturn's rings were first seen in the 17th century, and Christiaan Huygens, in 1656, explained them; previously Saturn had even been regarded as a triple planet.

There are two bright rings (A and B) and a fainter inner ring (C) which was discovered in 1850, and is usually known as the Crêpe or Dusky Ring because it is semi-transparent. The bright rings are separated by a gap known as Cassini's Division in honour of G. D. Cassini, who discovered it in 1675. Various fainter rings, both inside and outside the main system, had been reported before the Space Age, but there had been no definite confirmation. The main system is relatively close to the planet, and lies well within the Roche limit – that is to say, the minimum distance at which a fragile body can survive without being gravitationally disrupted; the outer edge of Ring A lies at 135,200 kilometres (84,000 miles) from Saturn's centre, while Mimas, the innermost of the satellites known before the space missions, is much further out at 185,600 kilometres (116,000 miles).

The full diameter of the ring system is about 270,000 kilometres (169,000 miles), but the thickness is no more than a few tens of metres (60 to 100 feet). Represent the full spread of the rings by the diameter of a cricket or baseball field, and the thickness will be no more than that of a piece of cigarette paper. This means that when the rings are edgewise-on to us they almost disappear. Edgewise presentations occur at intervals of 13 years 9 months and 15 years 9 months alternately, as in 1966, 1980 and 1995. This inequality is due to Saturn's orbital eccentricity.

During the shorter interval, the south pole is tilted sunwards – in other words, it is summer in the southern hemisphere – and part of the northern hemisphere is covered up by the rings; during this time Saturn passes through perihelion, and is moving at its fastest. During the longer interval the north pole is turned sunwards, so that parts of the southern hemisphere are covered up; Saturn passes through aphelion, and is moving at its slowest.

The rings are at their most obscure when the Earth is passing through the main plane or when the Sun is doing so. It is wrong to claim that they vanish completely; they can be followed at all times with powerful telescopes, but they cannot be seen with smaller instruments, and at best they look like very thin, faint lines of light.

No solid or liquid ring could exist so close to Saturn (if, indeed, such a ring could ever be formed in the first place). It has long been known that the rings are made up of small particles, all moving round the planet in the manner of tiny moons. There is no mystery about their composition; they are made up of ordinary water ice.

◀ ***Aspects of the rings*** *as seen from Earth. Intervals between edgewise views are 13 years 9 months and 15 years 9 months.*

▲ ***Saturn's rings from Voyager 1****. Taken by Voyager 1 on 13 November 1980, the range was 1.5 million km (938,000 miles). The bright limb of Saturn is clearly visible through the rings; the image was exposed to bring out the ring detail, so that the illuminated crescent of the planet is over-exposed. The Crêpe C ring scatters light in a way that makes it look more blue than the A and B rings.*

Of the two main rings, B is the brighter. The Cassini Division is very conspicuous when the system is favourably tilted to the Earth, and before the Pioneer and Voyager missions several minor divisions had been reported, though only one (Encke's Division, in Ring A) had been confirmed, and it was believed that the other divisions were mere 'ripples' in an otherwise fairly regular and homogeneous flat ring.

In 1907 the French observer G. Fournier announced the discovery of a dim ring outside the main system; though at that time confirmation was lacking, and it became known as Ring F. There were also reports of a faint ring between the Crêpe Ring and the cloud-tops, and this was usually referred to as Ring D, though again positive confirmation was lacking.

The Cassini Division was thought to be due mainly to the gravitational pull of the 400-kilometre (250-mile) satellite Mimas, which had been discovered by William Herschel as long ago as 1789. A particle moving in the Division would have an orbital period exactly half that of Mimas, and cumulative perturbations would drive it away from the 'forbidden zone'. No doubt there is some substance in this, though the Voyager revelations showed that there must be other effects involved as well. The rings turned out to be completely different from anything which had been expected.

DISTANCES AND PERIODS OF RINGS AND INNER SATELLITES

	Distance from centre of Saturn, km	Period, h
Cloud-tops	60,330	10.66
Inner edge of 'Ring' D	67,000	4.91
Inner edge of Ring C	73,200	5.61
Inner edge of Ring B	92,200	7.93
Outer edge of Ring B	117,500	11.41
Middle of Cassini Division	119,000	11.75
Inner edge of Ring A	121,000	11.92
Encke Division	133,500	13.82
Pan	133,600	14
Outer edge of Ring A	135,200	14.14
Atlas	137,670	14.61
Prometheus	139,350	14.71
Ring F	140,600	14.94
Pandora	141,700	15.07
Epimetheus	151,420	16.65
Janus	151,420	16.68
Inner edge of Ring G	165,800	18
Outer edge of Ring G	173,800	21
Inner edge of Ring E	180,000	22
Mimas	185,540	22.60
Enceladus	238,040	32.88
Tethys	294,760	1.88 **d**
Dione	377,420	2.74
Outer edge of Ring E	480,000	4
Rhea	527,040	4.52

▼ Saturn from Voyager
1. *This image was taken on 18 October 1980 at a range of 34 million km (21.1 million miles) from the planet. This photograph was taken on the last day Saturn and its rings could be captured within a single narrow-angle camera frame as the spacecraft closed in on the planet for its nearest approach on 12 November 1980. Dione, one of Saturn's inner satellites, appears as three colour spots just below the planet's south pole.*

Details of Saturn's Rings

When the first missions to Saturn were planned, it was quite naturally thought that stray ring particles might present a serious hazard. The initial foray was made by Pioneer 11, which passed within 21,000 kilometres (13,000 miles) of the cloud-tops. Estimates of its survival ranged from 99 per cent down to only 1 per cent, and it was a relief when the probe emerged unscathed. The Voyagers did not approach so closely to Saturn (124,200 kilometres (77,000 miles) and 101,300 kilometres (63,000 miles) respectively, and they also were undamaged. The scan platform of Voyager 2 jammed during the outward journey from Saturn, and for a while it was thought that a collision with a ring particle might have been responsible, but the problem turned out to be one of insufficient lubrication. During the Uranus and Neptune encounters the scan platforms worked perfectly.

The main surprise was that the rings proved to be made up of thousands of ringlets and narrow divisions; there are even rings inside the Cassini and Encke gaps. Some sort of wave effect may be involved, though it is fair to say that even now we do not fully understand the dynamics of the system.

The innermost or D region of the system is not a true ring, as there is no sharp inner edge, and the particles may spread down almost to the cloud-tops. The C or Crêpe Ring particles seem on average to be about 2 metres (7 feet) in diameter; in the B Ring the particle sizes range from 10 centimetres to about a metre (4 to 40 inches), with temperatures of −180°C in sunlight down to −200°C in shadow. Here we find strange, darkish radial 'spokes'; they had been glimpsed earlier by Earth-based observers such as Antoniadi, but the Voyagers gave the first clear views of them. Logically they ought not to exist, because, following Kepler's Laws, the orbital speeds of the particles decrease with increasing distance from the planet, and the difference in period between the inner and outer edges of Ring B is over three hours – yet the spokes persisted for hours after emerging from the shadow of the globe, and when they broke up they were replaced by new ones coming from out of the shadow. Presumably they are due to particles elevated away from the ring-plane by magnetic or electrostatic forces. The spokes are confined entirely to Ring B.

Ring A is made up of particles ranging from fine 'dust' to larger blocks up to about 10 metres (over 30 feet) across. The main division in it, Encke's Division, was found to contain some discontinuous, irregular ringlets along with a tiny satellite, now named Pan. Another satellite, Atlas, moves close to the outer edge of Ring A, and is responsible for its sharp border.

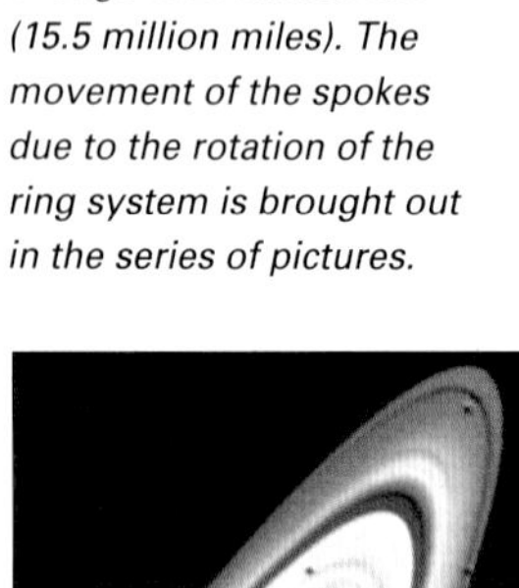

▼ ***'Spokes' in Saturn's rings**. These images were obtained from Voyager 1 on 24 October 1980, from a range of 25 million km (15.5 million miles). The movement of the spokes due to the rotation of the ring system is brought out in the series of pictures.*

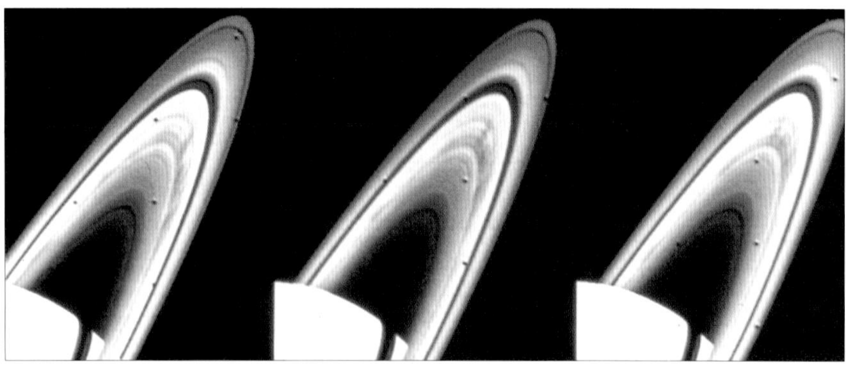

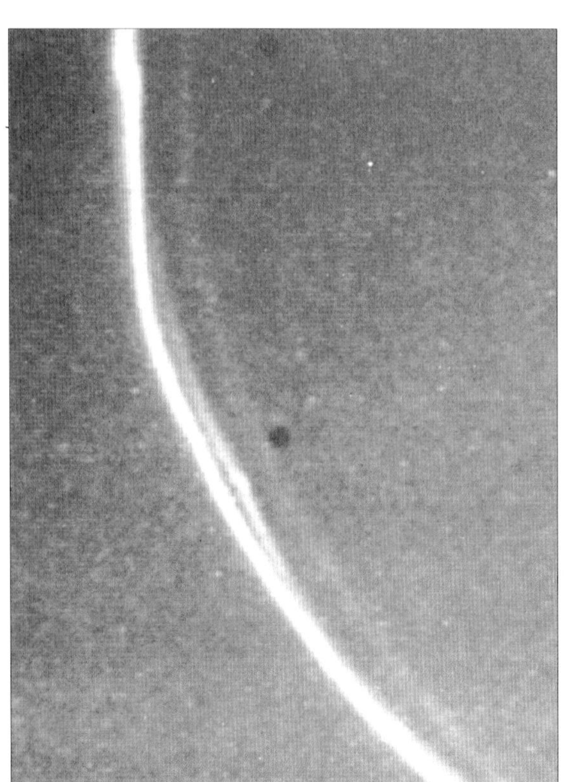

◀ ***The 'Braided' F Ring** as imaged from Voyager 2 from a range of 103,000 km (64,000 miles). The complex structure of the ring was unexpected, and seems to be due to the gravitational effects of the small satellites Prometheus and Pandora.*

▶ ***False colour Saturn's rings**, from Voyager 2 on 20 August 1981. Sunlight is seen coming through the Cassini Division. The resolution is down to 56 m (180 feet).*

Outside the main system comes Ring F, which is faint and complex. It is stabilized by two more small satellites, Prometheus and Pandora, which act as 'shepherds' and keep the ring particles in place. Prometheus, slightly closer to Saturn than the ring, moves faster than the ring particles, and will speed up a particle if it moves inwards, so returning it to the main ring zone; Pandora, on the far side, will be moving more slowly, and will drag any errant particle back.

The outer rings (G and E) are very tenuous indeed. The brightest part of Ring E is just inside the orbit of the icy satellite Enceladus, and it has even been suggested that material ejected from Enceladus may have been concerned in the formation of the ring. It is difficult to say where Ring E ends; traces of it may extend out to as far as the orbit of the larger satellite Rhea, more than 500,000 kilometres (312,000 miles) from Saturn.

There has been considerable discussion about the origin of the rings. According to one theory, they represent the debris of an icy satellite which wandered too close to Saturn and paid the supreme penalty, though on the whole it seems more likely that the rings are formed from material which never condensed into a larger body. At any rate, they are there for our inspection, and for sheer beauty Saturn is unrivalled in the Solar System.

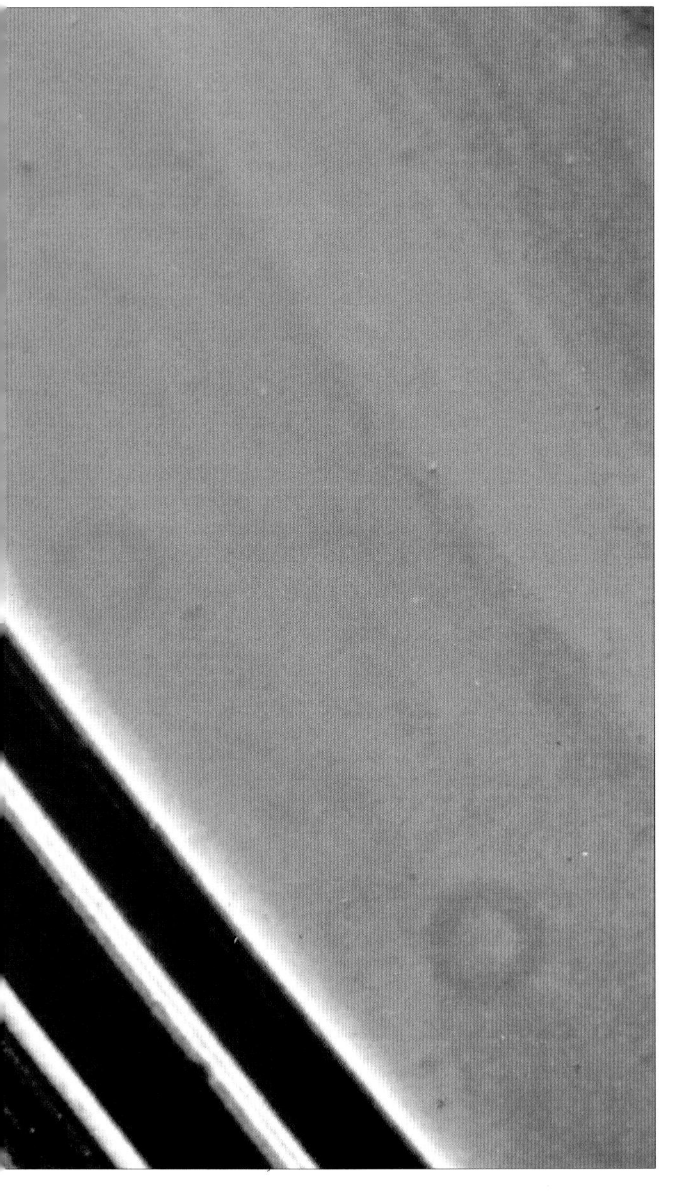

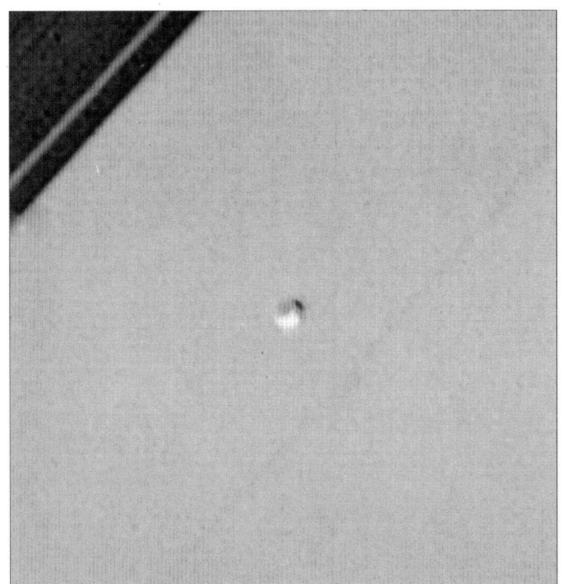

▲ ***Voyager 2's image*** *of Saturn's F-ring and its small inner shepherding satellite Prometheus from a range of 1.2 million km (740,000 miles) taken 25 August 1981. Prometheus is more reflective than Saturn's clouds, suggesting that it is an icy, bright-surfaced object like the larger satellites and the ring particles themselves.*

▲ ***Rings of Saturn****. The C-ring (and to a lesser extent, the B-ring at top and left) is seen in this false-colour image made from three separate pictures taken through ultra-violet, clear, and green filters. Voyager 2 obtained this image on 23 August 1981 from a distance of 2.7 million kilometres (1.7 million miles). More than 60 bright and dark ringlets are evident here. Colour differences between the C-ring (blue in this picture) and the B-ring indicate differing surface compositions of the material that comprises these complex structures.*

Missions to Saturn

▶ ***Saturn from the two Voyagers****. These pictures are shown in false colour – that is to say, colour has been added in order to help in analysis. The left-hand image shows Saturn from Voyager 1, in October 1980; the right-hand image shows Saturn from Voyager 2, August 1981. There are differences in detail, though the basic aspect is the same.*

Three spacecraft have passed by Saturn. The first encounter, by Pioneer 11 in September 1979, was in the nature of a brief preliminary reconnaissance; Pioneer had not originally been planned to go on Saturn after its rendezvous with Jupiter, but when it became clear that this was a possibility full advantage was taken of it. Pioneer did indeed send back useful information, but the main results have come from Voyager 1 (1980) and Voyager 2 (1981).

Voyager 1 was scheduled to survey not only Saturn itself, but also Titan, the largest of the satellites, which was known to have an atmosphere and to be a world of exceptional interest. Had Voyager 1 failed, then Voyager 2 would have had to study Titan – and this would have meant that it would have been unable to continue on to Uranus and Neptune. Therefore, there was great relief when Voyager 1 proved to be a success.

Saturn is a much blander world than Jupiter. The cloud structure is of the same type, but the lower temperature means that ammonia crystals form at higher levels, producing the generally hazy appearance. There are none of the vivid colours so striking on Jupiter.

The main belts are usually obvious enough, though there are long periods when a large part of one or the other hemisphere is hidden by the rings. Spots are usually inconspicuous, but there are major outbreaks now and then. Bright white spots were seen in 1876, 1903, 1933 (discovered by W. T. Hay – perhaps better remembered by most people as Will Hay, the actor), 1960 and 1990. The most prominent of these have been the spots of 1933, which persisted for some weeks, and of 1990, which were well imaged by the Hubble Space Telescope and were clearly due to an uprush of material from below. The time intervals between these white spots have been 27, 30, 27 and 30 years respectively. This is close to Saturn's orbital period of 29½ years, which may or may not be significant; at any rate, observers will be watching out for a new white spot around the year 2020. The spots are important because they tell us a good deal about conditions below the visible surface, and also help in measuring rotation periods.

The Voyager missions confirmed that Saturn, again like Jupiter, has a surface which is in constant turmoil (even if not in the way that Proctor had supposed in 1882) and that windspeeds are very high. There is a wide equatorial jet-stream, 80,000 kilometres (50,000 miles) broad and stretching from about latitude 35°N to 35°S, where the winds reach 1800 kilometres per hour (1120 miles per hour), much faster than any on Jupiter. A major surprise was that the wind zones do not follow the light and dark bands, but instead are symmetrical with the equator. One prominent 'ribbon' at latitude 47°N was taken to be a wave pattern in a particularly unstable jet-stream.

A careful search was made for spots. There is nothing remotely comparable with the Great Red Spot on Jupiter, but one relatively large oval feature in the southern hemisphere did appear to be somewhat coloured (it was first noted by Anne Bunker, and became known as Anne's Spot), and there were other, smaller markings of the same kind, some of which were noted by Voyager 1 and were still present when Voyager 2 made its fly-by – though it is not likely that any of them are really long-lived.

Saturn's seasons are very long, and this means that there are measurable temperature differences between the two hemispheres. The Sun crossed into the northern hemisphere of the planet in 1980, but there is a definite 'lag' effect, and during the Voyager encounters the northern hemisphere was still the colder of the two; the difference between the two poles amounted to 10 degrees C.

▼ ***Computer-generated image of Saturn's rings****, from data obtained by the first Saturn probe, Pioneer 11, which bypassed the planet at 21,400 km (13,300 miles) on 1 September 1979. This is not a photograph, but shows ring data taken at 6° above the plane of the rings as though it were seen from 90° above the rings (or directly over Saturn's north pole). The graphic shows the rings as though the observer were 1 million km (620,000 miles) above the north pole. The resolution of the ring bands is 500 km (300 miles); the area occupied by the planet has been filled by a selected portion of a cloud-top picture.*

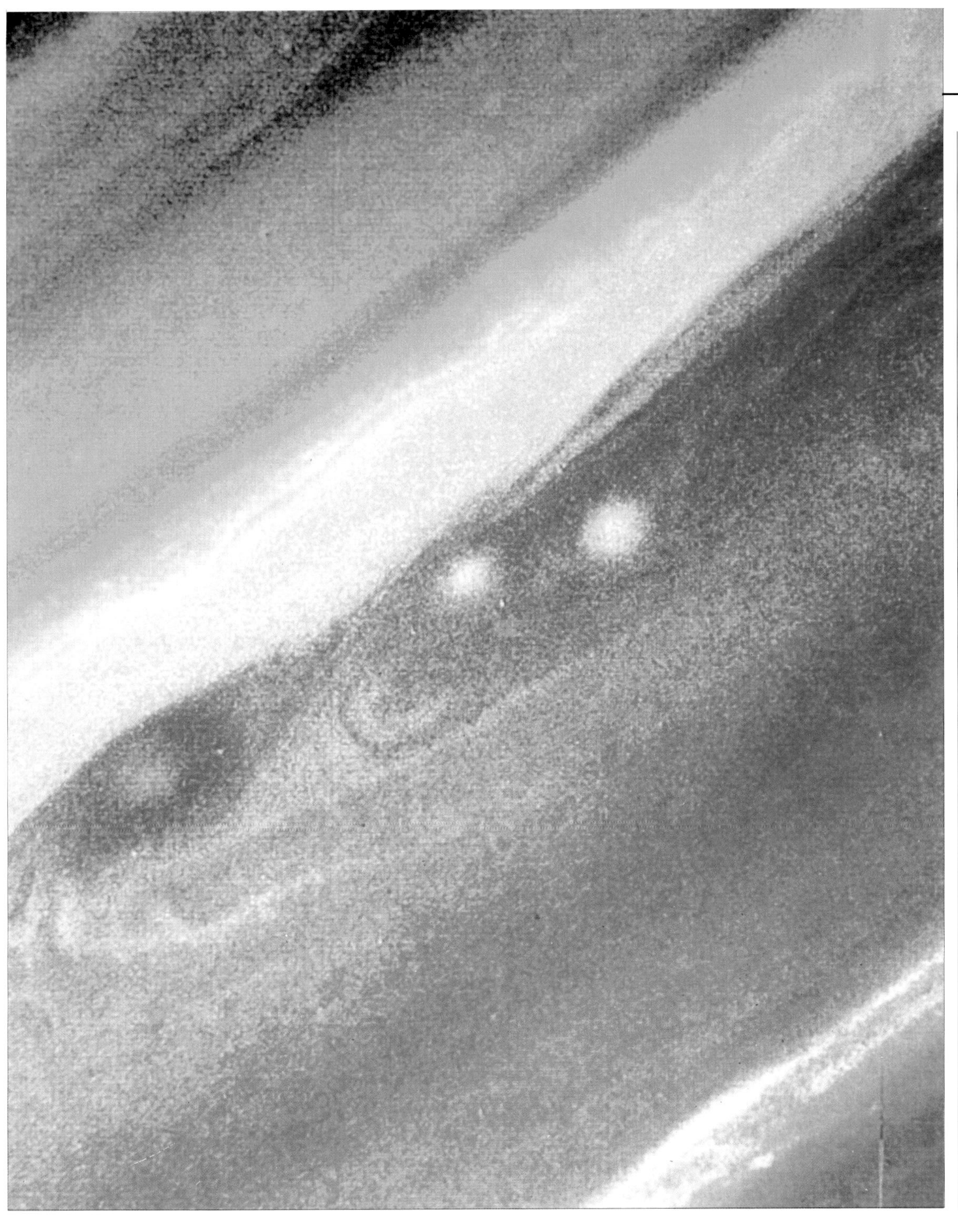

◀ ***Three spots*** *in Saturn's northern hemisphere from Voyager 2, 19 August 1981. The largest is 3000 km (over 1800 miles) in diameter. The colour has been enhanced. The delicate disk features recorded by the Voyagers are beyond the range of Earth-based telescopes, though high-quality images are now being obtained by the Hubble Space Telescope. We may hope for more information when the Cassini probe reaches Saturn in 2004. Meanwhile, amateur observers as well as professionals will keep watch – there is always the chance of finding a new major white spot.*

◀ ***The north polar region*** *from Voyager 2, 25 August 1981. Note the obvious difference in atmospheric structure in the polar zone.*

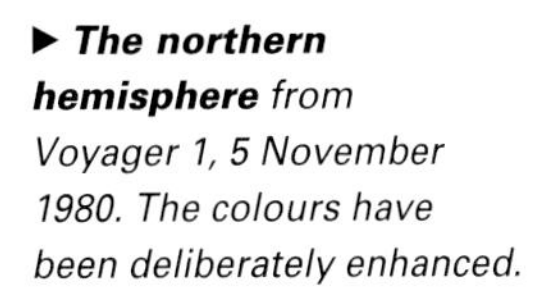

▶ ***The northern hemisphere*** *from Voyager 1, 5 November 1980. The colours have been deliberately enhanced.*

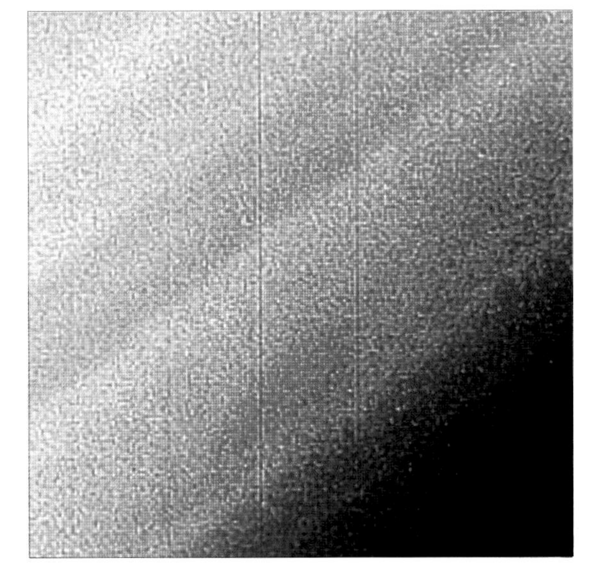

Satellites of Saturn

Saturn's satellite family is quite different from that of Jupiter. Jupiter has four large attendants and a dozen small ones; Saturn has one really large satellite (Titan) and seven which are medium-sized, together with the remote Phoebe, which has retrograde motion and is almost certainly a captured asteroid. Eight new satellites, all very small, have been found on the Voyager images.

Titan, with a diameter of over 5000 kilometres (3201 miles), is the largest satellite in the Solar System apart from Ganymede, and is actually larger than the planet Mercury, though less massive. It is also unique among satellites in having a dense atmosphere; all that Voyager was able to see was the top part of a layer of orange 'smog'. The atmosphere was found to be made up chiefly of nitrogen, with a good deal of methane.

Of the icy satellites, Rhea and Iapetus are around 1500 kilometres (940 miles) in diameter, Dione and Tethys around 1110 kilometres (690 miles), and Enceladus, Hyperion and Mimas between 220 kilometres and 320 kilometres (between 140 and 320 miles), though Hyperion is decidedly irregular in shape (it has been likened, rather unromantically, to a cosmic hamburger). The globes appear to be made up of a mixture of rock and ice, though Tethys in particular has a mean density only just greater than that of water, so that rock may be a very minor constituent. Not much is known about Phoebe, which was unfavourably placed during both the Voyager passes, but all the other main members of the family were well surveyed.

All the icy satellites have their own special points of interest. They are not alike; for example Rhea and Mimas are very heavily cratered, while Enceladus has a surface which looks much younger, and Tethys shows a tremendous trench which reaches three-quarters of the way round the globe. Iapetus has one hemisphere which is bright and one which is dark; evidently dark material has welled up from below and covered the icy surface. This had been suspected long before the Voyager missions, because Iapetus is so variable in brightness. When west of the planet, with its reflective area turned towards us, it is easy to see with a small telescope; when east of Saturn, with its blacker side displayed, owners of small telescopes will find it very elusive.

The 'new' satellites found on the Voyager images are also, presumably, icy and cratered. Pan actually moves inside the Encke Division in Ring A, while Prometheus and Pandora act as 'shepherds' to the F Ring. Epimetheus and Janus seem to take part in what may be called a game of musical chairs; their paths are almost the same, and

▼ ***Mimas*** *(Voyager 1). The surface is dominated by one very large crater, now named Herschel, which has a diameter of 130 km (80 miles) – one-third that of Mimas itself – with walls which rise to 5 km (3 miles) above the floor, the lowest part of which is 10 km (6 miles) deep and includes a massive central mountain.*

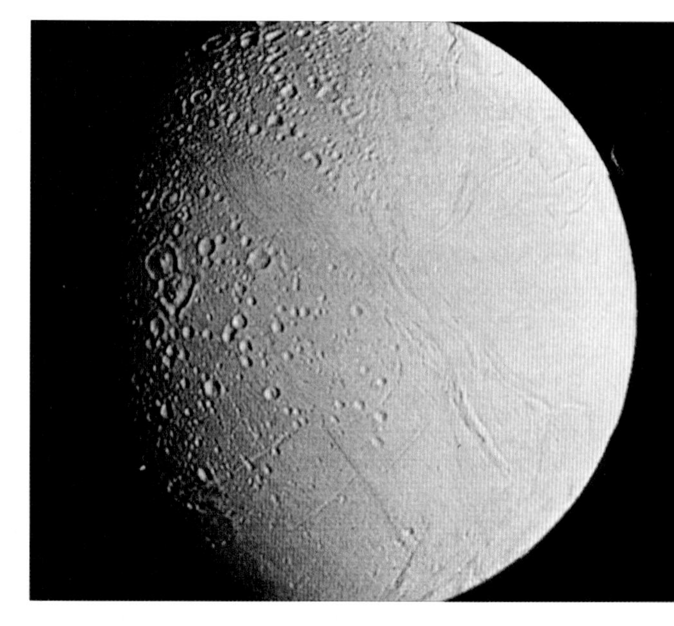

► ***Enceladus*** *(Voyager 1). Enceladus is quite different from Mimas; there are several completely different types of terrain. Craters exist in many areas, but appear 'sharp' and relatively young, while there is also an extensive plain which is almost crater-free. It has been suggested that there are periods when water wells up from below the surface, obliterating existing features.*

SATELLITES OF SATURN

Name	Distance from Saturn, km	Orbital period, days	Orbital incl.,°	Orbital ecc.	Diameter, km	Mean opp. mag.
Pan	133,600	0.57	0	0	20?	20?
Atlas	137,670	0.602	0.3	0.002	37 × 34 × 27	18.1
Prometheus	139,350	0.613	0.8	0.004	48 × 100 × 68	16.5
Pandora	141,700	0.629	0.1	0.004	110 × 88 × 62	16.3
Epimetheus	151,420	0.694	0.3	0.009	194 × 190 × 154	14.5
Janus	151,470	0.695	0.1	0.007	138 × 110 × 110	15.5
Mimas	185,540	0.942	1.52	0.020	194 × 190 × 154	12.9
Enceladus	238,040	1.370	0.07	0.004	421 × 395 × 395	11.8
Tethys	294,670	1.888	1.86	0.000	1046	10.3
Telesto	294,670	1.888	2	0	30 × 25 × 15	19.0
Calypso	294,670	1.888	2	0	30 × 16 ×16	18.5
Dione	377,420	2.737	0.02	0.002	1120	10.4
Helene	377,420	2.737	0.2	0.005	35	18.5
Rhea	527,040	4.518	0.35	0.001	1528	9.7
Titan	1,221,860	15.495	0.33	0.029	5150	8.4
Hyperion	1,481,100	21.277	0.43	0.104	360 × 280 × 225	14.2
Iapetus	3,561,300	79.331	7.52	0.028	1436	10 (var.)
Phoebe	12,954,000	550.4	175	0.163	30 × 220 × 210	16.5

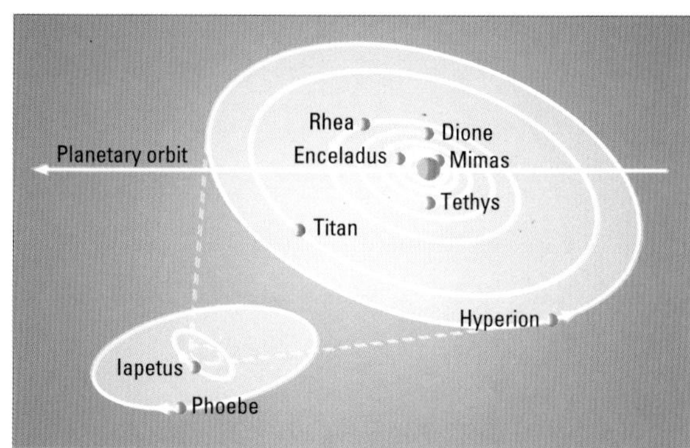

▲ ***Satellite orbits*** *of Saturn's nine larger satellites. In addition there are at least eight smaller moons, some in very unusual orbits. For instance, two tiny satellites, Telesto and Calypso, move at the same distance as Tethys, one about 60° ahead of, and the other 60° behind Tethys in its orbit, in the same way that the Trojan asteroids travel in Jupiter's orbit.*

periodically they actually exchange orbits. They do not collide – otherwise they could not continue to exist as separate bodies – but they may come within a few kilometres (a mile or two) of each other. Both are very irregular, and are unquestionably the fragments of a larger object which met with disaster in the remote past.

Telesto and Calypso move in the same orbit as Tethys, just as the Trojan asteroids do with respect to Jupiter; they oscillate around the Lagrangian points 60 degrees ahead and 60 degrees behind. Dione has one 'Trojan' satellite, Helene, and another has been suspected.

Titan is visible in almost any telescope, and very keen-sighted people can glimpse it with good binoculars. A 7.5-centimetre (3-inch) refractor will show Iapetus (when west of Saturn) and Rhea easily, and Tethys and Dione with more difficulty. The other pre-Voyager satellites require larger apertures, though all except Phoebe are within the range of a 30-centimetre (12-inch) reflector.

In 1904 W. H. Pickering, discoverer of Phoebe, reported another satellite, moving between the orbits of Titan and Hyperion. It was even given a name – Themis – but has never been confirmed, and probably does not exist. On the other hand there may be several more small inner satellites awaiting discovery.

◄ ***Tethys*** *(Voyager 2). Tethys seems to be made up of almost pure ice. There is one huge crater, Odysseus, with a diameter of 400 km (250 miles) – larger than the whole of Mimas! There is also Ithaca Chasma, a tremendous trench 2000 km (1250 miles) long, running from near the north pole across the equator and along to the south pole.*

◄ ***Dione*** *(Voyager 2). The surface is not uniform; the trailing hemisphere is relatively dark. There are large craters, such as Dido, Penelope and Aeneas, and a curious feature, Amata, which may be either a crater or a basin.*

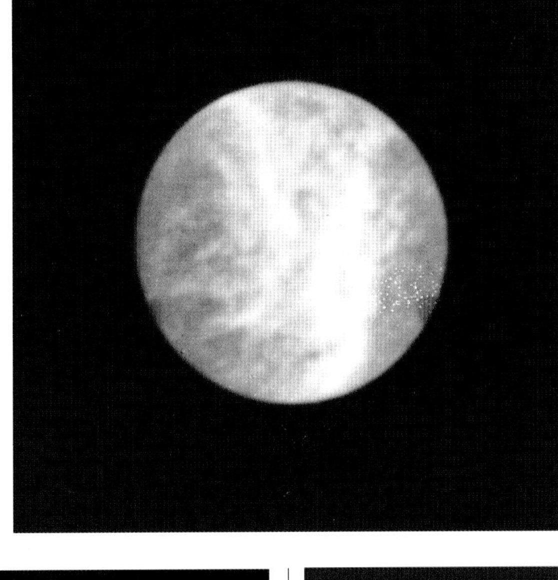

► ***Rhea*** *(Voyager 1). Rhea is heavily cratered, but with few really large formations. As with Dione, the trailing hemisphere is darkish, with wispy features which are not unlike those in Dione, though less prominent. Rhea seems to be made up of a mixture of rock and ice in almost equal amounts.*

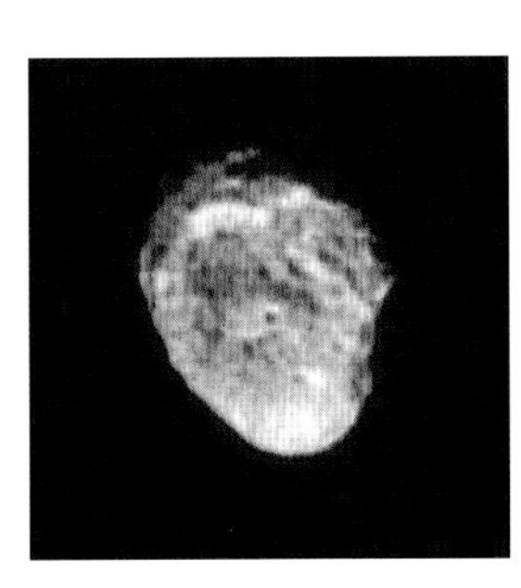

▼ ***Hyperion*** *(Voyager 2). It has been said that Hyperion is shaped like a hamburger! Its rotation is 'chaotic' rather than synchronous, and the longer axis does not point towards Saturn, as it might have been expected to do if its rotation was more settled. There are many craters, and one long scarp (Bons-Lassell) running for 300 km (190 miles). The surface is in general less reflective than those of the other icy satellites.*

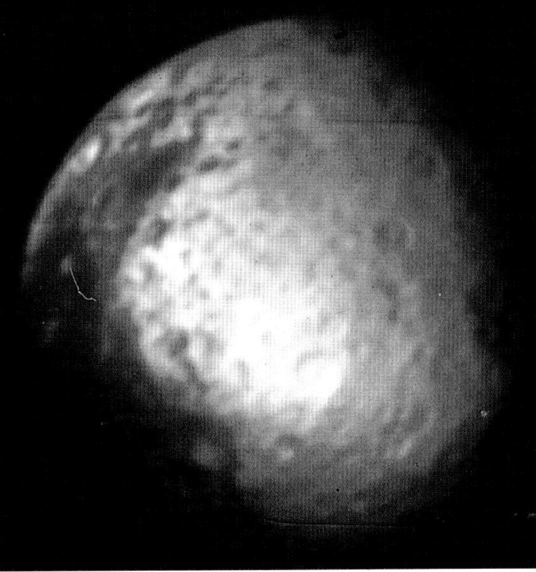

► ***Iapetus*** *(Voyager 2). Here there are two hemispheres of very different albedo. In this picture the sunlight is coming not from the left, as might be expected, but from the right! This is why Iapetus is variable in brightness; when west of the planet, its brighter side faces us. However, the topography seems to be the same over both regions.*

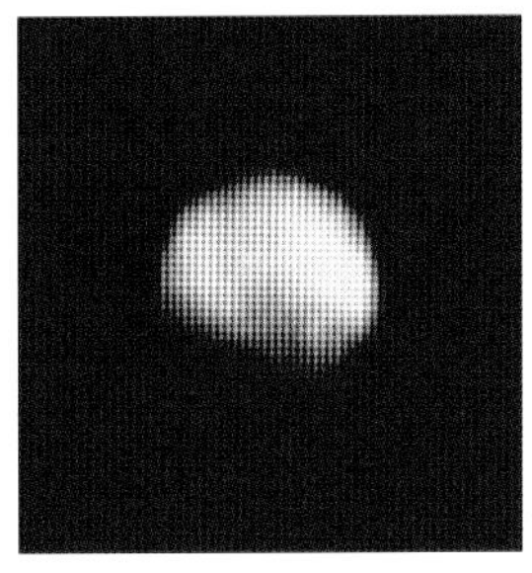

▲ ***Phoebe*** *(Voyager 2) was the only pre-Voyager satellite not well imaged; the only picture came from Voyager 2 at 1,473,000 km (915,000 miles). Revolution is retrograde with a period of 9.4 hours; the surface is darkish. Phoebe is almost certainly a captured asteroid.*

Maps of Saturn's Icy Satellites

Apart from planet-sized Titan, Saturn's satellites are small and icy, but they each have their unique characteristics. The eight known before the Voyager missions are mapped here with data from those missions. The Voyagers also found eight more satellites, even smaller ones.

Mimas *The globe seems to be composed mainly of ice, though there must be some rock as well. The surface is dominated by the huge crater now named Herschel, which is 130 kilometres (80 miles) in diameter – one third the diameter of Mimas itself – and has a massive central mountain rising to 6 kilometres (nearly 4 miles) above the floor. There are many other craters of lesser size, together with grooves (chasma) such as Oeta and Ossa.*

Enceladus *Enceladus is quite different from Mimas. Craters exist in many areas, but look 'young', and it may be that the interior is flexed by the pull of the more massive Dione, so that at times soft ice wells out from below and covers older formations. There are also ditches (fossae) and planitia (plains) such as Diyar and Sarandib.*

Tethys *There is one huge crater, Odysseus, which is 400 kilometres (250 miles) across (larger than Mimas) but not very deep. The main feature is Ithaca Chasma, a tremendous trench running from the north pole across the equator to the region of the south pole. Its average width is 100 kilometres (over 60 miles), and is 4 to 5 kilometres (about 3 miles) deep, with a rim which rises to half a kilometre (1600 feet) above the outer surface. Other craters include Penelope, Anticleia and Eumaeus.*

Dione *Dione is much denser and more massive than Tethys, so that its globe presumably contains less ice and more rock. The trailing hemisphere is darkish; the leading hemisphere is bright. The most prominent feature is Amata, 240 kilometres (150 miles) in diameter, which may be either a crater or a basin, and is associated with bright wispy*

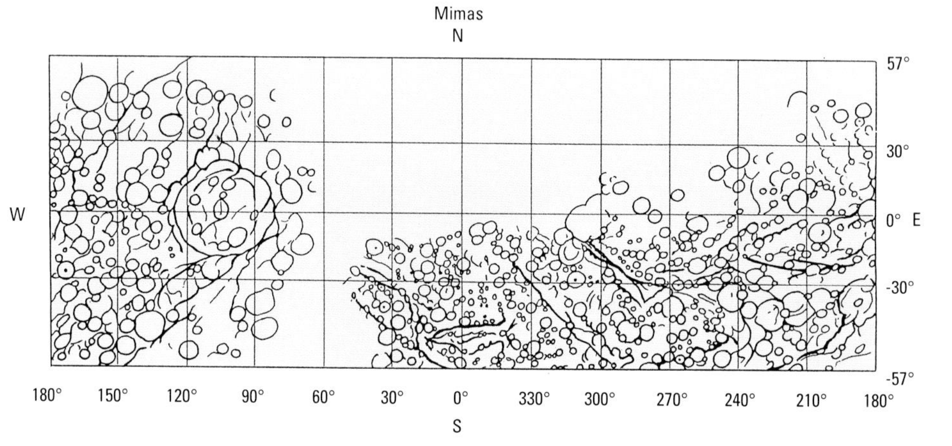

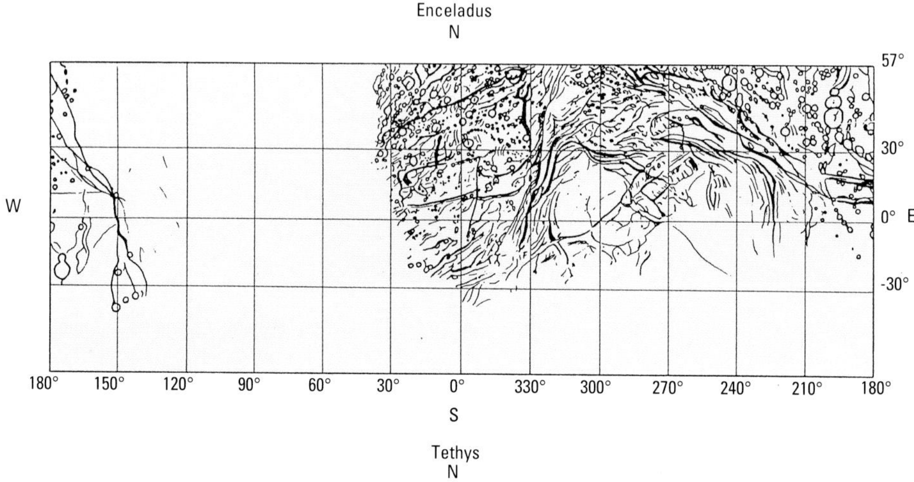

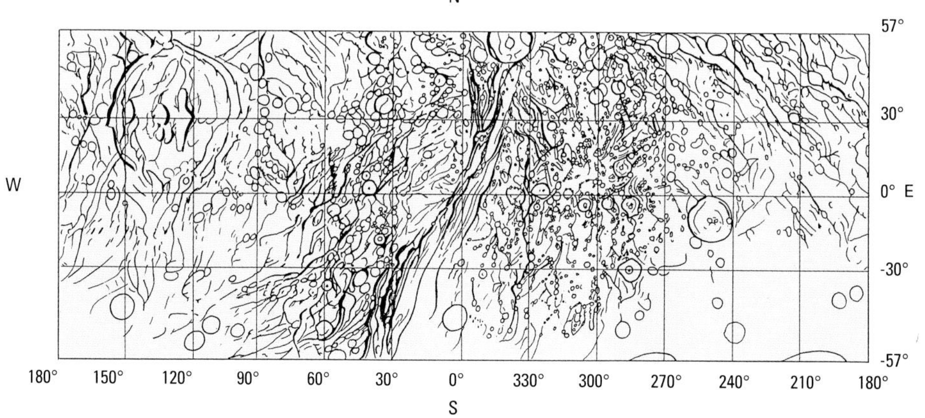

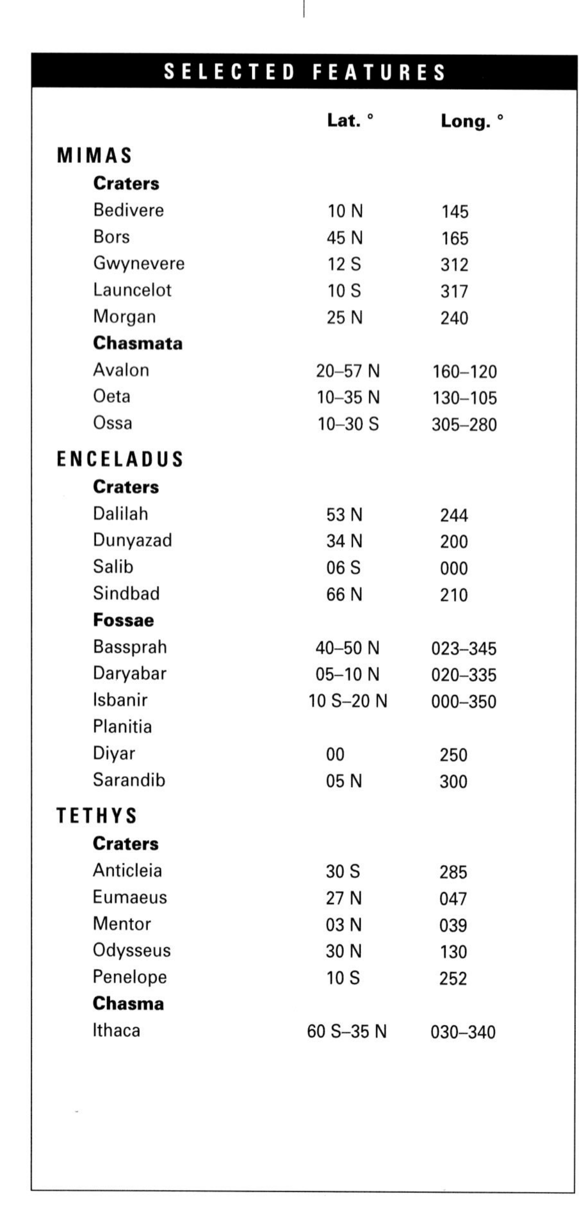

SELECTED FEATURES

	Lat. °	Long. °
MIMAS		
Craters		
Bedivere	10 N	145
Bors	45 N	165
Gwynevere	12 S	312
Launcelot	10 S	317
Morgan	25 N	240
Chasmata		
Avalon	20–57 N	160–120
Oeta	10–35 N	130–105
Ossa	10–30 S	305–280
ENCELADUS		
Craters		
Dalilah	53 N	244
Dunyazad	34 N	200
Salib	06 S	000
Sindbad	66 N	210
Fossae		
Bassprah	40–50 N	023–345
Daryabar	05–10 N	020–335
Isbanir	10 S–20 N	000–350
Planitia		
Diyar	00	250
Sarandib	05 N	300
TETHYS		
Craters		
Anticleia	30 S	285
Eumaeus	27 N	047
Mentor	03 N	039
Odysseus	30 N	130
Penelope	10 S	252
Chasma		
Ithaca	60 S–35 N	030–340

features which extend over the trailing hemisphere. Other major craters are Aeneas and Dido. There are also chasma (Larissa, Tibur) and linea (Carthage, Palatine).

Rhea *A very ancient, cratered surface. The most prominent crater is Izanagi, but there are few others of great size, and they tend to be irregular in shape. As with Dione, the trailing hemisphere is darkish, with wispy features which are not unlike those on Dione but are less prominent.*

Hyperion *This is one of the few medium-sized satellites which does not have synchronous rotation, and the rotation period is indeed 'chaotic' and variable. Hyperion is less reflective than the other icy satellites, so there may be a 'dirty' layer covering wide areas. There are several craters, such as Helios, Bahloo and Jarilo, as well as a long ridge or scarp, Bond-Lassell.*

Iapetus *Here the leading hemisphere is as black as a blackboard, with an albedo of no more than 0.05, while the trailing hemisphere is bright: albedo 0.5. The line of demarcation is not abrupt, and there is a transition zone 200 to 300 kilometres (125 to 190 miles) wide. Some craters in the bright region (Roncevaux Terra) have dark floors, but we do not know whether the dark floor material is the same as the dark area (Cassini Regio). Craters include Otho and Charlemagne.*

Phoebe *Whether this should be classed as an icy satellite is not clear; apparently it has a darkish surface. Unfortunately it was not closely surveyed by either Voyager. It may be very similar to the strange asteroid Chiron, and it is worth noting that in 1664* BC *Chiron approached Saturn to within a distance of 16 million kilometres (10 million miles) which is not much greater than the distance between Saturn and Phoebe. Like Hyperion, the rotation period is not synchronous, and amounts to only 9.4 hours.*

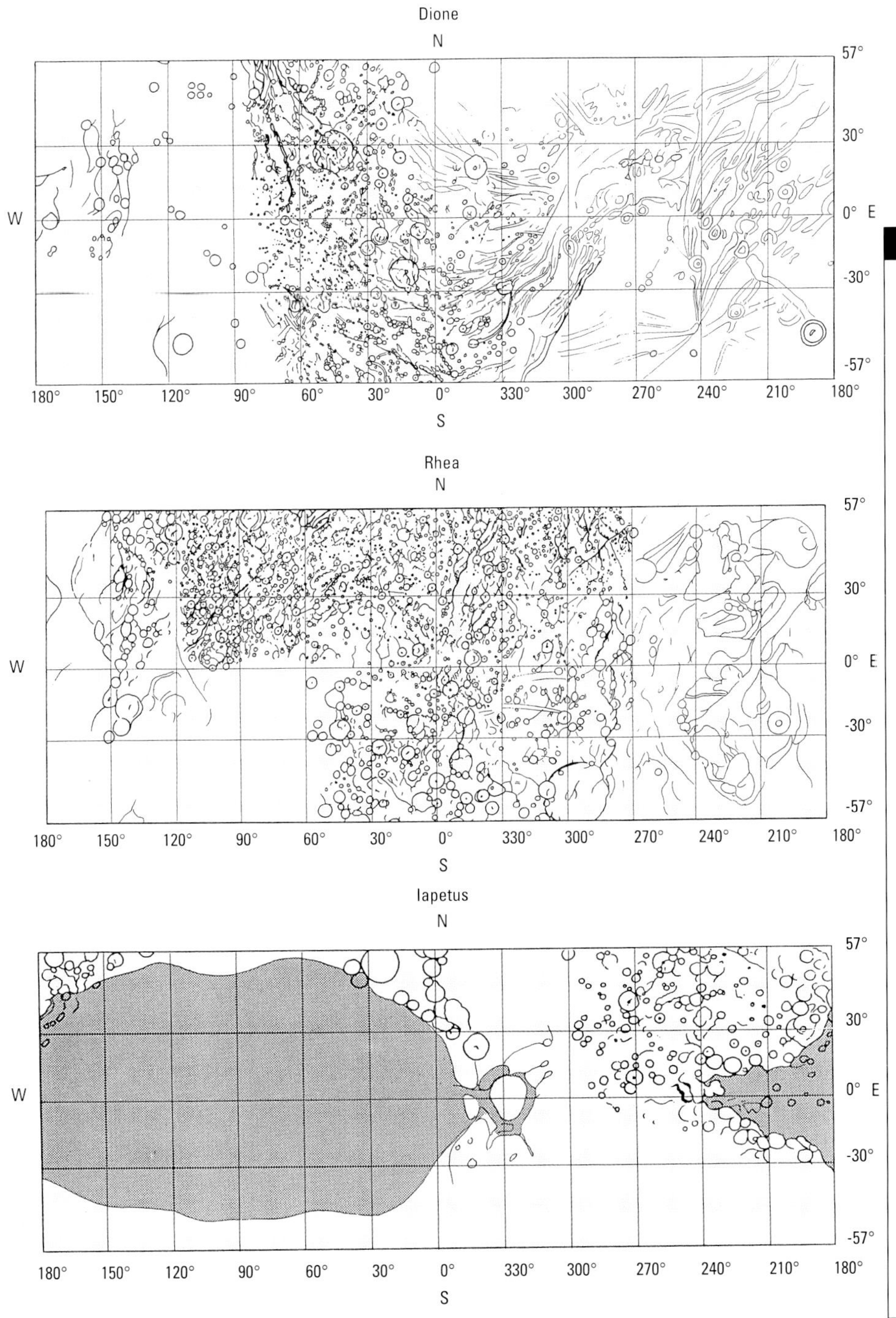

SELECTED FEATURES

	Lat. °	Long. °
DIONE		
Craters		
Aeneas	26 N	047
Amata	07 N	287
Adrastus	64 S	040
Dido	22 S	015
Italus	20 S	076
Lausus	38 N	023
Chasmata		
Larissa	20–48 N	015–065
Latium	03–45 N	064–075
Palatine	55–73 S	075–230
Tibur	48–80 N	060–080
Linea		
Carthage	20–40 N	337–310
Padua	05 N–40 S	245
Palatine	10–55 S	285–320
RHEA		
Craters		
Izanagi	49 S	298
Izanami	46 S	310
Leza	19 S	304
Melo	51 S	006
Chasmata		
Kun Lun	37–50 N	275–300
Pu Chou	10–35 N	085–115
Tapetus		
IAPETUS		
Craters		
Charlemagne	54 N	266
Hamon	10 N	271
Othon	24 N	344
Regio		
Cassini	48 S–55 N	210–340
Terra		
Roncevaux	30 S–90 N	300–130

Titan

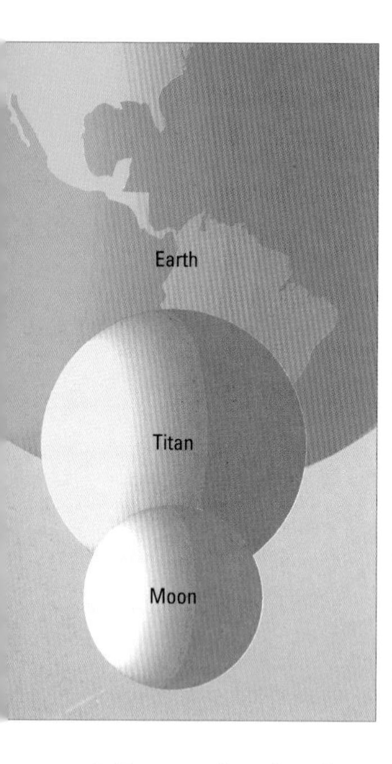

▲ ***Comparative size of Titan*** *as against that of the Earth and the Moon. Unlike our Moon, Titan has an atmosphere.*

Apart from the Galileans, Titan was the first planetary satellite to be discovered – by Christiaan Huygens, in 1656. It became of special interest in 1944, when G. P. Kuiper showed spectroscopically that it is surrounded by an extensive atmosphere; and, not unnaturally, it was scheduled to be a prime target for Voyager 1.

Various facts had been established. The globe is certainly solid, and there is presumably a rocky core; rock makes up perhaps 55 per cent of the total mass. Two models had been proposed. On the first, the rocky core was surrounded by a mantle of liquid water with some dissolved ammonia and methane; on the second, the core was surrounded by ice layers with differing crystal structures. The atmosphere was assumed to be made up of methane.

The Voyager 1 results came as a surprise. Titan's atmosphere hid the surface completely, and no detail could be seen apart from a slight difference in between the two hemispheres. This was understandable, because there is every reason to suppose that Titan's rotational axis is aligned with Saturn's, so that the seasons are very long indeed; during Voyager 1's pass, the northern hemisphere had just emerged from a 'night' lasting for 7½ Earth years. But instead of being composed of methane, the atmosphere turned out to be at least 90 per cent nitrogen, leaving methane as a minor constituent. Moreover, the atmosphere is dense, giving a ground pressure 1.5 times as great as that of the Earth's air at sea level. Haze was seen above the limb, reaching to a height of 200 kilometres (125 miles), with a more tenuous layer a further 100 kilometres (over 60 miles) above.

What lies below the orange shield? We have to admit that as yet we do not know. The surface temperature of −168 degrees C is close to the triple point of methane – that is to say, methane could exist as a solid, a liquid or gas, just as H_20 can do on Earth as ice, liquid water or water vapour. Consequently, there could be an extensive methane (or ethane) ocean, which Carl Sagan suggested might be 350 metres (over 1000 feet) deep. There is one piece of supporting evidence. Methane gas can be broken up by sunlight, and the process is irreversible, so that the atmospheric methane must be continuously replenished, presumably from Titan's surface. On the other hand, radar results indicate that a chemical ocean cannot cover the entire surface. It would be a poor reflector of radar pulses, but in fact the reflectivity over different parts of the surface is not uniform, so that there must be at least some 'land'. There could be masses of water ice together with solid carbon dioxide, silicates and tars.

Certainly Titan is an extraordinary world, unlike any other in the Solar System. We may know more in 2004, when a special probe arrives there. The Cassini mission carries a Titan lander, appropriately named in honour of Christiaan Huygens, which will – we hope – make a gentle landing at a speed of only a few kilometres per hour. Whether it will touch down on land or liquid remains to be seen. There will not be much time to find out; Huygens cannot operate for more than a few minutes at most before the orbiting Cassini probe loses contact with it.

All the ingredients for life exist on Titan, but it seems virtually certain that the very low temperature has prevented life from appearing there. There have been suggestions that in the far future, when the Sun swells out and becomes much more luminous than it is now, Titan could become habitable. Unfortunately, there is a fatal objection to this idea. Titan has a low escape velocity – only 2.4 kilometres (1.5 miles) per second, almost exactly the same as that of the Moon – and it can hold on to its dense atmosphere only because it is so cold; low temperatures slow down the movements of atoms and molecules. Raise the temperature, and Titan's atmosphere will promptly escape. In its way, Titan is as fascinating and puzzling a place as any world in the Sun's family.

◄ ***Crescent Titan*** *from Voyager 2; range 906,111 km (563,000 miles). The crescent is prolonged beyond the semi-circle, because of the presence of a dense atmosphere.*

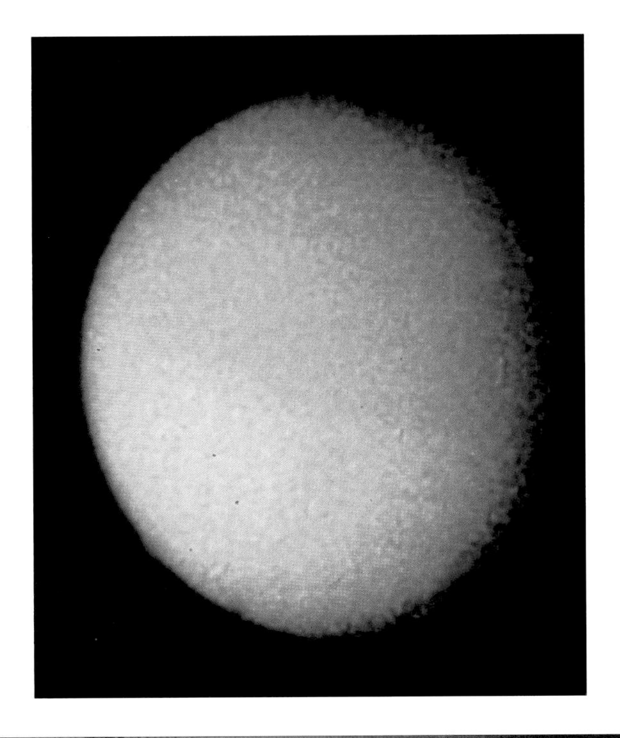

◀ ***Titan from Voyager 1***, *9 November 1980, from 7000 km (4350 miles). The southern hemisphere is lighter than the northern; the difference between the two hemispheres is quite possibly seasonal.*

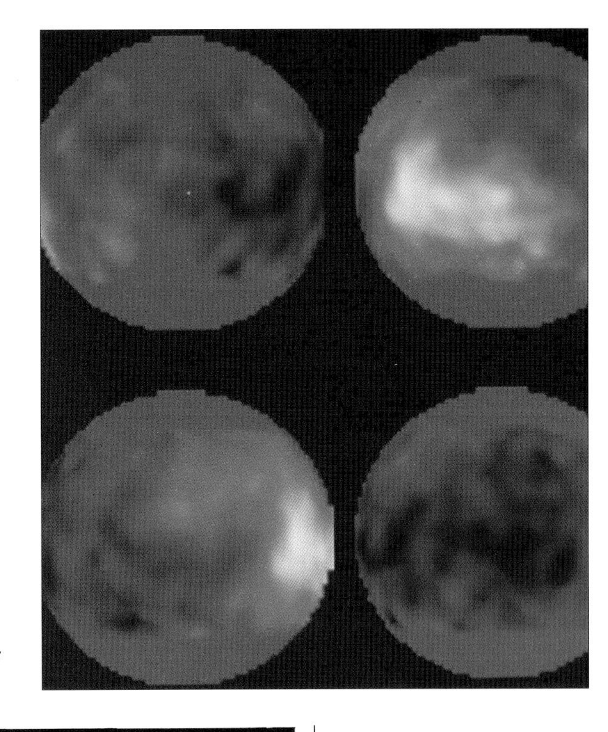

▶ ***Surface details on Titan***, *shown in infra-red from the Hubble Space Telescope. The bright area is about the size of Australia; its nature is uncertain.*

◀ ***Haze layers above Titan*** *from Voyager 1, 1 November 1980, range 22,000 km (13,700 miles). This is a false colour picture; the haze layers are shown in blue. We cannot see through the atmosphere and examine the surface; we must await the arrival of the Huygens space probe.*

Uranus

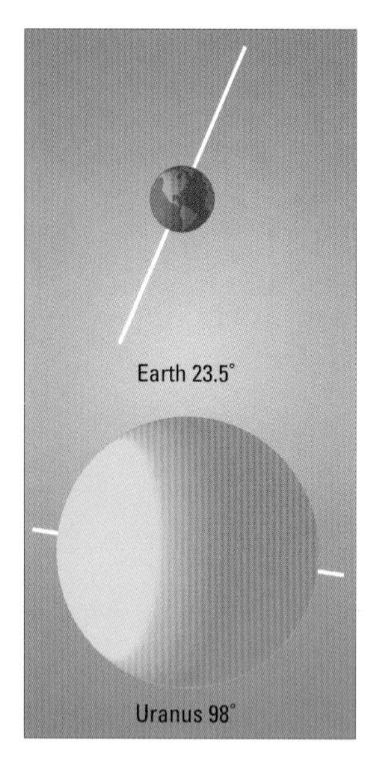

▲ ***Axial inclination of Uranus**. The planets' inclinations have a wide range; 2° (Mercury), 178° (Venus), 24° (Mars), 3° (Jupiter), 26.5° (Saturn), 98° (Uranus), 29° (Neptune) and 122° (Pluto). Uranus thus differs from all the other planets – discounting Pluto.*

Uranus, third of the giant planets, was discovered by William Herschel in 1781. Herschel was not looking for a planet; he was engaged in a systematic 'review of the heavens' with a home-made reflecting telescope when he came across an object which was certainly not a star. It showed a small disk, and it moved slowly from night to night. Herschel believed it to be a comet, but calculations soon showed it to be a planet, moving far beyond the orbit of Saturn. After some discussion it was named Uranus, after the mythological father of Saturn.

Uranus is just visible with the naked eye, and it had been seen on several occasions before Herschel's discovery. John Flamsteed, England's first Astronomer Royal, even included it in his star catalogue, and gave it a number: 34 Tauri. However, a small telescope will show its tiny, greenish disk. The equatorial diameter is 51,118 kilometres (31,770 miles), rather less than half that of Saturn; the mass is over 14 times that of the Earth, and the visible surface is made up of gas, mainly hydrogen together with a considerable amount of helium.

Irregularities in the movements of Uranus led to the tracking down of the outermost giant, Neptune, in 1846. In size and mass the two are near-twins, so that in some ways they may be considered together even though there are marked differences between them. As a pair, moreover, Uranus and Neptune are very different from Jupiter and Saturn, quite apart from being much smaller and less massive; it has been suggested that they are intermediate in type between the hydrogen- and helium-rich Jupiter and Saturn on the one hand, and the oxygen-rich metallic planets on the other. According to the so-called three-layer model of Uranus, there is a silicate core surrounded by an ocean of liquid water which is in turn overlaid by the atmosphere; on the more convincing two-layer model there is a core surrounded by a deep layer in which gases are mixed with 'ices', mainly water, ammonia and methane. Above this comes the predominantly hydrogen atmosphere, together with around 15 per cent of helium and smaller quantities of other gases. It is not easy to decide just where the 'atmosphere' ends and the real body of the planet begins; neither is it certain whether there is a sharp boundary to the core.

What is certain is that Uranus, unlike Jupiter, Saturn and Neptune, has no appreciable source of internal heat. This means that the temperature at the cloud-tops is much the same as that of Neptune, even though Neptune is so much further from the Sun.

Uranus is a slow mover; it takes 84 years to orbit the Sun. The rotation period is 17 hours 14 minutes, though, as with the other giants, the planet does not spin in the way that a rigid body would do. The most extraordinary feature is the tilt of the axis, which amounts to 98 degrees; this is more than a right angle, so that the rotation is technically retrograde. The Uranian calendar is very curious. Sometimes one of the poles is turned towards the Sun, and has a 'day' lasting for 21 Earth years, with a corresponding period of darkness at the opposite pole; sometimes the equator is presented. In total, the poles receive more heat from the Sun than does the equator. The reason for this exceptional tilt is not known. It is often thought that at an early stage in its evolution Uranus was hit by a massive body, and literally knocked sideways. This does not sound very likely, but it is hard to think of anything better. Significantly, the satellites and the ring system lie virtually in the plane of Uranus's equator.

(*En passant*, which is the 'north' pole and which is the 'south'? The International Astronomical Union has decreed that all poles above the ecliptic, i.e. the plane of the Earth's orbit, are north poles, while all poles below the ecliptic are south poles. In this case it was the south pole which was in sunlight during the Voyager 2 pass of 1986. However, the Voyager team reversed this, and referred to the sunlit pole as the north pole. Take your pick!)

No Earth-based telescope will show definite markings on the disk of Uranus. Before the Voyager mission, five satellites were known – Miranda, Ariel, Umbriel, Titania and Oberon; Voyager added ten more, all close to the planet.

On 10 March 1977, Uranus passed in front of a star, and hid or occulted it. This gave astronomers an excellent chance of measuring Uranus's apparent diameter – which is not easy by sheer visual observation, because the edge of the disk is not sharp, and the slightest error in measurement will make a tremendous difference to the final value. Therefore the phenomenon was carefully observed, with surprising results. Both before and after the actual occultation the star 'winked' several times, and this could be due only to a system of rings surrounding the planet. Subsequently D. A. Allen, at Siding Spring in Australia, managed to photograph the rings in infra-red light. However, our knowledge of Uranus and its system remained decidedly meagre, and a detailed survey had to await the fly-by of Voyager 2 in January 1986.

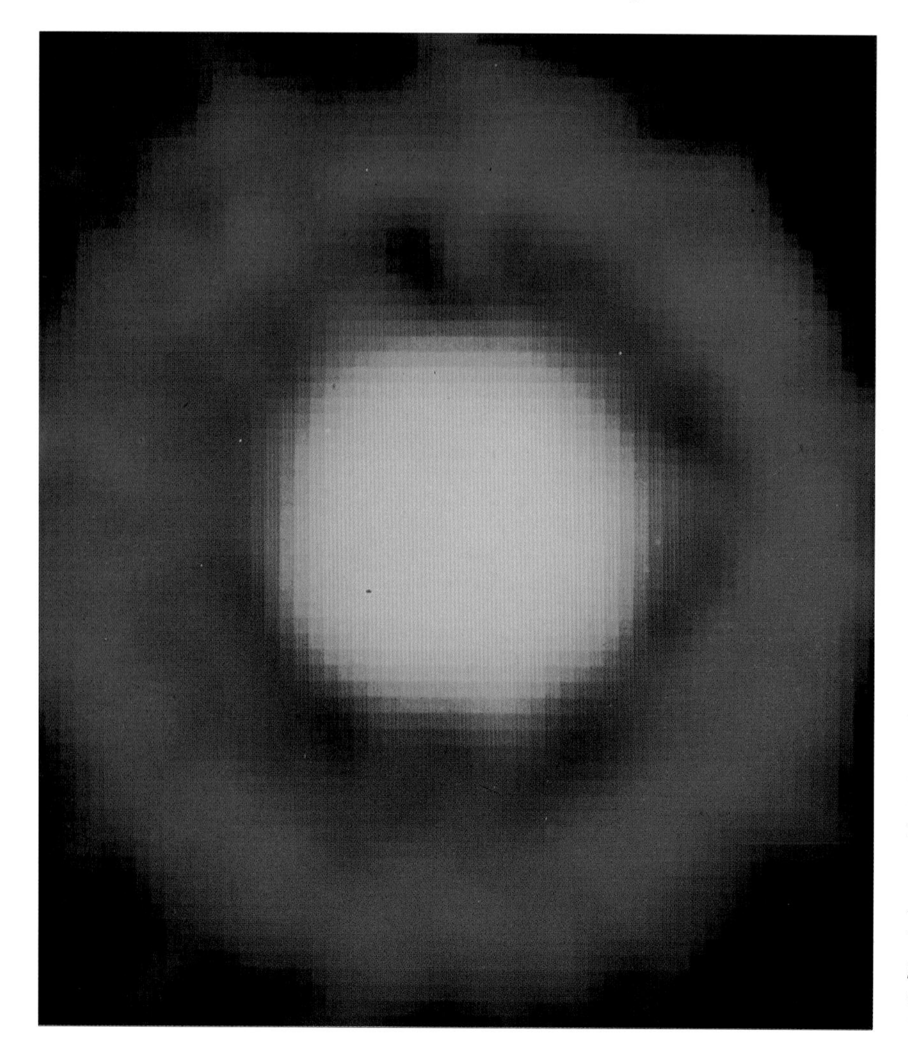

◀ ***The rings in infra-red**. This was the first really good picture of the rings; it was obtained by D. A. Allen in 1985, over 200 years since its discovery, using the Anglo-Australian Telescope at Siding Spring. Obviously the details of the rings are not shown.*

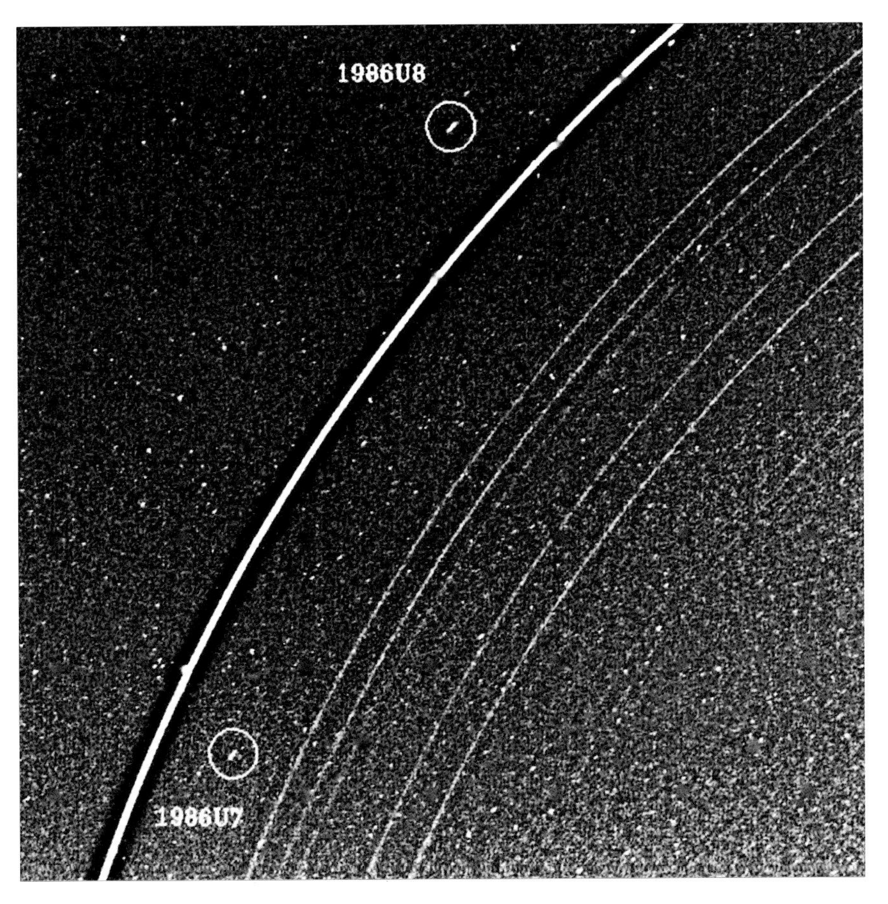

◀ ***Shepherd satellites*** *of the rings of Uranus. Two minor satellites, Cordelia and Ophelia, guard the orbit of particles in the Epsilon ring of the planet. No shepherd satellites were found for the other rings.*

▲ ***Uranus****, 24 August 1991 – a drawing I made with a magnification of 1000 on the Palomar 60-inch (152-cm) reflector. Even with this giant telescope, no surface details could be made out; all that could be seen was a greenish disk.*

▶ ***Discovery of the rings of Uranus****. On 10 March 1977 Uranus occulted the star SAO 158687, magnitude 8.9, and observations from South Africa and from the Kuiper Airborne Observatory, flying over the Indian Ocean, established the existence of a ring system – confirmed by subsequent observations.*

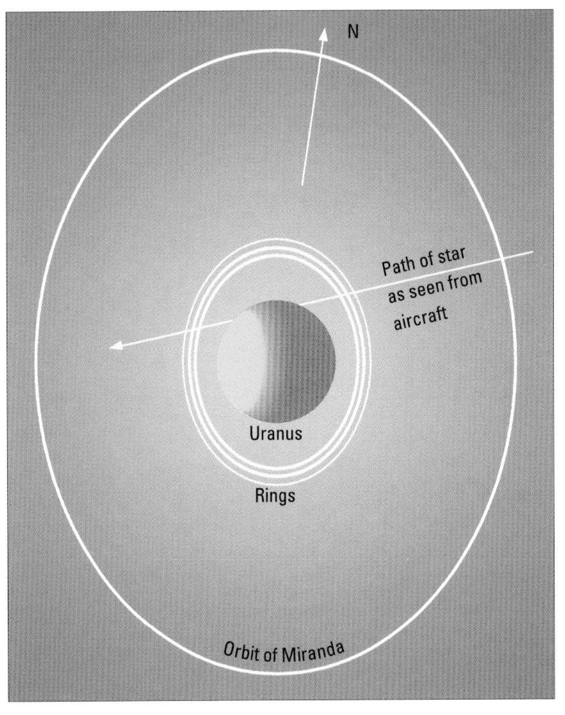

PLANETARY DATA – URANUS	
Sidereal period	30,684.9 days
Rotation period	17.2 hours
Mean orbital velocity	6.80 km/s (4.22 miles/s)
Orbital inclination	0.773°
Orbital eccentricity	0.047
Apparent diameter	max. 3.7″, min. 3.1″
Reciprocal mass, Sun = 1	22,800
Density, water = 1	1.27
Mass, Earth = 1	14.6
Volume, Earth = 1	67
Escape velocity	22.5 km/s (14.0 miles/s)
Surface gravity, Earth = 1	1.17
Mean surface temperature	−214°C
Oblateness	0.24
Albedo	0.35
Maximum magnitude	+5.6
Diameter (equatorial)	51,118 km (31,770 miles)

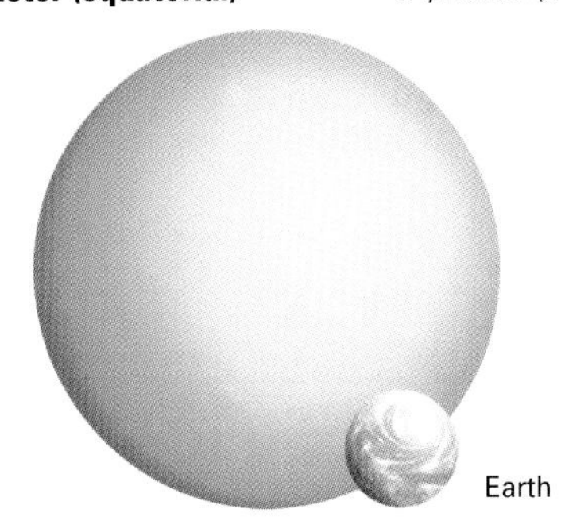

▼ ***The changing presentation of Uranus****. Sometimes a pole appears in the middle of the disk as seen from Earth; sometimes the equator is presented. Adopting the International Astronomical Union definition, it was the south pole which was in sunlight during the Voyager 2 pass in 1986.*

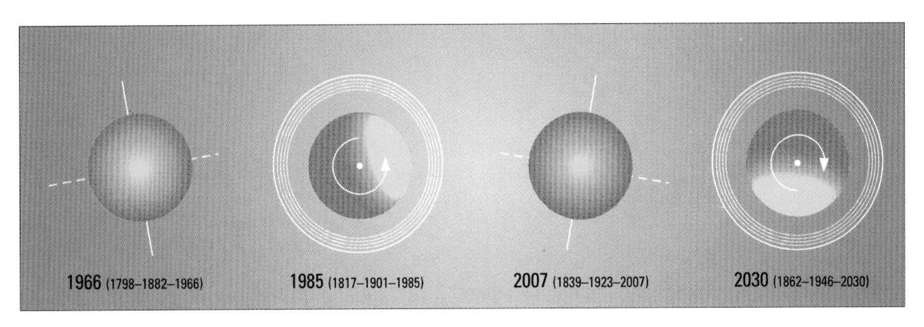

Missions to Uranus

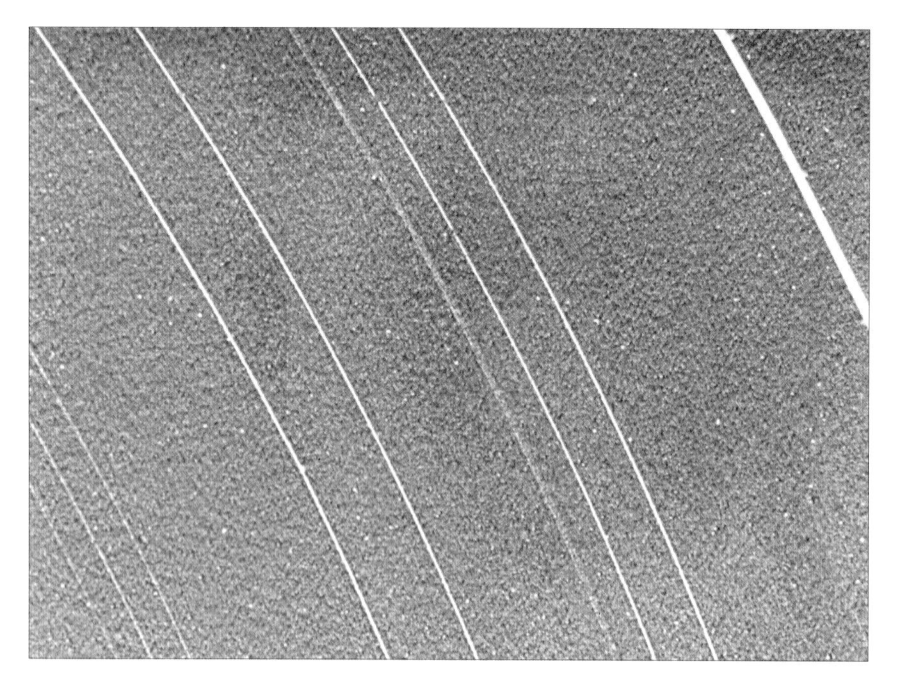

▲ ***Complete ring system of Uranus*** *(Voyager 2). In addition to the ten rings, there is a broad sheet of material closer-in than Ring 6, extending from 39,500 to 23,000 km from Uranus.*

▼ ***Uranus on 25 January 1986.*** *As Voyager 2 left and set forth on its cruise to Neptune, the spacecraft used its wide-angle camera to photograph this crescent view of Uranus. Voyager 2 was then 1 million km (about 600,000 miles) beyond Uranus. The picture, a composite of images taken in blue, green and orange, resolves features 140 km (90 miles) across.*

Only one spacecraft has so far encountered Uranus. On 24 January 1986, Voyager 2 flew past the planet at a distance of 80,000 kilometres (50,000 miles), and told us more in a few hours than we had been able to find out all through the whole of scientific history.

Several new inner satellites were discovered during Voyager's approach, but little could be seen on the disk itself; of course the planet was being seen pole-on, so that the equator lay round the rim of the disk (it was rather like aiming at the bull's eye of a dart board). Ten new satellites were discovered in all, all within the orbit of Miranda. Finally some cloud formations were made out, mainly in the range of latitudes from 20 to 45 degrees where sunlight can penetrate to slightly warmer levels, but all the clouds are very obscure, and in general Uranus appears almost featureless even from close range. Windspeeds could be measured, and, to general surprise, seem to be strongest at high levels in the atmosphere; there is a westwards air current at low latitudes, and a meandering eastwards jet-stream further from the equator.

Uranus has a decidedly bluish-green hue. This is because of the abundance of methane in the upper clouds; methane absorbs red light and allows the shorter wavelengths to be reflected. In the Uranian atmosphere it seems that water, ammonia and methane condense in that order to form thick, icy cloud-layers. Methane freezes at the lowest temperature, and so forms the top layer, above which comes the hydrogen-rich atmosphere. Aurorae were seen on the planet's night side, and on the day side ultraviolet observations showed strong emissions, producing what is termed the electroglow – the origin of which is still unclear.

As expected, Uranus is a source of radio waves, and there is a fairly strong magnetic field. The surprising fact about this is that the magnetic axis is displaced by 58.6 degrees from the axis of rotation. Moreover, the magnetic axis does not even pass through the centre of the globe; it is displaced by more than 7,500 kilometres (4,700 miles), and the polarity is opposite to that of the Earth. The reason for the tilt of the magnetic axis is not known. Initially it was believed to be connected in some way with the 98-degree tilt of the axis of rotation, but since Neptune has since been found to share the same peculiarity we must think again. The windsock-shaped magnetosphere is so extensive that all the members of the satellite family are engulfed by it.

Voyager was able to make a detailed survey of the ring-system. Ten individual rings have been identified, plus a broad sheet of material closer-in than the main system; the nomenclature is frankly chaotic, and one can only hope that in the future it will be revised. All the rings are very thin, with remarkably sharp borders; their thickness cannot be more than a few tens of metres (40 to 100 feet), and they are probably made up of boulders a metre or two (3 to 7 feet) in diameter. There are not many smaller centimetre-sized objects.

All the rings of Uranus are not alike. The outer or Epsilon ring is not symmetrical. It is variable in width. The part of it closest to Uranus is around 20 kilometres (12.5 miles) wide, while the part furthest from the planet has a maximum width of around 100 kilometres (about 60 miles). All the other rings are much narrower, and some of them show definite structure. The satellites Cordelia and Ophelia act as 'shepherds' to the Epsilon ring; a close search was made for shepherd satellites of the other rings, but without success. The rings of Uranus are as black as coal-dust, and are totally unlike the magnificently colourful icy rings that surround Saturn.

◀ ***A true-colour photograph*** *on 17 January 1986 by the narrow-angle camera of Voyager 2, 9.1 million km (5.7 million miles) from the planet, seven days before closest approach. The blue-green colour is due to absorption of red light by methane gas in Uranus's deep, cold and remarkably clear atmosphere.*

THE RINGS OF URANUS

Ring	Distance from Uranus, km	Width, km
6	41,800	1–3
5	42,200	2–3
4	42,600	2
(Alpha)	44,700	4–11
(Beta)	45,700	7–11
(Eta)	47,200	2
(Gamma)	47,600	1–4
(Delta)	48,300	3–9
(Lambda)	50,000	1–2
(Epsilon)	51,150	20–96

The broad sheet of material closer-in than Ring 6, extending from 39,500 km to 23,000 km from Uranus, is sometimes regarded as a ring.

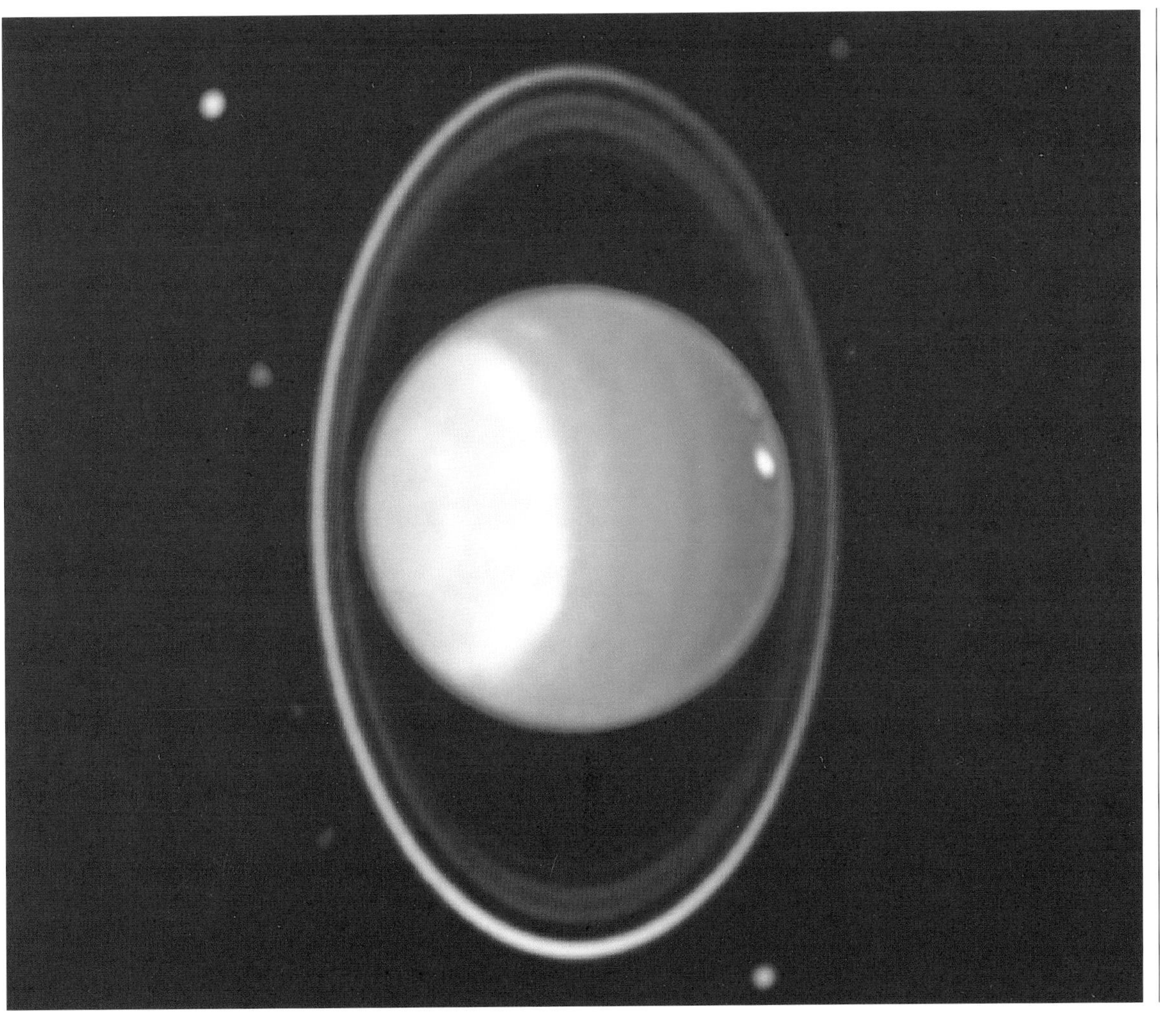

◀ ***Bright clouds on Uranus.*** *This false-colour image was generated by Erich Karkoschka using data taken on 8 August 1998 with Hubble's Near Infrared Camera and Multi-Object Spectrometer. The orange clouds near the prominent bright band move at over 500 km/h (300 mph). Colours indicate altitude; green and blue show that the atmosphere is clear, so that sunlight can penetrate the atmosphere deeply. In yellow and grey regions the sunlight reflects from a higher haze or cloud layer. Orange and red colours indicate high clouds. [Reproduced by kind permission of Erich Karkoschka (University of Arizona) and NASA.]*

Satellites of Uranus

► ***Umbriel***. *The surface of Umbriel is much darker and more subdued than that of Ariel. The largest crater, Skynd, is 110 km (68 miles) in diameter, with a bright central peak. Wunda, diameter 140 km (87 miles), lies near Umbriel's equator; its nature is uncertain, but it is the most reflective feature on the satellite. (Remember that owing to the pole-on view, the equator lies round the limb in this picture.)*

Uranus, like all the giant planets, has an extensive satellite family. The two outer members, Titania and Oberon, were discovered by William Herschel in 1787. Herschel also announced the discovery of four more satellites, but three of these are non-existent and must have been faint stars; the fourth may have been Umbriel, but there is considerable doubt. Umbriel and Ariel were found in 1851 by the English amateur William Lassell. All the four first-discovered satellites are between 1100 and 1600 kilometres (700 to 1000 miles) in diameter, so that they are comparable with the medium-sized icy satellites in Saturn's system, but their greater distance makes them rather elusive telescopic objects.

During the 1890s W. H. Pickering (discoverer of Phoebe, the outermost satellite of Saturn) searched for further members of the system, but without success. The fifth moon, Miranda, was discovered by G. P. Kuiper in 1948; it is much fainter and closer-in than the original four. Voyager 2 found another ten satellites, all moving inside

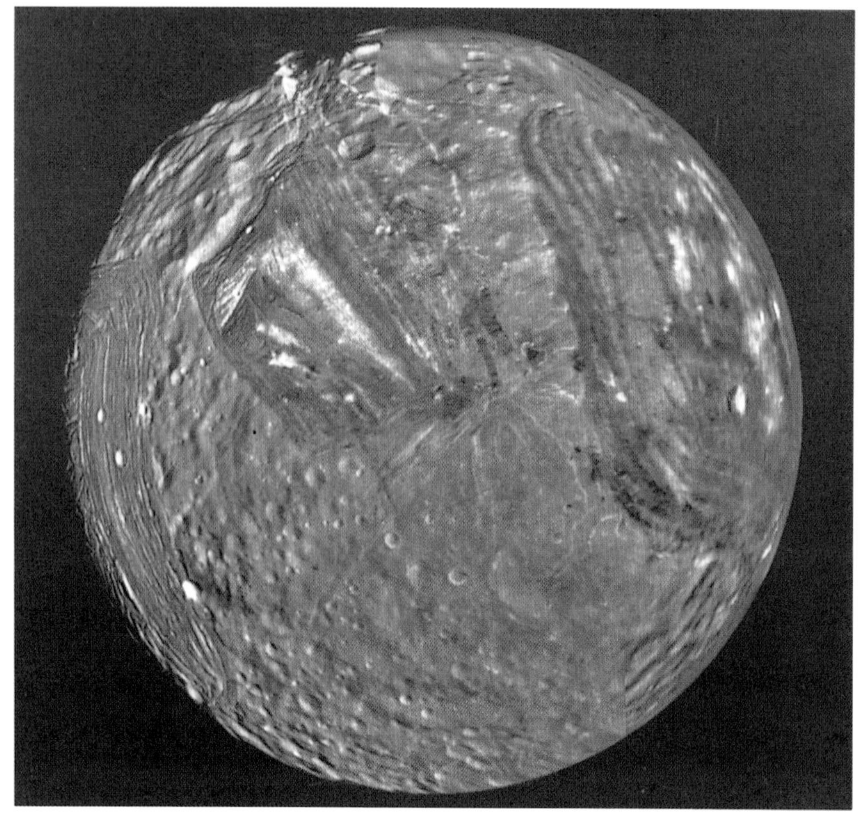

◄ ***Miranda***. *The innermost of Uranus's large satellites is seen at close range in this image from Voyager 2, taken from a distance of 35,000 km (22,000 miles). Scarps, ice-cliffs and craters are visible.*

▲ ***Oberon***. *Voyager 2 took this picture from around 660,000 km (413,000 miles), with a resolution of 11 km (7 miles). Note the high peak, which is about 6 km high (3.8 miles), projecting from the lower left limb.*

SATELLITES OF URANUS

Name	Distance from Uranus, km	Orbital period, days	Orbital inclination, °	Orbital eccentricity	Diameter, km	Density, water = 1	Escape velocity, km/s	Magnitude
Cordelia	49,471	0.330	0.14	0.0005	26	?	very low	24.2
Ophelia	53,796	0.372	0.09	0.0101	32	?	very low	23.9
Bianca	59,173	0.433	0.16	0.0009	42	?	very low	23.1
Cressida	51,777	0.463	0.04	0.0001	62	?	very low	22.3
Desdemona	62,676	0.475	0.16	0.0002	54	?	very low	22.5
Juliet	64,352	0.493	0.04	0.0002	84	?	very low	21.7
Portia	66,085	0.513	0.09	0.0002	106	?	very low	21.1
Rosalind	69,941	0.558	0.08	0.0006	54	?	very low	22.5
Belinda	75.258	0.622	0.03	0.0001	66	?	very low	22.1
1986 U10	75.258	0.62	low	low	40	?	very low	23
Puck	86,000	0.762	0.31	0.0001	154	?	very low	20.4
Miranda	129,400	1.414	4.22	0.0027	481 × 466 × 466	1.3	0.5	16.3
Ariel	191,000	2.520	0.31	0.0034	1158	1.6	1.2	14.2
Umbriel	256,300	4.144	0.36	0.0050	1169	1.4	1.2	14.8
Titania	435,000	8.706	0.014	0.0022	1578	1.6	1.6	13.7
Oberon	583,500	13.463	0.10	0.0008	1523	1.5	1.5	13.9
Caliban	7,775,000	654	146	0.2	60	?	very low	22.3
Sycorax	8,846,000	795	154	0.34	120	?	very low	20.7
1999 U1	10,000,000	950	–	–	40	?	very low	24
1999 U2	25,000,000	3773	–	–	40	?	very low	24

Miranda's orbit. The only newcomer to exceed 100 kilometres (60 miles) in diameter is Puck, which was imaged from a range of 500,000 kilometres (312,000 miles) and found to be dark and roughly spherical; three craters were seen, and given the rather bizarre names of Bogle, Lob and Butz.

Incidentally, it may be asked why the names of the Uranian moons come from literature, not mythology. The names Titania and Oberon were suggested by Sir John Herschel, and the later satellites were also given names coming either from Shakespeare or from Pope's poem *The Rape of the Lock*. This is certainly a departure from the norm, and arguably an undesirable one, but the names are now well established, and all have been ratified by the International Astronomical Union.

The nine innermost satellites are presumably icy, but nothing is known about their physical make-up. Cordelia and Ophelia act as shepherds to the Epilson ring. A careful search was made for similar shepherds inside the main part of the ring system, but without success; if any such shepherds exist, they must be very small indeed.

The four largest members of the family are not alike. In general they are denser than the icy satellites of Saturn, and so must contain more rock and less ice; the proportion of rocky material is probably between 50 and 55 per cent. All have icy surfaces, but there are marked differences between them. Umbriel is the darkest of the four, with a rather subdued surface and one bright feature, Wunda, which lies almost on the equator – so that with the pole-on view it appears near the edge of the disk; it may be a crater, but its nature is uncertain. Umbriel is fainter than the other major satellites, and in pre-Voyager days was assumed to be the smallest, though in fact it is marginally larger than Ariel. Oberon is heavily cratered, and some of the craters such as Hamlet, Othello and Falstaff have dark floors, due perhaps to a mixture of ice and carbonaceous material erupted from the interior; on the limb, near the crater Macbeth, there is a high mountain. Titania is distinguished by high ice-cliffs, and there are broad, branching and interconnected valleys, so that there seems to have been more past internal activity than on Oberon. Ariel also has very wide, branching valleys which look as though they have been cut by liquid – though, needless to say, all the satellites are far too lightweight to retain any trace of atmosphere.

Miranda has an amazingly varied surface. There are regions of totally different types – some cratered, some relatively smooth; there are ice-cliffs up to 20 kilometres (over 12 miles) high, and large trapezoid-shaped areas or 'coronae' which were initially nicknamed 'race-tracks'. The three main coronae (Arden, Elsinore and Inverness) cover much of the hemisphere which was imaged by Voyager 2. It has been suggested that during its evolution Miranda has been broken up by collision, perhaps several times, and that the fragments have subsequently re-formed. This may or may not be true, but certainly it would go some way to explaining the jumble of surface features now seen.

Two new outer satellites were discovered in 1998, and have been named Caliban and Sycorax. Both are small and reddish, so that they are presumably captured bodies. Another two small outer satellites were detected in 1999, bringing the grand total to 20. As yet (2000) little is known about them.

▲ ***Titania***. *Voyager 2 took this picture 24 January 1986 from 483,000 km (302,000 miles). It shows details down to 9 km (5.6 miles). The surface is generally cratered, with ice-cliffs and trench-like features; there is considerable evidence of past tectonic activity.*

▼ ***Satellite orbits***. *Orbits of the five larger satellites. Voyager discovered ten more small moons, within the orbit of Miranda.*

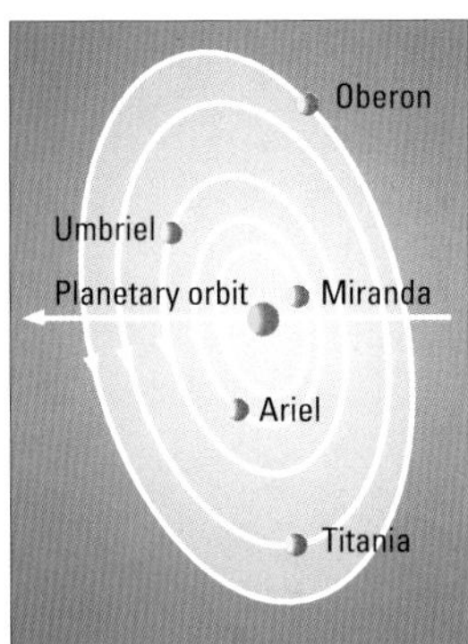

◄ ***Puck***. *Discovered on 30 December 1985 and imaged on 24 January 1986 from a range of 500,000 km (300,000 miles). Resolution is 10 km (6 miles). Three craters were recorded: Bogle, Lob and Butz. Puck is roughly spherical, with a darkish surface.*

► ***Ariel***. *From 169,000 km (106,000 miles), the resolution is 3.2 km (2 miles). The surface is cratered, with fault-scarps and graben, suggesting considerable past tectonic activity, and there is evidence of erosion.*

Maps of the Satellites of Uranus

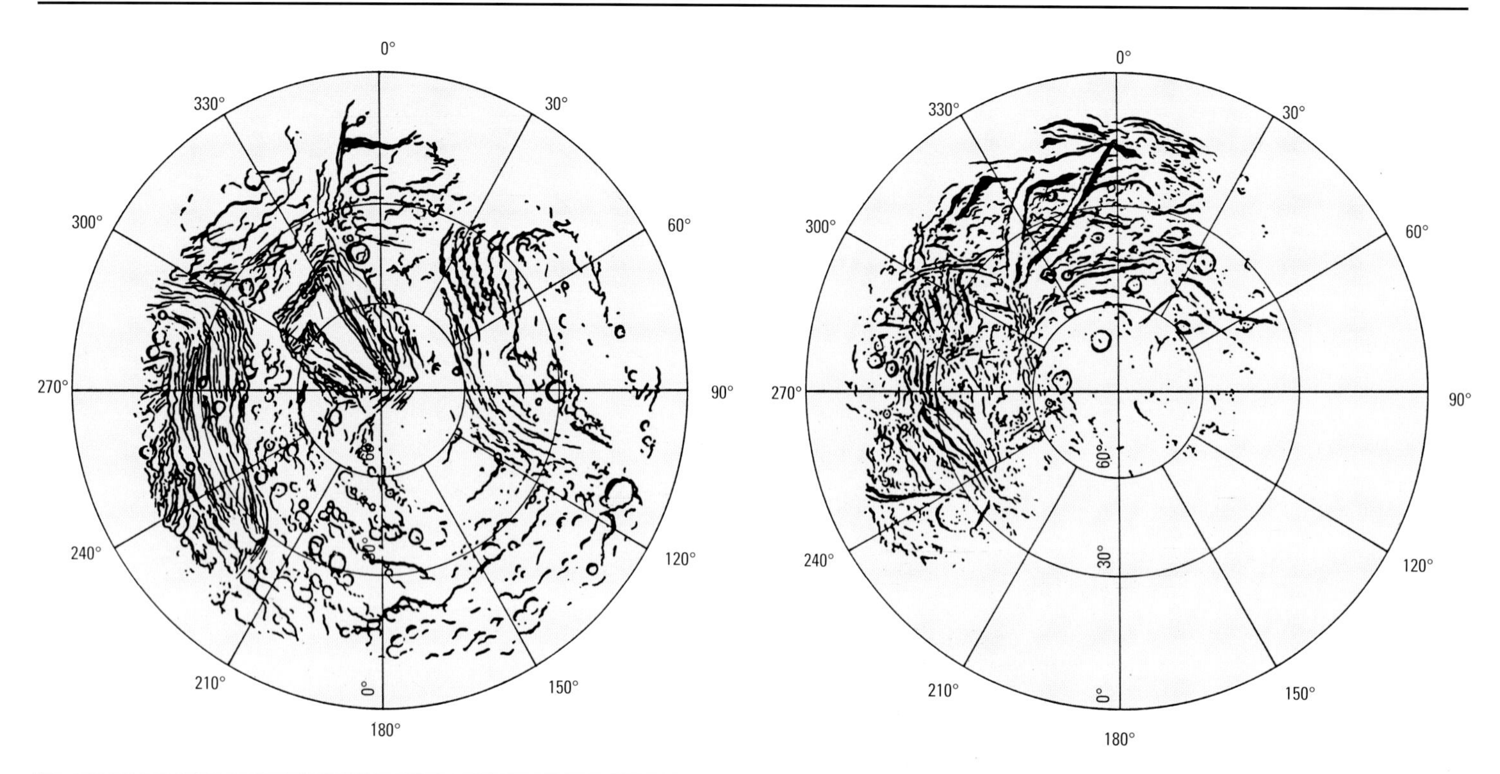

MIRANDA	Lat. °S	Long. °E
Arden Corona	10–60	30–120
Dunsinane Regio	20–75	345–65
Elsinore Corona	10–42	215–305
Ferdinand	36	208
Gonzalo	13	75
Inverness Corona	38–90	0–350
Mantua Regio	10–90	75–300
Prospero	35	323
Sicilia Regio	10–50	295–340
Trinculo	67	168

ARIEL	Lat. °S	Long. °E
Ataksak	53	225
Brownie Chasma	5–21	325–357
Domovoy	72	339
Kachina Chasma	24–40	210–280
Kewpie Chasma	15–42	307–335
Korrigan Chasma	25–46	328–353
Kra Chasma	32–36	355–002
Laica	22	44
Mab	39	353
Sylph Chasma	45–50	328–015
Yangoor	68	260

When Voyager 2 flew by Uranus in January 1986, it allowed detailed maps of the planet's satellites to be made for the first time.

Miranda *The landscape is incredibly varied, and something of a jumble. The main features are the three coronae: Elsinore, Arden ('the Race-Track') and Inverness ('the Chevron'). There are few large craters. Voyager 2 passed Miranda at only 3000 kilometres (1880 miles), and the pictures sent back gave a resolution down to 600 metres (2000 feet), so that the views of Miranda are more detailed than those of any other world except those upon which spacecraft have actually landed.*

Ariel *Ariel was imaged from 130,000 kilometres (81,000 miles), giving a resolution down to 2.4 kilometres (1.5 miles). There are many craters, some with bright rims and ray-systems, but the main features are the broad, branching, smooth-floored valleys such as Korrigan Chasma and Kewpie Chasma. There are also grooves, sinuous scarps, and faults. Ariel's surface seems to be younger than those of the other major satellites.*

Umbriel *The most detailed picture of the Umbriel's darkish, rather subdued surface was taken from a range of 537,000 kilometres (335,000 miles), giving a resolution of about 10 kilometres (6 miles). The most prominent crater is Skynd, on the terminator; it is 110 kilometres (68 miles) in diameter, with a bright central or near-central peak. The other bright feature, Wunda, is much more puzzling. It seems to be a ring about 140 kilometres (87 miles) across, but is so badly placed that its form cannot be made out, though it is probably a crater.*

Titania *Like Ariel, Titania seems to have experienced considerable tectonic activity in the past. On the best Voyager view, obtained from a range of 369,000 kilometres (230,000 miles), many craters are shown, together with linear troughs and fault valleys. The 200-kilometre (125-mile) crater Ursula is cut by a fault valley over 100 kilometres (62 miles) wide; the largest formation, Gertrude, may be more in the nature of a basin than a true crater. There are ice-cliffs and valleys such as Messina Chasma, which is 1500 kilometres (940 miles) long.*

Oberon *Oberon was imaged from 660,000 kilometres (412,000 miles), giving a resolution down to 12 kilometres (7.5 miles). There are many craters, some of which, such as Hamlet, Othello and Falstaff, have dark floors. One interesting feature is a lofty mountain, about 6 kilometres (3.75 miles) high, shown on the best Voyager picture exactly at the edge of the disk, near Macbeth, so that it protrudes from the limb (otherwise it might not be identifiable). Whether or not it is exceptional, we do not know. Only new observations will tell.*

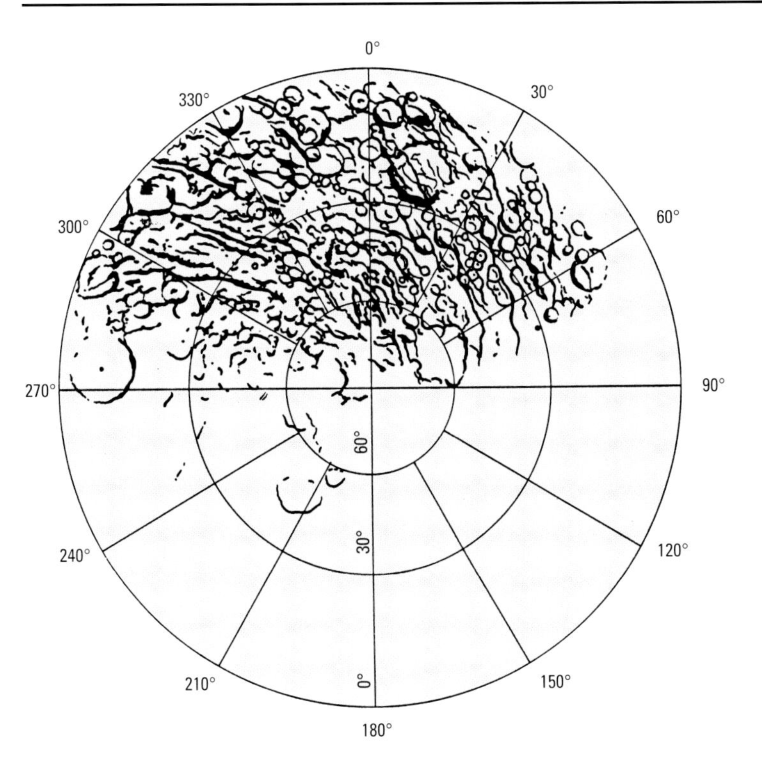

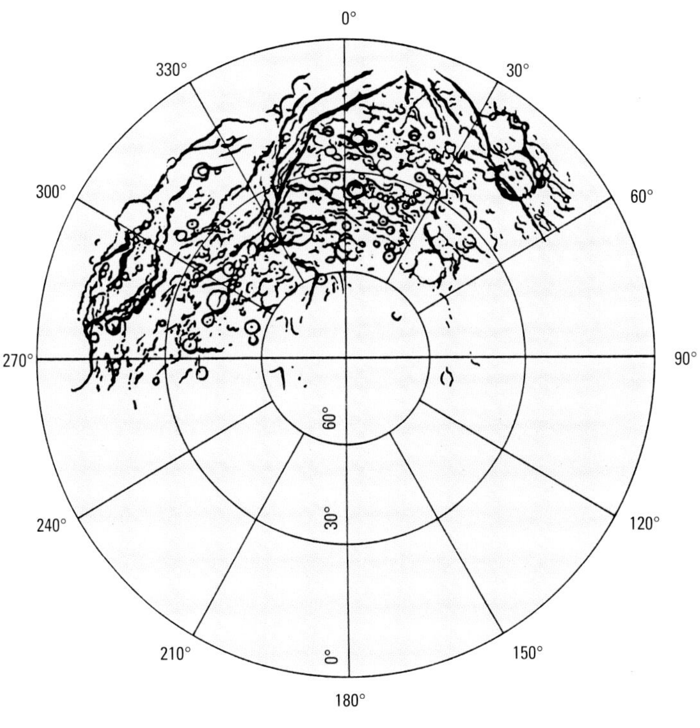

UMBRIEL

	Lat. °S	Long. °E
Kanaloa	11	351
Malingee	22	13
Setibos	31	350
Skynd	1 (N)	335
Vuver	2	311
Wunda	6	274
Zlyden	24	330

TITANIA

	Lat. °S	Long. °E
Belmont Chasma	4–25	25–35
Gertrude	15	288
Lucetta	9	277
Messin Chasma	8–28	325–005
Rousillon Rupes	7–25	17–38
Ursula	13	44
Valeria	34	40

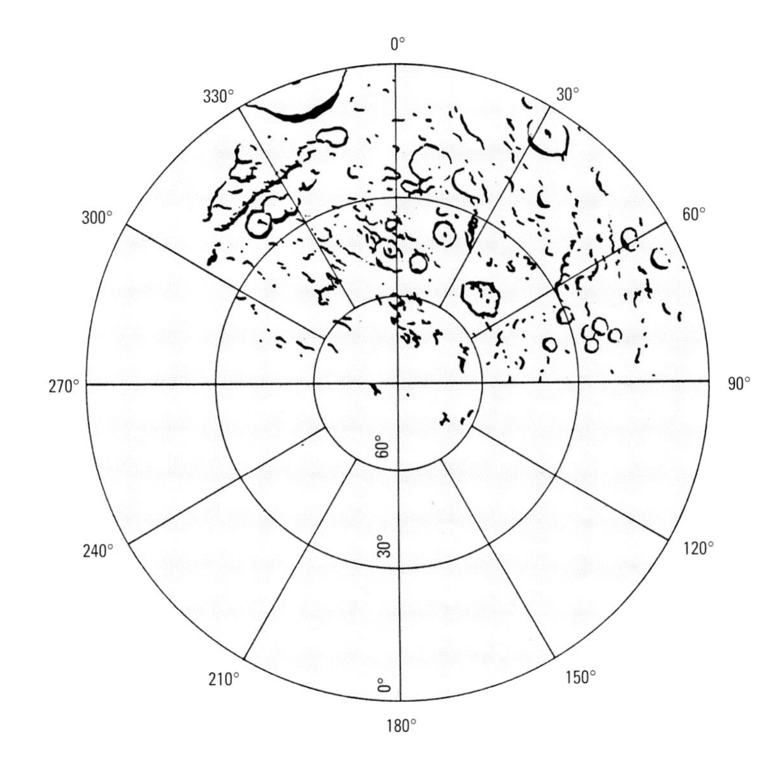

OBERON

	Lat. °S	Long. °E
Coriolanus	11	345
Falstaff	22	19
Hamlet	46	45
Lear	5	31
Macbeth	59	112
Mommur Chasma	16–20	240–343
Othello	65	44
Romeo	28	88

Neptune

▲ ***Neptune's blue-green atmosphere*** *seen by Voyager at a distance of 16 million km (10 million miles). The Great Dark Spot at the centre is about 13,000 × 6600 km (8000 × 4100 miles). 'Cirrus-type' clouds are higher.*

Some years after Uranus was discovered, it became clear that it was not moving as had been expected. The logical cause was perturbation by an unknown planet at a greater distance from the Sun. Two mathematicians, U. J. J. Le Verrier in France and J. C. Adams in England, independently calculated the position of the new planet, and in 1846 Johann Galle and Heinrich D'Arrest, at the Berlin Observatory, identified it. After some discussion it was named Neptune, after the mythological sea-god.

Neptune is too faint to be seen with the naked eye; its magnitude is 7.7, so that it is within binocular range. Telescopes show it as a small, bluish disk. In size it is almost identical with Uranus, but it is appreciably more massive. The orbital period is almost 165 years. Like all the giants, it is a quick spinner, with an axial rotation period of 16 hours 7 minutes. Neptune does not share Uranus's unusual inclination; the axis is tilted by only 28 degrees 48 minutes to the perpendicular.

Though Uranus and Neptune are near-twins, they are not identical. Unlike Uranus, Neptune has a strong source of internal heat, so that the temperature at the cloud-tops is almost the same as that of Uranus, even though Neptune is over 1600 million kilometres (1000 million miles) further from the Sun. In composition Neptune is presumably dominated by planetary 'ices', such as water-ice; there may be a silicate core surrounded by the mantle, but it is

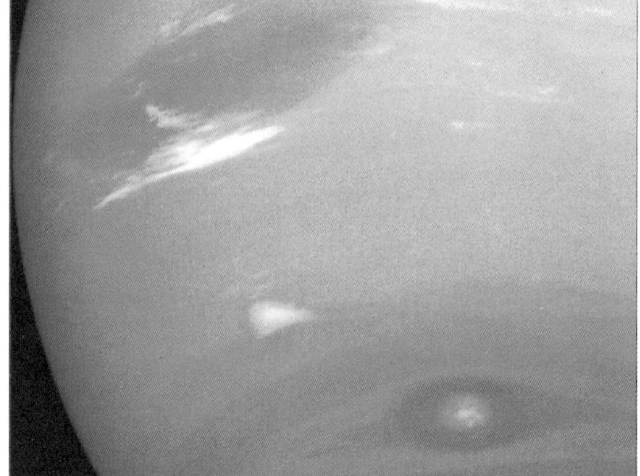

▶ ***Three prominent features*** *reconstructed from two Voyager images. At the north (top) is the Great Dark Spot. To the south is the 'Scooter' which rotates around the globe faster than other features. Still further south is the feature called 'Dark Spot 2'. Each moves eastwards at a different velocity.*

▼ ***Neptune's clouds*** *two hours before Voyager 2's closest approach. In this view, reminiscent of Earth from an airliner, fluffy white clouds are seen high above Neptune. Cloud shadows have not been seen on any other planet.*

PLANETARY DATA – NEPTUNE	
Sidereal period	60,190.3 days
Rotation period	16h 7m
Mean orbital velocity	5.43 km/s (3.37 miles/s)
Orbital inclination	1° 45′ 19.8″
Orbital eccentricity	0.009
Apparent diameter	max. 2.2″, min. 2.0″
Reciprocal mass, Sun = 1	19,300
Density, water = 1	1.77
Mass, Earth = 1	17.2
Volume, Earth = 1	57
Escape velocity	23.9 km/s (14.8 miles/s)
Surface gravity, Earth = 1	1.2
Mean surface temperature	−220°C
Oblateness	0.02
Albedo	0.35
Maximum magnitude	+7.7
Diameter	50,538 km (31,410 miles)

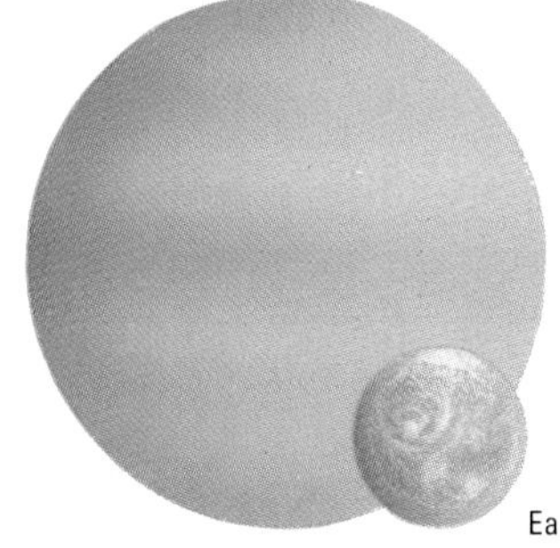

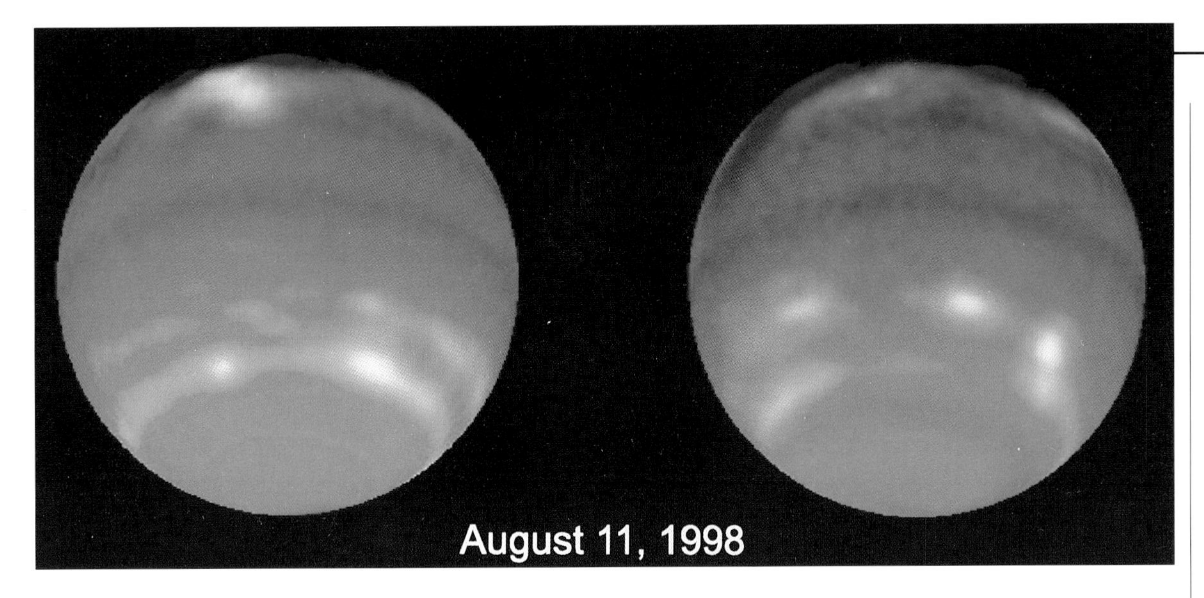

◀ ***Views of Neptune's weather**, on opposite hemispheres. Taken on 11 August 1998 with Hubble's Wide Field Planetary Camera 2, these composite images show Neptune's blustery weather. The predominant blue colour is a result of the absorption of red and infra-red light by its methane atmosphere. Clouds elevated above most of the methane absorption appear white, while the very highest clouds tend to be yellow-red. Neptune's powerful equatorial jet – where winds blow at nearly 1500 km/h (900 mph) – is centred on the dark blue belt just south of Neptune's equator. Further south, the green belt indicates a region where the atmosphere absorbs blue light.*

quite likely that the core is not sharply differentiated from the ice components.

Almost all our detailed knowledge of Neptune has been provided by one spacecraft, Voyager 2, which flew past the planet on 25 August 1989 – at 4425 million kilometres (2750 million miles) from the Earth. Voyager passed over the darkened north pole at a relative velocity of just over 17 kilometres (10 miles) per second; at that time the southern hemisphere was having its long 'summer'.

Well before Voyager closed in, the images showed that Neptune is a far more dynamic world than Uranus. The most conspicuous feature on the blue surface was a huge oval, the Great Dark Spot, at latitude 8 degrees 28 minutes south; it had a rotation period of over 18 hours, so that it drifted westwards relative to the nearby clouds at 30 metres (100 feet) per second. It rotated in an anti-clockwise direction, and showed more or less predictable changes in shape and orientation. Above it lay wispy clouds made of methane crystals ('methane cirrus') and between these and the main cloud deck there was a 50-kilometre (31-mile) clear zone. Further south (latitude 42 degrees S) was a smaller, very variable feature with a bright centre, which had a shorter rotation period and was nicknamed the 'Scooter'; still further south (latitude 55 degrees S) was a second dark spot.

Neptune is a windy place. At the equator the winds blow westwards (retrograde) at up to 450 metres (1500 feet) per second; further south the winds slacken, and beyond latitude 50 degrees they become eastwards, reaching 300 metres (1000 feet) per second but decreasing once more near the south pole. Temperature measurements show that there are cold mid-latitude regions with a warmer equator and pole.

The upper atmosphere is made up chiefly of hydrogen (85 per cent), with a considerable amount of helium and a little methane. There are various cloud layers, above which lies the general methane haze.

Neptune is a source of radio waves, which was only to be expected, but the magnetic field proved to be very surprising. The magnetic axis makes an angle of 47 degrees with the axis of rotation, so that in this respect Neptune resembles Uranus more than Jupiter or Saturn; here also the magnetic axis does not pass through the centre of the globe, but is displaced by 10,000 kilometres (6200 miles). The magnetic field itself is weaker than those of the other giants. Aurorae were confirmed, though they are of course brightest near the magnetic poles.

Voyager confirmed that Neptune has a ring system, though it is much less evident than those of the other giants. Altogether there seem to be five separate rings, plus the so-called 'plateau', a diffuse band of material made up of very small particles. There may also be 'dust' extending down almost to the cloud-tops.

The rings have been named in honour of astronomers who were involved in Neptune's discovery. The Adams ring is the most pronounced and is 'clumpy' with three brighter arcs which may be due to the gravitational pull of Galatea, one of the newly-discovered small satellites. The ring at 62,000 kilometres (38,750 miles) is close to Galatea's orbit. The rings are dark and ghostly, and the fainter sections were only just above the threshold of visibility from Voyager.

Details on Neptune are now within the range of the Hubble Space Telescope – and, to general surprise, the images taken in August 1996 (and subsequently) showed no trace of the Great Dark Spot. Smaller features were seen, and there seems no escape from the conclusion that the Great Dark Spot has disappeared. Other spots have been recorded, and it is becoming clear that Neptune is much more variable than had been expected.

◀ ***Neptune's two main rings**, about 53,000 km (33,000 miles) and 63,000 km (39,000 miles) from the centre of the planet, were backlit by the Sun as Voyager 2 swept past. Neptune's rings appear bright as microscopic ring particles scatter sunlight towards the camera. Particle-size distribution in Neptune's rings is quite different from that in Uranus's rings.*

THE RINGS OF NEPTUNE

Name	Distance from centre of Neptune, km	Width, km
Galle	41,900	50
Le Verrier	53,200	50
'Plateau'	53,200–59,100	4000
—	62,000	30
Adams	62,900	50

Satellites of Neptune

▶ ***Neptune's small moon*** *Proteus was discovered in June 1989, early enough for Voyager personnel to target it. The image was taken on 25 August 1989 from a distance of 146,000 km (90,500 miles). Seen here, about half-illuminated, it has an average diameter of over 400 km (250 miles), slightly smaller than Uranus's moon Miranda. It is dark (reflecting only 6 per cent of light it receives) and spectrally grey. Hints of craters and grooves can be discerned.*

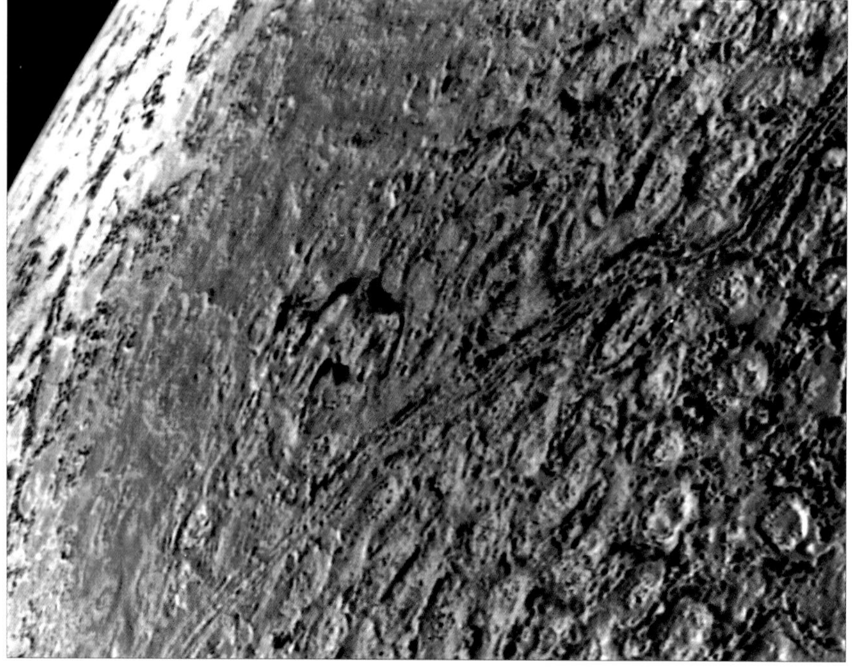

▲ ***A detailed view of Triton*** *taken by Voyager 2 on 25 August 1989 from 40,000 km (25,000 miles). The frame is about 220 km (140 miles) across and shows details as small as 750 m (0.5 miles). Most of the area is covered by roughly circular depressions separated by rugged ridges. This terrain, which covers Triton's northern hemisphere, is unlike anything seen elsewhere in the Solar System.*

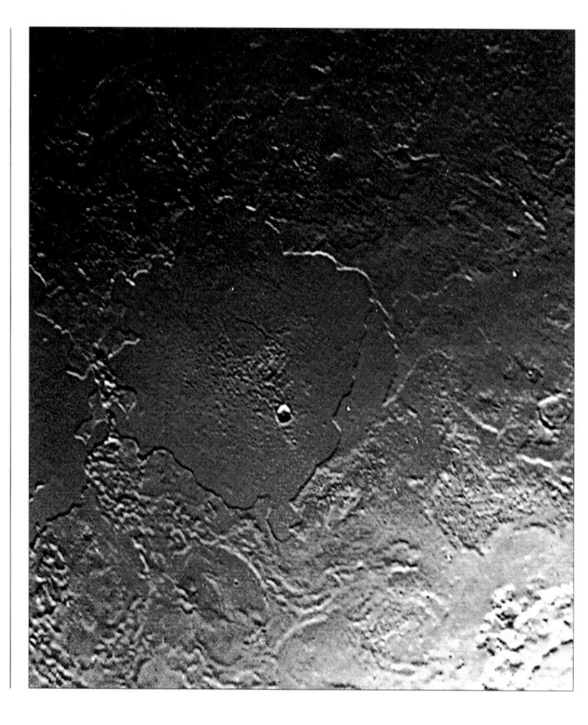

▶ ***Triton*** *on 25 August 1989. This view is about 500 kilometres (300 miles) across. It encompasses two depressions that have been extensively modified by flooding, melting, faulting, and collapsing.*

Two satellites of Neptune were known before the Voyager fly-by: Triton and Nereid. Each was exceptional in its own way. Triton, discovered by Lassell a few weeks after Neptune itself had been found, is large by satellite standards but has retrograde motion; that is to say, it moves round Neptune in a sense opposite to that in which Neptune rotates. This makes it unique among major satellites, since all other attendants with retrograde motion (the four outermost members of Jupiter's system, and Phoebe in Saturn's) are asteroidal. Nereid is only 240 kilometres (140 miles) across; and though it moves in the direct sense its eccentric orbit is more like that of a comet than a satellite; the distance from Neptune varies by over 8 million kilometres (5 million miles), and the revolution period is only one week short of an Earth year, so that obviously the axial rotation is not synchronous.

Voyager discovered six new inner satellites, one of which (Proteus) is actually larger than Nereid, but is virtually unobservable from Earth because of its closeness to Neptune. Proteus and one of the other new discoveries, Larissa, were imaged from Voyager, and both turned out to be dark and cratered; Proteus shows a major depression in its southern hemisphere, with a rugged floor. No doubt the other inner satellites are of the same type. Nereid was in the wrong part of its orbit during the Voyager pass, and only one very poor image was obtained, but Triton more than made up for this omission.

Estimates of the diameter of Triton had been discordant, and at one time it was even thought to be larger than Mercury, with an atmosphere dense enough to support clouds similar to those of Titan. Voyager proved otherwise. Triton is smaller than the Moon, and is well over twice as dense as water, so that its globe is made up of more rock than ice. The surface temperature is around −236°C, so that Triton is the chilliest world so far encountered by a spacecraft.

The escape velocity is 1.4 kilometres (0.9 miles) per second, and this is enough for Triton to retain a very tenuous atmosphere, made up chiefly of nitrogen with an appreciable amount of methane. There is considerable haze, seen by Voyager above the limb and which extends to at least 6 kilometres (3.7 miles) above the surface; it is probably composed of tiny particles of methane or nitrogen ice. Winds in the atmosphere average about 5 metres (16 feet) per second in a westwards direction.

The surface of Triton is very varied, but there is a general coating of water ice, overlaid by nitrogen and methane ices. There are very few craters, but many flows which are probably due to ammonia-water fluids; surface relief is very muted, and certainly there are no mountains. The most striking feature is the southern polar cap, which is pink and makes Triton look quite different from any other planet or satellite. The pink colour must be due to nitrogen ice and snow. The long Tritonian season means that the south pole has been in constant sunlight for over a century now, and along the borders of the cap there are signs of evaporation. North of the cap there is an 'edge' which looks darker and redder, perhaps because of the action of ultra-violet light upon methane, and running across this region is a slightly bluish layer, caused by the scattering of incoming light by tiny crystals of methane.

The surface imaged from Voyager 2 is divided into three main regions: Uhlanga Regio (polar), Monad Regio (eastern equatorial) and Bubembe Regio (western equatorial). It is in Uhlanga that we find the remarkable nitrogen geysers. According to the most plausible explanation, there is a layer of liquid nitrogen 20 or 30 metres (65 to 100 feet) below the surface. If for any reason this liquid

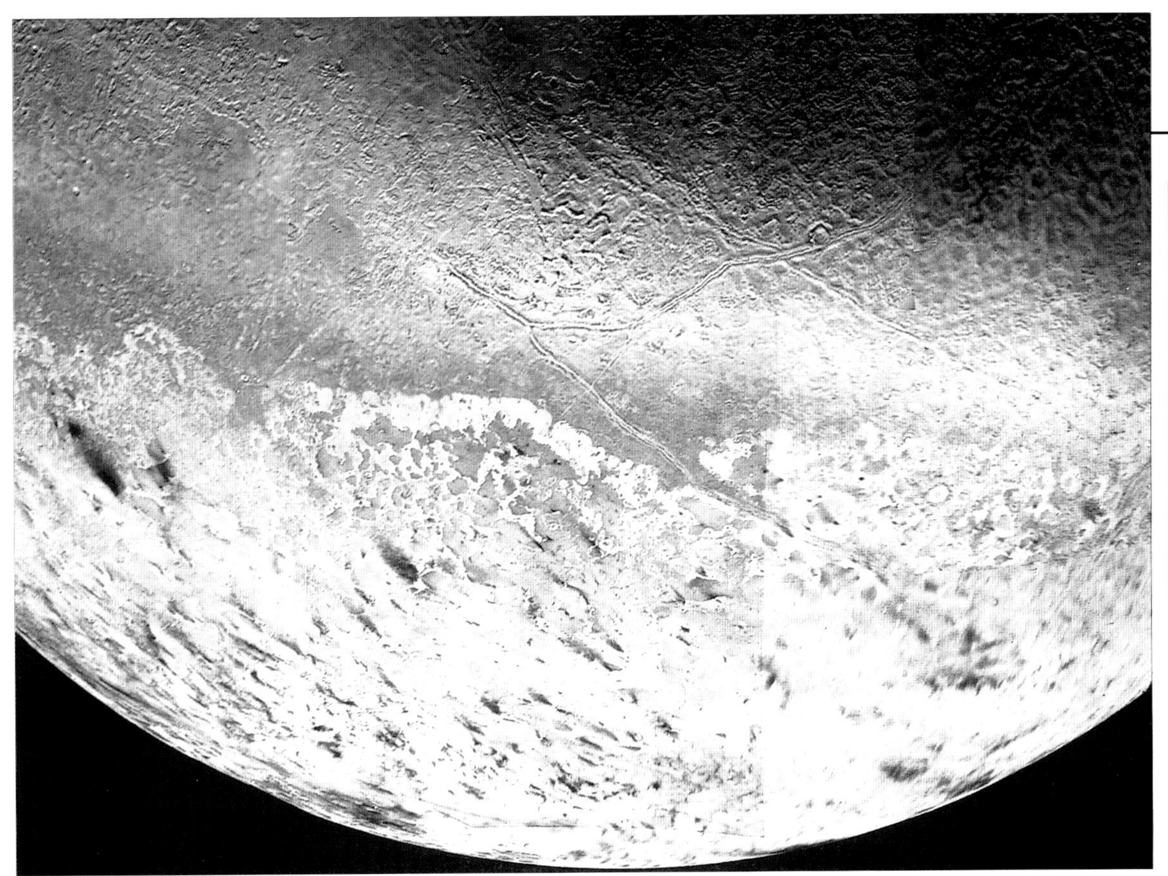

◄ Triton photomosaic *made up of 12 Voyager 2 images – the light south-polar cap lies at the bottom.*

▼ Orbits of Triton and Nereid. *Triton has an almost circular orbit, with retrograde motion; Nereid has direct motion but a highly eccentric orbit. All the six satellites discovered from Voyager 2 are closer-in than Triton; all have direct motion.*

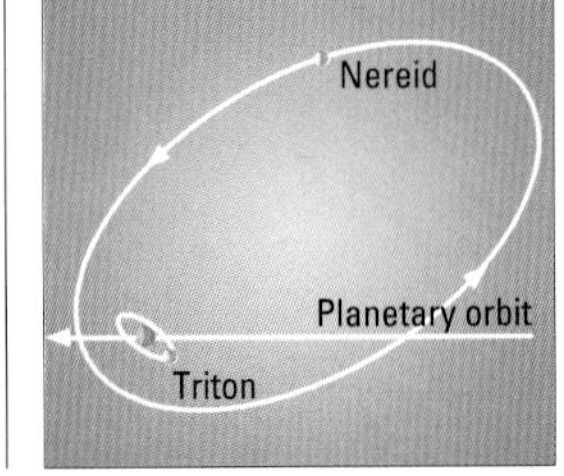

SATELLITES OF NEPTUNE

Name	Dist. from Neptune, km	Orbital period, days	Orbital incl.,°	Orbital eccentricity	Diameter, km	Mag.
Naiad	48,000	0.296	4.5	0	54	26
Thalassa	50,000	0.312	0	0	80	24
Despina	52,500	0.333	0	0	180	23
Galatea	62,000	0.429	0	0	150	23
Larissa	73,500	0.544	0	0	192	21
Proteus	117,600	1.121	0	0	416	20
Triton	354,800	5.877	159.9	0.0002	2705	13.6
Nereid	1,345,500–9,688,500	360.15	27.2	0.749	240	18.7

SELECTED FEATURES ON TRITON

Name	Lat. °	Long. E °
Abatos Planum	35–8S	35–81
Akupara Maculae	24–31S	61–65
Bin Sulci	28–48S	351–14
Boyenne Sulci	18–4S	351–14
Bubembe Regio	25–43S	285–25
Medamothi Planitia	16S–17N	50–90
Monad Regio	30°S 45′N	330–90
Namazu Macula	24–28S	12–16
Ob Sulci	19–14N	325–37
Ruach Planitia	24–31S	20–28
Ryugu Planitia	3–7S	25–29
Tuonela Planitia	36–42N	7–19
Uhlanga Regio	60–0S	285–0
Viviane Macula	30–32S	34–38
Zin Maculae	21–27S	65–72

migrates towards the upper part of the crust, the pressure will be relaxed, and the nitrogen will explode in a shower of ice and gas, travelling up the nozzle of the geyser-like vent at a rate of up to 150 metres (500 feet) per second – fast enough to send the material up for many kilometres before it falls back. The outrush sweeps dark debris along with it, and this debris is wafted downwind, producing plumes of dark material such as Viviane Macula and Namazu Macula. Some of these plumes are over 70 kilometres (40 miles) long.

Monad Regio is part smooth and in part hummocky, with walled plains or 'lakes' such as Tuonela and Ruach; these have flat floors, and water must be the main material from which they were formed, because nitrogen ice and methane ice are not rigid enough to maintain surface relief over long periods. Bubembe Regio is characterized by the so-called 'cantaloupe terrain', a name given because of the superficial resemblance to a melon-skin. Fissures cross the surface, meeting in huge X or Y junctions, and there are subdued circular pits with diameters of around 30 kilometres (19 miles).

It may well be that there will be marked changes in Triton's surface over the coming decades, because the seasons there are very long indeed, and the pink snow may migrate across to the opposite pole – which was in darkness during the Voyager pass. Unfortunately we cannot hope for another space mission before southern midsummer on Neptune and Triton, which is not due until the year 2006.

▼ Map of Triton *showing the main features recorded from Voyager.*

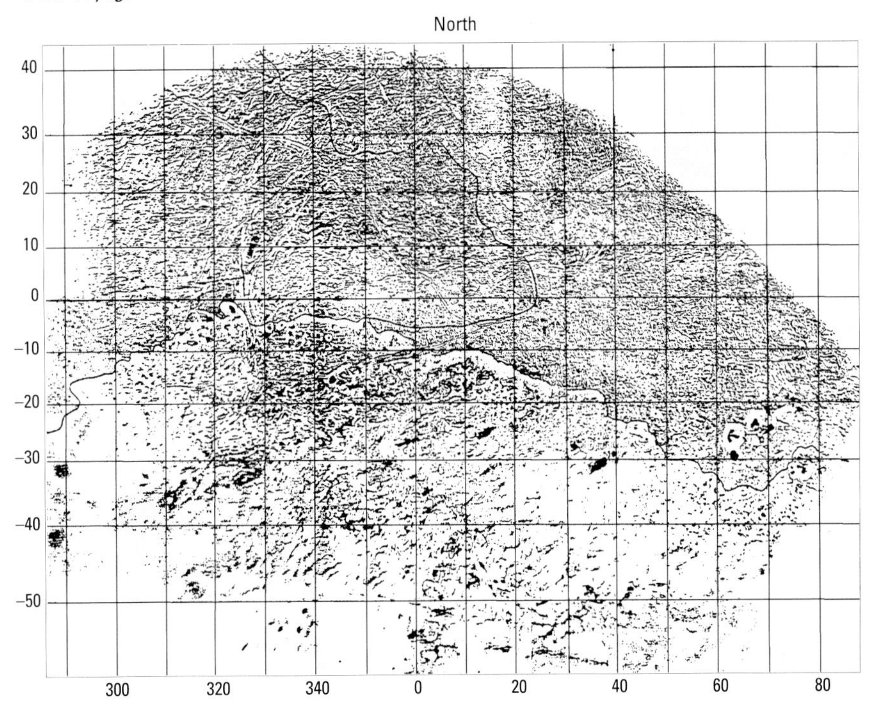

Pluto

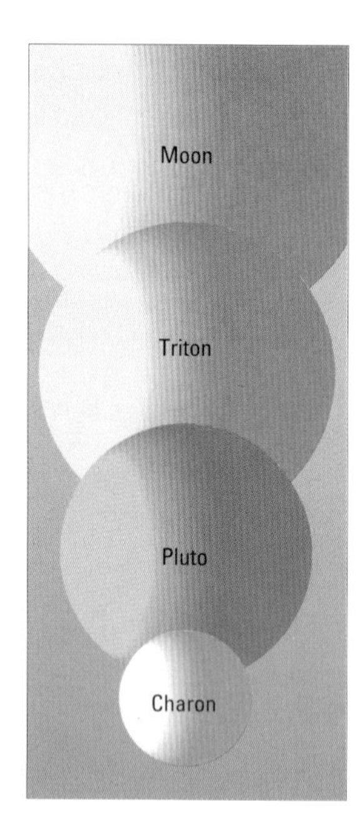

▲ ***Size of Pluto**. Pluto is shown here compared with the Moon, Triton and Charon. It is clear the Pluto–Charon pair cannot be regarded as a planet-and-satellite system; Charon has more than half the diameter of Pluto.*

Even after the discovery of Neptune, there were still tiny irregularities in the moments of the outer giants which led Percival Lowell (of Martian canal fame) to make fresh calculations in the hope of tracking down yet another planet. In 1930, 16 years after Lowell's death, Clyde Tombaugh used a specially-obtained telescope at the Lowell Observatory to identify a new planetary object only a few degrees from the predicted place. After some discussion it was named Pluto – a suitable name, since Pluto was the god of the Underworld, and the planet named after him is a gloomy place even though sunlight there would still be 1500 times brighter than full moonlight on Earth.

Pluto has a curiously eccentric orbit, and when closest to the Sun it moves well within the orbit of Neptune, though since Pluto's path is inclined by as much as 17 degrees there is no fear of collision. The last perihelion passage fell in 1989, and not until 1999 was Pluto's distance from the Sun again greater than that of Neptune. The revolution period is almost 248 years; the axial rotation period is 6 days 9 hours, and the axis of rotation is inclined by 122 degrees to the perpendicular, so that the calendar there is very complicated indeed.

The main puzzle about Pluto is its small size and mass. The diameter is a mere 2324 kilometres (1444 miles), which is less than that of the Moon or several other planetary satellites, including Triton. The mass is no more than 0.002 that of the Earth, and obviously Pluto can have no measurable effect upon the motions of giants such as Uranus and Neptune. Either Lowell's reasonably accurate prediction was sheer luck (which is hard to believe), or else the real planet for which he was hunting remains to be discovered.

The density is over twice that of water, so that there must be a fairly high percentage of rock in its globe; there could be a silicate core surrounded by a thick mantle of ice, but we have no definite information, because no spacecraft has been anywhere near Pluto. One thing we do know is that there is a thin but extensive atmosphere. When Pluto passes in front of a star, and hides or occults it, the star fades appreciably well before it is covered up, so that for a brief period its light is coming to us by way of Pluto's atmosphere. The atmosphere may be methane, nitrogen or a mixture. When Pluto moves out to the far part of its orbit the temperature will become so low that the atmosphere may freeze out on the surface, so for part of the Plutonian 'year' there is no gaseous surround at all. The next aphelion is due in 2114, but the atmosphere will probably condense out long before that.

In 1977 it was found that Pluto is not a solitary traveller in space. It is associated with a secondary body, which has been named Charon in honour of the somewhat sinister boatman who used to ferry departed souls across the River Styx on their way to the Underworld. Photographs taken with the Hubble Space Telescope show the two bodies separately, even though they are less than 20,000 kilometres (12,500 miles) apart. Charon has a diameter of 1270 kilometres (790 miles), more than half that of Pluto itself; the mass is one-twelfth that of Pluto, and when the two are shining together Charon contributes only 20 per cent of the total light. Its orbital period is 6.3 days, which is the same as the axial rotation period of Pluto, so that the two are 'locked' and an observer on Pluto would see Charon hanging motionless in the sky.

By a fortunate chance there were mutual eclipse and occultation phenomena during the late 1980s – a state of affairs which will not recur for 120 years. When Charon passed behind Pluto it was completely hidden, and Pluto's spectrum could be examined alone; when Charon passed in front of Pluto the two spectra were seen together, and that of Pluto could be subtracted. Pluto appears to have a surface coated with methane ice, perhaps with some ice of

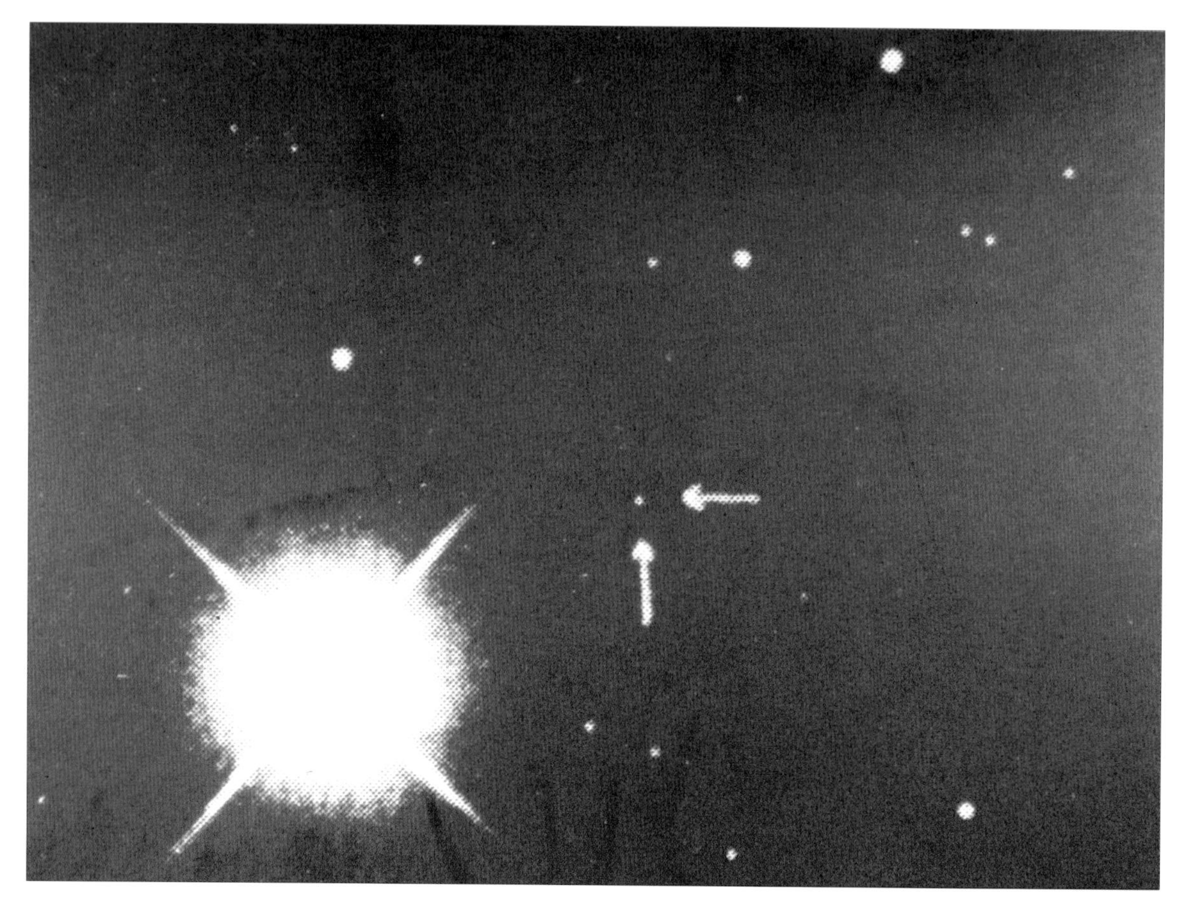

▲ ***Clyde Tombaugh**, discoverer of Pluto; this photograph was taken in 1980 at the 50th anniversary of the discovery.*

◄ ***Discovery plate of Pluto**, taken by Clyde Tombaugh in 1930. Pluto is indicated by the arrows; it looks exactly like a star, and was identified only because of its motion from night to night. The very over-exposed star image is that of Delta Geminorum, magnitude 3.5; the magnitude of Pluto was below 14.*

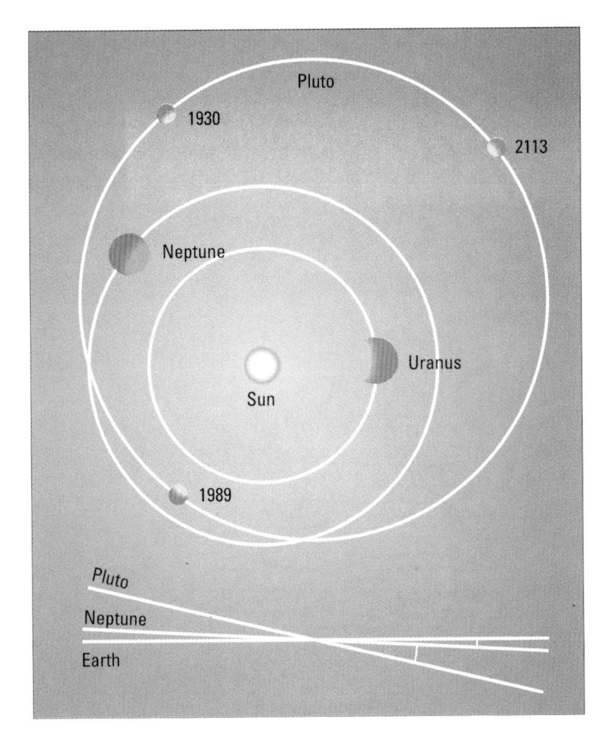

▲ ***Orbit of Pluto**. Pluto's eccentric path brings it within the orbit of Neptune, but its orbital inclination of 17° means that there is no fear of a collision occurring. Perihelion was passed in 1989, and aphelion will occur in 2114.*

nitrogen as well, while Charon shows signs of water ice; it has no detectable atmosphere. The occultation results even gave some clues as to surface markings. Pluto has a darkish equatorial band and extensive polar caps, while Charon may have a darkish patch in one hemisphere and a brighter band in the other.

The nature of Pluto is uncertain. It is not a normal planet; it may be an exceptional asteroid; but is more probably a planetesimal, representing material 'left over', so to speak, when the main planets were formed from the original solar nebula. Its magnitude is 14, so that a moderate-sized telescope will show it as a star-like point. Until it is surveyed by a spacecraft we can hardly hope to find out much more, but it may well be that the surface features are of the same type as those on Triton.

▼ ***Dome of the telescope** used to discover Pluto – then at the Lowell Observatory, now at the outstation (Anderson Mesa).*

▲ ***Pluto and Charon**: Hubble Space Telescope, 21 February 1994.*

◀ ***The 9-inch refractor** used to discover Pluto.*

PLANETARY DATA – PLUTO

Sidereal period	90,465 days
Rotation period	6d 9h 17m
Mean orbital velocity	4.7 km/s (2.9 miles/s)
Orbital inclination	17.2°
Orbital eccentricity	0.248
Apparent diameter	< 0.25″
Reciprocal mass, Sun = 1	< 4,000,000
Mass, Earth, = 1	0.0022
Escape velocity	1.18 km/s (0.7 miles/s)
Mean surface temperature	about −220°C
Albedo	about 0.4
Maximum magnitude	14
Diameter	2324 km (1444 miles)

Earth

The Surface of Pluto

▶ ***Pluto from Hubble****. The two smaller inset pictures at the top are actual images from Hubble (north at top). Each square pixel is more than 160 km (100 miles) across. At this resolution, Hubble discerns roughly 12 major 'regions' where the surface is either bright or dark. The larger images (below) are from a global map constructed through computer image-processing performed on the Hubble data. The tile pattern is an artefact of the image enhancement technique. Opposite hemispheres of Pluto are seen in these two views. Some of the variations across Pluto's surface may be caused by topographic features such as basins, or fresh impact craters. However, most of the surface features unveiled by Hubble, including the prominent northern polar cap, are likely produced by the complex distribution of frosts that migrate across Pluto's surface with its orbital and seasonal cycles and chemical byproducts deposited out of Pluto's nitrogen–methane atmosphere.*

The Hubble Space Telescope has allowed astronomers to view the surface of Pluto as never before. Pluto had always appeared as nothing more than a dot of light in even the largest Earth-based telescopes because its disk is much smaller than can be resolved from beneath the Earth's atmosphere.

Hubble imaged nearly the entire surface of Pluto, as it rotated through its 6.4-day period, in late June and early July 1994. These images, which were made in blue light, show that Pluto is an unusually complex object, with more large-scale contrast than any planet, except Earth. Pluto itself probably shows even more contrast and perhaps sharper boundaries between light and dark areas than is shown in the images here, but Hubble's resolution tends to blur edges and blend together small features sitting inside larger ones.

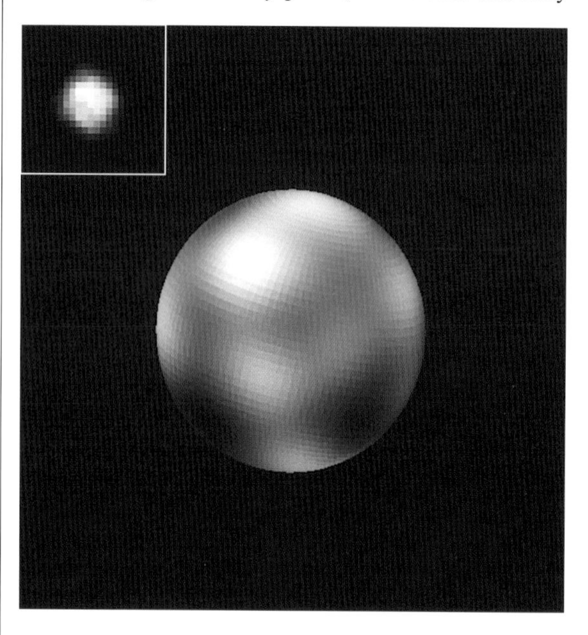

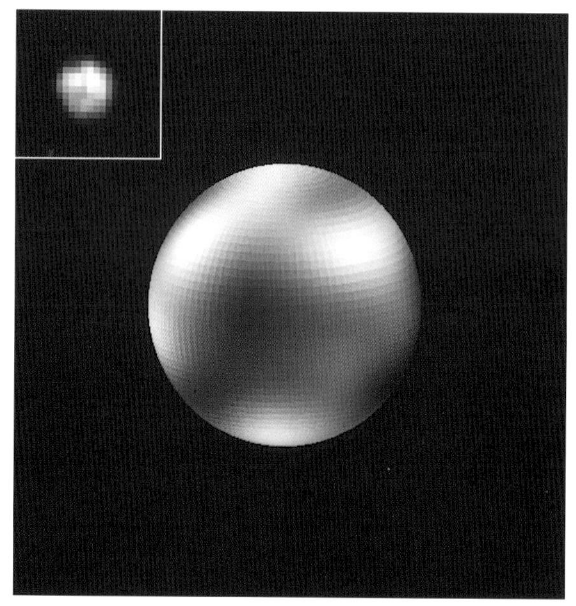

▼ ***Map of Pluto's surface****, assembled by computer image-processing software from four separate images of Pluto's disk taken with the European Space Agency's (ESA) Faint Object Camera (FOC) aboard the Hubble Space Telescope. Hubble imaged nearly the entire surface, as Pluto rotated on its axis in late June and early July 1994. The map, which covers 85 per cent of the planet's surface, confirms that Pluto has a dark equatorial belt and bright polar caps, as inferred from ground-based light curves obtained during the mutual eclipses that occurred between Pluto and its satellite Charon in the late 1980s. The brightness variations in this may be due to topographic features such as basins and fresh impact craters. The black strip across the bottom corresponds to the region surrounding Pluto's south pole, which was pointed away from Earth when the observations were made, and could not be imaged.*

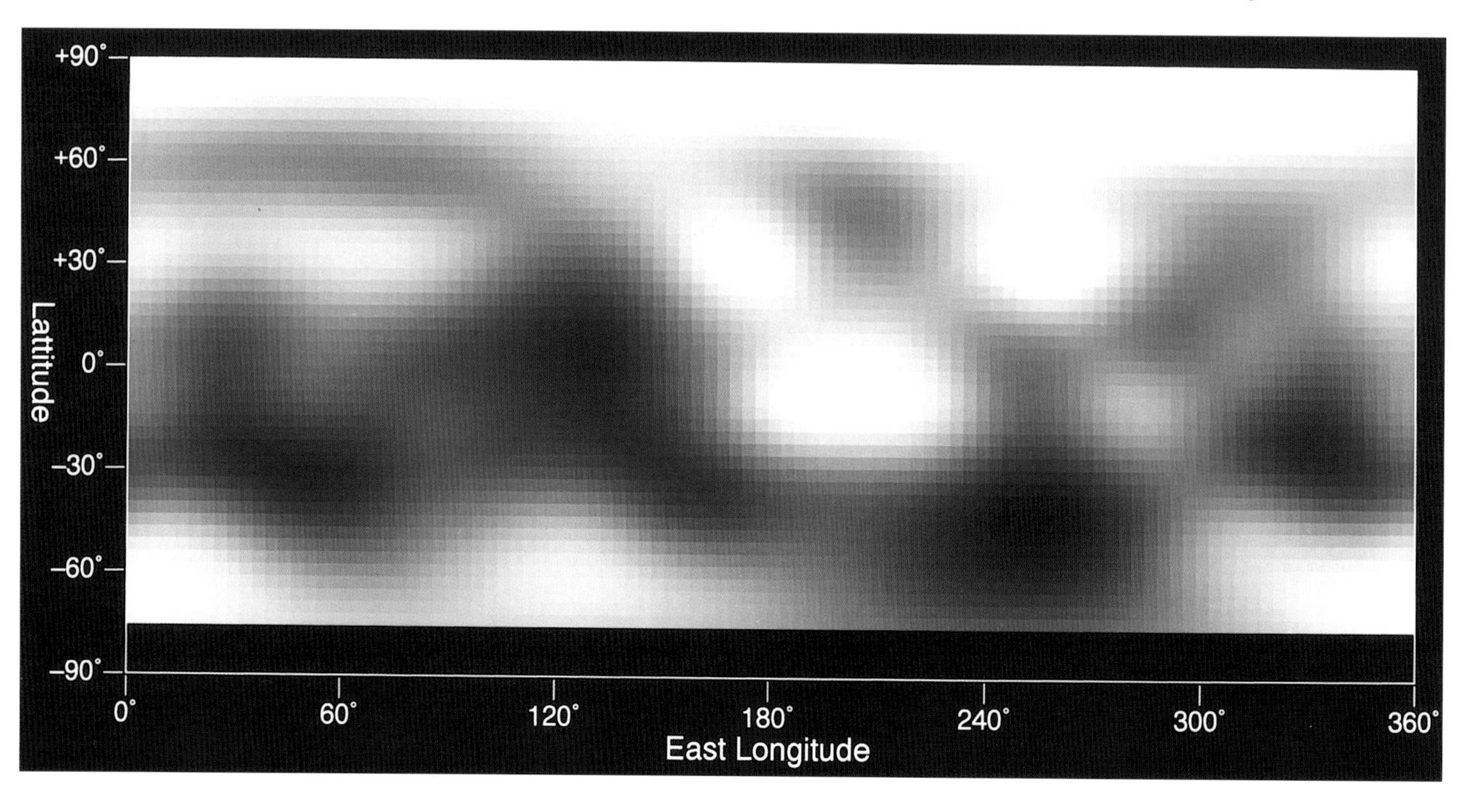

Boundaries of the Solar System

It is not easy to define where the Solar System 'ends'. The nearest star beyond the Sun is over four light-years away, corresponding to a distance of over 40 million million kilometres (25 million million miles), so that the region where the Sun's influence is dominant may extend out to roughly half this amount, but there can be no sharp boundary.

Recently there have been observations of a swarm of bodies moving round the Sun some way beyond the orbit of Neptune. This has been named the Kuiper Belt, in honour of the late Gerard Kuiper, who first suggested its existence. The first Kuiper Belt object was discovered in 1992 by David Jewitt and Jane Luu, from the Mauna Kea Observatory in Hawaii, and has been given the provisional designation of 1992 QB_1. It moves round the Sun in an orbit which carries it from 5100 million kilometres (3170 million miles) out to 6600 million kilometres (4100 million miles), and has a period 296 years; the diameter is probably of the order of 150 kilometres (below 100 miles), but may be more, as we have no idea of the value of the albedo. Since then other similar bodies have been found, and there are also other asteroidal bodies, such as 5145 Pholus, which have very eccentric orbits; that of Pholus crosses the paths of Saturn, Uranus and Neptune.

There are also asteroidal objects which wander out to immense distances. In September 1996 Eleanor Helin discovered 1996 RQ_{20}, which has a period of 327 years and ranges between 33 and 61 astronomical units from the Sun; 1996 TL_{66}, found by Jane Luu in October of that year, has a period of almost 800 years, and moves out to 84 astronomical units – not far short of 13,000 million kilometres (8000 million miles). It may be at least 500 kilometres (300 miles) across, comparable with Vesta or Pallas. No doubt many more of these curious bodies await discovery.

Finally, there is always the chance that another large planet exists in the depths of the Solar System. Searches for it have been made from time to time, but whether it really exists is a matter for debate.

What, then, of the comets? Short-period comets may come from the Kuiper Belt, but long-period comets come from the so-called Oort Cloud, a swarm of these icy bodies orbiting the Sun at a distance of more than a light-year. If one of the members of the Cloud is perturbed for any reason, it may start to fall inwards towards the Sun, and eventually it will invade the inner part of the Solar System. One of several things may happen. The comet may simply swing round the Sun and return to the Oort Cloud, not to be back for many centuries – or even thousands or millions of years. It may fall into the Sun, and be destroyed. It may be perturbed by a planet (usually Jupiter) and either thrown out of the Solar System altogether, or else forced into a short-period orbit which brings it back to perihelion after a few years. Or it may collide with a planet, as Comet Shoemaker–Levy 9 did in July 1994, when it impacted Jupiter. One comet, Pizarro, moves wholly within the asteroid belt – but it is definitely a comet, as it has a pronounced tail. But really brilliant comets have periods so long that we cannot predict their appearance, and they are always apt to arrive without warning and take us by surprise.

Amateurs have a fine record of cometary discovery. The hunter may discover a comet only a night or two after starting work, but he may have to wait for years. Luck plays a part, of course, but the main essential in searching for comets is a truly encyclopaedic knowledge of the sky.

▲ ***Pholus**. Asteroid 5145, discovered in 1992 by D. L. Rabinowitz from Kitt Peak Observatory in Arizona. The magnitude was then 17. The diameter may be around 150 km (below 100 miles). The revolution period is 93 years. The distance from the Sun ranges between 1305 million and 4800 million km (810 million and 2980 million miles). In colour it is red. It may well be a planetesimal from the Kuiper Belt.*

◀ ***1992 QB_1**. These images were taken by Alain Smette and Christian Vanderriest, using the 3.5-m (138-inch) New Technology Telescope at La Silla. The magnitude was 23; the faint image of the object is circled. The distance was then over 6000 million km (3700 million miles) from the Sun, beyond the orbit of Pluto.*

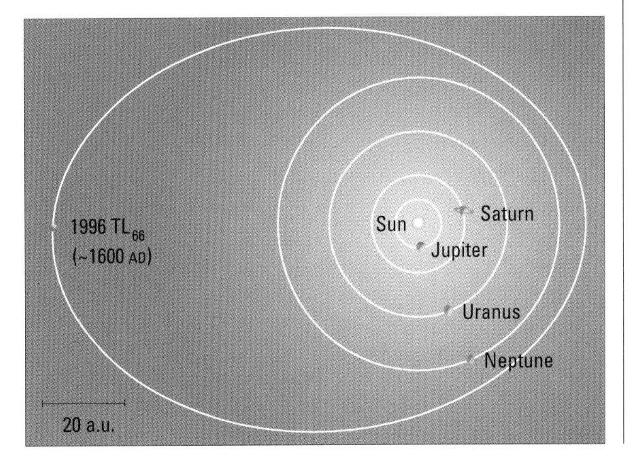

◀ ***The orbit of 1996 TL_{66}** – discovered by Jane Luu in October 1996.*

Comets

Comets are the least predictable members of the Solar System. All in all, a comet is a wraith-like object. The only substantial part is the nucleus, which has been aptly described as a dirty ice-ball and is never more than a few kilometres across. When the comet is heated, as it nears the Sun, the ices in the nucleus start to evaporate, so that the comet develops a head or coma, which may be huge; the coma of the Great Comet of 1811 was larger than the Sun. There may be one or more tails, though many small comets never produce tails of any sort.

A cometary nucleus is composed of rocky fragments held together with ices such as frozen ammonia, methane and water. Tails are of two kinds. A gas or ion tail is produced by the pressure of sunlight, which drives very small particles out of the head, while a dust tail is due to the pressure of the solar wind; in general an ion tail is straight, while a dust tail is curved. Tails always point more or less away from the Sun, so that when a comet is travelling outwards it moves tail-first.

Each time a comet passes through perihelion it loses material to produce a coma and (in some cases) tails; for example Halley's Comet, with a period of 76 years, loses about 300 million tons of material at each return to the Sun. This means that by cosmical standards, comets must be short-lived. Some short-period comets which used to be seen regularly have now disappeared; such are the comets of Biela, Brorsen and Westphal. (Comets are usually named after their discoverers, though occasionally after the mathematician who first computed the orbit – as with Halley's Comet.) A prefix P/ indicates that the comet is periodical. A comet leaves a 'dusty' trail behind it as it moves along, and when the Earth plunges through one of these trails the result is a shower of shooting-stars. Most meteor showers have known parent comets; for instance the main annual shower, that of early August, is associated with Comet P/Swift–Tuttle, which has a period of 130 years and last returned to perihelion in 1992.

Because a comet is so flimsy and of such low mass, it is at the mercy of planetary perturbations, and orbits may be drastically altered from one cycle to another. The classic case is that of Lexell's Comet of 1770, which became a bright naked-eye object. A few years later it made a close approach to Jupiter, and its orbit was completely changed, so that we have no idea where the comet is now.

▲ ***West's Comet**, one of the brightest of modern times even though not a 'great comet'. This comet was discovered by Richard West, who took the photograph shown here. The comet was a bright naked-eye object for several mornings in March 1975; as it receded from the Sun it showed signs of disruption in its nucleus, so that it may not be so bright when it next returns to perihelion in approximately 553,000 years!*

▶ ***Bennett's Comet of 1970**. This was one of the brighter comets of recent times, and its nucleus became as bright as the Pole Star. It was discovered by the South African amateur astronomer Jack Bennett, one of the best known comet-hunters.*

Short-period Comets

It is generally believed that large comets come from the Oort Cloud, a swarm of icy bodies orbiting the Sun at a distance over a light-year, while short-period comets come from the Kuiper Cloud, which lies not far beyond the orbit of Neptune. It is a pity that all the comets with periods of less than half a century are faint. No doubt they were much more imposing when they first plunged sunwards, but by now they are mere ghosts of their former selves. Encke's Comet, the first to be identified, is a case in point. It was originally found in 1786 by the French astronomer Pierre Méchain, when it was of the fifth magnitude, and had a short tail. It was seen again in 1795 by Caroline Herschel, William Herschel's sister, and yet again in 1805 by Thulis, from Marseilles. In 1818 it turned up once more, and was detected by Jean Louis Pons, whose grand total of comet discoveries amounted to 37. (His story is unusual, because he began his career as an observatory doorkeeper and ended it as an observatory director.) The orbit was calculated by J. F. Encke, of Berlin, who concluded that the comets of 1786, 1795, 1805 and 1818 were one and the same; he gave the period as 3.3 years, and predicted a return for 1822. The comet duly appeared just where Encke had expected, and, very appropriately, was named after him. Since then it has been seen at every return except that of 1945, when it was badly placed and when most astronomers had other things on their minds.

At some returns during the last century Encke's Comet was quite prominent; in 1829 it reached magnitude 3.5, with a tail 18 minutes of arc in length. Nowadays it does not achieve such eminence, and although it is hard to be sure – estimating comet magnitudes is far from easy – it does seem to have faded. Whether it will survive into the 22nd century remains to be seen.

Encke's Comet has a small orbit; at perihelion it ventures just inside the orbit of Mercury, while at its furthest from the Sun it moves out into the asteroid zone. Modern instruments can follow it all around its path; its period is the shortest known. In 1949 a new comet, Wilson–Harrington, was believed to have a period of only 2.4 years, but it was not seen again until 1979, when it was recovered – this time as an asteroid, designated No. 4015! There is little doubt that it has changed its status, and it may well be that many of the small close-approach asteroids, such as Phaethon, are ex-comets which have lost all their volatiles.

Biela's Comet met with a sad fate. It was discovered in 1772 by Montaigne from Limoges, recovered by Pons in 1805 and again by an Austrian amateur, Wilhelm von Biela, in 1826. The period was given as between six and seven years, and it returned on schedule in 1832, when it was first sighted by John Herschel. (It was unwittingly responsible for a major panic in Europe. The French astronomer Charles Damoiseau had predicted that the orbit of the comet would cut that of the Earth; he was quite right, but at that time the comet was nowhere near the point of intersection.) Biela's Comet was missed in 1839 because of its unfavourable position in the sky, but it came back once more in 1846, when it astonished astronomers by splitting in two. The pair returned in 1852, were missed in 1859 again because they were badly placed, and failed to appear at the expected return of 1866 – in fact they have never been seen again. When they ought to have returned, in 1872, a brilliant meteor shower was seen coming from that part of the sky where the comet had been expected, and there is no doubt that the meteors represented the funeral pyre of the comet. The shower was repeated in 1885, 1892 and 1899, but no more brilliant displays have been seen since then; to all intents and purposes the shower has ceased, so that we really have, regretfully, seen the last of Biela's Comet.

Other periodical comets have been 'mislaid', only to be found again after a lapse of many years; thus Holmes' Comet, which reached naked-eye visibility in 1892 and had a period of nearly seven years, was lost between 1908 and 1965; it has since been seen at several returns, but is excessively faint. Comet Brooks 2 made a close approach to Jupiter in 1886, when it actually moved inside the orbit of Io and was partially disrupted, spawning four minor companion comets which soon faded away. During the Jupiter encounter, the orbit was changed from 29 years to its present value of seven years.

Comet Schwassmann–Wachmann 1 is of unusual interest. Its orbit lies wholly between those of Jupiter and Saturn, and normally it is very faint, but sometimes it shows sudden outbursts which bring it within the range of small telescopes. Large instruments can follow it all round its orbit, as is also the case with a few other comets with near-circular paths, such as Smirnova–Chernykh and Gunn. Oterma's Comet used to have a period of 7.9 years, but an encounter with Jupiter in 1973 altered this to 19.3 years, and the comet now comes nowhere near the Earth, so that its future recovery is very doubtful.

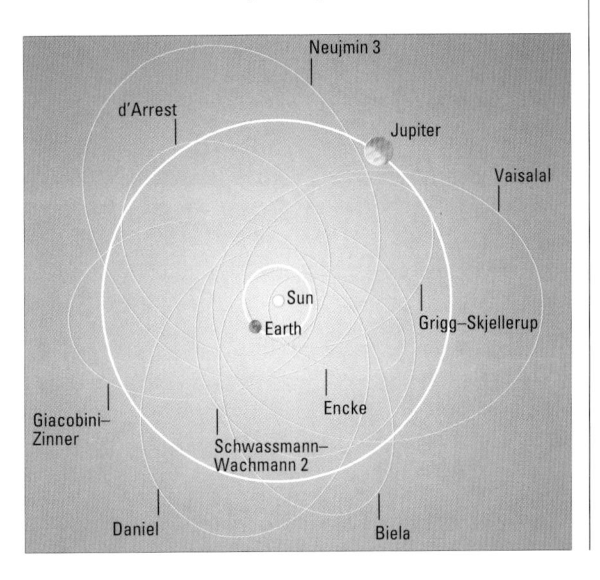

◀ ***Orbits of some short period comets.*** *Many have their aphelia at around Jupiter's orbit. Some, such as Encke's, are observable all round their orbits with modern equipment.*

PERIODICAL COMETS WHICH HAVE BEEN OBSERVED AT TEN OR MORE RETURNS

Name	Year of discovery	Period, years	Eccentricity	Incl.	Dist. from Sun, astr. units min.	max.
Encke*	1786	3.3	0.85	12.0	0.34	4.10
Grigg–Skjellerup	1902	5.1	0.66	21.1	0.99	4.93
Tempel 2	1873	5.3	0.55	12.5	1.38	4.70
Pons–Winnecke	1819	6.4	0.64	22.3	1.25	5.61
D'Arrest	1851	6.4	0.66	16.7	1.29	5.59
Kopff	1906	6.4	0.55	4.7	1.58	5.34
Schwassmann–Wachmann 2	1929	6.5	0.39	3.7	2.14	4.83
Giacobini–Zinner	1900	6.6	0.71	13.7	1.01	6.00
Borrelly	1905	6.8	0.63	30.2	1.32	5.83
Brooks 2	1889	6.9	0.49	5.6	1.85	5.41
Finlay	1886	7.0	0.70	3.6	1.10	6.19
Faye	1843	7.4	0.58	9.1	1.59	5.96
Wolf 1	1884	8.4	0.40	27.3	2.42	5.73
Tuttle	1790	13.7	0.82	54.4	1.01	10.45
Schwassmann–Wachmann 1*	1908	15.0	0.11	9.7	5.45	6.73
Halley	240 BC	76.0	0.97	162.2	0.59	34.99

(* = comets which can be followed throughout their orbits)

Halley's Comet

▲ ***Halley's Comet as seen from Christchurch, New Zealand**, 16 March 1986. Photograph by Peter Carrington.*

Of all the comets in the sky
There's none like Comet Halley.
We see it with the naked eye,
And periodically.

Nobody seems to know who wrote this piece of doggerel, but certainly Halley's Comet is in a class of its own. It has been seen at every return since that of 164 BC, and the earliest record of it, from Chinese sources, may date back as far as 1059 BC. Note that the interval between successive perihelion passages is not always 76 years; like all of its kind, Halley's Comet is strongly affected by the gravitational pulls of the planets.

Edmond Halley, later to become Astronomer Royal, observed the return of 1682. He calculated the orbit, and realized that it was strikingly similar to those of comets previously seen in 1607 and 1531, so that he felt confident in predicting a return for 1758. On Christmas night of that year – long after Halley's death – the comet was recovered by the German amateur astronomer Palitzsch, and it came to perihelion in March 1759, within the limits of error given by Halley. This was the first predicted return of any comet; previously it had been thought by most astronomers that comets travelled in straight lines.

Halley's Comet has a very elliptical orbit. At its closest it is about 88 million kilometres (55 million miles) from the Sun, within the orbit of Venus; at aphelion it recedes to 5250 million kilometres (3260 million miles), beyond the orbit of Neptune and possibly near the Kuiper Belt. At its brightest recorded return, that of AD 837, it passed by the Earth at only 6 million kilometres (3.75 million miles), and contemporary reports tell us that its head was as brilliant as Venus, with a tail stretching 90 degrees across the sky. Another bright return was that of 1066, before the Battle

▼ ***Giotto**, the British-built spacecraft which encountered Halley's Comet and subsequently went on to an encounter with Comet P/Grigg–Skjellerup.*

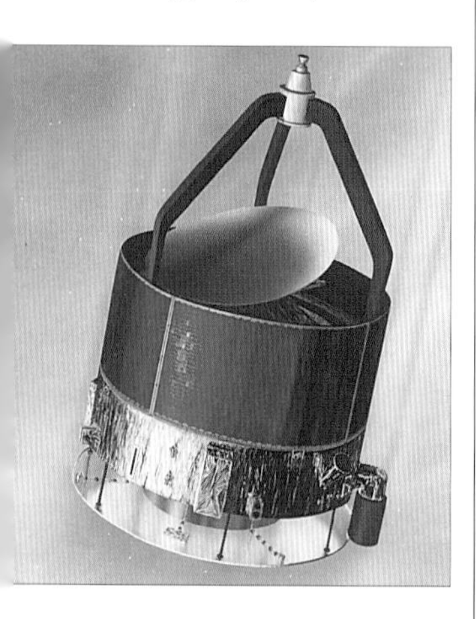

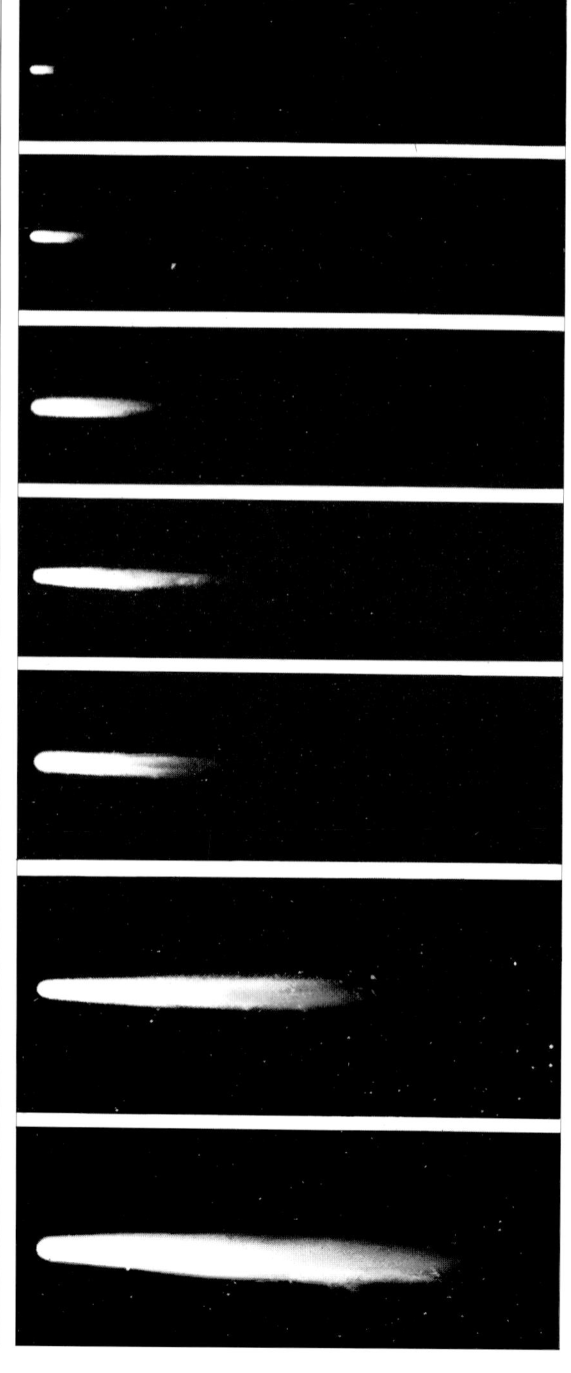

▶ ***Halley's Comet** taken during the 1910 return. When far from the Sun, a comet has no tail; the tail starts to develop when the comet draws inward and is heated, so that the ices in its nucleus begin to evaporate. This series shows the tail increasing to a maximum.*

of Hastings; the comet caused great alarm among the Saxons, and it is shown on the Bayeux Tapestry, with King Harold toppling on his throne and the courtiers looking on aghast. In 1301 it was seen by the Florentine artist Giotto di Bondone, who used it as a model for the Star of Bethlehem in his picture *The Adoration of the Magi* – even though Halley's Comet was certainly not the Star of Bethlehem; it returned in 12 BC, years before the birth of Christ. At the return of 1456 the comet was condemned by Pope Calixtus III as an agent of the Devil. It was prominent in 1835 and in 1910, but unfortunately not in 1986, when it was badly placed and never came within 39 million kilometres (24 million miles) of the Earth, and though it became an easy naked-eye object it was by no means spectacular. The next return, that of 2061, will be no better. We must wait until 2137, when it will again be a magnificent sight.

The fact that Halley's Comet can still become brilliant shows that it came in fairly recently from the Oort Cloud. It loses about 250 million tonnes of material at each perihelion passage, but it should survive in more or less its present form for at least 150,000 years to come.

The comet was first photographed at the 1910 return, after which it remained out of range until 16 October 1982, when it was recovered by D. Jewitt and E. Danielson, at the Palomar Observatory, only six minutes of arc away from its predicted position; it was moving between the orbits of Saturn and Uranus. As it drew in towards perihelion, four spacecraft were sent to it: two Japanese, two Russian and one European. The European probe Giotto, named after the painter, invaded the comet's head, and on the night of 13–14 March 1986 passed within 605 kilometres (376 miles) of the nucleus. Giotto's camera functioned until about 14 seconds before closest approach, when the spacecraft was struck by a particle about the size of a rice-grain and contact was temporarily broken; in fact the camera never worked again, and the closest image of the nucleus was obtained from a range of 1675 kilometres (about 1000 miles).

The nucleus was found to be shaped rather like a peanut, measuring 15 × 8 × 8 kilometres (9 × 5 × 5 miles), with a total volume of over 500 cubic kilometres (120 cubic miles) and a mass of from 50,000 million to 100,000 million tonnes (it would need 60,000 million comets of this mass to equal the mass of the Earth). The main constituent is water ice, insulated by an upper layer of black material which cracks in places when heated by the Sun, exposing the ice below and resulting in dust jets. Jet activity was very marked during the Giotto pass, though the jets themselves were confined to a small area on the sunward side of the nucleus. The central region of the nucleus was smoother than the ends; a bright 1.5-kilometre (one-mile) patch was presumably a hill, and there were features which appeared to be craters with diameters of around a kilometre (3280 feet). The comet was rotating in a period of 55 hours with respect to the long axis of the nucleus.

Tails of both types were formed, and showed marked changes even over short periods. By the end of April the comet had faded below naked-eye visibility, but it provided a major surprise in February 1991, when observers using the Danish 154-cm (60-inch) reflector at La Silla in Chile found that it had flared up by several magnitudes. There had been some sort of outburst, though the reason for it is unclear.

Giotto survived the Halley encounter, and was then sent on to rendezvous with a much smaller and less active comet, P/Grigg–Skjellerup, in July 1992. Despite the loss of the camera, a great deal of valuable information was obtained. Unfortunately Giotto did not have enough propellant remaining for a third cometary encounter.

OBSERVED RETURNS OF HALLEY'S COMET

Year	Date of perihelion
1059 BC	3 Dec
240 BC	25 May
164 BC	12 Nov
87 BC	6 Aug
12 BC	10 Oct
AD 66	25 Jan
141	22 Mar
218	17 May
295	20 Apr
374	16 Feb
451	28 June
530	27 Sept
607	15 Mar
684	2 Oct
760	20 May
837	28 Feb
912	18 July
989	5 Sept
1066	20 Mar
1145	18 Apr
1222	28 Sept
1301	25 Oct
1378	10 Nov
1456	9 June
1531	26 Aug
1607	27 Oct
1682	15 Sept
1759	13 Mar
1835	16 Nov
1910	10 Apr
1986	9 Feb

◄ ***Halley through the Schmidt telescope****.* *One of the most beautiful photographs of Halley was taken on 9 January 1986 by Dr Birkle at Calar Alto in Southern Spain. The comet was then 200 million km (125 million miles) from the Earth with a tail 6 million km (3.75 million miles) long.*

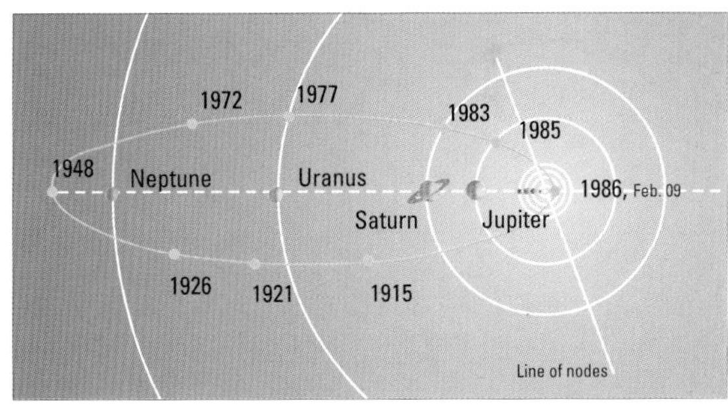

▲ ***Orbit of Halley's Comet****. Aphelion was reached in 1948; the comet passed perihelion in 1986, and has now receded once more as far as the orbit of Uranus. The next perihelion will be in 2061.*

◄ ***Nucleus of Halley's Comet****, from the Halley Multi-colour Camera carried in Giotto. The range was 20,000 km (12,500 miles). This is, of course, a false-colour picture.*

Great Comets

◀ ***The Great Comet of 1811***, *discovered by Honoré Flaugergues. This impression shows the comet on 15 October, from Otterbourne Hill, near Winchester in England.*

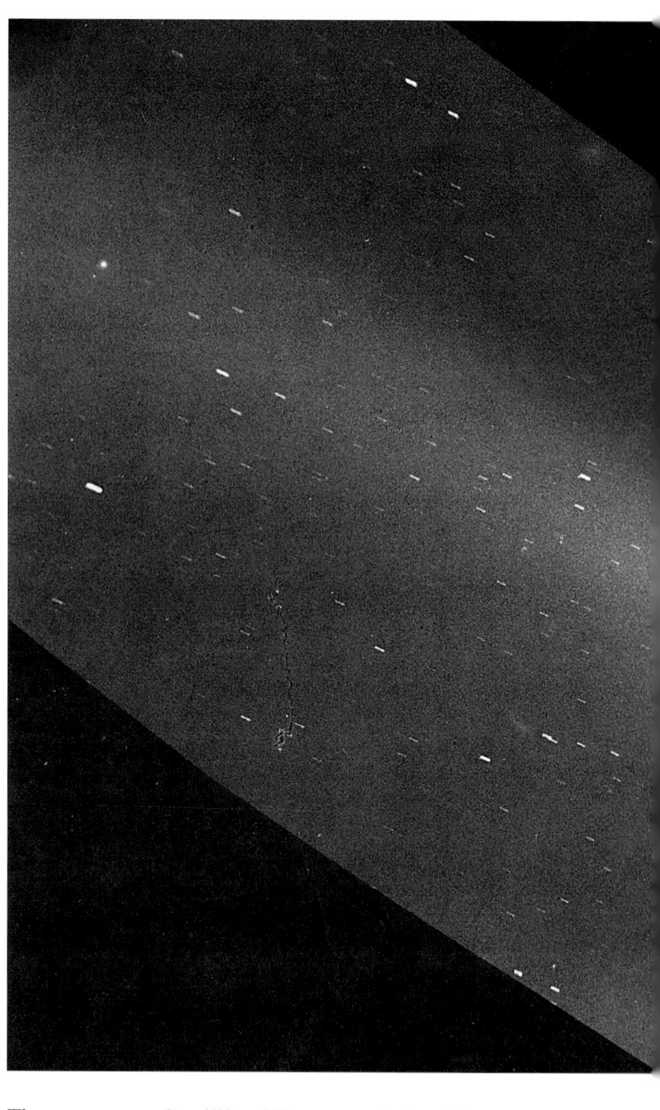

▶ ***The Daylight Comet of 1910***, *Lowell Observatory. 27 January. (Many people who claim to remember Halley's Comet in 1910 actually saw the Daylight Comet, which was considerably brighter!)*

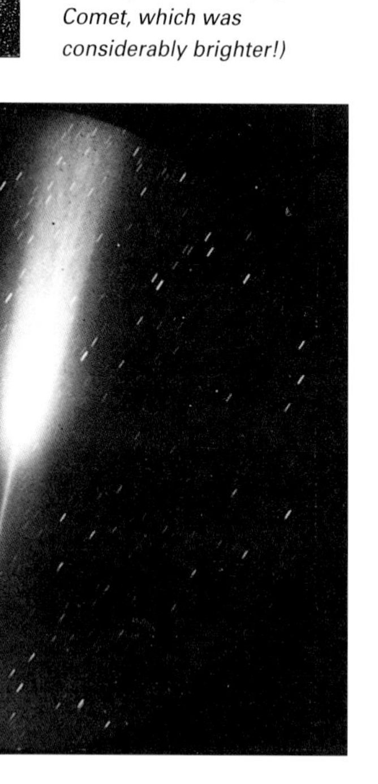

▶ ***Comet Arend–Roland***, *1957, as photographed by E. M. Lindsay from Armagh, Northern Ireland. This comet will never return; it has been perturbed into an open orbit.*

It is not surprising that ancient peoples were alarmed whenever a brilliant comet appeared. These so-called 'hairy stars' were regarded as unlucky; remember Shakespeare's lines in *Julius Caesar* – 'When beggars die, there are no comets seen; the heavens themselves blaze forth the death of princes.' There have been various comet panics, one of which was sparked off in 1736 by no less a person than the Rev. William Whiston, who succeeded Newton as Lucasian Professor of Mathematics at Cambridge. Mainly on religious grounds, Whiston predicted that the world would be brought to an end by a collision with a comet on 16 October of that year, and the alarm in London was so great that the Archbishop of Canterbury felt bound to issue a public disclaimer!

If the Earth were struck by a cometary nucleus a few kilometres in diameter there would undoubtedly be widespread damage, but the chances are very slight. A theory of a different type has been proposed in recent years by Sir Fred Hoyle and Chandra Wickramasinghe, who believe that comets can deposit viruses in the upper air and cause epidemics such as smallpox. It must be said, however, that these ideas have met with practically no support either from astronomers or from medical experts.

Great comets have been rare during the present century, but many have been seen in the past. For example, de Chéseaux Comet of 1744 developed multiple tails, and a contemporary drawing of it has been likened to a Japanese fan. Even more impressive was the comet of 1811, discovered by the French astronomer Honoré Flaugergues. The coma was 2 million kilometres (1.2 million miles) across, and the 16-million-kilometre (10-million-mile) tail stretched out to over 90 degrees, while the tail of the Great Comet of 1843 extended to 330 million kilometres (205 million miles), considerably greater than the distance between the Sun and Mars. It is not easy to remember that these huge bodies are so flimsy, and that their masses are absolutely negligible by planetary standards.

Donati's Comet of 1858 is said to have been the most beautiful ever seen, with its brilliant head, straight ion tail and curved dust-tail. Three years later came Tebbutt's Comet, discovered by an Australian amateur, which came within 2 million kilometres (1.2 million miles) of the Earth; we may even have passed through the tip of the tail, though nothing unusual was reported apart from a slight, unconfirmed yellowish tinge over the sky.

The Great Southern Comet of 1882 was bright enough to cast shadows, and to remain visible even when the Sun was above the horizon. This was the first comet to be properly photographed. Sir David Gill, at the Cape of Good Hope, obtained an excellent picture of it, and this led to an important development. Gill's picture showed so many stars that he realized that the best way to map the stellar sky was by photographic methods rather than by laborious visual measurement. The 1882 comet was a member of the Kreutz Sun-grazing group, distinguished by very small perihelion distance.

The Daylight Comet of 1910 appeared a few weeks before Halley's, and was decidedly the brighter of the two.

▲ ***The Great Comet of 1843**, as seen from the Cape of Good Hope on the evening of 3 March. This may have been the brightest comet for many centuries.*

▲ ***De Chéseaux's Comet of 1744**, with its multiple tail; this is a famous impression of it, but it did not remain brilliant for long, and is not well documented.*

▼ ***Donati's Comet of 1858**, often said to have been the most beautiful comet ever seen; it had tails of both types.*

It also was visible at the same time as the Sun, and it had a long, imposing tail. The orbit is elliptical, but we will not be seeing the comet again yet awhile, because the estimated period is of the order of 4 million years. Obviously we cannot be precise; we can measure only a very small segment of the orbit, and it is very difficult to distinguish between a very eccentric ellipse and a parabola.

Comet Skjellerup–Maristany of 1927 was also very brilliant, but its glory was brief, and it remained inconveniently close to the Sun in the sky. This was also true, though not to so great an extent, of Comet 1965 VIII, discovered independently by two Japanese observers, Ikeya and Seki. From some parts of the world it was brilliant for a while, but it soon faded, and will not be back for at least 880 years. Kohoutek's Comet of 1973 was a great disappointment. It was discovered on 7 March by Lubos Kohoutek at the Hamburg Observatory, and was expected to become extremely brilliant, but it signally failed to do so, and was none too conspicuous as seen with the naked eye. Perhaps it will make a better showing at its next return, about 75,000 years from now.

Of lesser comets, special mention should be made of Arend–Roland (1957), Bennett (1970) and West (1976). Arend–Roland was quite conspicuous in the evening sky for a week or two in April 1957, and showed a curious sunward spike which was not a reverse tail, but was due merely to thinly-spread material in the comet's orbit catching the sunlight at a favourable angle. Bennett's Comet was rather brighter, with a long tail; the period here is about 1700 years. West's Comet was also bright, but suffered badly as it passed through perihelion, and the nucleus was broken up. No doubt observers will be interested to see what has happened to it when it returns in around the year AD 559,000.

The only really bright comets of very recent years came in 1996 and 1997 – Comet Hyakutake and Comet Hale–Bopp. When the next will appear we do not know, but we hope it will not be too long delayed. At least the appearance of two bright comets so near the end of the 20th century is encouraging.

SELECTED LIST OF GREAT COMETS

Year	Name	Date of discovery	Greatest brightness	Mag.	Min. dist. from Earth, 10^6 km
1577		1 Nov	10 Nov	−4	94
1618		16 Nov	6 Dec	−4	54
1665		27 Mar	20 Apr	−4	85
1743	De Chéseaux	29 Nov	20 Feb 1744	−7	125
1811	Flaugergues	25 Mar	20 Oct	0	180
1843		5 Feb	3 Jul	−7	125
1858	Donati	2 June	7 Oct	−1	80
1861	Tebbutt	13 May	27 June	0	20
1874	Coggia	17 Apr	13 July	0	44
1882	Great Southern Comet	18 Mar	9 Sept	−10	148
1910	Daylight Comet	13 Jan	30 Jan	−4	130
1927	Skjellerup–Maristany	27 Nov	6 Dec	−6	110
1965	Ikeya–Seki	18 Sept	14 Oct	−10	135
1996	Hyakutake	30 Jan	1 May	−1	21
1997	Hale–Bopp	22 July 1995	30 Apr	−1.5	193

Millennium Comets

▲ ***Comet Hyakutake**, photographed in 1996 by Kent Blackwell (Virginia Beach, USA). Note the lovely green colour of the comet in this photograph.*

The closing years of the old century were graced by two bright comets. The first of these was discovered on 30 January 1996 by the Japanese amateur Yuji Hyakutake, using 25 × 150 binoculars; it was then of the 11th magnitude. It brightened steadily, and moved north in the sky; it reached perihelion on 1 May 1996, at 34 million kilometres (21 million miles) from the Sun. On 24 March it had passed Earth at 15 million kilometres (9,300,000 miles) – 40 times as far away as the Moon. At this time it was near Polaris in the sky; the magnitude was −1, and there was a long, gossamer-like tail extending for 100 degrees. The main feature of the comet was its beautiful green colour. It faded quickly during April; its period is around 15,000 years. In fact it was a small comet, with a nucleus estimated to be no more than 3.2 kilometres (2 miles) in diameter.

The second bright comet was discovered on 22 July 1995 independently by two American observers, Alan Hale and Thomas Bopp. It was by no means a faint telescopic object, but was 900 million kilometres (560 million miles) from the Sun, beyond the orbit of Jupiter. It brightened steadily; by the autumn of 1996 it had reached naked-eye visibility, and became brilliant in March and April 1997, with a magnitude exceeding -1. It passed Earth on 22 March 1997, at over 190 million kilometres (120 million miles); had it come as close as Hyakutake had done, it would have cast shadows. Perihelion was reached on 1 April, at over 125 million kilometres (over 80 million miles) from the Sun. It will return in about 3500 years.

Hale–Bopp was a very active comet, throwing off shells from its rotating nucleus; there was a curved reddish-brown dust tail and a very long, blue gas or ion tail. Unquestionably, it was the most striking comet of recent times, and possibly the best since the Daylight Comet of 1910. Astronomers everywhere were sorry to bid it farewell!

► ***Comet Hale–Bopp**, photographed by Kent Blackwell on 7 March 1997 (50 mm F2, exposure 2 minutes 6 seconds, Fuji Super G Plus 800).*

▼ ***Comet Hale–Bopp**, photographed by the author on 1 April 1997, at Selsey, West Sussex (Nikon F3, 50 mm, exposure 40 seconds, Fuji ISO 800).*

▲ ***Comet Hale–Bopp***, *photographed by Martin Mobberley on 13 March 1997.*

◀ ***Head of Hale–Bopp***, *photographed by the author on 1 April 1997 (15-inch reflector, exposure 1 minute).*

Meteors

▲ ***Woodcut of the Leonid Meteor Storm of 1833**, when it was said that meteors 'rained down like snowflakes'. Other major Leonid meteor storms were those of 1833, 1866 and 1966; we may well be treated to another in 1999.*

▼ ***Great Meteor of 7 October 1868**. Old painting by an unknown artist. The meteor was so brilliant that it attracted widespread attention, and seems to have been as bright as the Moon, lasting for several seconds and leaving a trail which persisted for minutes.*

Meteors are cometary debris. They are very small, and we see them only during the last seconds of their lives as they enter the upper atmosphere at speeds of up to 72 kilometres (45 miles) per second. What we actually observe, of course, is not the tiny particles themselves (known more properly as meteoroids) but the luminous effects which they produce as they plunge through the air. On average a 'shooting-star' will become visible at a height of about 115 kilometres (70 miles) above ground level, and the meteoroid will burn out by the time it has penetrated to 70 kilometres (45 miles), finishing its journey in the form of fine 'dust'. Still smaller particles, no more than a tenth of a millimetre across, cannot produce luminous effects, and are known as micrometeorites.

When the Earth moves through a trail of cometary debris we see a shower of shooting-stars, but there are also sporadic meteors, not connected with known comets, which may appear from any direction at any moment. The total number of meteors of magnitude 5 or brighter entering the Earth's atmosphere is around 75 million per day, so that an observer may expect to see something of the order of ten naked-eye meteors per hour, though during a shower the number will naturally be higher.

It is also worth noting that more meteors may be expected after midnight than before. During evenings, the observer will be on the trailing side of the Earth as it moves round the Sun, so that incoming meteors will have to catch it up; after midnight the observer will be on the leading side, so that meteors meet the Earth head-on, so to speak, and the relative velocities are higher.

The meteors of a shower will seem to issue from one particular point in the sky, known as the radiant. The particles are travelling through space in parallel paths, so that we are dealing with an effect of perspective – just as the parallel lanes of a motorway appear to 'radiate' from a point near the horizon.

The richness of a shower is measured by its Zenithal Hourly Rate (ZHR). This is the number of naked-eye meteors which could be seen by an observer under ideal conditions, with the radiant at the zenith. These conditions are never met, so that the observed rate is always appreciably lower than the theoretical ZHR.

Each shower has its own particular characteristics. The Quadrantids of early January have no known parent comet; the radiant lies in the constellation of Boötes (the Herdsman), the site of a former constellation, the Quadrant, which was rejected by the International Astronomical Union and has now disappeared from the maps. The ZHR can be very high, but the maximum is very brief. The April Lyrids are associated with Thatcher's Comet of 1861, which has an estimated period of 415 years; the ZHR is not usually very high, but there can be occasional rich displays, as last happened in 1982. Two showers, the Eta Aquarids of April–May and the Orionids of October, come from Halley's Comet, though they were not particularly rich around the time of the comet's last return in 1986. The October Draconids are associated with the periodical comet Giacobini–Zinner, and are sometimes referred to as the Giacobinids. Usually they are sparse, but they produced a major storm in 1933, when for a short time the rate of observed meteors reached 350 per minute. Ever since then, unfortunately, the Draconids have been very disappointing.

Two major showers occur in December: the Geminids and the Ursids. The Geminids have an unusual parent – the asteroid Phaethon, which is very probably a dead comet. The Ursids, with the radiant in the Great Bear, are associated with Tuttle's Comet and can sometimes be rich, as in 1945 and again in 1986.

Some showers appear to have decreased over the years. The Andromedids, as we have seen, are now almost extinct. The Taurids, associated with Encke's Comet, are not usually striking, though they last for well over a month; reports seem to indicate that in past centuries they were decidedly richer than they are now.

Probably the most interesting showers are the Perseids and the Leonids. The Perseids are very reliable, and last for several weeks with a sharp maximum on 12 August each year; if you look up into a clear, dark sky for a few minutes during the first fortnight in August, you will be very unlucky not to see several Perseids. The fact that the display never fails us shows that the particles have had time to spread all round the orbit of the parent comet, Swift–Tuttle, which has a period of 130 years and was last back to perihelion in 1992. The comet was not then conspicuous, but at its next return it will come very near the Earth – certainly within a couple of million kilometres, perhaps even closer – and there have been suggestions that it might hit us. In fact the chances of a collision are many hundreds to one against, but certainly Swift–Tuttle will be a magnificent spectacle. It is a pity that nobody born before the end of the 20th century will see it.

The Leonids are quite different. The parent comet, Tempel–Tuttle, has a period of 33 years, and it is when the comet returns to perihelion that we see major Leonid displays; the particles are not yet spread out all round the comet's orbit. Superb meteor storms were seen in 1799, 1833 and 1866. The expected displays of 1899 and 1933 were missed, because the swarm had been perturbed by Jupiter and Saturn, but in 1966 the Leonids were back with a vengeance, reaching a peak rate of over 60,000 per hour. Sadly, this lasted for only about 40 minutes, and it occurred during daylight in Europe, so the observers in the New World had the best view. Leonid showers have been traced back for many centuries, and indeed 902 was known as 'the Year of the Stars'. A few Leonids are always seen around 17 November.

▲ ***Fireball*** *(a brilliant meteor) photographed at 22.55 UT on 8 November 1991 by John Fletcher, from Gloucester, England. Exposure time 6 seconds; film 3M 1000; focal length 50 mm; f/2.8.*

▲ ***The 'radiant' principle.*** *I took this picture from Alaska in 1992; the parallel tracks seem to radiate from a point near the horizon.*

◄ ***The Leonid meteor storm*** *as seen from Arizona, 17 November 1966. It seems to have been just as rich as the storms of 1799, 1833 and 1866.*

▼ ***Comet Swift–Tuttle****, the parent comet of the Perseid meteors, photographed by Don Trombino at 23.35 UT on 12 December 1992. It never became bright at this return, but was widely observed.*

SELECTED ANNUAL METEOR SHOWERS

Shower	Begins	Max.	Ends	Max. ZHR	Parent comet	Notes
Quadrantids	1 Jan	4 Jan	6 Jan	60	–	Radiant in Boötes. Short, sharp max.
Lyrids	19 Apr	21 Apr	25 Apr	10	Thatcher	Occasionally rich, as in 1922 and 1982.
Eta Aquarids	24 Apr	5 May	20 May	35	Halley	Broad maximum.
Delta Aquarids	15 July	29 July 20 Aug	6 Aug	20	–	Double radiant. Faint meteors.
Perseids	23 Jul	12 Aug	20 Aug	75	Swift–Tuttle	Rich; consistent.
Orionids	16 Oct	22 Oct	27 Oct	25	Halley	Swift; fine trails.
Draconids	10 Oct	10 Oct	10 Oct	var.	Giacobini–Zinner	Usually weak, but occasional great displays, as in 1933 and 1946.
Taurids	20 Oct	3 Nov	30 Nov	10	Encke	Slow meteors. Fine display in 1988.
Leonids	15 Nov	17 Nov	20 Nov	var.	Tempel–Tuttle	Usually weak, but occasional storms, as in 1933, 1866 and 1966. Storm expected in 1999.
Andromedids	15 Nov	20 Nov	6 Dec	v. low	Biela	Now almost extinct.
Geminids	7 Dec	13 Dec	16 Dec	75	Phaethon (asteroid)	Rich, consistent.
Ursids	17 Dec	23 Dec	25 Dec	5	Tuttle	Can be rich, as in 1945 and 1986.

Meteorites

A meteorite is a solid particle which comes from space and lands on the Earth, sometimes making a crater. It is not simply a large meteor, and there is no connection between the two types of objects. Meteors, as we have seen, are the debris of comets. Meteorites come from the asteroid zone, and are associated neither with shooting-star meteors nor with comets. It is probably true to say that there is no difference between a large meteorite and a small asteroid.

Meteorites are divided into three main classes: irons (siderites), stony-irons (siderolites) and stones (aerolites). Irons are composed almost entirely of iron and nickel. Aerolites are of two sorts, chondrites and achondrites. Chondrites contain small spherical particles known as chondrules, which may be from one to ten millimetres (less than half an inch) across and are fragments of minerals, often metallic; achondrites lack these chondrules. Of special interest are the carbonaceous chondrites which contain not only carbon compounds but also organic materials. It was even suggested that one famous carbonaceous chondrite, the Orgueil Meteorite which fell in France on 14 May 1864, contained 'organized elements' which could have come from living material, though it seems much more likely that the meteorite was contaminated after it landed.

Most museums have meteorite collections; irons are more often on display than stones, but only because they are more durable and are more likely to be recovered in recognizable form. Areas such as Western Australia and, particularly, Antarctica are fruitful grounds for meteorite-hunters, because there has been relatively little human activity there. All known meteorites weighing more than 10 tonnes are irons (the largest aerolite, which fell in Manchuria in 1976, has a weight of only 1766 kilograms) but it is not always easy to identify a meteorite simply by its appearance, and often it takes a trained geologist to tell what is meteoritic and what is not. One test for an iron meteorite is to cut it and etch with dilute acid. Some irons will show the geometrical 'Widmanstätten patterns' not found in ordinary minerals.

▼ ▶ ***Tektites*** *may be meteoritic, but are more probably of terrestrial origin.*

▶ ***The Glatton Meteorite*** *which fell in Cambridgeshire on 5 May 1991. It weighed 767 grams.*

▶ ***Nickel-iron meteorites*** *found at the site of the Meteor Crater in Arizona, and now on display in the museum there.*

Meteorites have been known since very early times, though it was not until 1803 that a shower of stones, at L'Aigle in France, gave conclusive proof that they come from the sky. Some interesting specimens are found here and there. The Sacred Stone at Mecca is certainly a meteorite, and it is on record that as recently as the 19th century part of a South African meteorite was used to make a sword for the Emperor Alexander of Russia.

The largest known meteorite is still lying where it fell, in prehistoric times, at Grootfontein near Hoba West in Namibia. It weighs at least 60 tonnes. There are no plans to shift it, but not so long ago action had to be taken to protect it from being vandalized by troops of the United Nations peacekeeping force. Second in order of size is the Ahnighito ('Tent'), which was found in Greenland by the explorer Robert Peary in 1897, and is now in the Hayden Planetarium in New York.

Over 20 meteorites have been known to fall over the British Isles, and most have been recovered. The most celebrated of them shot over England on Christmas Eve in 1965 and broke up, showering fragments around the Leicestershire village of Barwell. The latest British meteorite – a small chondrite – fell at Glatton, in Cambridgeshire, on 5 May 1991, landing 20 metres from a retired civil servant who was doing some casual gardening. Incidentally, there is no known case of serious injury caused by a tumbling meteorite, though admittedly a few people have had narrow escapes.

Both the greatest falls during the 20th century have been in Siberia. On 30 June 1908 an object struck the Tunguska region, blowing pine trees flat over a wide area which was, mercifully, uninhabited. Owing to the disturbed state of Russia at that time no expedition reached the site until 1927, and though the pine trees were still flat there was no crater and no evidence of meteoritic material. It is possible that the impactor was icy, in which case it may have been a fragment of a comet, but we do not really know. There is no mystery about the second Siberian fall, in the Sikhote–Alin area on 12 February 1947; many small craters were found, and many pieces of the meteorite were salvaged.

There has been much discussion about the eight SNC meteorites, named after the regions in which they were found (Shergotty in India, Nakhla in Egypt and Chassigny in France). They seem to be much younger than most meteorites, and to be different in composition; it has been suggested that they have come from the Moon or even Mars. This is highly speculative, but is at least an intriguing possibility, though it is not easy to see how they could have arrived here. More enigmatic are the tektites, small glassy objects which seem to have been heated twice and are aerodynamically shaped; they are found only in localized areas, notably in Australasia and parts of the Czech and Slovak Republics. For many years, they were classed as unusual meteorites, but it now seems more probable that they are of terrestrial origin, presumably shot out from volcanoes.

One thing which we can do is to measure the ages of meteorites. Most seem to be about 4.6 thousand million years old, which is about the same as the age of the Solar System itself. Pick up a meteorite, and you are handling a piece of material which moved around between the planets for thousands of millions of years before coming to its final resting-place on the surface of our own world.

SOME LARGE METEORITES

Name	Weight, tonnes
Hoba West, Grootfontein, Namibia, Africa	Over 60
Ahnighito (The Tent), Cape York, West Greenland	34
Bacuberito, Mexico	27
Mbosi, Tanzania	26
Agalik, Cape York, West Greenland	21
Armanty, Outer Mongolia	20
Willamette, Oregon, USA	14
Chapuderos, Mexico	14
Campo del Cielo, Argentina	13
Mundrabilla, Western Australia	12
Morito, Mexico	11

◀ ***The Hoba West Meteorite**, photographed by Ludolf Meyer. This is the heaviest known meteorite.*

◀ ***Fragment of the Barwell Meteorite**, which was found in Leicestershire; the meteorite landed on 24 December 1965. It was widely observed as it passed across England, and broke up during the descent. It was the largest meteorite to fall in Britain in recorded times; the original weight may have been of the order of 46 kilograms.*

Meteorite Craters

Go to Arizona, not far from the town of Winslow, and you will come to what has been described as 'the most interesting place on Earth'. It is a huge crater, 1265 metres (4150 feet) in diameter and 175 metres (575 feet) deep; it is well preserved, and has become a well-known tourist attraction, particularly as there is easy access from Highway 99. There is no doubt about its origin; it was formed by the impact of a meteorite which hit the Arizonan desert in prehistoric times. The date of its origin is not known with certainty, and earlier estimates of 22,000 years ago may be too low. White men have known about it since 1871.

The crater is circular, even though the impactor came in at an angle. When the meteorite struck, its kinetic energy was converted into heat, and it became what was to all intents and purposes a very powerful bomb. What is left of the meteorite itself is very probably buried beneath the crater's south wall. Incidentally, the popular name is wrong. It is called Meteor Crater, but this should really be 'Meteorite' Crater.

A smaller but basically similar impact crater is Wolf Creek in Western Australia. There are various local legends about it. The Kjaru Aborigines call it Kandimalal, and describe how two rainbow snakes made sinuous tracks across the desert, forming Wolf Creek and the adjacent Sturt Creek, while the crater marks the spot where one of the snakes emerged from below the ground. It is much younger than the Arizona crater; the age cannot be more than 15 million years, and 2 million years is a more likely value. Wolf Creek is more difficult to reach than Meteor Crater, and the road from the nearest settlement, Halls Creek, is usually open for only part of the year, but it has now been well studied since aerial surveys first identified it in 1947. The wall rises at an angle of 15 to 35 degrees, and the floor is flat, 55 metres (180 feet) below the rim and 25 metres (80 feet) below the level of the surrounding plain. The diameter is 675 metres (2200 feet). Meteoritic fragments found in the area leave no doubt that it really is of cosmic origin.

Also in Australia there are other impact craters; one at Boxhole and a whole group at Henbury, both in Northern Territory. Equally intriguing is Gosse Bluff, which is at least 50,000 years old and very eroded, though there is the remnant of a central structure and indications of the old walls.

Lists of impact craters include structures in America, Arabia, Argentina, Estonia and elsewhere, but one must be wary of jumping to conclusions; for example, unbiased geologists who have made careful studies of the Vredefort Ring, near Pretoria in South Africa, are unanimous in finding that it is of internal origin. It is linked with local

▼ ***Site of the Siberian impact of 1908**, Tunguska; photographed by Don Trombino in 1991. No crater was produced, so presumably the projectile broke up before landing, but the results of the impact are still very evident.*

▼ ***Wolf Creek Crater** in Western Australia; an aerial photograph which I took in 1993. This is a very well-formed crater and possibly the most perfect example of an impact structure on the Earth, apart from the famous Meteor Crater in Arizona, USA.*

► ***'Saltpan'** near Pretoria, South Africa, was identified as an impact crater recently. Larger than the Arizona crater, the associated breccia are clearly seen. The water in the lake is salty. The surrounding wall is uniform in height. Photograph by Dr Kelvin Kemm, 1994.*

geology, and the form is not characteristic of collision. Note also that no crater is associated with the giant Hoba West Meteorite.

It has often been suggested that the Earth was struck by a large missile 65 million years ago, and that this caused such a change in the Earth's climate that many forms of life became extinct, including the dinosaurs. It has been claimed that the buried Chicxulub impact crater in the Yucatan Peninsula, Mexico, was the result of the meteorite fall which killed the dinosaurs.

No doubt further craters will be formed in the future; there are plenty of potential impactors moving in the closer part of the Solar System. Although the chances of a major collision are slight, they are not nil, which is partly why constant watch is now being kept to identify wandering bodies. It is even possible that if one of these bodies could be seen during approach, we might be able to divert it by nuclear warheads carried on ballistic missiles – though whether we would be given enough advance warning is problematical.

In January 2000, the British government set up a special committee to look into the whole question of danger from asteroidal or cometary impact. If there is such an impact, let us hope that we cope with the situation better than the dinosaurs did.

▲ ***Meteor Crater in Arizona, USA****. This is the most famous of all impact structures, though not now the largest one to be found. I took this aerial photograph in 1991.*

▼ ***Gosse's Bluff****, Northern Territory of Australia; photograph by Gerry Gerrard. Its impact origin is not in doubt, but it is very ancient, and has been greatly eroded.*

SOME IMPORTANT METEORITIC CRATERS

Name	Diameter, m	Date of discovery
Meteor Crater, Arizona	1265	1871
Wolf Creek, Australia	675	1947
Henbury, Australia	200 × 110	1931 (13 craters)
Boxhole, Australia	175	1937
Odessa, Texas, USA	170	1921
Waqar, Arabia	100	1932
Oesel, Estonia	100	1927

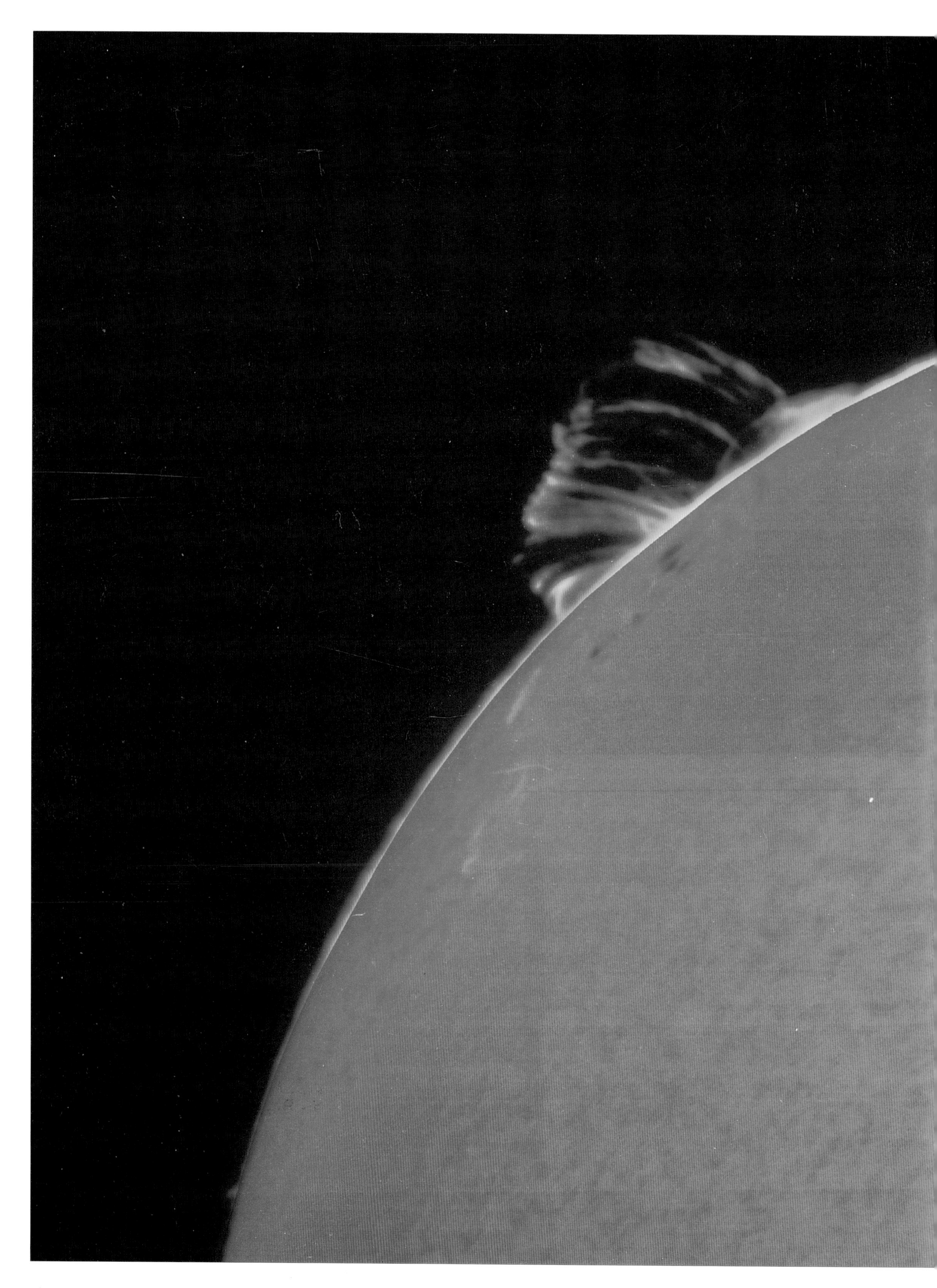

THE SUN

◄ ***Immense loop prominence*** *recorded by Don Trombino on 15 June 1992 using a Daystar 12-cm (4.5-inch) telescope. The black-and-white negative was converted to colour in the enlarger.*

Our Star: the Sun

Because the Sun appears so glorious in our sky, some people are disinclined to believe that it is only a star; indeed, astronomers relegate it to the status of a Yellow Dwarf! Its closeness to us means that it is the only star which we can examine in detail.

Its diameter is 1,392,000 kilometres (865,000 miles), and it could engulf over a million globes the volume of the Earth, but it is very much less dense, because it is made up of incandescent gas. At the core, where the energy is being produced, the temperature may be as high as 15,000,000 degrees C; even the bright surface which we can see – the photosphere – is at a temperature of 5500 degrees C. It is here that we see the familiar sunspots and the bright regions known as faculae. Above the photosphere comes the chromosphere, a layer of much more rarefied gas, and finally the corona, which may be regarded as the Sun's outer atmosphere.

The Sun is nowhere near the centre of the Galaxy; it is around 25,000 light-years from the nucleus. It is sharing in the general rotation of the Galaxy, moving at 220 kilometres (140 miles) per second, and taking 225 million years to complete one circuit – a period often called the cosmic year; one cosmic year ago, even the dinosaurs lay in the future!

The Sun is rotating on its axis, but it does not spin in the way that a solid body would do. The rotation period at the equator is 25.4 days, but near the poles it is about 34 days. This is easy to observe by the drift of the sunspots across the disk; it takes about a fortnight for a group to cross the disk from one limb to the other.

The greatest care must be taken when observing the Sun. Looking directly at it with any telescope, or even binoculars, means focusing all the light and (worse) the heat on to the observer's eye, and total and permanent blindness will result. Even using a dark filter is unsafe; filters are apt to shatter without warning, and in any case cannot give full protection. The only sensible method is to use the telescope as a projector, and observe the Sun's disk on a screen held or fastened behind the telescope eyepiece.

We know that the Earth is about 4600 million years old, and the Sun is certainly older than this. A Sun made up entirely of coal, and burning furiously enough to emit as much energy as the real Sun actually does, would be reduced to ashes in only 5000 years. In fact, the Sun's energy is drawn from nuclear transformations near its core, where the temperatures and pressures are colossal. Not surprisingly, the Sun consists largely of hydrogen (over 70 per cent), and near the core the nuclei of hydrogen atoms are combining to form nuclei of the next lightest element, helium. It takes four hydrogen nuclei to make one helium nucleus; each time this happens, a little energy is released and a little mass is lost. It is this energy which keeps the Sun shining, and the mass–loss amounts to 4 million tonnes per second. Fortunately there is no cause for immediate alarm; the Sun will not change dramatically for at least a thousand million years yet.

The photosphere extends down to about 300 kilometres (190 miles), and below this comes the convection zone, which has a depth of about 200,000 kilometres (125,000 miles); here, energy is carried upwards from below by moving streams and masses of gas. Next comes the radiative zone, and finally the energy-producing core, which seems to have a diameter of around 450,000 kilometres (280,000 miles). The theoretical models seem satisfactory enough, but there is one major problem out-

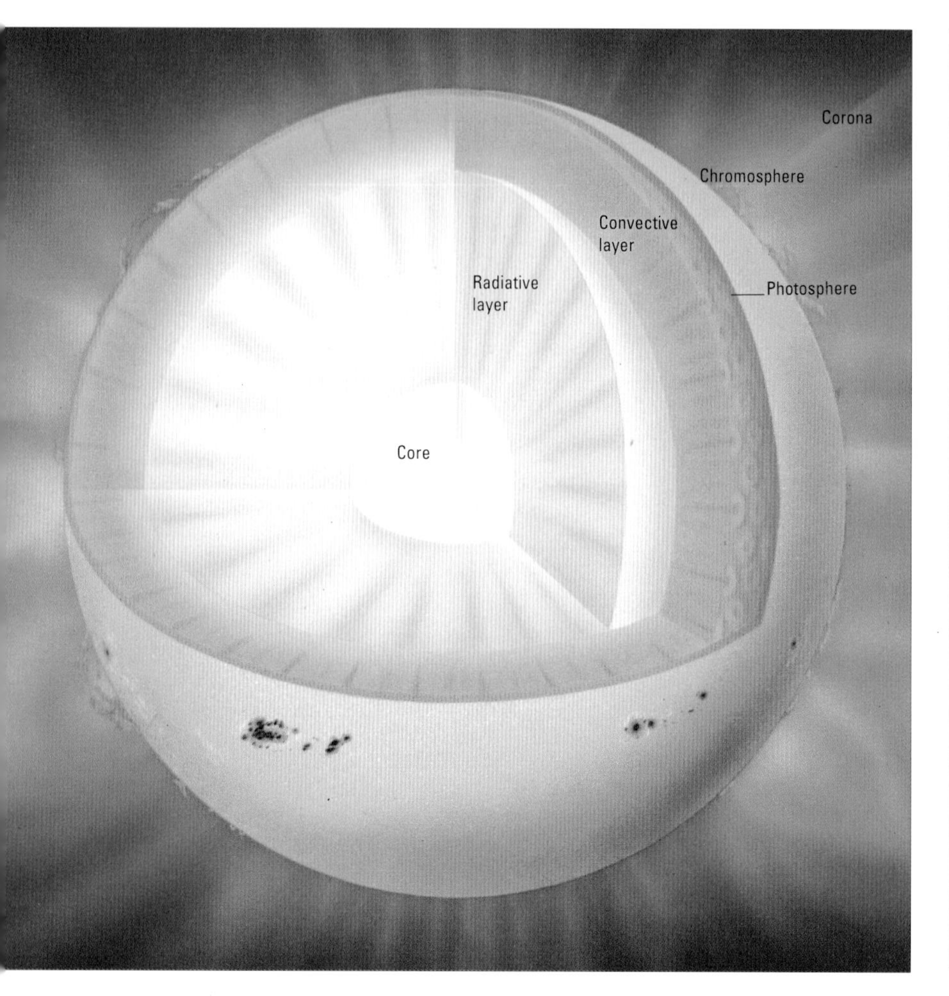

▼ ***Cross-section of the Sun*** *showing the core, radiative zone, convective zone, photosphere, chromosphere and corona.*

SOLAR DATA

Distance from Earth	149,597,893 km (92.970,000 mile or 1 astronomical unit)
Mean distance from centre of Galaxy	25,000 light-years
Velocity round centre of Galaxy	220 km/s (140 miles/s)
Revolution period round centre of Galaxy	225,000,000 years
Apparent diameter	max. 32′ 35″, mean 32′ 01″, min. 31′ 31″
Density, water = 1	1.409
Mass, Earth = 1	332,946
Mass	2 x 10^{27} tonnes
Volume, Earth = 1	1,303,600
Surface gravity, Earth = 1	27.9
Escape velocity	617.5 km/s (384 miles/s)
Mean apparent magnitude	−26.8 (600,000 Full Moons)
Absolute magnitude	+4.83
Spectrum	G2
Surface temperature	5500°C
Core temperature	about 15,000,000°C
Rotation period (equatorial)	25.4 days.
Diameter (equatorial)	1,392,000 km (865,000 miles)

Jupiter
Earth

standing. The Sun ought to send out vast numbers of strange particles called neutrinos, which are difficult to detect because they have no electrical charge and almost no mass – perhaps none at all. Yet the Sun appears to emit far fewer neutrinos than predicted.

If a neutrino scores a direct hit upon an atom of chlorine, the chlorine may be changed into a form of radioactive argon. Deep in Homestake Gold Mine in South Dakota, Ray Davies and his colleagues filled a large tank with over 450,000 litres of cleaning fluid, which is rich in chlorine; every few weeks they flushed out the tank to see how much argon had been produced by neutrino hits. In fact the numbers were strikingly less than they should have been, and similar experiments elsewhere have confirmed this. If the Sun's core temperature were reduced to 14 million degrees, the dearth of neutrinos could be explained; but this would introduce other difficulties, and at present the mystery remains. (It was essential to install the tank deep below the ground; otherwise the results would be affected by cosmic ray particles which, unlike neutrinos, cannot penetrate far below the Earth's surface.)

Like all other stars, the Sun began its career by condensing out of interstellar material, and at first it was not hot enough to shine. As it shrank, under the influence of gravity, it heated up, and when the core temperature had risen to 10 million degrees nuclear reactions were triggered off; hydrogen was converted into helium, and the Sun began a long period of steady emission of energy. As we have seen, it was not initially as luminous as it is now, and the increase in power may have had disastrous results for any life which may have appeared on Venus. But at the moment the Sun changes very little; the fluctuations due to its 11-year cycle are insignificant.

However, this will not last for ever. The real crisis will come when the supply of available hydrogen begins to become exhausted. The core will shrink and heat up as different types of reactions begin; the outer layers will expand and cool. The Sun will become a red giant star, and will be at least 100 times as luminous as at present, so that the Earth and the other inner planets are certain to be destroyed. Subsequently the Sun will throw off its outer layers, and the core will collapse, so that the Sun becomes a very small, incredibly dense star of the type known as a white dwarf. Eventually all its light and heat will leave it, and it will become a cold, dead globe – a black dwarf.

This may sound depressing, but the crisis lies so far ahead that we need not concern ourselves with it. In our own time, at least, there is no danger from the Sun.

▼ ***Homestake Mine**, in South Dakota, site of the world's most unusual 'telescope' – a large tank of cleaning fluid (tetrachloroethylene), rich in chlorine to trap solar neutrinos. The observed flux is only about one-third as great as predicted. The same has been found by investigators in Russia, using 100 tonnes of liquid scintillator and 144 photodetectors in a mine in the Donetsk Basin, and at Kamiokande in Japan.*

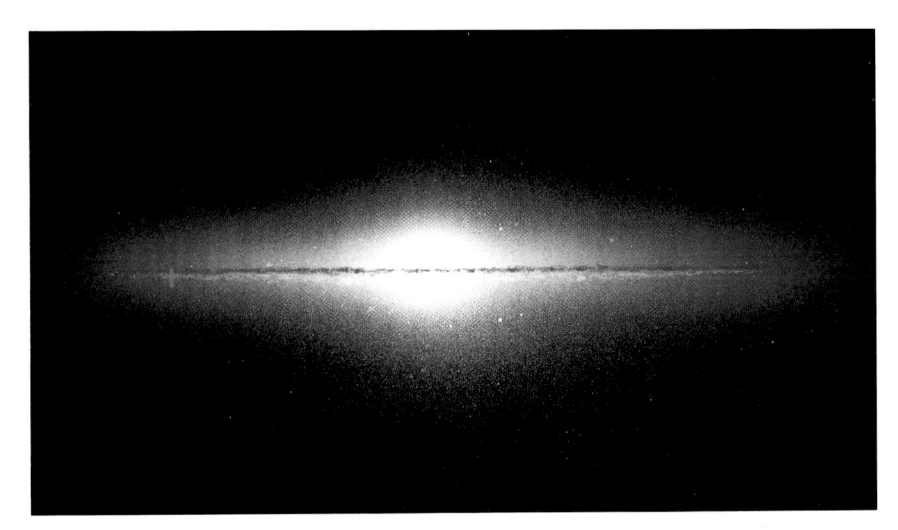

▲ ***The Sun in the Galaxy**. The Sun lies well away from the centre of the Galaxy; the distance from the centre is less than 30,000 light-years, and the Sun lies near the edge of one of the spiral arms.*

▶ ***Projecting the Sun**. The only safe way to view the Sun is to project it through a telescope on to a screen.*

The Surface of the Sun

Use a telescope to project the Sun's image, and you will see that the yellow disk is brightest at its centre and less brilliant at the edges; this is because towards the centre we are seeing into deeper and therefore hotter layers. There may be one or more darker patches which are known as sunspots. The spots are not genuinely black, but appear so because they are cooler than the surrounding regions of the photosphere.

A major spot is made up of a dark central portion or umbra, surrounded by a lighter penumbra. Sometimes the shapes are regular; sometimes they are very complex, with many umbrae contained in a single mass of penumbra. The temperature of the umbra is about 4500 degrees C, and of the penumbra 5000 degrees C, so that if a spot could be seen shining on its own, the surface brilliance would be greater than that of an arc-lamp.

Spots generally appear in groups. An 'average' two-spot group begins as a pair of tiny pores at the limit of visibility. The pores develop into proper spots, growing and separating in longitude; within two weeks the group has reached its maximum length, with a fairly regular leading spot and a less regular follower, together with many smaller spots spread around in the area. A slow decline then sets in, usually leaving the leader as the last survivor. Around 75 per cent of groups fit into this pattern, but there are many variations, and single spots are also common.

Sunspots may be huge; the largest on record, that of April 1947, covered an area of over 18,000 million square kilometres (7000 million square miles) when at its largest. Obviously they are not permanent. A major group may persist for anything up to six months, though very small spots often have lifetimes of less than a couple of hours.

Spots are essentially magnetic phenomena, and there is a fairly predictable cycle of events. Maxima, with many groups on view simultaneously, occur every 11 years or so; activity then dies down, until at minimum the disk may

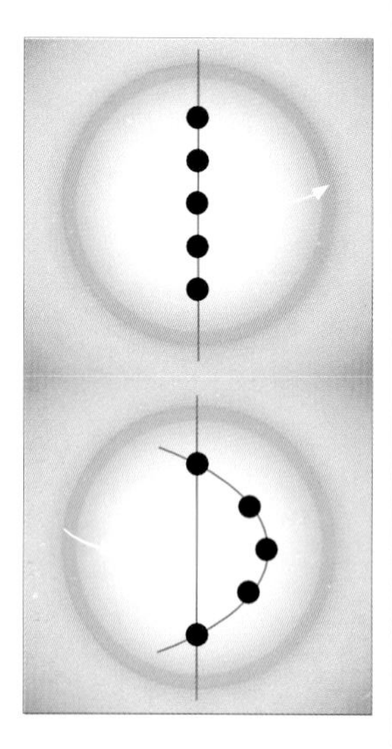

▲ ***Differential rotation****. The rotation period of the photosphere increases with increasing latitude. In the idealized situation shown here, if a row of sunspots lay along the Sun's central meridian, then, after one rotation, the spots would be spread out in a curve.*

▼ ***Solar rotation****. This sequence shows the giant sunspot group of 1947. After passing round the far side of the Sun, it reappeared to make a second crossing.*

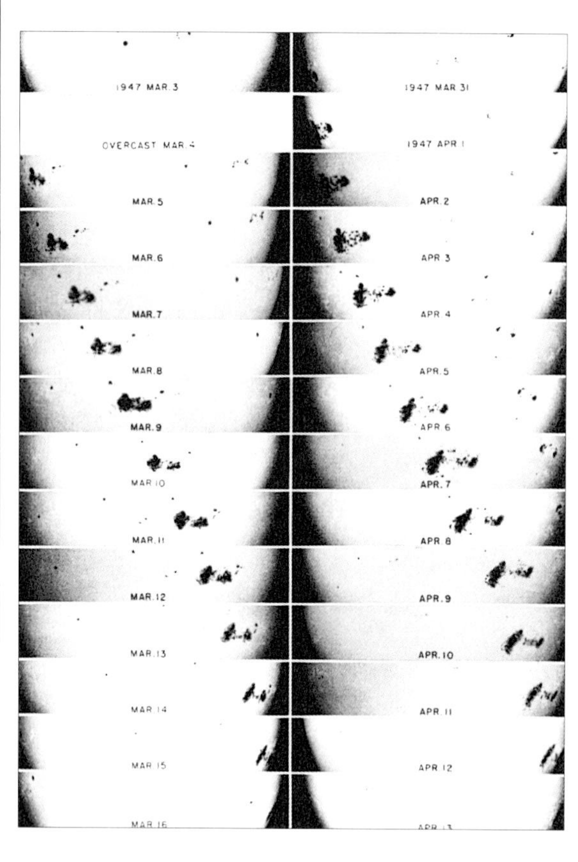

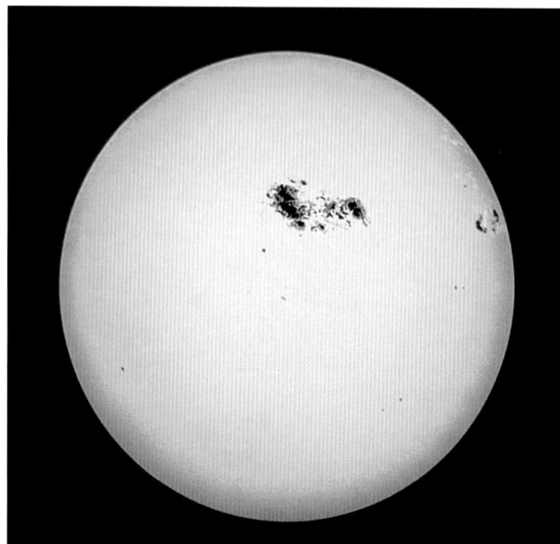

◀ ***The Great Sunspot of 1947*** *– the largest known. On April 8, it covered 18,000 million square km (7000 million square miles).*

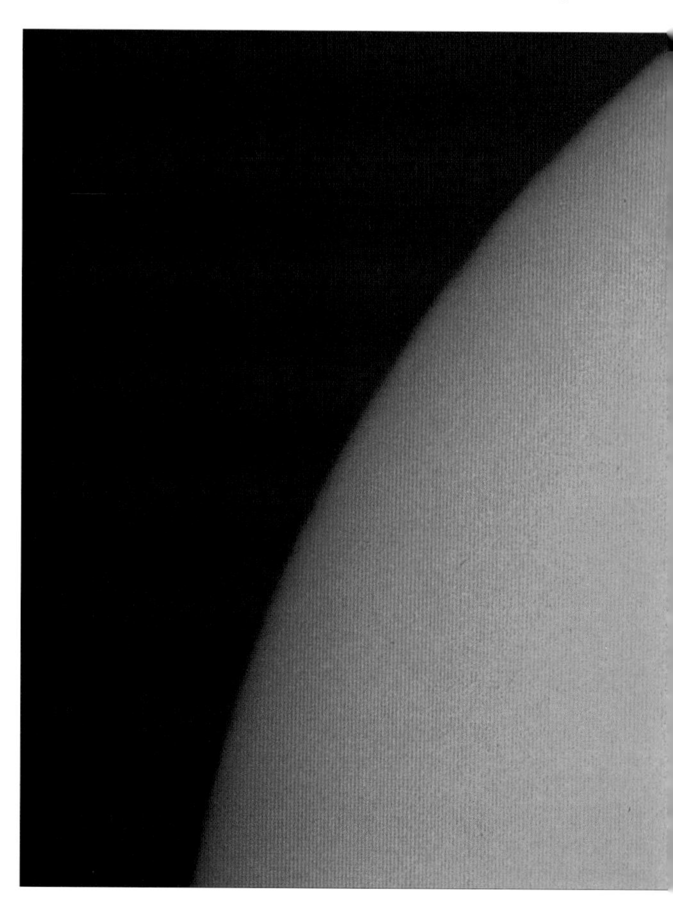

▶ ***Sunspots*** *photographed by H. J. P. Arnold.*

▼ ***The solar cycle****, 1650 to present. Not all maxima are equally energetic, and during the 'Maunder Minimum', 1645–1715, it seems that the cycle was suspended, though the records are incomplete. The vertical scale is the Zürich number, calculated from the number of groups and the number of spots.*

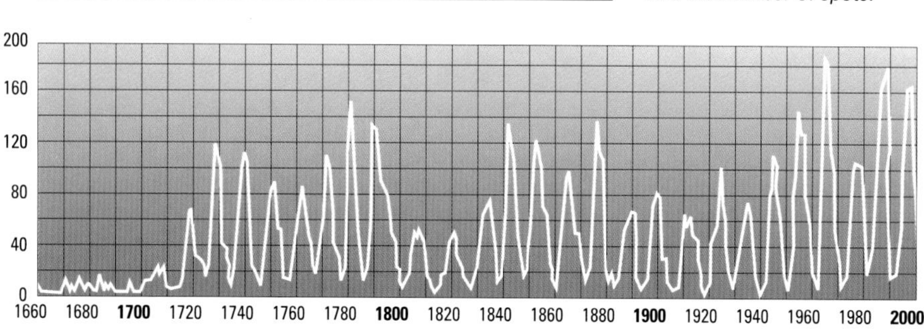

be free of spots for many consecutive days or even weeks, after which activity starts to build up once more towards the next maximum. The cycle is not perfectly regular, but 11 years is a good average length, so that there were maxima in 1957–8, 1968–9, 1979–80 and 1990–91.

The maxima are not equally energetic, and there seems to have been a long spell, between 1645 and 1715, when there were almost no spots at all, so that the cycle was suspended. This is termed the Maunder Minimum, after the British astronomer E. W. Maunder, who was one of the first to draw attention to it. Obviously the records at that time are not complete, but certainly there was a dearth of spots for reasons which are not understood. There is also evidence of earlier periods when spots were either rare or absent, and it may well be that other prolonged minima will occur in the future. Whether this has any effect upon the Earth's climate is a matter for debate, but it is true that the Maunder Minimum was a 'cold spell'; during the 1680s the River Thames froze over in most winters, and frost-fairs were held upon it.

There is a further peculiarity, first noted by the German amateur F. W. Spörer. At the start of a new cycle, the spots break out at latitudes between 30 and 45 degrees north or south of the solar equator. As the cycle progresses, new spots appear closer and closer to the equator, until at maximum the average latitude is only 15 degrees north or south. After maximum new spots become less common, but may break out at latitudes down to seven degrees. They never appear on the equator itself, and before the last spots of the old cycle die away the first spots of the new cycle appear at higher latitudes.

According to the generally accepted theory, proposed by H. Babcock in 1961, spots are due to the effects of the Sun's magnetic field lines, which run from one pole to the other just below the bright surface. The rotation period at the equator is shorter than that at higher latitudes, so that the field lines are dragged along more quickly, and magnetic 'tunnels' or flux tubes, each about 500 kilometres (300 miles) in diameter, are formed below the surface. These float upwards and break through the surface, producing pairs of spots with opposite polarities. At maximum the magnetic field lines are looped and tangled, but then rejoin to make a more stable configuration, so at the end of the cycle activity fades away and the field lines revert to their original state.

The polarities of leader and follower are reversed in the two hemispheres, and at the end of two cycles there is a complete reversal, so there are grounds for suggesting that the true length of a cycle is 22 years rather than 11.

Tracking sunspots is a fascinating pastime. A group takes slightly less than two weeks to cross the disk from one limb to the other, and after an equivalent period it will reappear at the following limb if, of course, it still exists. A spot is foreshortened when near the limb, and the penumbra of a regular spot appears broadened to the limbward side. This 'Wilson effect' indicates that the spot is a depression rather than a hump, but not all spots show it.

Many spots are associated with faculae (Latin for 'torches') which may be described as bright, cloudlike features at higher levels; they are often seen in regions where spots are about to appear, and persist for some time after the spots have died out. And even in non-spot zones, the surface is not calm. The photosphere has a granular structure; each granule is about 1000 kilometres (600 miles) in diameter with a lifetime of about eight minutes. They represent currents, and it is estimated that the surface includes about four million granules at any one time.

It would be idle to pretend that we have anything like a complete understanding of the Sun. Many problems have been solved, but we still have much to learn about our 'daytime star'.

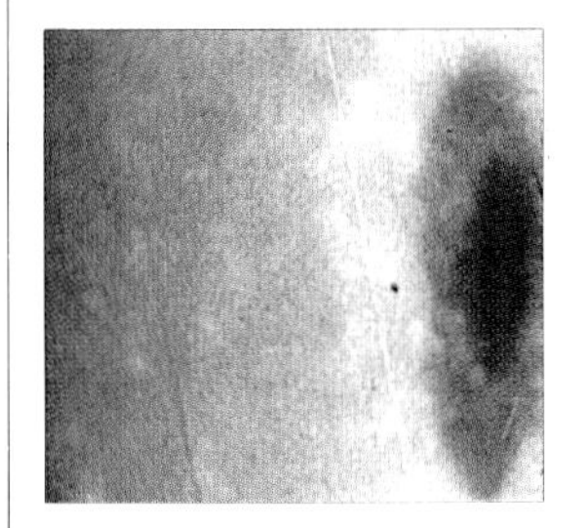

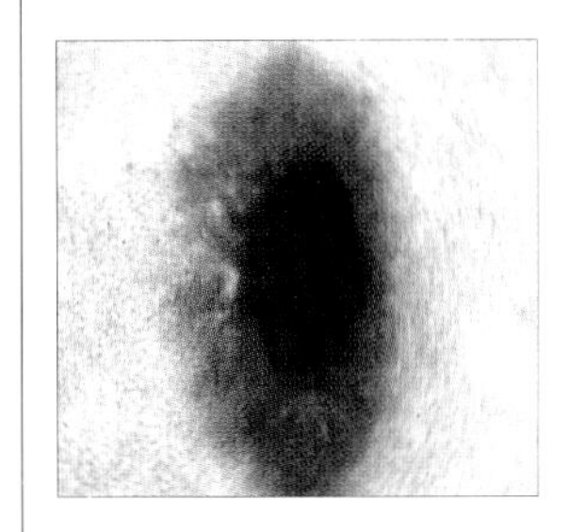

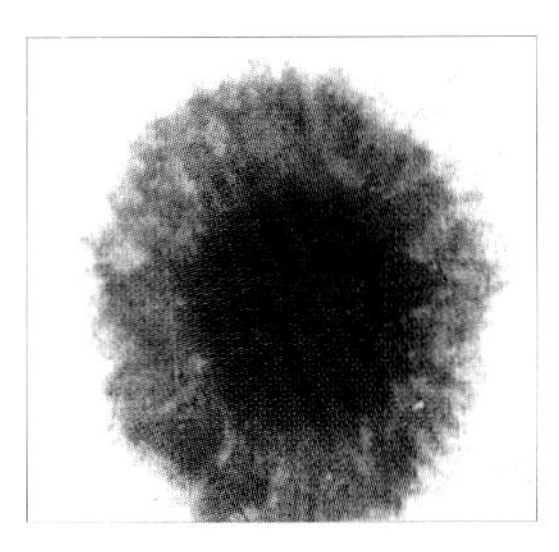

▲ ***The Wilson Effect.*** *As shown in these three pictures, many spots behave as though they were hollows; the penumbra to the inward side appears broadened when the spot is foreshortened. The original observations, by Scottish astronomer A. Wilson, were made in 1769.*

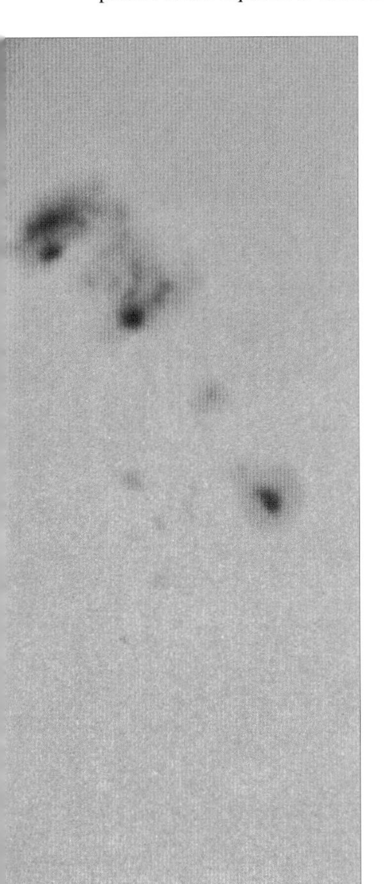

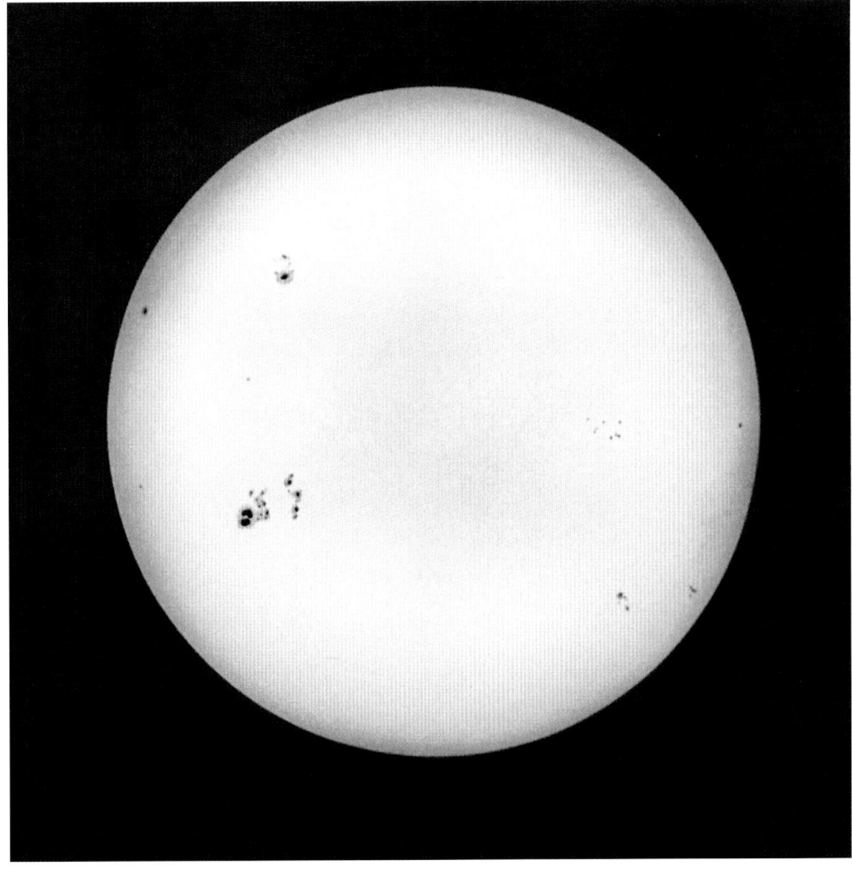

◄ ***Sunspots****, 26 May 1990; I made this sketch by projection with a 12.7-cm (5-inch) refractor. Faculae are shown to the upper left.*

The Solar Spectrum

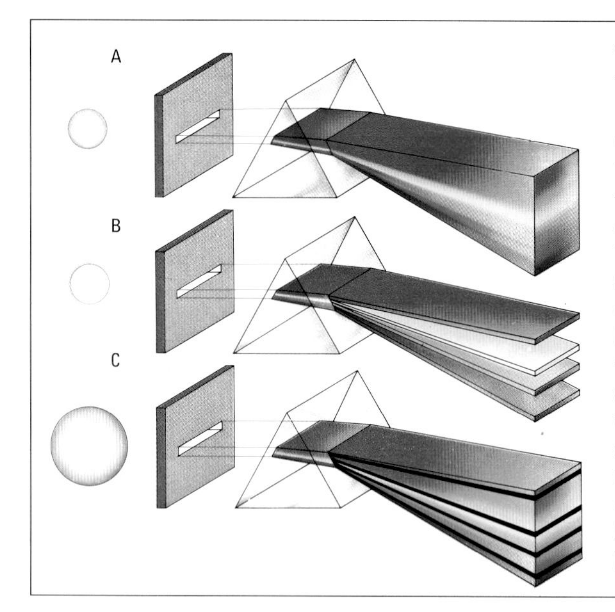

The solar spectrum The photosphere produces a rainbow or continuous spectrum from red at the long wavelength end to violet at the shortwave end (A). The solar atmosphere should produce an emission spectrum (B), but as light is radiated from the surface, gaseous elements in the atmosphere absorb specific wavelengths, so the spectrum observed on Earth has gaps (dark lines, called Fraunhofer lines) in it (C).

If we could do no more than examine the bright photosphere, and follow the changes in the spots, faculae and granules, our knowledge of the Sun would remain slender indeed. Luckily this is not the case, and we can turn to that other great astronomical instrument, the spectroscope.

Just as a telescope collects light, so a spectroscope splits it up. A beam of sunlight is made up of a mixture of colours, and a glass prism will bend or refract the various colours unequally; short wavelengths (blue and violet) are refracted most, long wavelengths (orange and red) least. The first experiments were made by Isaac Newton in 1666, but he never followed them up, perhaps because the prisms he had to use were of poor quality. In 1802 the English scientist W. H. Wollaston passed sunlight through a prism, via a slit in an opaque screen, and obtained a true solar spectrum, with red at one end through orange, yellow, green, blue and violet. Wollaston saw that the rainbow band was crossed by dark lines, but he mistakenly thought that these lines merely marked the boundaries between different colours. Twelve years later Josef Fraunhofer made a much more detailed investigation, and realized that the dark lines were permanent, keeping to the same positions and with the same intensities; he mapped 324 of them, and even today they are still often referred to as the Fraunhofer lines. In 1859, two German physicists, Gustav Kirchhoff and Robert Bunsen, interpreted them

▲ ***Solar telescope*** *at the Mount Wilson Observatory. The 46-m (150-foot) tower telescope at Mount Wilson Observatory, California, collects light using a mirror at the top of the tower, and sends it down in a fixed direction to the recording instruments.*

RATIO OF ELEMENTS IN THE SUN

Element	Number of atoms, the number of hydrogen atoms being taken as 1,000,000
Helium	63,000
Oxygen	690
Carbon	420
Nitrogen	87
Silicon	45
Magnesium	40
Neon	37
Iron	32
Sulphur	16
All others	Below 5

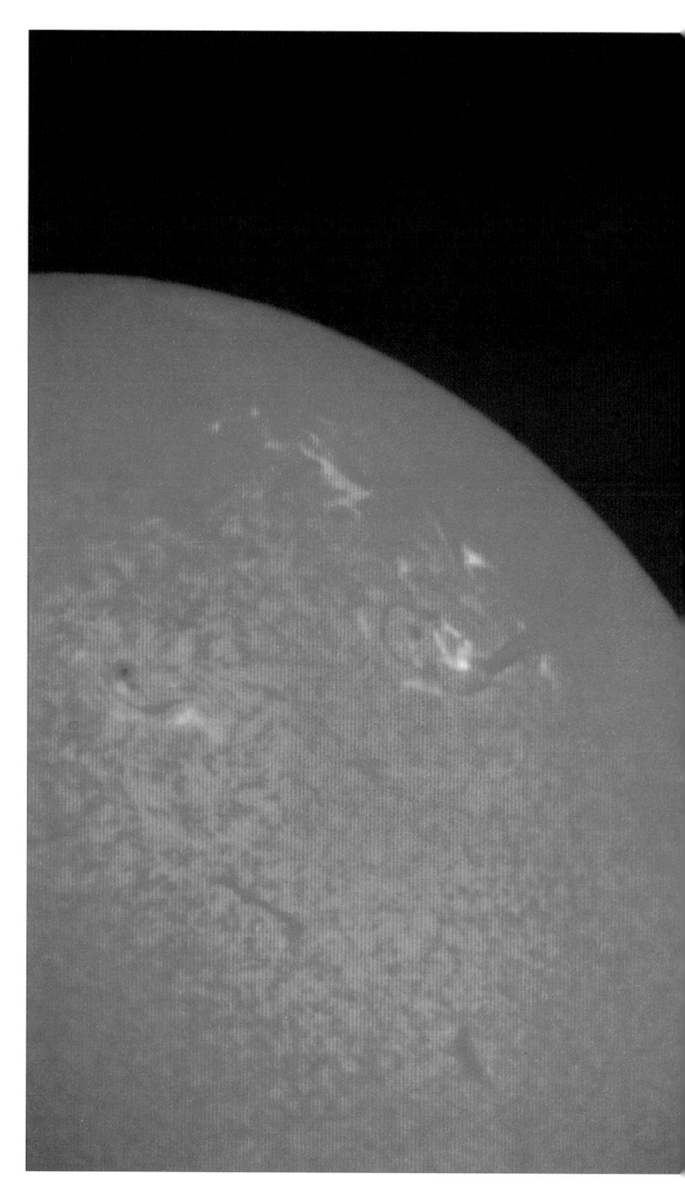

▶ ***The Sun*** *imaged in the light of hydrogen (H-alpha) by Don Trombino.*

correctly, and so laid the foundations of modern astrophysics.

An incandescent solid, liquid, or gas at high pressure, will yield a continuous spectrum, from red to violet. An incandescent gas at low pressure will produce a different spectrum, made up of isolated bright lines, each of which is characteristic of one particular element or group of elements; this is known as an emission spectrum. For example, incandescent sodium will produce a spectrum which includes two bright yellow lines; if these are seen, then sodium must be responsible, because nothing else can produce them. Many elements, such as iron, have spectra so complex that they include many thousands of lines in their unique fingerprints.

The Sun's photosphere yields a continuous spectrum. Above the photosphere lies the chromosphere, which is made up of low-pressure gas and produces an emission spectrum. Normally these lines would be bright; because they are silhouetted against the rainbow background they appear dark, but their positions and intensities are unaltered, so that there is no problem in identifying them. Two prominent dark lines in the yellow part of the band correspond exactly to the two famous lines of sodium, and therefore we can prove that there is sodium in the Sun.

It has been found that the most plentiful element in the Sun is hydrogen, which accounts for 71 per cent of the total mass; any other result would have been surprising, since in the universe as a whole the numbers of hydrogen atoms outnumber those of all the other elements combined. In the Sun, the next most plentiful element is helium, with 27 per cent. This does not leave much room for anything else, but by now most of the 92 elements known to occur in Nature have been identified in smaller quantities. Helium was actually identified in the solar spectrum before it was known on Earth; it was found by Lockyer in 1868, who named it after the Greek *helios* (Sun). Not until 1894 was it tracked down on our own world.

Many instruments of various kinds are based on the principle of the spectroscope. One such is the spectroheliograph, where two slits are used and it is possible to build up an image of the Sun in the light of one selected element only (the visual equivalent of the spectroheliograph is the spectrohelioscope). Similar results can be obtained by using special filters, which block out all the wavelengths except those which have been selected. Today, equipment of this sort is used by many amateur observers as well as professionals – and solar observation is always fascinating, if only because there is always something new to see; the Sun is always changing, and one can never tell what will happen next.

▲ ***The Kitt Peak Solar Telescope*** *in Arizona. The Sun's light is collected by the heliostat, a mirror at the top of the structure, and is directed down the slanted tunnel on to a curved mirror at the bottom; this in turn reflects the rays back up the tunnel to a flat mirror, which sends the rays down through a hole to the lab below.*

SOME IMPORTANT FRAUNHOFER LINES IN THE SOLAR SPECTRUM

Letter	Wavelength, Å	Identification
C (H-alpha)	6563	Hydrogen
D_1	5896	Sodium
D_2	5890	
b_1	5183	Magnesium
b_2	5173	
b_3	5169	
b_4	5167	
F (H-beta)	4861	Hydrogen
G	4308	Iron
g	4227	Calcium
h (H-delta)	4102	Hydrogen
H	3967	Ionized calcium
K	3933	

(One Ångström (Å), named in honour of the Swedish scientist Anders Ångström, is equal to one hundred-millionth part of a centimetre. The diameter of a human hair is roughly 500,000 Å. Another often-used unit is the nanometre. To convert Ångstroms into nanometres, divide by 10, so that, for instance, the wavelength of the H-alpha line is 656.3 nm.)

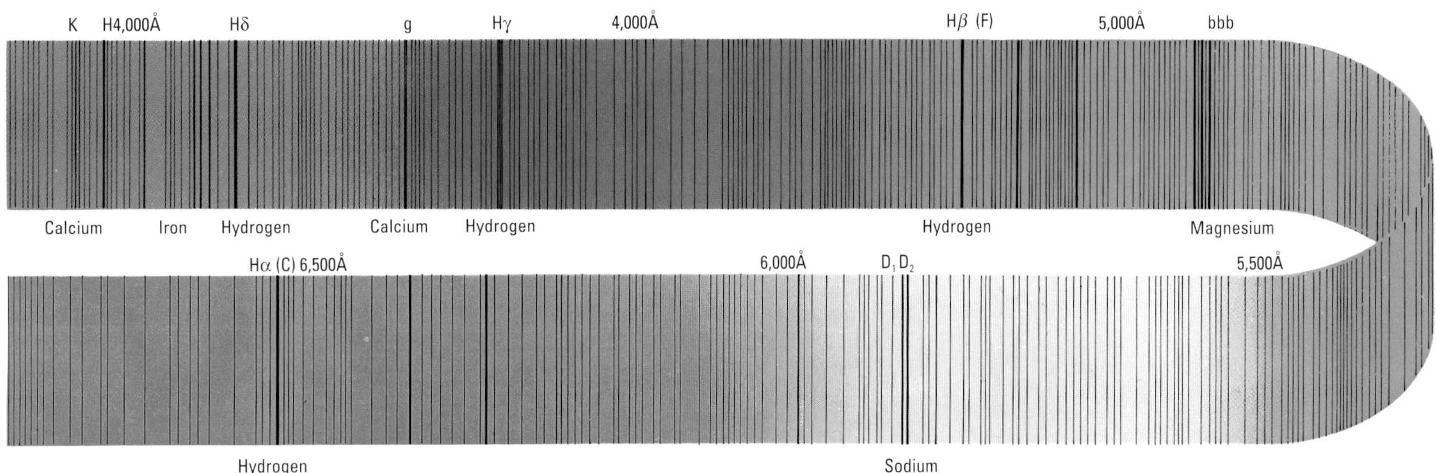

▼ ***The visible spectrum*** *of the Sun is very complex; more than 70 elements have been identified. The photosphere, a layer of dense gas, produces a rainbow or continuous spectrum. The rarefied gases in the chromosphere would yield bright lines if seen on their own but against the photosphere are 'reversed' and appear dark, although their positions and intensities are unaffected.*

Eclipses of the Sun

▼ ***The Moon's shadow** is divided, like any other, into two regions, the dark central 'umbra', and the lighter 'penumbra', within which part of the Sun remains visible. A total eclipse of the Sun occurs when the Earth passes into the shadow cast by the Moon. However, the eclipse only appears total from the limited region of the Earth's surface which is covered by the umbra; from inside the penumbra the eclipse is partial. An annular eclipse occurs when the Moon is near apogee, and its shadow cone does not reach the Earth. The angular size of the Moon as seen from Earth is therefore too small to cover the Sun's disk, so that a thin ring of light remains visible around the black disk of the Moon.*

The Moon moves round the Earth; the Earth moves round the Sun. Therefore, there must be times when the three bodies line up, with the Moon in the mid position. The result is what is termed a solar eclipse, though it should more properly be called an occultation of the Sun by the Moon.

Eclipses are of three types: total, partial and annular. At a total eclipse the photosphere is completely hidden, and the sight is probably the most magnificent in all Nature. As soon as the last segment of the bright disk is covered, the Sun's atmosphere flashes into view, and the chromosphere and corona shine out, together with any prominences which happen to be present. The sky darkens sufficiently for planets and bright stars to be seen; the temperature falls sharply, and the effect is dramatic by any standards. Unfortunately, total eclipses are rare as seen from any particular locality. The Moon's shadow can only just touch the Earth, and the track of totality can never be more than 272 kilometres (169 miles) wide; moreover, the total phase cannot last more than 7 minutes 31 seconds, and is generally shorter.

To either side of the main cone of shadow the eclipse is partial, and the glorious phenomena of totality cannot be seen; many partial eclipses are not total anywhere. Finally there are annular eclipses, when the alignment is perfect but the Moon is near its greatest distance from Earth; its disk is not then large enough to cover the photosphere completely, and a ring of sunlight is left showing round the dark mass of the Moon (Latin *annulus*, a ring).

For obvious reasons, a solar eclipse can happen only when the Moon is new, and thus lies on the Sun-side of the Earth. If the lunar orbit lay in the same plane as that of the Earth, there would be an eclipse every month, but in fact the Moon's orbit is tilted at an angle of just over five degrees, so that in general the New Moon passes unseen either above or below the Sun in the sky.

The points at which the Moon's orbit cuts the ecliptic are known as the nodes, so that to produce an eclipse the Moon must be at or very near a node. Because of the gravitational pull of the Sun, the nodes shift slowly but regularly. After a period of 18 years 11.3 days, the Earth, Sun and Moon return to almost the same relative positions, so that a solar eclipse is likely to be followed by another eclipse 18 years 11.3 days later – a period known as the Saros. It is not exact, but it was good enough for ancient peoples to predict eclipses with fair certainty. For example, the Greek philosopher Thales is said to have forecast the eclipse of 25 May 585 BC, which put an abrupt end to a battle being fought between the armies of King Alyattes of the Lydians and King Cyraxes of the Medes; the combatants were so alarmed by the sudden darkness that they made haste to conclude peace.

From any particular point on the Earth's surface, solar eclipses are less common than those of the Moon. This is because to see a solar eclipse, the observer has to be in just the right place at just the right time, whereas a lunar eclipse is visible from any location where the Moon is above the horizon. England had two total eclipses during the 20th century, those of 29 June 1927 and 11 August 1999. The track of the 1927 eclipse crossed North England, but at the 'return' at the end of the Saros (9 July 1945) the track missed England altogether, though it crossed Canada, Greenland and North Europe. The 11 August 1999 total eclipse crossed the Scilly Isles, Cornwall, South Devon and Alderney, and thence across Europe.

The main phenomena seen during totality are the chromosphere, the prominences and the corona. The chromosphere is from 2000 to 10,000 kilometres (1250 to 6250 miles) deep, with a temperature which reaches 8000 degrees C at an altitude of 1500 kilometres (950 miles) and then increases rapidly until the chromosphere merges with the corona. Prominences – once, misleadingly, called Red Flames – are masses of red, glowing hydrogen.

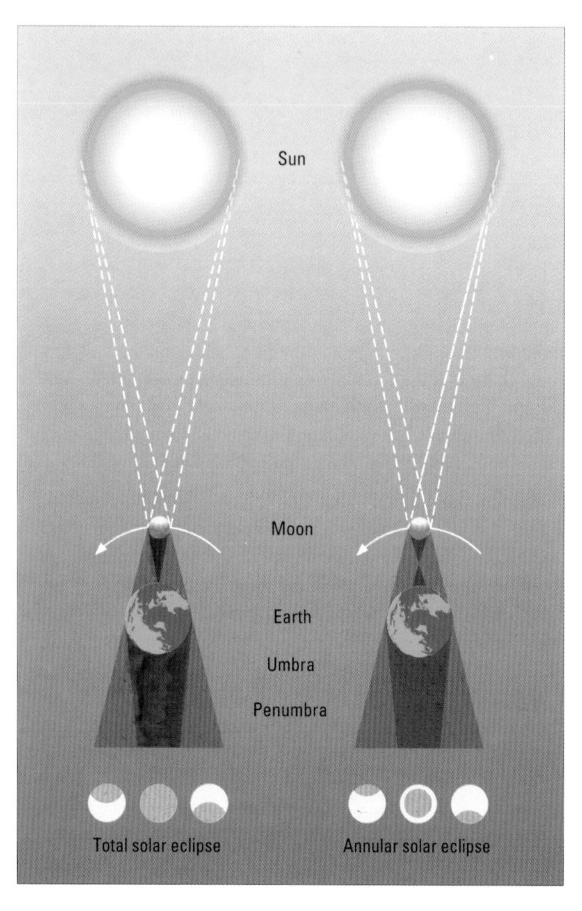

SOLAR ECLIPSES, 1999–2005

Date	GMT	Type	Duration (if total or annular) min.	sec.	% eclipsed (if partial)	Area
16 Feb 1999	07	A	1	19	–	Indian Ocean, Australia, Pacific
11 Aug 1999	11	T	2	23	–	Atlantic, England (Cornwall), France, Turkey, India
5 Feb 2000	13	P	–		56	Antarctic
31 July 2000	02	P	–		60	Arctic
25 Dec 2000	18	P	–		72	Arctic
21 Jun 2001	12	T	4	56	–	Atlantic, South Africa
14 Dec 2001	21	A	3	54	–	Central America, Pacific
10 Jun 2002	24	A	1	13	–	Pacific
4 Dec 2002	08	T	2	04	–	S. Africa, Indian Ocean, Australia
31 May 2003	04	A	3	37	–	Iceland
23 Nov 2003	23	T	1	57	–	Antarctic
19 Apr 2004	14	P	–		74	Antarctica
14 Oct 2004	03	P	–		93	Arctic
8 Apr 2005	21	T	0	42	–	Pacific, America, north of S. America
3 Oct 2005	11	A	4	32	–	Atlantic, Spain, Africa, Indian Ocean

Quiescent prominences may hang in the chromosphere for many weeks, but eruptive prominences show violent motion, often rising to thousands of kilometres; in some cases material is hurled away from the Sun altogether. They can be seen with the naked eye only during totality, but spectroscopic equipment now makes it possible for them to be studied at any time. By observing in hydrogen light, prominences may also be seen against the bright disk as dark filaments, sometimes termed flocculi. (Bright flocculi are due to calcium.)

Shadow bands are wavy lines seen across the Earth's surface just before and just after totality. They are due to effects in the atmosphere, and are remarkably difficult to photograph well; neither are they seen at every total eclipse.

During totality, the scene is dominated by the glorious pearly corona, which stretches outwards from the Sun in all directions; at times of spot-maxima it is reasonably symmetrical, but near spot-minimum there are long streamers. It is extremely rarefied, with a density less than one million-millionth of that of the Earth's air at sea level. Its temperature is well over a million degrees, but this does not indicate that it sends out much heat. Scientifically, temperature is measured by the speeds at which the various atoms and molecules move around; the greater the speeds, the higher the temperature. In the corona the speeds are very high, but there are so few particles that the heat is negligible. The cause of the high temperature seems to be linked with magnetic phenomena, though it is not yet fully understood.

Eclipse photography is fascinating, but there is one point to be borne in mind. Though it is quite safe to look directly at the totally-eclipsed Sun, the slightest trace of the photosphere means that the danger returns, and it is essential to remember that pointing an SLR camera at the Sun is tantamount to using a telescope. As always, the greatest care must be taken – but nobody should ever pass up the chance of seeing the splendour of a total solar eclipse.

▶ ***The partial eclipse*** *of 21 November 1966 photographed from Sussex by Henry Brinton with a 10-cm (4-inch) reflector.*

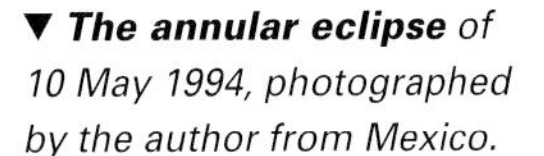

▼ ***The annular eclipse*** *of 10 May 1994, photographed by the author from Mexico.*

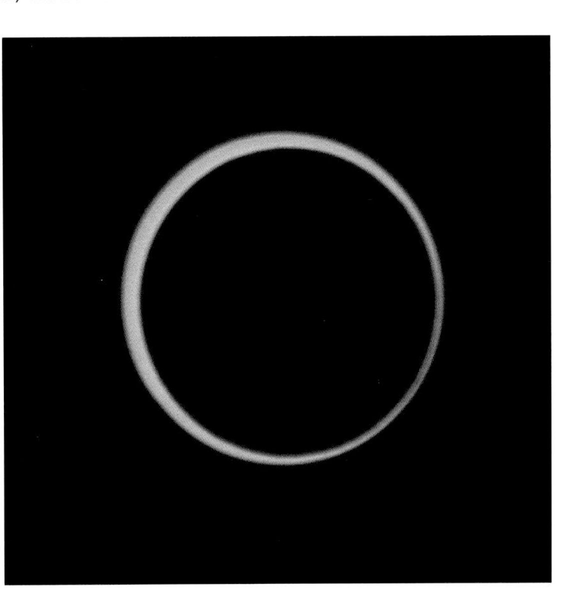

▼ ***The lovely Diamond Ring effect****, seen just before and just after totality. This photograph was taken from a transatlantic jet on 21 November 1966, at the end of totality; as the first segment of the Sun reappears it seems to flash out for a few seconds.*

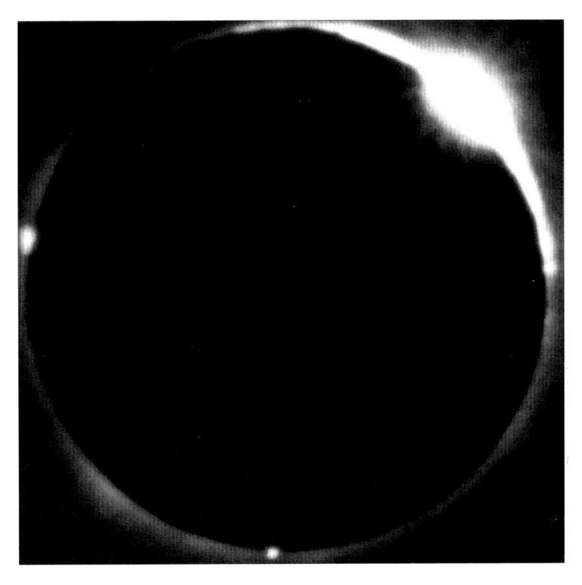

◀ ***Total eclipse, 1983****; photographed by Don Trombino, from Java. The corona was magnificently displayed. The shape of the corona varies according to the state of the solar cycle; near spot-maximum it is fairly regular, while near spot-minimum long streamers extend from the equatorial regions. During totality the sky darkens and planets and bright stars may be seen. Before the Space Age, total eclipses were of the utmost importance to physicists, because there were no other opportunities to observe the outer corona.*

The Sun in Action

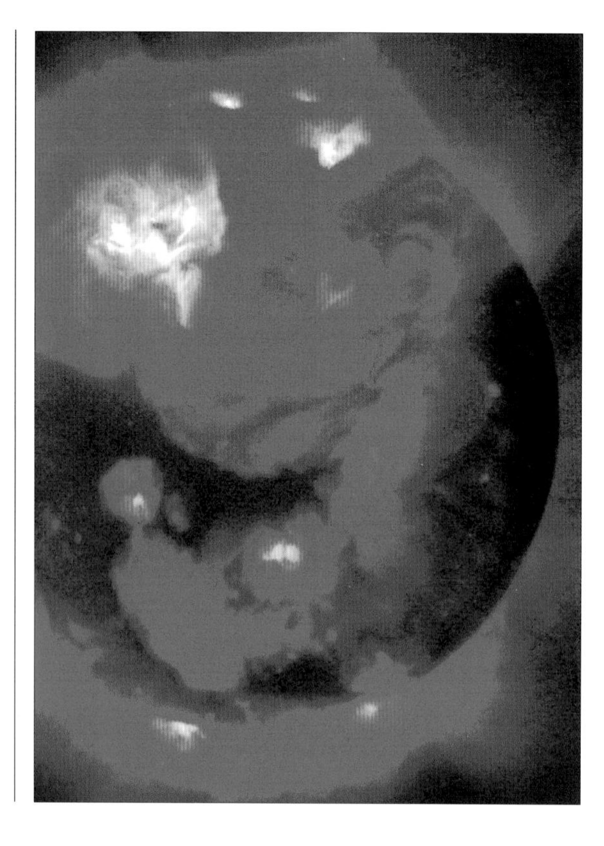

▶ ***The X-Ray Sun**, 25 November 1990; imaged by the Japanese X-ray satellite Yohkoh. This picture shows regions of different X-ray emission; there is clear evidence of coronal holes.*

The Sun is never calm. Even the photosphere is in a constant state of turmoil. In the chromosphere we have the prominences, some of which are violently eruptive, and there are also spicules, narrow vertical gas-jets which begin on the bright surface and soar to as much as 10,000 kilometres (over 6200 miles) into the chromosphere. They are always present, and at any one time there may be as many as a quarter of a million of them.

Even more dramatic are the flares, which usually, though not always, occur above active spot-groups; they are seldom seen in ordinary light, so that spectroscopic equipment has to be used to study them. They are short-lived, and generally last for no more than 20 minutes or so, though a few have been known to persist for several hours. They produce shock-waves in the chromosphere and the corona, and considerable quantities of material may be blown away from the Sun altogether; the temperatures may rocket to many millions of degrees. Flares are essentially magnetic phenomena, and it seems that rapid rearrangement of magnetic fields in active regions of the corona results in a sudden release of energy which accelerates and heats matter in the Sun's atmosphere. Radiations at all wavelengths are emitted, and are particularly strong in the X-ray and ultra-violet regions of the electromagnetic spectrum.

The solar wind is made up of charged particles sent out from the Sun at all times. It is made up of a plasma (that is to say, an ionized gas, made up of a mixture of electrons and the nuclei of atoms), and is responsible for repelling the ion tails of comets, making them point away from the Sun. When these charged particles reach the Earth they are responsible for the lovely displays of aurorae or polar lights – aurora borealis in the northern hemisphere, aurora australis in the southern.

The average velocity of the solar wind as it passes the Earth is 300 to 400 kilometres per second (190 to 250 miles per second); we are not sure how far it extends, but it is hoped that four of the current space probes (Pioneers 10 and 11, and Voyagers 1 and 2) will keep on transmitting until they reach the edge of the heliosphere, that is, the region where the solar wind ceases to be detectable.

The solar wind escapes most easily through coronal holes, where the magnetic field lines are open instead of looped. In 1990 a special spacecraft, Ulysses, was launched to study the polar regions of the Sun, which have never been well known simply because from Earth, and from all previous probes, we have always seen the Sun more or less broadside-on. Ulysses had to move well out of the ecliptic plane, which it did by going first out to Jupiter and using the powerful gravitational pull of the giant planet to put it into the correct path.

Many solar probes have been launched, and have provided an immense amount of information, particularly in the X-ray and ultra-violet regions of the electromagnetic spectrum. Detailed studies of the Sun were carried out by the three successive crews of the American space station Skylab, which was manned during 1973 and 1974, while the elaborate unmanned SMM or Solar Maximum Mission vehicle lasted from 1980 to 1989.

We have come a long way since William Herschel believed that the Sun might be inhabited, and we are learning more all the time, but we have to admit that we do not yet have anything like a full understanding of our own particular star.

◀ ***Launch of the Japanese X-Ray satellite Yohkoh.** The satellite was launched by a Japanese M-3SII-6 rocket. X-rays cannot penetrate the atmosphere, so all research has to be conducted from space.*

▶ ***Active region of the Sun**, imaged from the Skylab space station on 14 August 1973. The huge 'loop' has a weaker secondary image close beside it.*

▼ ***Solar prominence**; photograph by Don Trombino. This is an active prominence; in some cases material escapes from the Sun completely.*

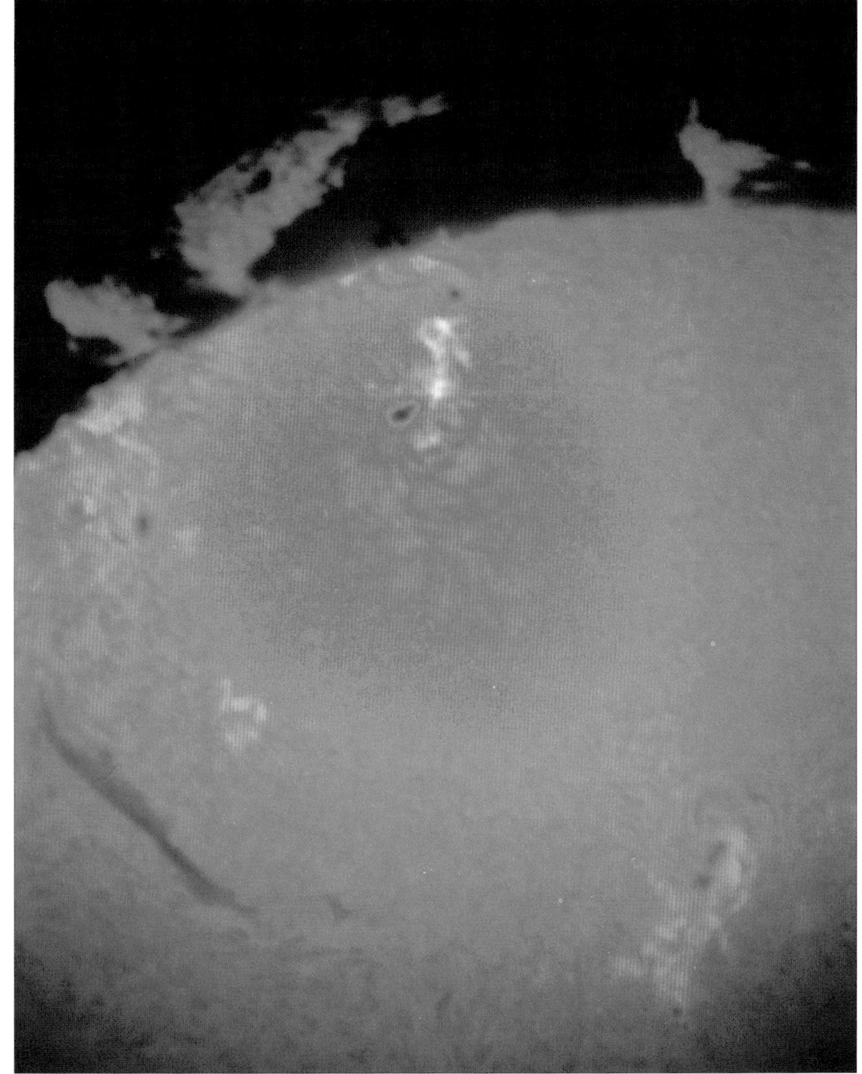

▼ ***Aurora borealis**, 13 March 1989; Paul Doherty, Stoke-on-Trent, England. This display was one of the most widespread for many years, and was associated with a particularly large and active sunspot group. From Stoke, the aurora was actually bright enough to cast shadows!*

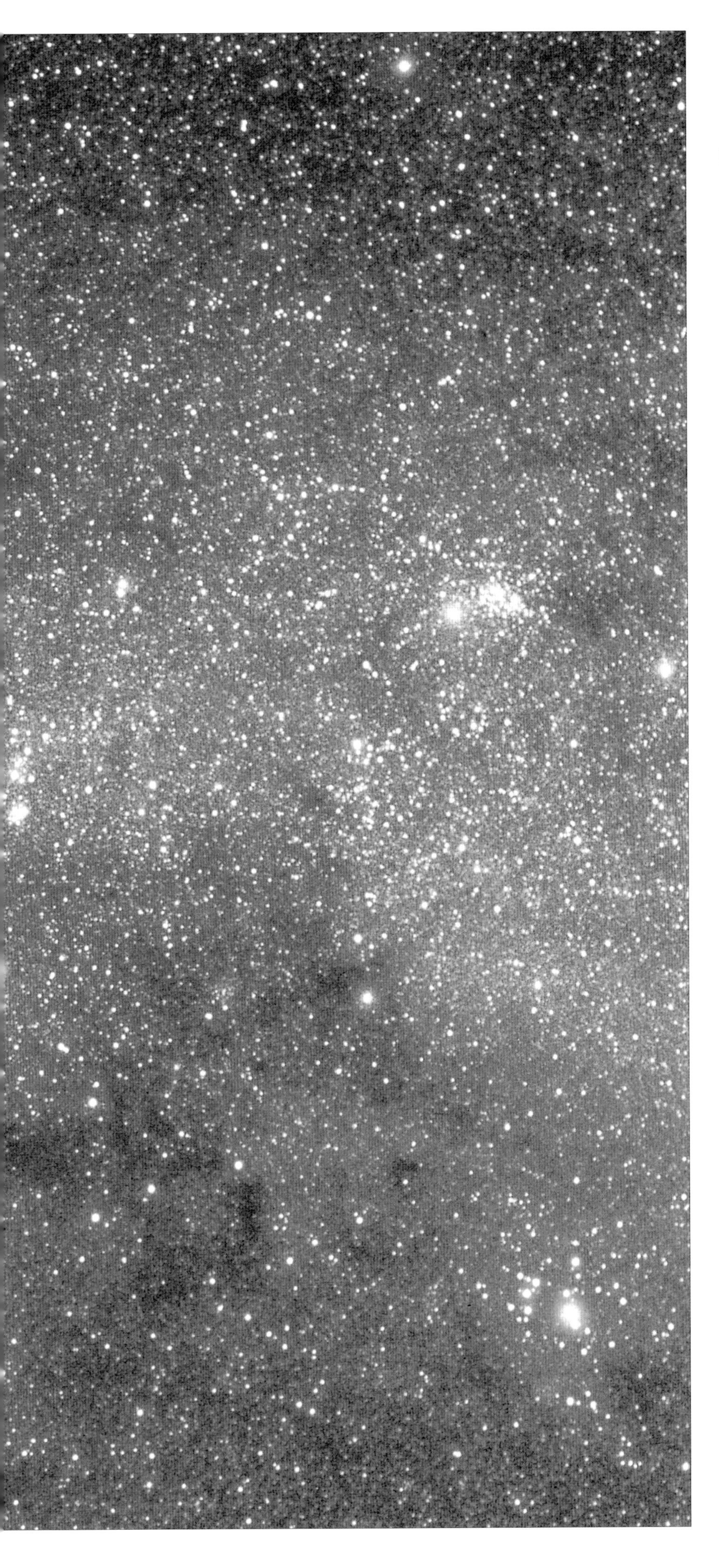

THE STARS

◀ ***The Southern Cross*** *embedded in the Milky Way. The colours are natural but enhanced by the light-gathering power of a large telescope.*

Introduction to the Stars

THE GREEK ALPHABET

α	Alpha
β	Beta
γ	Gamma
δ	Delta
ε	Epsilon
ζ	Zeta
η	Eta
θ	Theta
ι	Iota
κ	Kappa
λ	Lambda
μ	Mu
ν	Nu
ξ	Xi
ο	Omicron
π	Pi
ρ	Rho
σ	Sigma
τ	Tau
υ	Upsilon
φ	Phi
χ	Chi
ψ	Psi
ω	Omega

How many stars can you see with the naked eye on a clear, dark night? Many people will say, 'Millions', but this is quite wrong. There are roughly 5800 stars within naked-eye range. Only half these will be above the horizon at any one time, and faint stars which are low down will probably not be seen. This means that if you can see a grand total of 2500 stars, you are doing very well.

The ancients divided up the stars into groups of constellations, which were named in various ways. The Egyptians had one method, the Chinese another, and so on; the constellations we use today are those of the Greeks (admittedly with Latin names) and if we had used one of the other systems our sky-maps would look very different, though the stars themselves would be exactly the same. In fact, a constellation pattern has no real significance, because the stars are at very different distances from us, and we are dealing with nothing more than line of sight effects.

Ptolemy, last of the great astronomers of Classical times, listed 48 constellations, all of which are still given modern maps even though they have been modified in places. Some of the groups were named after mythological characters, such as Orion and Perseus; others after animals or birds, such as Cygnus (the Swan) and of course Ursa Major (the Great Bear), and there are a few inanimate objects, such as Triangulum (the Triangle). Other constellations have been added since, notably those in the far south of the sky which never rose above the horizon in Egypt, where Ptolemy seems to have spent the whole of his life.

Some of the new groups have modern-sounding names, such as Telescopium (the Telescope) and Octans (the Octant). During the 17th century various astronomers compiled star catalogues, usually by stealing stars from older groups. Some of the additions have survived (including Crux Australis, the Southern Cross), while others have been mercifully deleted; our maps no longer show constellations such as Globus Aerostaticus (the Balloon), Officina Typographica (the Printing Press) and Sceptrum Brandenburgicum (the Sceptre of Brandenburg). Today we recognize a total of 88 constellations. They are very unequal in size and importance, and some of them are so obscure that they seem to have little claim to separate identity. One can sympathize with Sir John Herschel, who once commented that the constellation patterns seemed to have been drawn up so as to cause the maximum possible inconvenience and confusion.

Very bright stars such as Sirius, Canopus, Betelgeux and Rigel have individual names, most of which are Arabic, but in other cases a different system is used. In 1603 Johann Bayer, a German amateur astronomer, drew up a star catalogue in which he took each constellation and gave its stars Greek letters, starting with Alpha for the brightest star and working through to Omega. This proved to be very satisfactory, and Bayer's letters are still in use, though in many cases the proper alphabetical sequence has not been followed; thus in Sagittarius (the Archer), the brightest stars are Epsilon, Sigma and Zeta, with Alpha and Beta Sagittarii very much 'also rans'. Later in the century John Flamsteed, the first Astronomer Royal, gave numbers to the stars, and these too are still in use; thus Sirius, in Canis Major (the Great Dog) is not only Alpha Canis Majoris but also 19 Canis Majoris.

The stars are divided into classes or magnitudes depending upon their apparent brilliance. The scheme works rather in the manner of a golfer's handicap, with the more brilliant performers having the lower values; thus magnitude 1 is brighter than 2, 2 brighter than 3, and so on. The faintest stars normally visible with the naked eye

▲ ***Ursa Major*** *(The Great Bear). This is the most famous of all the northern constellations. Seven main stars make up the 'Plough' or 'Dipper' pattern. Six of them are white; the seventh, Dubhe, is orange.*

► ***Orion****. This brilliant constellation is crossed by the celestial equator, and can therefore be seen from every inhabited country. Its pattern is unmistakable; two of the stars (Rigel and Betelgeux) are of the first magnitude, and the remainder between $1\frac{1}{2}$ and 2. Like Ursa Major, Orion is a 'guide' to the constellations, though it is out of view for part of the year when it is close to the Sun and above the horizon only during daylight.*

are of magnitude 6, though modern telescopes using electronic equipment can reach down to at least 28. On the other end of the scale, there are few stars with zero or even negative magnitudes; Sirius is −1.46, while on the same scale the Sun would be −26.8. In fact, the scale is logarithmic, and a star of magnitude 1.0 is exactly a hundred times as bright as a star of magnitude 6.0.

Note that the apparent magnitude of a star has nothing directly to do with its real luminosity. A star may look bright either because it is very close on the cosmic scale, or because it is genuinely very large and powerful, or a combination of both. The two brightest stars are Sirius (−1.46) and Canopus (−0.73), so that Sirius is over half a magnitude the more brilliant of the two; yet Sirius is 'only' 26 times as luminous as the Sun, while the much more remote Canopus could match 200,000 Suns. Appearances can often be deceptive.

There is one curious anomaly. It is customary to refer to the 21 brightest stars as being of the first magnitude; they range from Sirius down to Regulus in Leo, whose magnitude is 1.35. Next in order comes Adhara in Canis Major; its magnitude is 1.50, but even so it is not included among the élite.

The stars are not genuinely fixed in space. They are moving about in all sorts of directions at all sorts of speeds, but they are so far away that their individual or proper motions are very slight; the result is that the constellation patterns do not change appreciably even over periods of many lifetimes and they look virtually the same today as they must have done in the time of Julius Caesar or even the builders of the Pyramids. It is only our nearer neighbours, the members of the Solar System, which move about from one constellation into another. The nearest star beyond the Sun lies at a distance of 4.2 light-years, a light-year being the distance travelled by a ray of light in one year – over 9 million million kilometres (around 6 million million miles).

This, of course, is why the stars appear relatively small and dim; no normal telescope will show a star as anything but a point of light. Yet some stars are huge; Betelgeux in Orion is so large that it could contain the entire orbit of the Earth round the Sun. Other stars are much smaller than the Sun, or even the Earth, but the differences in mass are not so great as might be expected, because small stars are denser than large ones. It is rather like balancing a cream puff against a lead pellet.

There is a tremendous range in luminosity. We know of stars which are more than a million times as powerful as the Sun, while others have only a tiny fraction of the Sun's power. The colours, too, are not the same; our Sun is yellow, while other stars may be bluish, white, orange or red. These differences are due to real differences in surface temperature. The hottest known stars have temperatures of up to 80,000 degrees C while the coolest are so dim that they barely shine at all.

Many stars, such as the Sun, are single (though they may well be the centres of planetary systems). Others are double or members of multiple systems. There are stars which are variable in light; there are clusters of stars, and there are also vast clouds of dust and gas which are termed nebulae. Our star system or Galaxy contains about 100,000 million stars, and beyond we come to other galaxies, so remote that their light takes millions, hundreds of millions or even thousands of millions of years to reach us. Look at these distant systems today, and you see them not as they are now, but as they used to be when the universe was young – long before the Earth or even the Sun came into existence. It is a sobering thought.

*▼ **Crux Australis**, the most famous southern constellation, is shaped like a kite rather than an X. Two of the stars in the main pattern are of the first magnitude, a third 1½ and the fourth just above 3. Two more brilliant stars, Alpha and Beta Centauri, point to it. Of the four chief stars in Crux, three are hot and bluish white; the fourth (Gamma Crucis) orange-red. Crux also includes a famous dark nebula, the Coal Sack, and the lovely Jewel Box open cluster. These three photographs (Ursa Major, Orion, Crux) were taken by the author with the same camera and exposure.*

The Celestial Sphere

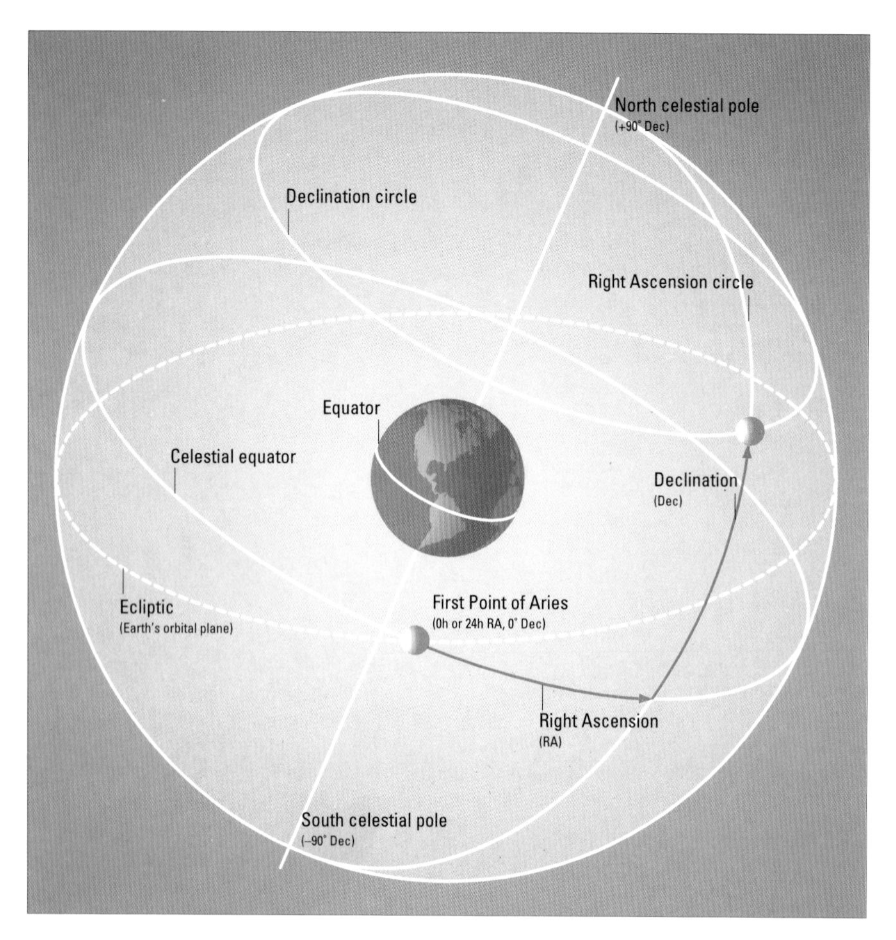

▲ ***The celestial sphere.*** *For some purposes it is still convenient to assume that the sky really is solid, and that the celestial sphere is concentric with the surface of the Earth. We can then mark out the celestial poles, which are defined by the projection of the Earth's axis on to the celestial sphere; the north pole is marked closely by the bright star Polaris in Ursa Minor, but there is no bright south pole star, and the nearest candidate is the dim Sigma Octantis. Similarly, the celestial equator is the projection of the Earth's equator on to the celestial sphere; it divides the sky into two hemispheres. Declination is the angular distance of a body from the celestial equator, reckoned from the centre of the Earth (or the centre of the celestial sphere, which is the same thing); it therefore corresponds to terrestrial latitude. Sirius is shown; here its angular distance from the celestial equator is 16° 39′ south, so that its declination is – 16° 39′.*

Ancient peoples believed the sky to be solid, with the stars fixed on to an invisible crystal sphere. This is a convenient fiction, so let's assume that the celestial sphere really exists, making one revolution round the Earth in 24 hours and carrying all the celestial bodies with it.

The north pole of the sky is simply the point on the celestial sphere which lies in the direction of the Earth's axis; it is marked within one degree by the second-magnitude star Polaris, in the Little Bear (Ursa Minor). Of course there is also a south celestial pole, but unfortunately there is no bright star anywhere near it, and we have to make do with the obscure Sigma Octantis, which is none too easy to see with the naked eye even under good conditions.

Just as the Earth's equator divides the world into two hemispheres, so the celestial equator divides the sky into two hemispheres – north and south. The celestial equator is defined as the projection of the Earth's equator on to the celestial sphere as shown in the diagram above.

To define a position on Earth, we need to know two things – one's latitude, and one's longitude. Latitude is the angular distance north or south of the equator, as measured from the centre of the globe; for example, the latitude of London is approximately 51 degrees N, that of Sydney 34 degrees S. The latitude of the north pole is 90 degrees N, that of the south pole 90 degrees S. The sky equivalent of latitude is known as declination, and is reckoned in precisely the same way; thus the declination of Betelgeux in Orion is 7 degrees 24 minutes N, that of Sirius 16 degrees 43 minutes S. (Northern values are given as + or positive, southern values as – or negative.)

All this is quite straightforward, but when we consider the celestial equivalent of longitude, matters are less simple. On Earth, longitude is defined as the angular distance of the site east or west of a particular scientific instrument, the Airy Transit Circle, in Greenwich Observatory. Greenwich was selected as the zero for longitude over a century ago, when international agreement was much easier to obtain; there were very few dissenters apart from France.

We need a 'celestial Greenwich', and there is only one obvious candidate: the vernal equinox, or First Point of Aries. To explain this, it is necessary to say something about the way in which the Sun seems to move across the sky.

Because the Earth goes round the Sun in a period of one year (just over 365 days), the Sun appears to travel right round the sky in the same period. The apparent yearly path of the Sun against the stars is known as the ecliptic, and passes through the 12 constellations of the Zodiac (plus a small part of a 13th constellation, Ophiuchus, the Serpent-bearer). The Earth's equator is tilted to the orbital plane by 23½ degrees, and so the angle between the ecliptic and the celestial equator is also 23½ degrees. Each year, the Sun crosses the equator twice. On or about 22 March – the date is not quite constant, owing to the vagaries of our calendar – the Sun reaches the equator, travelling from south to north; its declination is then 0 degrees, and it has reached the Vernal Equinox or First Point of Aries, which is again unmarked by any bright star. The Sun then spends six months in the northern hemisphere of the sky. About 22 September it again reaches the equator, this time moving from north to south; it has reached the autumnal equinox or First Point of Libra, and spends the next six months in the southern hemisphere.

The celestial equivalent of longitude is termed right ascension. Rather confusingly, it is measured not in degrees, but in units of time. As the Earth spins, each point in the sky must reach its highest point above the horizon once every 24 hours; this is termed culmination. The right ascension of a star is simply the time which elapses between the culmination of the First Point of Aries, and the culmination of the star. Betelgeux culminates 5 hours 53 minutes after the First Point has done so; therefore its right ascension is 5h 53m.

Oddly enough, the First Point is no longer in the constellation of Aries, the Ram; it has moved into the adjacent constellation of Pisces, the Fishes. This is because of the phenomenon of precession. The Earth is not a perfect sphere; the equator bulges out slightly. The Sun and Moon pull on this bulge, and the effect is to make the Earth's axis wobble slightly in the manner of a gyroscope which is running down and has started to topple. But whereas a gyroscope swings round in a few seconds, the Earth's axis takes 25,800 years to describe a small circle in the sky. Thousands of years ago, when the Egyptians were building their Pyramids, the north celestial pole lay not close to Polaris, but near a much fainter star, Thuban in the constellation of the Dragon; in 12,000 years' time we will have a really brilliant pole star, Vega in Lyra (the Lyre).

For the moment, let us suppose that Polaris, our present pole star, lies exactly at the pole instead of being rather less than one degree away (its exact declination is +89 degrees 15 minutes 51 seconds). To an observer standing at the North Pole of the Earth, Polaris will have an altitude of 90 degrees; in other words it will lie at the zenith or overhead point. From the Earth's equator the altitude of Polaris will be 0 degrees; it will be on the horizon, while from southern latitudes it will never be seen at all. When observed from the northern hemisphere, the altitude of Polaris is always the same as the latitude of the observer. Thus from London, latitude 51 degrees N, Polaris will be 51 degrees above the horizon. From Sydney, the altitude of the dim Sigma Octantis will be 34 degrees.

A star which never sets, but merely goes round and round the pole without dipping below the horizon, is said

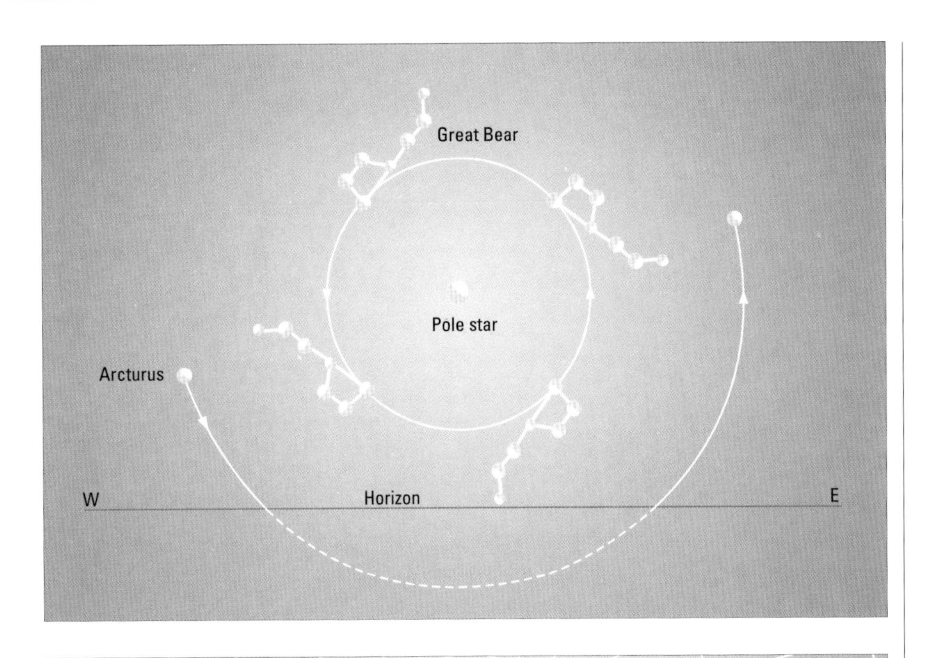

▶ ***Circumpolar and non-circumpolar stars****.* *In this diagram Ursa Major is shown, together with Arcturus in Boötes; it is assumed that the observer's latitude is that of England. Ursa Major is so close to the north celestial pole that it never sets, but Arcturus drops below the horizon for part of its diurnal circuit. Ursa Major is therefore circumpolar from England, while Arcturus is not.*

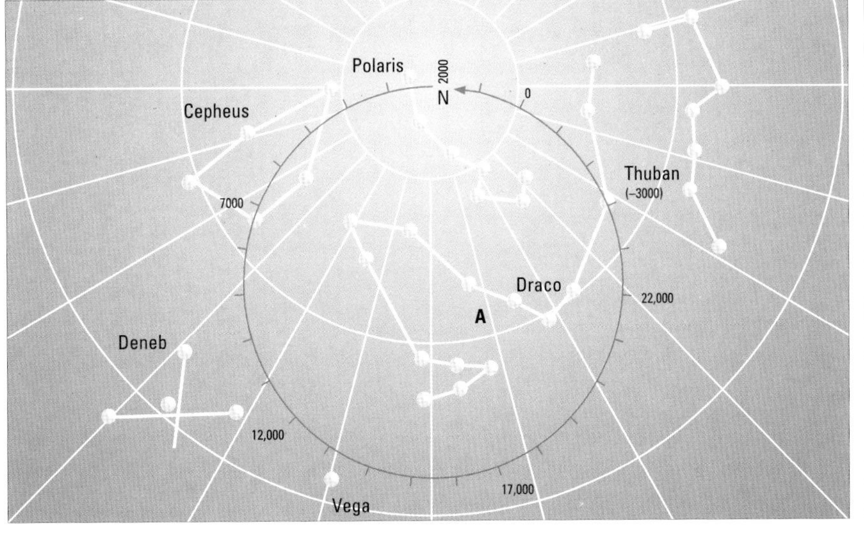

▶ ***Precession****. The precession circle, 47° in diameter, showing the shift in position of the north celestial pole around the pole of the ecliptic (A). In Egyptian times (c. 3000* BC*) the polar point lay near Thuban or Alpha Draconis; it is now near Polaris in Ursa Minor (declination +89° 15′); in* AD *12,000 it will be near Vega. The south celestial pole describes an analogous precession circle.*

▼ ***Star trails****. This photograph, with an exposure of two hours, shows the stars near the south celestial pole (the picture was taken from New Zealand). The pole itself is at the bottom of the picture.*

to be circumpolar. To decide which stars are circumpolar and which are not, simply subtract the latitude of the observing site from 90. In the case of London, 90 − 51 = 39; it follows that any star north of declination +39° will never set, and any star south of declination −39° will never rise. Thus constellations such as Ursa Major (the Great Bear) and Cassiopeia are circumpolar from anywhere in the British Isles, but not from the southern Mediterranean.

Another useful example concerns the Southern Cross, which is as familiar to Australians and New Zealanders as the Great Bear is to Britons. The declination of Acrux, the brightest star in the Cross, is −63 degrees. 90 − 63 = 27, so that Acrux can never be seen from any part of Europe, though it does rise in Hawaii, where the latitude is 20 degrees N. To have a reasonable view of the Cross, there is no need to travel as far south as the equator.

Incidentally, it was this sort of calculation which gave an early proof that the Earth is round. Canopus, the second brightest star in the sky, has a declination of −53 degrees; therefore it can be seen from Alexandria (latitude 31 degrees N) but not from Athens (38 degrees N), where it grazes the horizon. The Greeks knew this, and realized that such a situation could arise only if the Earth is a globe rather than a flat plane. From Wellington, in New Zealand, Canopus is circumpolar, so that it can always be seen whenever the sky there is sufficiently dark and clear.

Distances and Movement of the Stars

▶ ***Trigonometrical parallax.*** *A represents the Earth in its position in January; the nearby star X, measured against the background of more remote stars, appears at X_1. Six months later, by July, the Earth has moved to position B; as the Earth is 150 million km (93 million miles) from the Sun, the distance A–B is twice 150 = 300 million km (186 million miles). Star X now appears at X_2. The angle AXS can therefore be found, and this is known as the parallax. Since the length of the baseline A–B is known, the triangle can be solved, and the distance (X–S) of the star can be calculated.*

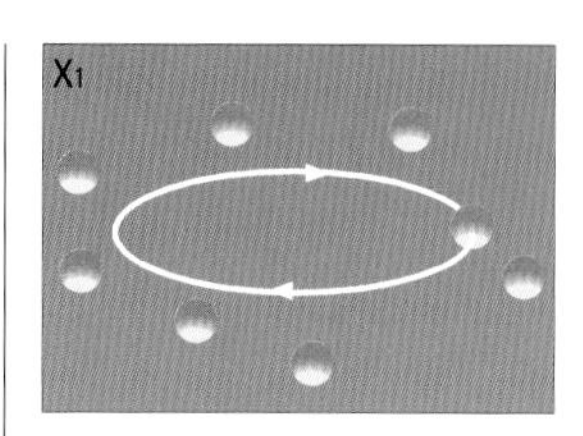

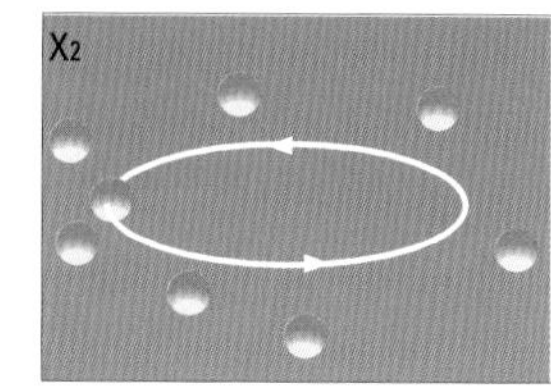

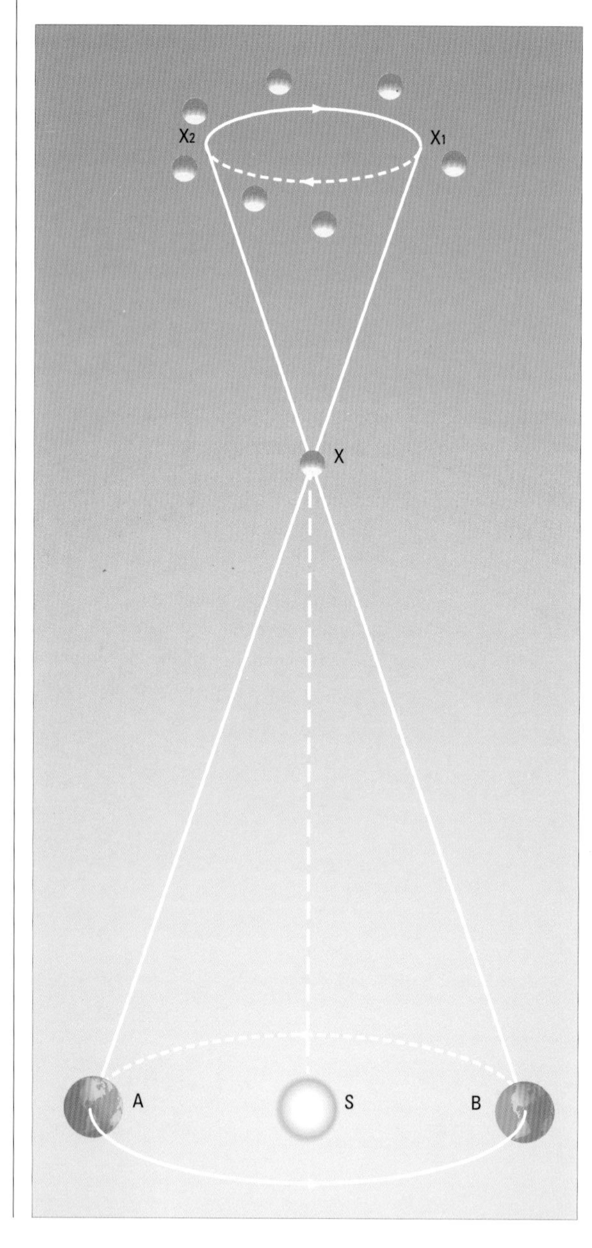

It is very easy to take an attractive star picture. All you need is a camera capable of giving a time exposure. Using a reasonably fast film, open the shutter and point the camera skywards on a dark, clear night. Wait for half an hour or so – the exact timing is not important – and end the exposure. You will find that you have a picture showing the trails left by the stars as they crawl across the field of view by virtue of the Earth's rotation, and you may be lucky enough to catch a meteor or an artificial satellite. Perhaps the most rewarding effects are obtained by pointing the camera at the celestial pole.

The first successful measurement of the distance of a star was made in 1838 by the German astronomer Friedrich Bessel. His method was that of parallax; the principle was the same as that used by a surveyor who wants to find the distance of some inaccessible object such as a mountain top. He measures out a baseline and then observes the direction of the target from its opposite ends. From this he can find the angle at the target, half of which is termed the parallax. He knows the length of the baseline, and simple trigonometry will enable him to work out the distance to the target, which is what he needs to know.

With the stars, a much longer baseline is needed, and Bessel chose the diameter of the Earth's orbit. A now represents the position of the Earth in January and B the position of the Earth in June, when it has moved round to the other side of its orbit; since the Earth is 150 million kilometres (93 million miles) from the Sun (S) the distance A–B is twice this, or 300 million kilometres (186 million miles). X now represents the target star, 61 Cygni in the constellation of the Swan, which Bessel calculated because he had reason to believe that it might be comparatively close. He worked out that the parallax is 0.29 of a second of arc, corresponding to a distance of 11.2 light-years.

The parallax method works well out to a few hundreds of light-years, but at greater distances the annual shifts become so small that they are swamped by unavoidable errors of observation, and we have to turn to less direct methods. What is done is to find out how luminous the star really is, using spectroscopic analysis. Once this has been found, the distance follows, provided that many complications have been taken into account – such as the absorption of light in space.

A star at a distance of 3.26 light-years would have a parallax of one second of arc, so that this distance is known as a parsec; professional astronomers generally use it in preference to the light-year. In fact, no star (apart from the Sun, of course) is within one parsec of us, and our nearest neighbour, the dim southern Proxima Centauri, has an annual parallax of 0.76 of a second of arc, corresponding to a distance of 4.249 light-years. Another term in common use is absolute magnitude, which is the apparent magnitude that a star would have if it could be viewed from a standard distance of 10 parsecs or 32.6 light-years. The absolute magnitude of the Sun is +4.8, so that from the standard distance it would be a dim naked-eye object. Sirius has an absolute magnitude of +1.4, but the absolute magnitude of Rigel in Orion is −7.1, so that if it could be seen from the standard distance it would cast strong shadows.

(*En passant*, the distances and luminosities of remote stars are bound to be rather uncertain, and different catalogues give decidedly different values. In this *Atlas*, I have followed the authoritative Cambridge catalogue. At least we may be sure that Rigel qualifies as a cosmic searchlight.)

Though the individual or proper motions of the stars are very slight, because of the tremendous distances involved, they can be measured. The 'speed record' is held by a dim red dwarf, Barnard's Star, which is 5.8 light-years

▼ ***Proper motion of Proxima Centauri.*** *Proxima, the nearest star beyond the Sun (4.249 light-years) has the very large proper motion of 3".75 per year. These two pictures, one taken in 1897 and the other in 1940, show the shift very clearly (Proxima is arrowed).*

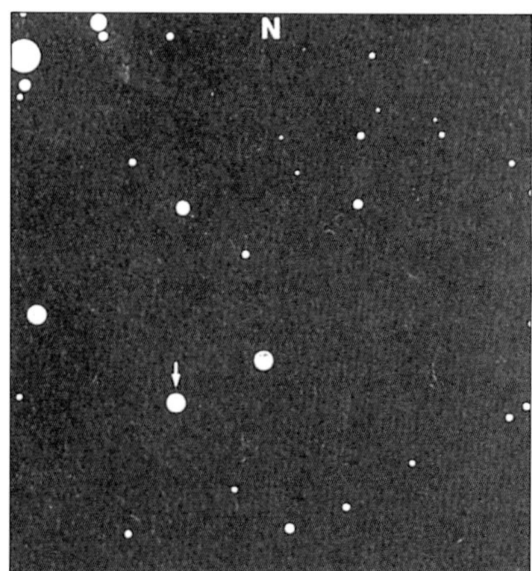

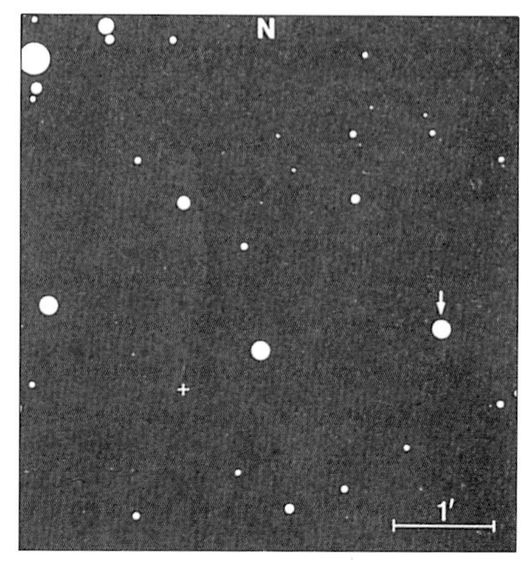

away and is our nearest neighbour apart from the three members of the Alpha Centauri group; the annual proper motion is 10.27 seconds of arc, so that in about 190 years it will crawl across the background by a distance equal to the apparent diameter of the Full Moon. (It has only 0.0005 the luminosity of the Sun, so that it is very feeble by stellar standards.)

Over sufficiently long periods, the shapes of the constellation patterns will change. For example, in the Great Bear (Ursa Major) we have the familiar seven-star pattern often called the Plough or, in America, the Big Dipper. Five of the stars are moving through space in much the same direction at much the same rate, so that presumably they had a common origin, but the other two – Alkaid and Dubhe – are moving in the opposite direction, so that in, say, 100,000 years' time the Plough pattern will have been distorted beyond all recognition. Neither are the Plough stars equally distant. Of the two 'end' stars, Mizar is 59 light-years away, Alkaid 108, so that Alkaid is very nearly as far away from Mizar as we are. There is another reminder that a constellation pattern is nothing more than a line of sight effect, and has no real significance. If we were observing from a different vantage point, Mizar and Alkaid could well be on opposite sides of the sky.

61 Cygni, the first star to have its distance measured, has an annual proper motion of over 4 seconds of arc, and it is also a wide binary, made up of two components which are gravitationally linked. That is why Bessel thought that it must be relatively close to us – and he was right.

We also have to deal with the radial or towards-or-away movements of the stars, which can be worked out by using the spectroscope. As we have seen, the spectrum of the Sun is made up of a rainbow background crossed by the dark Fraunhofer lines, and this is also true of all normal stars, though the details differ widely.

If the star is approaching us, all the lines in the spectrum will be shifted over to the blue or short-wave end of the rainbow band, while if the star is receding the shift will be to the red or long-wave end (this is the famous Doppler shift, about which much more will be said when we come to discuss galaxies).

By measuring the shifts of the lines we can find out whether the star is approaching or receding; the apparent proper motion of the star in the sky is a combination of the transverse and radial motions. (Conveniently, radial velocities are listed as negative if the star is approaching, positive if it is receding.) At the moment Barnard's Star is coming towards us at 108 kilometres (67 miles) per second, but it will not continue to do so indefinitely, and I can assure you that there is absolutely no fear of an eventual collision with our planet.

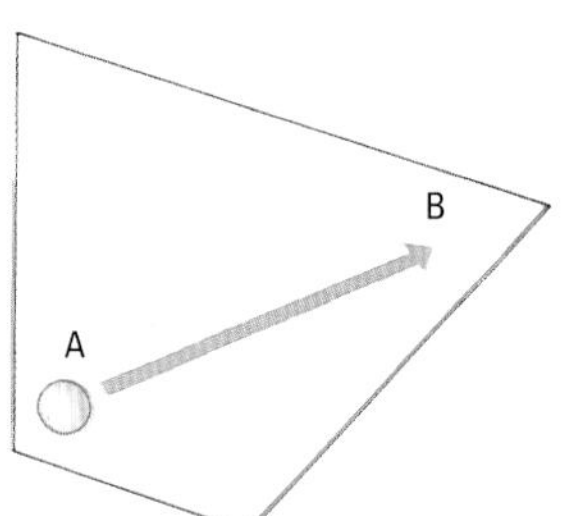

Actual motion *of a star in space when observed from Earth (A–B) is a combination of radial and transverse motion against the background of distant stars.*

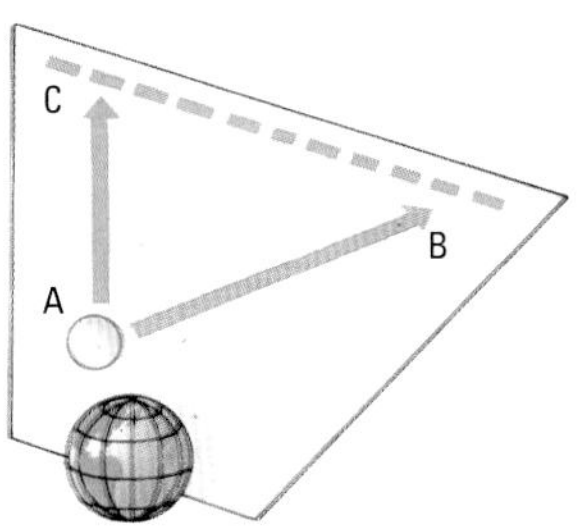

Radial motion *(A–C) is the velocity towards the Earth or away from the Earth; it is positive if the star is receding, negative if it is approaching.*

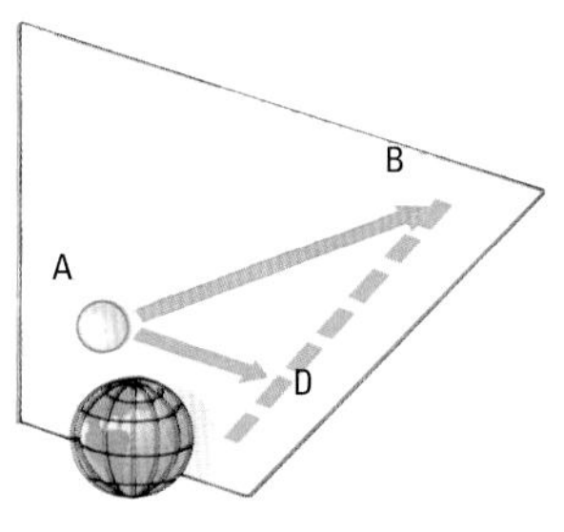

Proper motion *(A–D) is the transverse movement, or the motion across the sky. Barnard's Star (10".31 per year) has the greatest proper motion.*

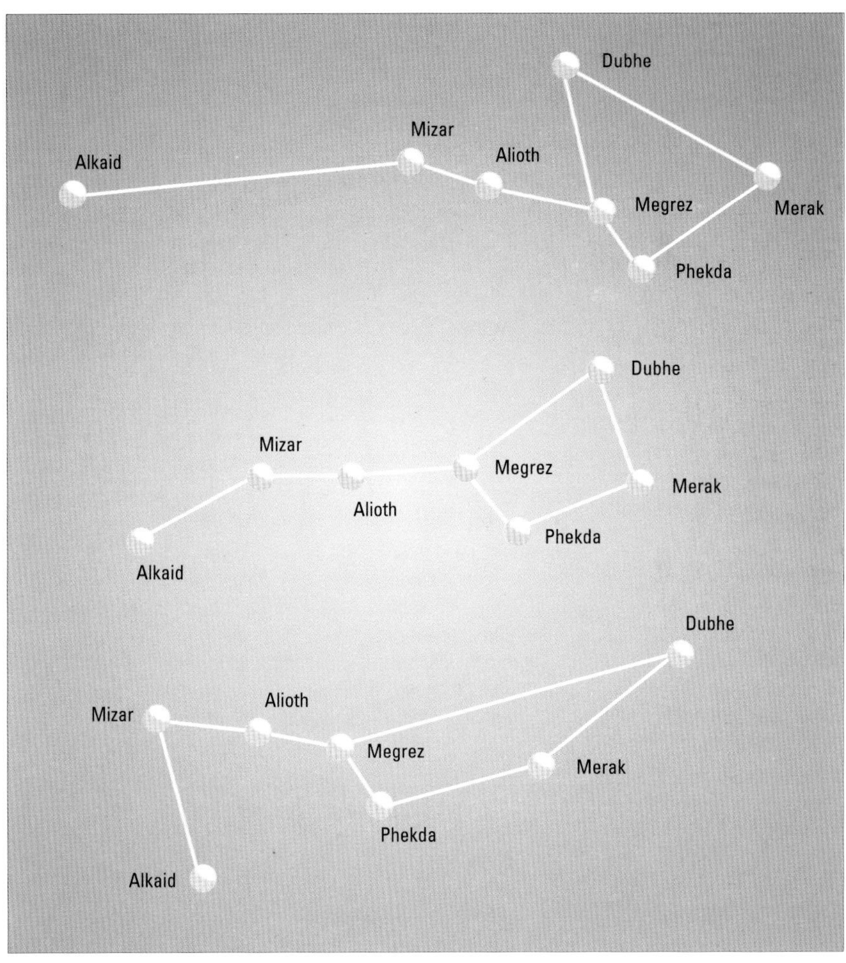

▲ ***Long term effect of proper motion****. The series of diagrams shows the movement of the seven main stars in Ursa Major (the Great Bear) that make up the famous pattern known as the Plough or Big Dipper. The upper diagram shows the arrangement of the stars as they were 100,000 years ago, the centre diagram gives the present appearance, and the bottom diagram shows the Bear as it will appear in 100,000 years' time.*

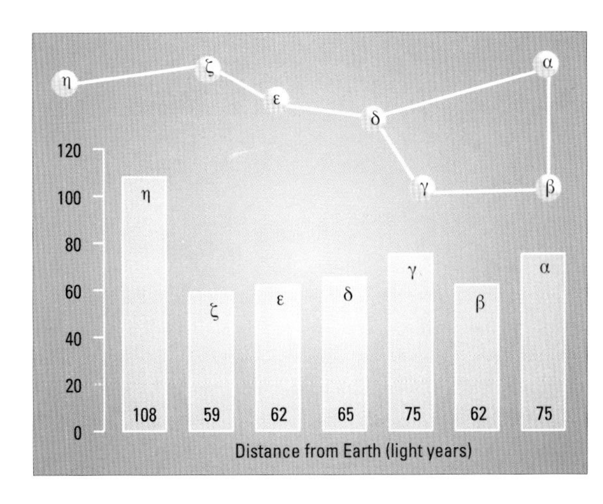

▶ ***Distances of the stars in the Plough****. The diagram shows the seven chief stars of the Plough, in Ursa Major, at their correct relative distances from the Earth. Alkaid, at 108 light-years, is the furthest away; Mizar, at 59 light-years, is the nearest. Therefore, Alkaid is almost as far away from Mizar as we are!*

Different Types of Stars

Ask any professional astronomer to name the most valuable scientific instrument at his disposal, and he will probably reply: 'The spectroscope.' Of course spectroscopes cannot be used without telescopes to collect the light in the first place, but without them our knowledge of the stars would indeed be meagre.

Since the stars are suns, it is logical to expect them to show spectra of the same type as our Sun. This is true, but there are very marked differences in detail. For example, the spectra of white stars such as Sirius are dominated by lines due to hydrogen, while the cool orange-red stars produce very complex spectra with many bands due to molecules.

Pioneering efforts were made during the last century to classify the stars into various spectral types. Finally the system adopted was that drawn up at Harvard, where each type of star was given a letter according to its spectrum. In order of decreasing surface temperature, the accepted types are W, O, B, A, F, G, K, M and then R, N and S, whose surface temperatures are much the same (nowadays types R and N are often classed together as type C). The original scheme began with A, B ... but so many complications became evident that the final sequence was alphabetically chaotic. There is a famous mnemonic to help in getting the order right: 'Wow! O, Be A Fine Girl Kiss Me Right Now Sweetie.' Each type is again subdivided; thus a star of type A5 is intermediate between A0 and F0. Our Sun is of type G2.

In 1908, the Danish astronomer Ejnar Hertzsprung drew up a diagram in which he plotted the luminosities of the stars against their spectral types (plotting absolute magnitude against surface temperature comes to the same thing). Similar work was carried out in America by Henry Norris Russell, and diagrams of this sort are now known as Hertzsprung–Russell or HR Diagrams. Their importance in astrophysics cannot be overestimated. You can see at once that most of the stars lie along a band running from the top left to the bottom right of the diagram; this makes up what is termed the Main Sequence. Our Sun is a typical Main Sequence star. To the upper right lie giants and supergiants of tremendous luminosity, while to the lower left there are the white dwarfs, which are in a different category and were not known when HR Diagrams were introduced. Note also that most of the stars belong to types B to M. The very hottest types (W and O) and the very coolest (R, N and S) are relatively rare.

It is also obvious that the red and orange stars (conventionally, though misleadingly, referred to as 'late' type) are of two definite kinds; very powerful giants and very feeble dwarfs, with virtually no examples of intermediate luminosity. The giant-and-dwarf separation is less marked for the yellow stars, though it is still perceptible; thus Capella and the Sun are both of type G, but Capella is a giant, while the Sun is ranked as a dwarf. The distinction does not apply to the white or bluish stars, those of 'early' type.

The stars show a tremendous range in size, temperature and luminosity. The very hottest stars are of type W; they are often called Wolf–Rayet stars, after the two French astronomers who made careful studies of them over a century ago, and have surface temperatures of up to 80,000 degrees C. Their spectra show many bright emission lines, and they are unstable, with expanding shells moving outwards at up to 3000 kilometres (over 1800 miles) per second. O-type stars show both emission and dark lines, and have temperatures of up to 40,000 degrees C. At the other end of the scale we have the cool red giants of types R, N and S, where the surface temperatures are no more than 2600 degrees C; almost all stars of this type are variable in output.

Some supergiants are powerful by any standards; S Doradûs, in the Large Cloud of Magellan – one of the closest of the external galaxies, at a distance of 169,000 light-years – is at least a million times as luminous as the Sun, though it is too far away to be seen with the naked eye. Even more powerful is the strange, erratic variable Eta Carinae, which may equal 6 million Suns and has a peculiar spectrum which cannot be put into any regular type. On the other hand a dim star known as MH 18, identified in 1990 by M. H. Hawkins at the Royal Observatory Edinburgh, has only 1/20,000 the luminosity of the Sun. The range in mass is not so great; the present holder of the 'heavyweight' record seems to be Plaskett's Star in Monoceros (the Unicorn), which is a binary system with two O7-type components, each of which is about 55 times as massive as the Sun.

Direct measurements of star diameters are very difficult. The star with the greatest apparent diameter is probably Betelgeux in Orion, with a value of 50 milliarc seconds – about the same as the apparent diameter of a metre ruler placed on the Moon and observed from the Earth. New direct measurements are being made by an Australian team led by John Davis, who has built SUSI, the Sydney University Stellar Interferometer; this is made up of a number of relatively small telescopes working together, and can measure the width of a human hair from a distance of some 100 kilometres (over 60 miles). It has even become possible to detect surface details on a few stars.

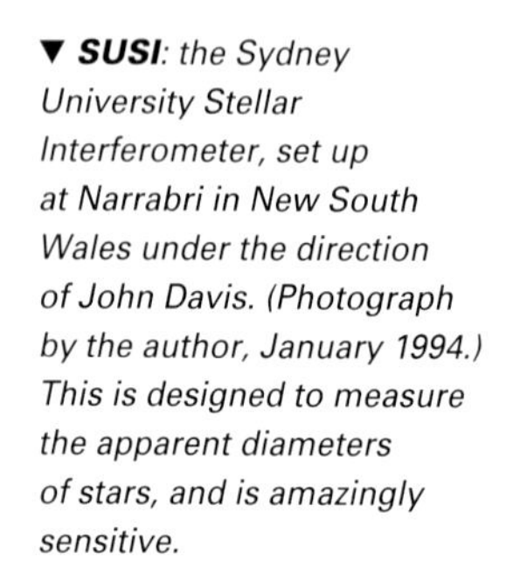

▼ ***SUSI**: the Sydney University Stellar Interferometer, set up at Narrabri in New South Wales under the direction of John Davis. (Photograph by the author, January 1994.) This is designed to measure the apparent diameters of stars, and is amazingly sensitive.*

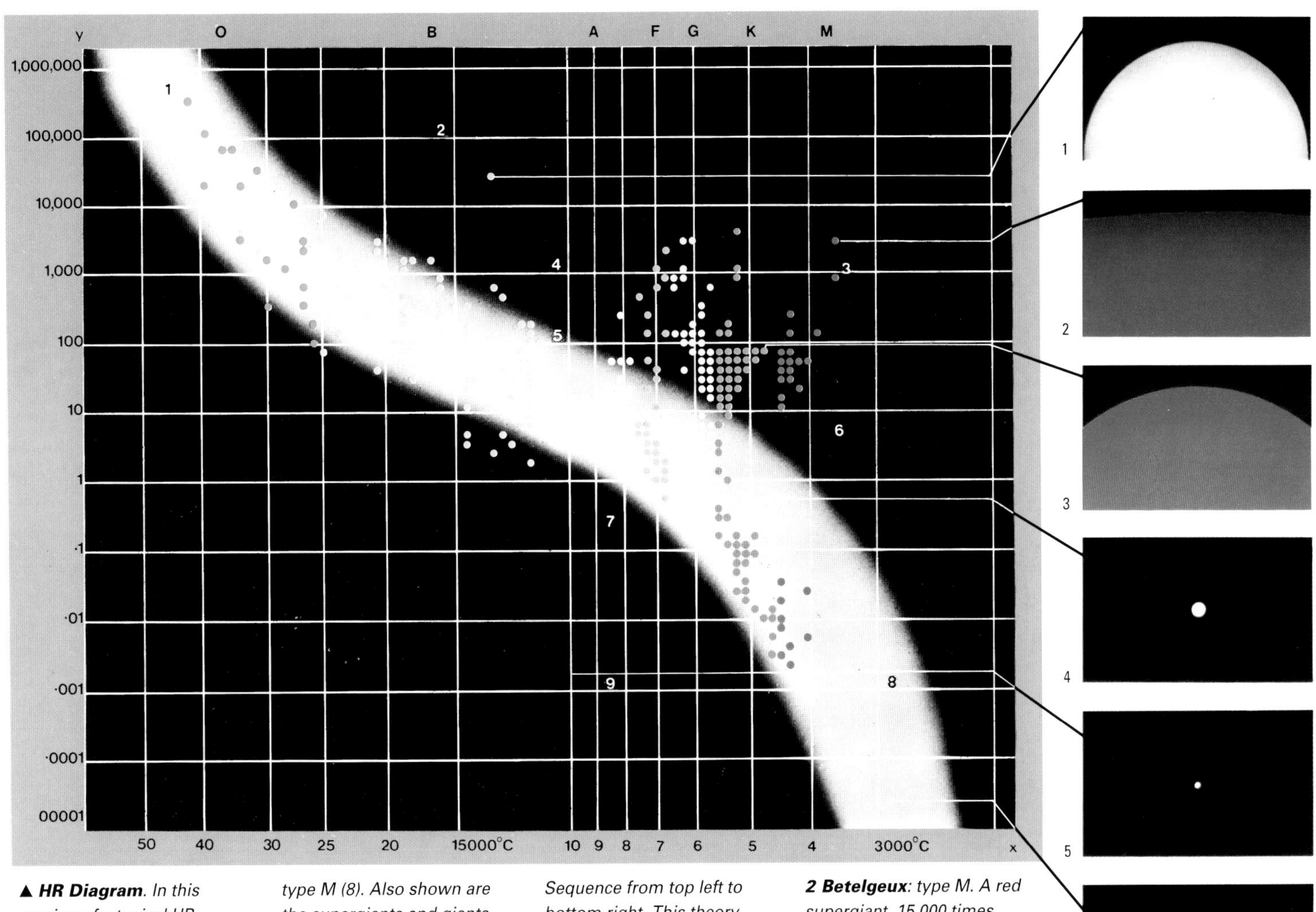

▲ ***HR Diagram**. In this version of a typical HR diagram, the stars are plotted according to their spectral types and surface temperatures (horizontal axis, x) and their luminosities in terms of the Sun (vertical axis, y). The Main Sequence is obvious at first glance, from the hot and powerful W and O stars (1), through to the dim red dwarfs of type M (8). Also shown are the supergiants and giants (2, 3); Cepheid variables (4); RR Lyrae variables (5); subgiants (6); subdwarfs (7); and white dwarfs (9). Originally it was believed that a star began its career as a large, cool red giant, and then heated up to join the Main Sequence; it then cooled and shrank as it passed down the Main Sequence from top left to bottom right. This theory has been found to be completely wrong. The red giants are at an advanced stage in their evolution.*

***1 Rigel**: type B8. A very massive and luminous star, at the upper end of the Main Sequence. It is 60,000 times as luminous as the Sun, and is white, with a temperature of more than 12,000°C.*

***2 Betelgeux**: type M. A red supergiant, 15,000 times as luminous as the Sun, with a greater diameter than that of the Earth's orbit. It is surrounded by a very tenuous 'shell' of potassium.*

***3 Aldebaran**: type K. A giant star, orange in colour, not as large as Betelgeux, although it is 100 times as powerful as the Sun and has a diameter believed to be at least 50 million km (30 million miles).*

***4 The Sun**: type G2. A typical Main Sequence star. It is officially ranked as a dwarf, while Capella, also of type G (G8), is a giant. (Capella is not a single star, but a close binary.)*

***5 Sirius B**: A white dwarf which has used up all its nuclear 'fuel'. It has a diameter of 40,000 km (25,000 miles), smaller than that of Uranus, but it is amazingly dense, and is as massive as the Sun.*

***6 Wolf 339**: type M. A dim red dwarf, with a surface temperature of 3000°C but a luminosity only 0.00002 that of the Sun. Yet its spectral type is the same as that of Betelgeux.*

STELLAR SPECTRA

Type	Spectrum	Surface temperature, °C	Example
W	Many bright lines. Divided into WN (nitrogen sequence) and WC (carbon sequence). Rare.	up to 80,000	γ Velorum (WC7)
O	Both bright and dark lines. Rare.	40,000–35,000	ζ Orionis (09.5)
B	Bluish-white. Prominent lines due to helium.	25,000–12,000	Spica, β Crucis
A	White. Prominent hydrogen lines.	10,000–8000	Sirius, Vega
F	White or very slightly yellowish. Calcium lines very prominent.	7500–6000	Canopus, Polaris
G	Yellowish; weaker hydrogen lines, many metallic lines.	Giants 5500–4200 Dwarfs 6000–5000	Capella, Sun
K	Orange. Strong metallic lines.	Giants 4000–3000 Dwarfs 5000–4000	Arcturus, Aldebaran ε Eridani, τ Ceti
M	Orange-red. Complicated spectra, with many bands due to molecules.	Giants 3400 Dwarfs 3000	Betelgeux, Antares Proxima Centauri
R	Reddish.	2600	T Lyrae
N	Reddish; strong carbon lines.	2500	R Leporis
S	Red; prominent bands of titanium oxide and zirconium oxide.	2600	χ Cygni, R Cygni

The Lives of the Stars

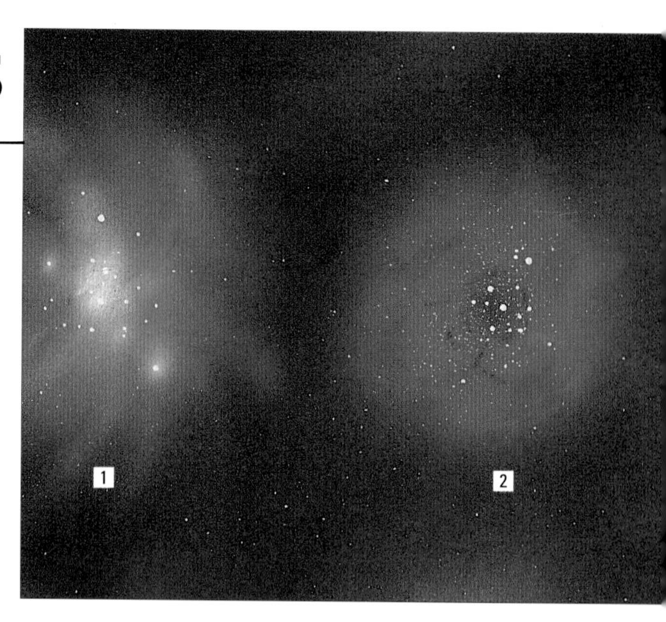

When we try to work out the life-story of a star, we are faced with the initial difficulty that we cannot – usually – see a star change its condition as we watch, just as an observer in a city street will not notice a boy changing into a man. All we can do is decide which stars are young and which are old, after which we can do our best to trace the sequence of events. Earlier theorists, through no fault of their own, picked wrong.

The mistake lay in a faulty interpretation of the HR Diagram. It was supposed that a star began as a very large, cool red giant such as Betelgeux; that it heated up and joined the Main Sequence at the upper left of the Diagram, and then slid down towards the lower right corner, cooling steadily and becoming a dim red dwarf before fading out. Certainly this would explain the giant-and-dwarf divisions, but we now know that red giants such as Betelgeux are not young at all; they are far advanced in their evolution, so that they rank as stellar old-age pensioners.

According to current theory, a star begins by condensing out of the tenuous material making up a nebula. Chance condensations lead to the appearance of non-luminous masses called globules, many of which can be seen in nebulae because they blot out the light of stars beyond. Gravity causes the mass to shrink, and as it does so it heats up near its centre. When the temperature has risen sufficiently, the mass begins to glow and turns into a protostar.

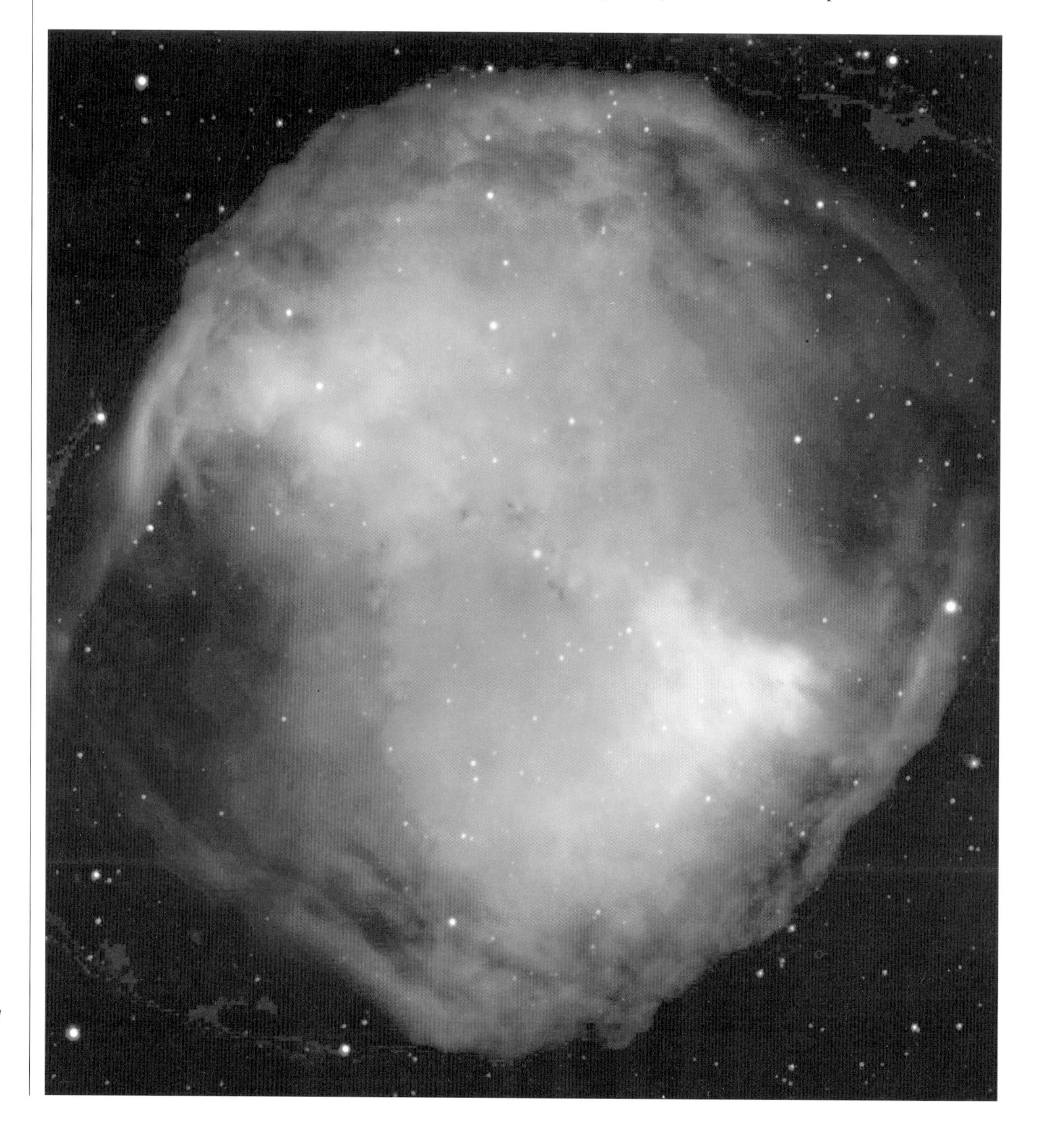

▶ ***M27 (NGC 6853)*** – *a planetary nebula in Vulpecula always known as the Dumbbell Nebula. It is just under 1000 light-years away. This image was taken in October 1998 with Antu, the first unit of the VLT (Very Large Telescope) at Cerro Paranal in Chile. The Antu mirror is 8.2 metres (323 inches) in diameter.*

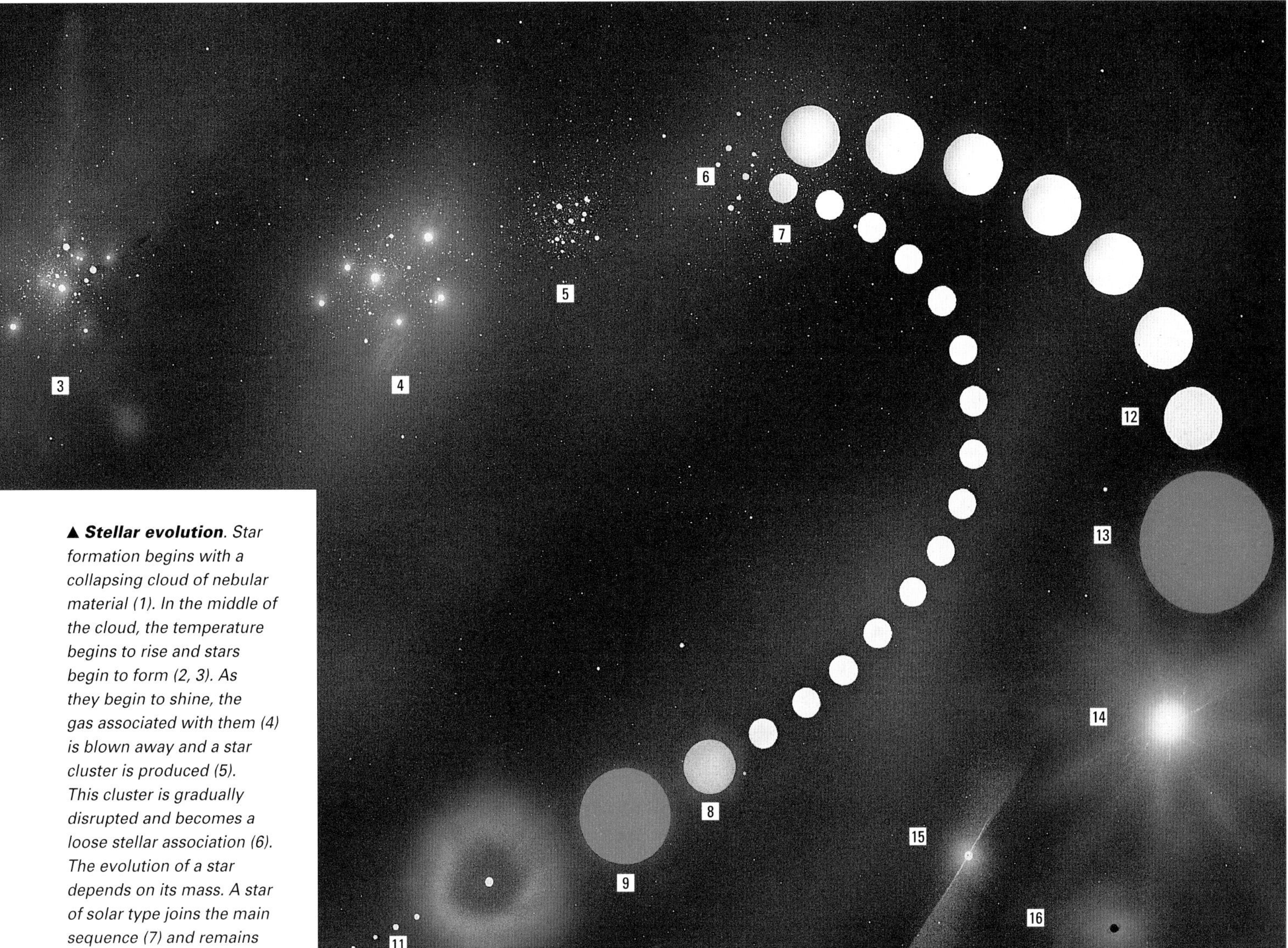

▲ ***Stellar evolution**. Star formation begins with a collapsing cloud of nebular material (1). In the middle of the cloud, the temperature begins to rise and stars begin to form (2, 3). As they begin to shine, the gas associated with them (4) is blown away and a star cluster is produced (5). This cluster is gradually disrupted and becomes a loose stellar association (6). The evolution of a star depends on its mass. A star of solar type joins the main sequence (7) and remains on it for a very long period. When its hydrogen 'fuel' begins to run low, it expands (8) and becomes a red giant (9). Eventually the outer layers are lost, and the result is the formation of a planetary nebula (10). The 'shell' of gas expands and is finally dissipated, leaving the core of the old star as a white dwarf (11). The white dwarf continues to shine feebly for an immense period before losing the last of its heat and becoming a cold, dead black dwarf. With a more massive star the sequence of events is much more rapid. After its Main Sequence period (12) the star becomes a red supergiant (13) and may explode as a supernova (14). It may end as a neutron star or pulsar (15), although if its mass is even greater it may produce a black hole (16).*

What happens next depends mainly upon the star's initial mass. If it is less than one-tenth that of the Sun, the core will never become hot enough for nuclear reactions to begin, and the star will simply glow feebly for a very long period before losing its energy.

If the mass is between 0.1 and 1.4 times that of the Sun, the story is very different. The star goes on shrinking, and fluctuates irregularly; it also sends out a strong stellar wind, and eventually blows away its original cocoon of dust. This is the so-called T Tauri stage, which in the case of the Sun may have lasted for around 30 million years. When the core temperature soars to 10 million degrees C, nuclear reactions are triggered off; the hydrogen-into-helium process begins (known, misleadingly, as 'hydrogen burning'), and the star joins the Main Sequence. Hydrogen burning will last for around 10,000 million years, but at last the supply of hydrogen 'fuel' must run low, and the star is forced to change its structure. The core temperature becomes so high that helium starts to 'burn', producing carbon; around this active core there is a shell where hydrogen is still producing energy. The star becomes unstable, and the outer layers swell out, cooling as they do so. The star becomes a red giant.

This is as far as the nuclear process can go, because the temperature does not increase sufficiently to trigger off carbon-burning. The star's outer layers are thrown off, and for a cosmically brief period – no more than about 100,000 years – we have the phenomenon of what is termed a planetary nebula. Finally, when the outer layers have dissipated in space, we are left with a white dwarf star; this is simply the original core, but now 'degenerate', so that the atoms are broken up and packed closely together with almost no waste of space. The density is amazingly high. If a spoonful of white dwarf material could be brought to Earth, it would weigh as much as a steam-roller. The best-known white dwarf is the dim companion of Sirius, which is only about 40,000 kilometres (25,000 miles) in diameter – smaller than a planet such as Uranus or Neptune – but is as massive as the Sun.

Bankrupt though it is, a white dwarf still has a high surface temperature when it is first formed; up to 100,000 degrees C in some cases, and it continues to radiate. Gradually it fades, and must end up as a cold, dead black dwarf; but at the moment no white dwarf with a surface temperature of below 3000 degrees C has been found, and it may be that the universe is not yet old enough for any black dwarfs to have been formed.

With stars of greater initial mass, everything happens at an accelerated rate. The core temperatures become so high that new reactions occur, producing heavier elements. Finally the core is made up principally of iron, which cannot 'burn' in the same way. There is a sudden collapse, followed by an explosion during which the star blows most of its material away in what is called a supernova outburst, leaving only a very small, super-dense core made up of neutrons – so dense that a thousand million tonnes of it could be crammed into an eggcup. If the mass is greater still, the star cannot even explode as a supernova; it will go on shrinking until it is pulling so powerfully that not even light can escape from it. It has produced a black hole.

We have learned a great deal about stellar evolution during the past decades, but many uncertainties remain.

Double Stars

Look at Mizar, the second star in the tail of the Great Bear, and you will see a much fainter star, Alcor, close beside it. Use a telescope, and Mizar itself is seen to be made up of two components, one rather brighter than the other. The two Mizars are genuinely associated, and make up a physically-connected or binary system, while Alcor is also a member of the group even though it is a long way from the bright pair. Binary systems are very common in the Galaxy; surprisingly, they seem to be more plentiful than single stars such as the Sun.

Many double stars are within the range of small telescopes, and some pairs are even separable with the naked eye; Alcor is by no means a difficult naked-eye object when the sky is reasonably clear and dark. Yet not all doubles are true binaries. In some cases one component is simply seen more or less in front of the other, so that we are dealing with nothing more significant than a line of sight effect. Alpha Capricorni, in the Sea-Goat, is a good case of this (Map 14 in this book). The two components are of magnitudes 3.6 and 4.2 respectively, and any normal-sighted person can see them separately without optical aid. The fainter member of the pair is 1600 light-years away, and over 5000 times as luminous as the Sun; the brighter component is only 117 light-years away, and a mere 75 Sun-power. There is absolutely no connection between the two. As so often happens, appearances are deceptive.

The components of a binary system move together round their common centre of gravity, much as the two bells of a dumbbell will do when twisted by the bar joining them. If the two members are equal in mass, the centre of gravity will be midway between them; if not, the centre of gravity will be displaced towards the 'heavier' star. However, the stars do not show nearly so wide a range in mass as they do in size or luminosity, so that in general the centre of gravity is not very far from the mid position. With very widely separated pairs, the orbital periods may be millions of years, so that all we can really say is that the components share a common motion in space. This is true of Alcor with respect to the Mizar pair, while the estimated period of the two bright Mizars round their common centre of gravity is of the order of

▲ ▶ ***Double stars (drawings by Paul Doherty).*** *(Top) Mizar and Alcor, the most famous of all naked-eye doubles; telescopically Mizar itself is seen to be made up of two components. (Centre) Albireo (Beta Cygni). This is almost certainly the most beautiful coloured double in the sky; the primary (magnitude 3.1) is golden yellow, the secondary (magnitude 5.1) vivid blue. The separation is almost 35" of arc. (Right) Almaak, or Gamma Andromedae. The primary is an orange K-type star of magnitude 2.2. The companion is of magnitude 5.0, and is a white star of type A. It is a close binary with a period of 61 years, and a separation of about 0".5.*

10,000 years. The real separation seems to be about 60,000 million kilometres (over 37,000 million miles).

With the Mizar group there is another complication. In 1889 E. C. Pickering, at the Harvard Observatory, examined the spectrum of the brighter component (Mizar A), and found that the spectral lines were periodically doubled. At once he realized that he was dealing with a binary whose two components were much too close to be seen separately. The revolution period is 20.5 days, and the two stars are about equal in brightness. There are times when one component is approaching us, and will show a blue shift, while the other is receding and will show a red shift; therefore the lines will be doubled. When the orbital motion is transverse, the lines will be single. Mizar A was the first-known spectroscopic binary; later it was found that both Mizar B and Alcor are also spectroscopic binaries. The eighth-magnitude star between Alcor and the bright pair is more remote, and not one of the group.

The position angle or P.A. of a double star – either a binary or an optical pair – is measured according to the angular direction of the secondary (B) from the primary (A), reckoned from 000 degrees at north round by 090 degrees at east, 180 degrees at south, and 270 degrees at west back to north. In general, it may be said that a 3-inch (7.6-cm) telescope will separate a pair 1.8 seconds of arc apart provided that the two components are equal; a 6-inch (15.2-cm) will reach down to 0.8 second of arc, and a 12-inch (30.5-cm) will reach to 0.4 second of arc.

Arich or Gamma Virginis (Map 6) is a good example of a binary which has changed its appearance over the years. The components are exactly equal at magnitude 3.5, and the orbital period is 171.4 years. Several decades ago it was very wide and easy, but it is now closing up, and by 2016 the star will appear single except with giant telescopes. This does not mean that the components are actually approaching each other, but only that we are seeing them from a less favourable angle. With Zeta Herculis (Map 9) the period is only 34 years, so that both the separation and the position angle alter quite quickly; so also with Alpha Centauri, the brighter of the two Pointers to the Southern Cross, where the period is 79.9 years. In 1995 the separation is 17.3 seconds and the P.A. is 218 degrees; by 2005 the separation will have decreased to 10.5 seconds and the P.A. will have increased to 230 degrees. (Alpha Centauri is the nearest bright star beyond the Sun. The dim red dwarf Proxima, more than a degree away from Alpha, is slightly closer to us; it has always been regarded as a member of the group, though there are suggestions that it is merely 'passing by'.)

In many cases the two components of a binary are very unequal. Sirius has its dwarf companion, only 1/10,000 as bright as the primary – though it must once have passed through the red giant stage and been much more luminous than it is now. Then there are pairs with beautiful contrasting colours; Albireo or Beta Cygni (Map 8) is a yellow star with a companion which is vivid blue, while some red supergiants, notably Antares (Map 11) and Alpha Herculis (Map 9) have companions which look greenish by contrast.

Multiple stars are also found. A famous case is that of Epsilon Lyrae, near Vega (Map 8). The two main components are of magnitude 4.7 and 5.1 respectively, and can be separated with the naked eye; a telescope shows that each is again double, so that we have a quadruple system. Theta Orionis, in the Great Nebula (Map 16), has its four main components arranged in the pattern which has led to the nickname of the Trapezium. Castor in Gemini, the senior though fainter member of the Twins (Map 17), is an easy telescopic pair; each component is a spectroscopic binary, and there is a much fainter member of the group which also is a spectroscopic binary.

It used to be thought that a binary system was formed when a rapidly-spinning star broke up, but this attractive theory has now fallen from favour, and it seems much more likely that the components of a binary were formed from the same cloud of material in the same region of space, so that they have always remained gravitationally linked. If their initial masses are different they will evolve at different rates, and in some cases there may even be exchange of material between the two members of the pair.

▼ ***Gemini*** *photographed by S. Andrew. The two bright stars are the 'Twins', Castor (upper) and Pollux (lower). Pollux is a single star; Castor, which appears single in this photograph, is actually a very complicated system made up of two bright pairs and one faint pair of stars.*

SELECTED DOUBLE STARS

Name	Mag.	Sep., "	P.A., °	Map	Notes
γ Andromedae	2.3, 5.0	9.4	064	12	Yellow, blue. B is double.
ζ Aquarii	4.3, 4.5	2.0	196	14	Widening.
γ Arietis	4.8, 4.8	7.6	000	12	Very easy.
α Canum Venaticorum	2.9, 5.5	19.6	228	1	Yellow, bluish.
α Centauri	0.0, 1.2	17.3	218	20	Very easy. Period 80 years.
γ Centauri	2.9, 2.9	1.2	351	20	Period 84 years.
δ Cephei	var, 7.5	41	192	3	Very easy.
α Crucis	1.4, 1.9	4.2	114	20	Third star in field.
β Cygni	3.1, 5.1	34.1	054	18	Yellow, blue.
γ Delphini	4.5, 5.5	9.3	267	18	Yellowish, bluish.
ν Draconis	4.9, 4.9	62	312	2	Naked-eye pair.
θ Eridani	3.4, 4.5	8.3	090	22	Both white.
α Geminorum	1.9, 2.9	3.5	072	17	Widening.
α Herculis	var, 5.4	4.6	106	9	Red, greenish.
ζ Herculis	2.9, 5.5	1.4	261	9	Period 34 years.
ε Lyrae	4.7, 5.1	207	173	18	Both double.
ζ Lyrae	4.3, 5.9	44	149	18	Fixed, easy.
β Orionis	0.1, 6.8	9.5	202	16	Not difficult.
ζ Orionis	1.9, 4.0	2.4	162	16	Split with 7.5 cm.
β Phoenicis	4.0, 4.2	1.5	324	21	Widening.
α Scorpii	1.2, 5.4	2.7	274	11	Red, greenish.
ν Scorpii	4.3, 6.4	42	336	11	Both double.
θ Serpentis	4.5, 4.5	22	104	10	Very easy.
β Tucanae	4.4, 4.8	27	170	21	Both double.
ζ Ursae Majoris	2.3, 4.0	14.4	151	1	Naked-eye pair with Alcor.
γ Virginis	3.5, 3.5	2.2	277	6	Period 171 years. Closing.

Variable Stars

Variable stars are very common in the Galaxy – and, for that matter, in other galaxies too. They are of many types. Some behave in a completely predictable manner, while others are always liable to take us by surprise.

First there are the eclipsing binaries, which are not truly variable at all. The prototype is Algol or Beta Persei, which, appropriately enough, lies in the head of the mythological Gorgon, and has long been nicknamed the 'Demon Star'. Normally it shines at magnitude 2.1; but every 2.9 days it begins to fade, dropping to magnitude 3.4 in just over four hours. It remains at minimum for a mere 20 minutes, after which it brightens up again.

Algol is in fact a binary system. The main component (Algol A) is of type B, and is a white star 100 times as luminous as the Sun; the secondary (Algol B) is a G-type subgiant, larger than A but less massive. When B passes in front of A, part of the light is blocked out, and the magnitude falls; when A passes in front of B there is a much shallower minimum not detectable with the naked eye. The eclipses are not total, and if Algol could be observed from a different vantage point in the Galaxy there would be no variation at all.

Incidentally, we have here a good example of what is called mass transfer. The G-type component was originally the more massive of the two, so that it left the Main Sequence earlier and swelled out. As it did so, its gravitational grip on its outer layers was weakened, and material was 'captured' by the companion, which is now the senior partner. The process is still going on, and radio observations show material streaming its way from B to A.

Other naked-eye Algol stars are Lambda Tauri (Map 17) and Delta Librae (Map 6). Beta Lyrae, near Vega (Map 8) is of different type. The period is almost 13 days, and there are alternate deep and shallow minima; the components are much less unequal than with Algol, and variations are always going on. Apparently the two components are almost touching each other, and each must be drawn out into the shape of an egg. Different again is Epsilon Aurigae, close to Capella (Map 18). The primary is a particularly luminous supergiant; the eclipsing secondary has never been seen, but is probably a smallish, hot star surrounded by a cloud of more or less opaque material. Eclipses occur only every 27 years; the next is due in 2011. Close beside Epsilon is Zeta Aurigae, also an eclipsing binary with a period of 972 days. Here we have a red supergiant together with a smaller, hot companion; it is when the hot star is eclipsed that we see a drop in brightness, but the amplitude is small (magnitude 3.7 to 4.2).

Pulsating stars are intrinsically variable. Much the most important are the Cepheids, named after the prototype star Delta Cephei (Map 3) in the far north of the sky. They are yellow supergiants in a fairly advanced stage of evolution, so that they have exhausted their available hydrogen and helium 'fuel' and have become unstable, swelling and shrinking. The period of pulsation is the time needed for a vibration to travel from the star's surface to the centre and back again, so that large luminous stars have longer periods than stars which are smaller and less powerful. There is a definite link between a Cepheid's period and its real luminosity – which in turn gives a clue to the distance, so that Cepheids are useful as 'standard candles', particularly since they are powerful enough to be seen over an immense range. W Virginis stars are not unlike Cepheids, but are less luminous; the brightest example is Kappa Pavonis in the southern hemisphere (Map 21). We also have RR Lyrae stars, which have short periods and small amplitude; all seem to be about 90 times as powerful as the Sun.

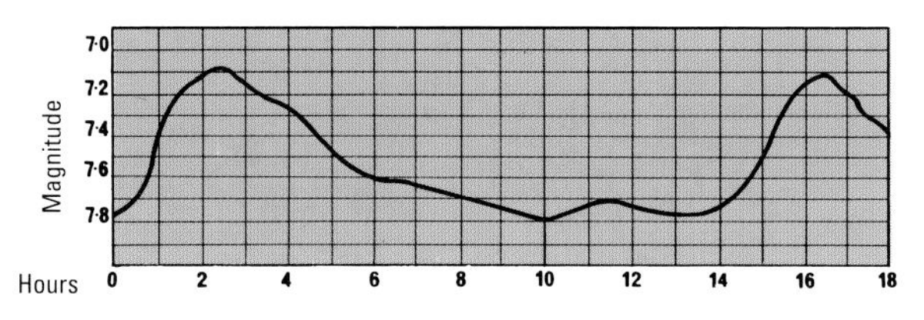

*► **RR Lyrae variables**. All have short periods and there are three main groups. In the first, the periods are about 0.5 days and the rise to maximum is sharp, followed by a slower decline; RR Lyrae is of this kind (see light-curve). Variables of the second class are similar, but the amplitudes are smaller and the rise to maximum is slower. Stars of the third class have symmetrical light-curves and periods of some 0.3 days.*

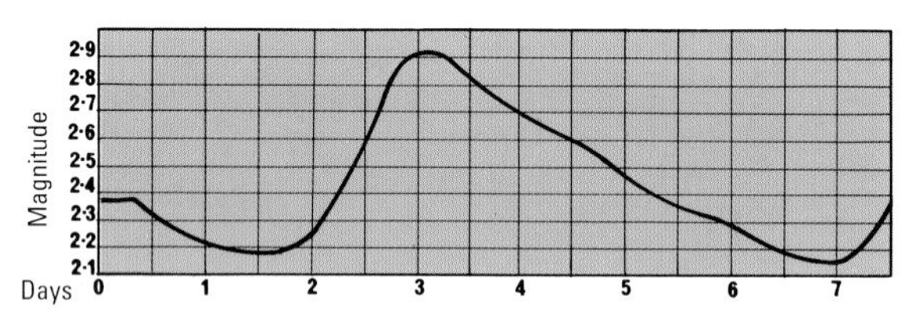

*► **Cepheid variables** have periods of from three days to over 50 days. Their light-curves are regular and are related to their absolute magnitudes. The light-curve shown here is for Delta Cephei, the prototype star. Delta Cephei belongs to Population I: there are also Cepheid variables of Population II, which are of similar kind but are less luminous; these are now known as W Virginis variables.*

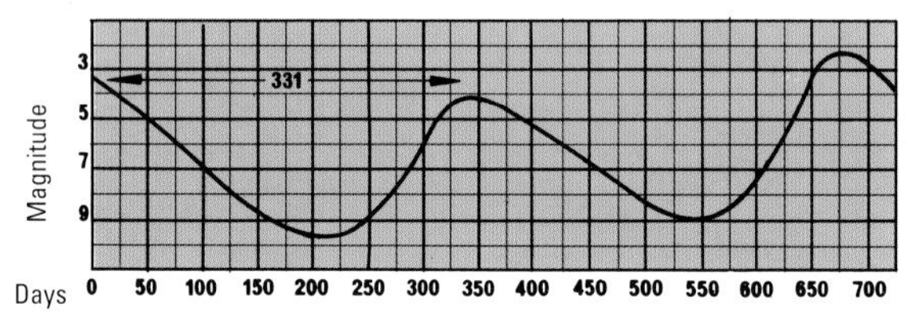

*► **Long-period variables**. With long-period variables, of which the prototype is Mira Ceti (light-curve shown here), neither the periods nor the maximum and minimum magnitudes are constant. For instance, Mira may at some maxima attain magnitude 2: at other maxima the magnitude never exceeds 4. Most long-period variables are red giants of spectral type M or later: and there is no Cepheid-type period–luminosity law.*

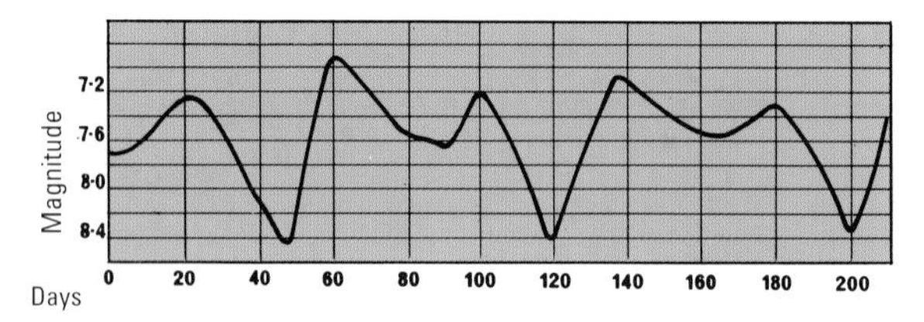

*► **RV Tauri variables**. The RV Tauri variables are very luminous and are characterized by light-curves which show alternate deep and shallow minima. There are considerable irregularities: two deep minima may occur in succession, and at times there is no semblance of regularity. RV Tauri stars are rare; the light-curve of a well-known member of the class, AC Herculis, is shown in the diagram here.*

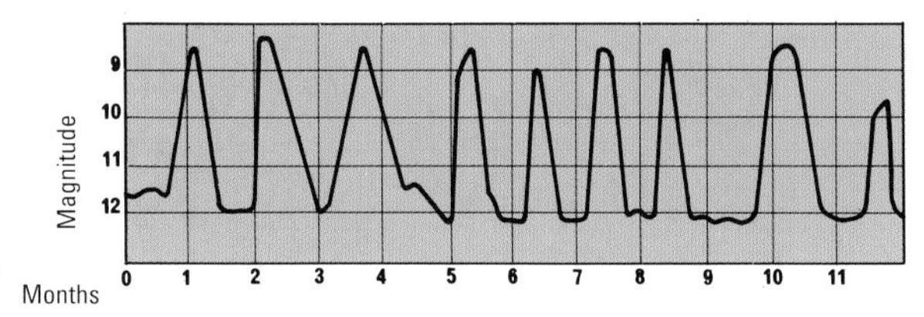

*► **SS Cygni or U Geminorum variables**. The so-called 'dwarf novae' are characterized by periodical outbursts; for most of the time they remain at minimum brightness. Outbursts may occur at fairly regular intervals or may be unpredictable. The prototype stars are U Geminorum and SS Cygni (light-curve shown here) which has outbursts approximately every 40 days, when its magnitude increases from 12 to 8.25.*

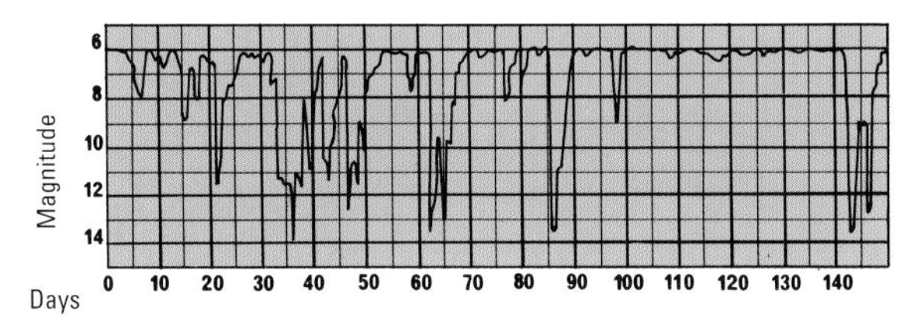

*► **R Coronae Borealis variables**. The most striking feature of the light-curve of the typical R Coronae Borealis variable shown here is that its magnitude remains more or less constant, but then at unpredictable intervals the magnitude plunges sharply to a deep, but brief, minimum. R Coronae, for example, is normally of around magnitude 6, but it may drop to below even magnitude 14; for long periods it may remain at its maximum. R Coronae variables are rare. Only six of them so far observed – R Coronae itself, UW Centauri, RY Sagittarii, SU Tauri and RS Telescopii – can exceed the 10th magnitude at their maximum brightness.*

BASIC CLASSIFICATION OF VARIABLE STARS

Symbol	Type	Example	Notes
EA	Algol	Algol	Periods 0.2d–27y. Maximum for most of the time.
EB	Beta Lyrae	Beta Lyrae	Periods over 1 day. Less unequal components. Continuous variation.
EW	W Ursae Majoris	W UMa	Dwarfs; periods usually less than 1 day.
PULSATING			
M	Mira	Mira	Long-period red giants. Periods 80–1000 days. Periods and amplitudes vary from cycle to cycle.
SR	Semi-regular	Ë Geminorum	Red giants. Periods and amplitudes very rough.
RV	RV Tauri	R Scuti	Red supergiants; alternate deep and shallow minima. Marked irregularities.
CEP	Cepheids	‰ Cephei	Regular; periods 1–135 days; spectra F to K.
CW	W Virginis	ÎPavonis	Population II Cepheids.
RR	RR Lyrae	RR Lyrae	Regular; short periods, 0.2–1.2 days; all of equal luminosity.
ERUPTIVE			
GCAS	Á Cassiopeiae	Á Cassiopeiae	Shell stars: rapid rotators; small amplitudes.
IT	T Tauri	T Tauri	Very young, irregularly-varying stars.
RCB	R Coronae Borealis	RCrB	Unpredictable deep minima. Large amplitude. Highly luminous.
SDOR	S Doradûs	S Doradûs	Very luminous supergiants with expanding shells.
CATACLYSMIC			
UG	U Geminorum	SS Cygni	Dwarf novae or SS Cygni.
UG2	Z Camelopardalis	Z Cam.	Dwarf novae with occasional standstill.
N	Novae	DQ Herculis	Violent outburst.
SN	Supernovae	B Cassiopeiae	Violent outburst. Type I; destruction of the white dwarf component of a binary system. Type II; collapse of a supergiant star.

▲ ***Eta Carinae*** *photographed with the Wide Field and Planetary Camera of the Hubble Space Telescope, January 1994. Eta Carinae has a mass about 150 times that of the Sun, and may be the most luminous star known; it will eventually explode as a supernova. The ghostly red glow surrounding the star is composed of fast-moving material ejected during the last-century outburst, when for a time Eta Carinae outshone every star apart from Sirius.*

Next come the longer-period Mira stars, named after Mira or Omicron Ceti, the first-discovered and brightest member of the class (Map 15). Unlike the Cepheids, the Mira stars – all of which are red giants or supergiants – are not perfectly regular; their periods vary somewhat from the mean value (332 days in the case of Mira itself) and so do the amplitudes. At some maxima Mira never becomes brighter than the fourth magnitude, while at others it has been known to reach 1.7. In general the amplitudes are large – over 10 magnitudes in the case of Chi Cygni, in the Swan (Map 8). Several Mira stars reach naked-eye visibility at maximum, but at minimum all of them fade below binocular range.

Semi-regular variables are not unlike the Mira stars, but have smaller amplitudes, and their periods are very rough. Betelgeux in Orion is the best-known example (Map 16). At times it may equal Rigel, while at others it is no brighter than Aldebaran in Taurus. RV Tauri stars have alternate deep and shallow minima, interspersed with spells of complete irregularity; R Scuti, in the Shield (Map 8) is the only brightish example.

Eruptive variables are unpredictable. Some, such as Gamma Cassiopeiae (Map 3) occasionally throw off shells of material; T Tauri stars are very young and have not yet joined the Main Sequence, so that they are varying irregularly. R Coronae Borealis stars remain at maximum for most of the time, but undergo sudden drops to minimum – because they accumulate clouds of soot in their atmospheres, and fade until the soot is blown away. These stars are very rare, and R Coronae itself (Map 4) is much the brightest member of the class.

Cataclysmic variables remain at minimum when at their normal brightness, but show outbursts which may be roughly periodical – as with the SS Cygni or U Geminorum stars – or else quite unexpected, as with classical novae. All these are binary systems. One component is a white dwarf, which pulls material away from its Main Sequence companion; when enough material has accumulated the situation becomes unstable, and a short-lived outburst results.

Supernovae, the most colossal outbursts known in nature, are best described separately (see pages 180–1). Mention should also be made of the unique Eta Carinae, in the Keel of the Ship (Map 19). For a time during the 19th century it was the brightest star in the sky apart from Sirius, but for over a hundred years now it has been just below naked-eye visibility; it is associated with nebulosity, and when seen through a telescope looks quite unlike a normal star. At its peak it must have been six million times as powerful as the Sun, making it the most powerful star known to us. It is highly unstable, and in the near future – cosmically speaking – it will probably explode as a supernova.

There are so many variable stars in the sky that professional astronomers cannot hope to keep track of them all, so that amateurs can do very valuable work. Accurate measurements can be made with sophisticated equipment such as photoelectric photometers, but a great deal can be done by making eye-estimates through an ordinary telescope. The procedure is to compare the variable with nearby stars of constant brightness. At least two comparison stars are needed. For example, if star A is known to be of magnitude 6.8 and star B of 7.2, and in brightness the variable is midway between them, its magnitude must be 7.0. With practice, estimates can be made to an accuracy of a tenth of a magnitude, and variable star work has now become one of the most important branches of modern amateur astronomy.

Novae

▼ ***Nova (HR) Delphini 1967**, discovered by the English amateur George Alcock. It reached naked-eye visibility. This photograph was taken on 10 August 1967 by Commander H. R. Hatfield. The quadrilateral of Delphinus is shown; HR is the brightish star near the top of the picture. The maximum was unusually prolonged, as the light-curve shows; fading was gradual, and at present (1997) the magnitude has returned to its pre-outburst value of about 13.*

Occasionally a bright star will flare up where no star has been seen before. Naturally enough, this is known as a 'nova', from the Latin for 'new', but the name is misleading; a nova is not really new at all. What has happened is that a formerly dim star has suffered an outburst and brightened up to many thousands of times its normal state. Its glory does not last for long; in a few days, weeks, or a few months at most it will fade back into its previous obscurity.

It now seems certain that a nova is the result of an outburst in the white dwarf component of a binary system. The other member of the pair is a normal star, which has not yet evolved to the white dwarf condition, and is of relatively low density. The white dwarf has a very powerful pull, and draws material away from its companion; as time goes by, a ring or 'accretion disk' builds up around the white dwarf. As more and more material arrives in the accretion disk, the temperature rises. In the lower part of the disk mild nuclear reactions are going on, but are 'blanketed', so to speak, by the non-reacting material above. This cannot last indefinitely; eventually the temperature builds up to such an extent that there is a violent nuclear explosion, and material is hurled outwards at speeds of up to 1500 kilometres (over 900 miles) per second. At the end of the outburst, the system reverts to its original state. Though the outburst releases a tremendous amount of energy – perhaps equivalent to a thousand million million nuclear bombs – the white dwarf loses only a tiny fraction of its mass.

Some novae may become very brilliant. GK Persei of 1901 reached magnitude 0.0; at its peak it must have been 200,000 times as luminous as the Sun, and it remained a naked-eye object for four months. As it faded, it was seen to be surrounded by nebulosity which gave every impression of expanding at a speed equal to that of light, though in fact the material had been there all the time and was merely being illuminated by the brilliance of the nova. It was only later that the actual nebulosity associated with the outburst became visible; the present magnitude of the star is about 13 – the same value as it had been before the explosion. In 1918, Nova Aquilae flared up abruptly, and outshone every star apart from Sirius; it was discovered on 8 June, and did not fade below the sixth magnitude until the following March. Spectroscopic research showed that it threw off shells of gas, and nebulosity became visible; this gradually expanded and became fainter, finally disappearing. At present, the old nova is of the 12th magnitude, and seems to be smaller but denser than the Sun.

DQ Herculis 1934 was discovered by an amateur, J. P. M. Prentice. It ranks as a 'slow nova', and was a naked-eye object for several months; it is now an eclipsing binary system with a period of only 4 hours 39 minutes. HR Delphini of 1967 was even slower, and took years to revert to its original magnitude of about 12. On the other hand V1500 Cygni of 1975 rocketed up to prominence in only a few hours, and had faded below naked-eye visibility in a few nights; it is now excessively faint.

A few stars have been known to show more than one outburst; these are known as recurrent novae. Thus the 'Blaze Star' T Coronae Borealis, in the Northern Crown (Map 4) is usually of about the tenth magnitude, but flared up to the second in 1866, and again brightened in 1946. The interval between the explosions was 80 years, and astronomers will be keeping a careful watch on it around 2026 to see if it will provide a repeat performance.

There is a definite link between novae and cataclysmic variables of the SS Cygni or U Geminorum type which are often termed dwarf novae. The exceptionally luminous and unstable P Cygni, in the Swan (Map 8) flared up to the third magnitude in 1600, and was once classed as a nova, but now seems to be a special type of variable star; for many years now its magnitude has hovered around 5, though it may well brighten again at any moment.

Novae show interesting changes in their spectra, and it is very important to start observing them as soon as possible after the start of the outburst. This is where amateurs come into their own, because they know the night sky far better than most professionals, and have a fine record of nova discovery. For example, an English schoolmaster, George Alcock, has now discovered five novae (as well as five comets); he uses powerful binoculars, and can identify some 30,000 stars on sight, so that he can recognize a newcomer at once.

Telescopic novae are not uncommon. Most of them appear in or near the Milky Way, and enthusiastic nova-hunters concentrate upon these regions of the sky. One never knows when a brilliant new star may burst forth without the slightest warning.

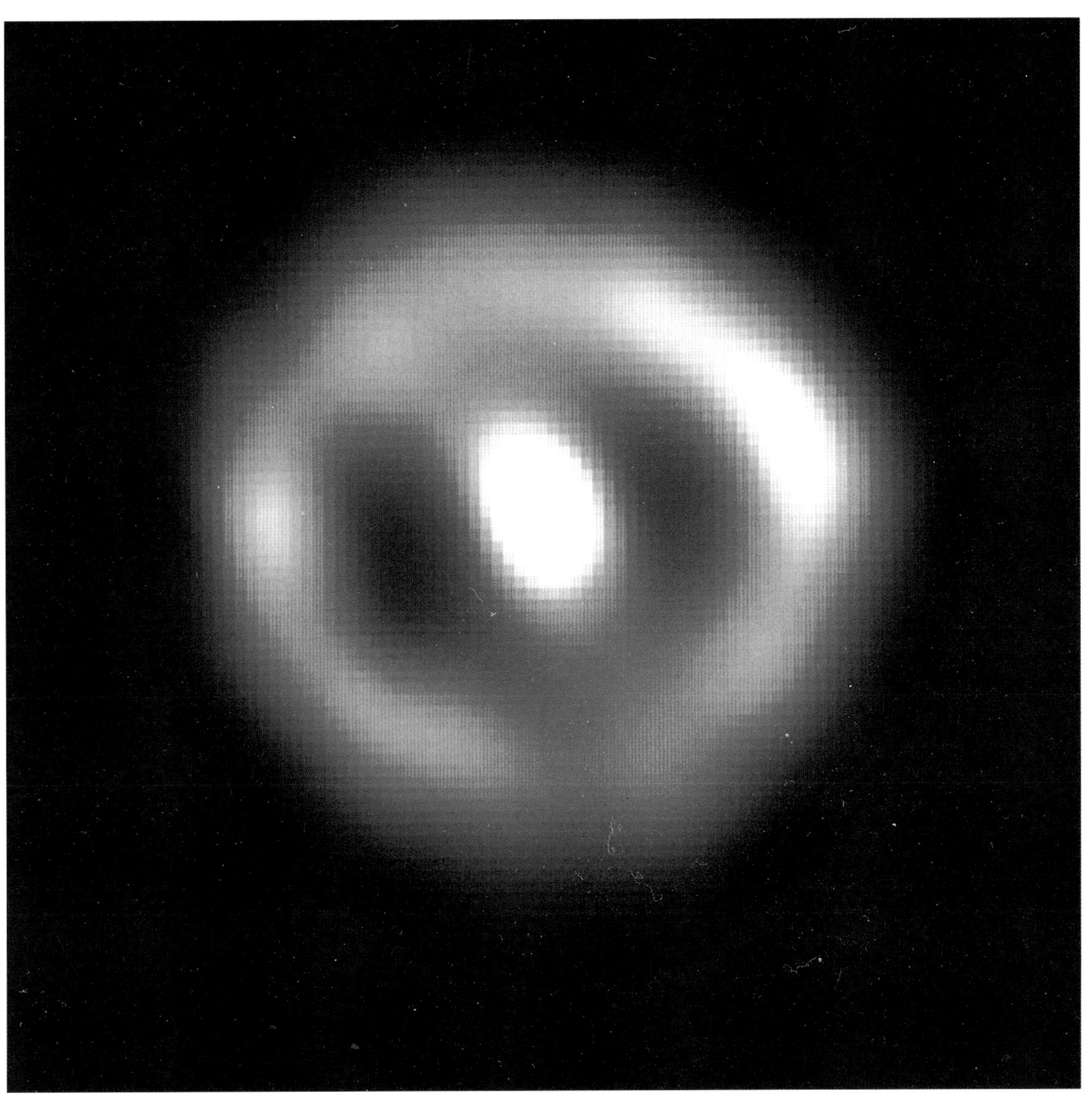

▲ ***Gas shell around Nova Cygni 1992****. A Hubble Space Telescope (HST) image of a rapidly ballooning bubble of gas blasted off a star. The shell surrounds Nova Cygni 1992, which erupted on 19 February 1992, and the image was taken on 31 May 1993, 467 days after the event. The shell is so young that it still contains a record of the initial conditions of the explosion.*

▼ ***Light curve for a nova****. HR Delphini observed by the author from discovery in 1967 to 1974.*

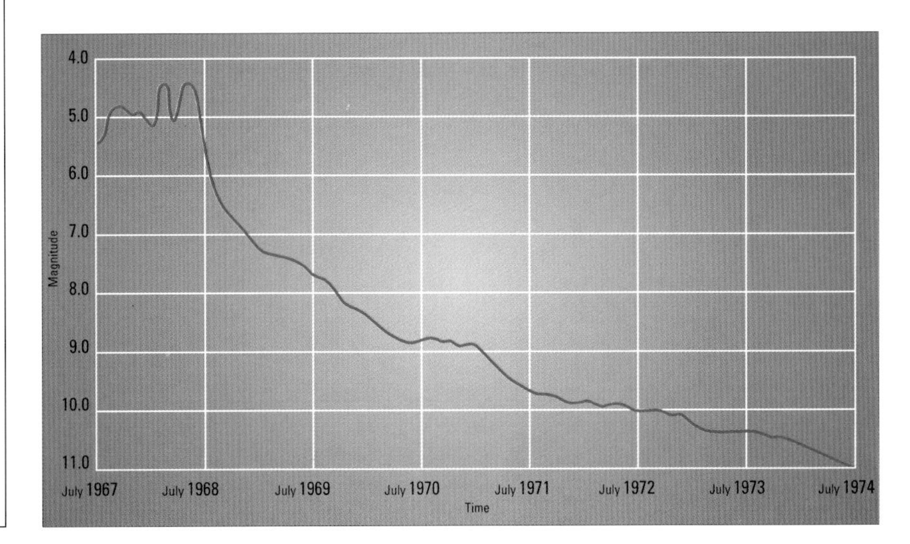

BRIGHT NOVAE, 1600–1999

The following list includes all classical novae which have reached magnitude 4.5 or brighter:

Year	Star	Discoverer	Max. magnitude
1670	CK Vulpeculae	Anthelm	3
1848	V841 Ophiuchi	Hind	4
1876	Q Cygni	Schmidt	3
1891	T Aurigae	Anderson	4.2
1901	GK Persei	Anderson	0.0
1912	DN Geminorum	Enebo	3.3
1918	V603 Aquilae	Bower	−1.1
1920	V476 Cygni	Denning	2.0
1925	RR Pictoris	Watson	1.1
1934	DQ Herculis	Prentice	1.2
1936	V630 Sagittarii	Okabayasi	4.5
1939	BT Monocerotis	Whipple and Wachmann	4.3
1942	CP Puppis	Dawson	0.4
1963	V533 Herculis	Dalgren and Peltier	3.2
1967	HR Delphini	Alcock	3.7
1970	FH Serpentis	Honda	4.4
1975	V1500 Cygni	Honda	1.8
1992	V1974 Cygni	Collins	4.3
1993	V705 Cassiopeiae	Kanatsu	5.4
1999	V382 Velorum	Williams and Gilmore	2.5
1999	V1994 Aquilae	Pereira	3.6

Supernovae

Stars have long lives. Even cosmic searchlights, such as Rigel and Canopus, go on pouring out their energy for tens of millions of years before disaster overtakes them. Yet sometimes we can witness a stellar death – the literal destruction of a star. We call it a supernova.

Supernovae are not merely very brilliant novae; they come into an entirely different category, and are of two distinct classes. A Type I supernova is a binary system, where one component (A) is initially more massive than its companion (B) and therefore evolves more quickly into the red giant stage. Material from it is pulled over to B, so that B grows in mass while A declines; eventually B becomes the more massive of the two, while A has become a white dwarf made up mainly of carbon. The situation then goes into reverse. B evolves to become a giant, and starts to lose material back to the shrunken A, with the result that the white dwarf builds up a gaseous layer made up mainly of hydrogen which it has stolen from B. However, there is a limit, once the mass of the white dwarf becomes greater than 1.4 times that of the Sun (a value known as the Chandrasekhar limit, after the Indian astronomer who first worked it out), the carbon detonates, and in a matter of a few seconds the white dwarf blows itself to pieces. There can be no return to the old state; the star has been completely destroyed. The energy released is incredible, and the luminosity may peak at at least 400,000 million times that of the Sun, greater than the combined luminosity of all the stars in an average galaxy. For some time afterwards wisps of material may be left, and can be detected because they send out radio radiation, but that is all.

A Type II supernova is very different, and is the result of the sudden collapse of a very massive supergiant – at least eight times as massive as the Sun – which has used up its nuclear fuel, and has produced a nickel-iron core which will not 'burn'. The structure of the star has been compared with that of an onion. Outside the iron-rich core is a zone of silicon and sulphur; next comes a layer of neon and magnesium; then a layer of carbon, neon and oxygen; then a layer of helium, and finally an outer-region of hydrogen. When all energy production stops, the outer layers crash down on to the core, which collapses; the protons and electrons are forced together to make up neutrons, and a flood of neutrinos is released, travelling right through the star and escaping into space. The temperature is now 100,000 million degrees C, and there is a rebound so violent that most of the star's material is blown away, leaving the neutron-star so dense that at least 2,500 million tonnes of its material could be packed inside a matchbox. The peak luminosity may be around 5000 million times that of the Sun.

▶ ***The Crab Nebula,*** *the remnants of the 1054 supernova. The Nebula itself was discovered by John Bevis in 1731, and independently by Messier in 1758. It is 6000 light-years away and radiates at almost all wavelengths, from the long radio waves down to the ultra-short X-rays and gamma-rays. Its 'power-house', the pulsar or neutron star in the centre, was detected optically as a very faint, flashing object in 1969 by observers in Arizona, at the Steward Observatory. It flashes 30 times per second, and is the quickest-spinning of the 'normal' pulsars.*

▲ ***Supernova in Messier 81****. M81 is a spiral galaxy in Ursa Major, 8.5 million light-years away (not so very far beyond the Local Group); it is a very easy telescopic object. In 1993 a supernova flared up in it. This picture was taken by R. W. Arbour when the supernova (arrowed) was at maximum brightness.*

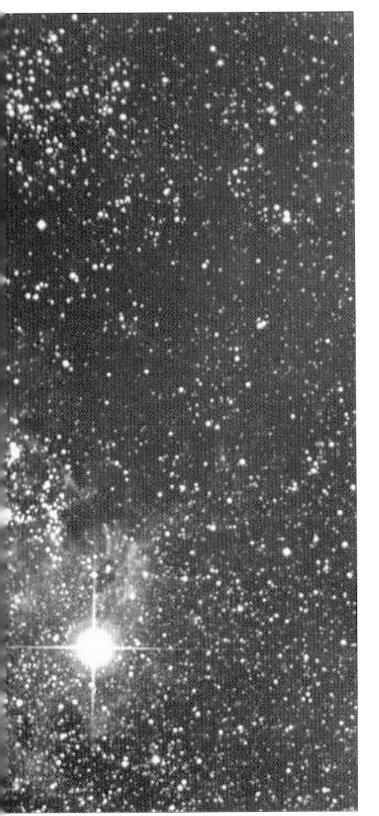

◄ ***Supernova 1987A in the Large Cloud of Magellan****. This photograph was taken from South Africa when the supernova (lower right) was near its maximum brightness.*

A neutron star is an amazing object. Its diameter may be no more than a few kilometres, but its mass will be equal to that of the Sun. The gravitational pull is very strong (objects would weigh a hundred thousand million times more on the surface of a neutron star than they would on the surface of the Earth), and so is the magnetic field. The rate of rotation is very fast, and beams of radio radiation come out from the magnetic poles, which are not coincident with the poles of rotation. If a radio beam sweeps across the Earth, we receive a pulse of radio emission; the effect may be likened to the beam of a rotating lighthouse illuminating an onlooker on the seashore. It is this which has led to neutron stars being known as pulsars.

Many supernovae have been seen in outer galaxies, but in our own Galaxy only four have been seen during the last thousand years; all these became brilliant enough to be seen with the naked eye in broad daylight. The brightest of all was seen in 1006 in the constellation of Lupus, the Wolf (Map 20); it is not well documented, but appears to have been as bright as the quarter-moon. We know more about the supernova of 1054, in Taurus (Map 17), because it has left the gas-patch known as the Crab Nebula, which contains a pulsar spinning round 30 times a second; this is one of the few pulsars to have been optically identified with a very faint, flashing object. The Crab is 6000 light-years away, so that the outburst actually occurred before there were any astronomers capable of observing it scientifically.

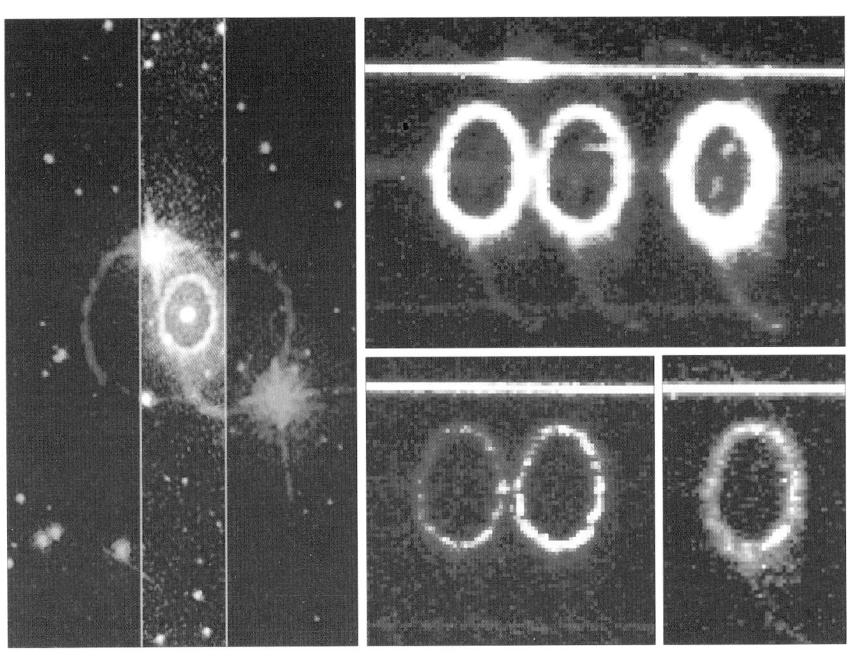

▲ ***Ring round a supernova****. These images from the Hubble Space Telescope show a light-year wide ring of glowing gas round the Supernova 1987A, in the Large Cloud of Magellan. The HST spectrograph viewed the entire ring system, and produced a detailed image of the ring in each of its constituent colours. Each colour represents light from a specific element: oxygen (single green ring), nitrogen and hydrogen (triple orange rings), and sulphur (double red rings). The ring formed 30,000 years before the star exploded, and so is a fossil record of the final stages of the star's existence. The light from the supernova heated the gas in the ring so that it now glows at temperatures from 5000 to 25,000°C. [Credit G. Sonneborn (GSFC) and NASA. WFPC image J. Pun (NOAO) and SINS Collaboration.]*

The supernova of 1572, in Cassiopeia (Map 3), is known as 'Tycho's Star', because it was carefully studied by the great Danish astronomer. The distance is 20,000 light-years; there is no pulsar, but radio emissions can be picked up from the wisps of gas which have been left. This is also true of the 1604 star, observed by Johannes Kepler. The radio source Cassiopeia A seems to be the remnant of a supernova which flared up in the late 17th century, but was not definitely observed because it was obscured by interstellar material near the plane of the Galaxy.

There have been two particularly notable supernovae since then. In 1885 a new star was seen in the Great Spiral in Andromeda (Map 12), which is over two million light-years away; it reached the fringe of naked-eye visibility, and is remembered as S Andromedae. Unfortunately nobody appreciated its true nature, because at that time it was not even generally believed that the so-called 'starry nebulae' were external systems.

Then, in 1987, came a flare-up in the Large Magellanic Cloud, which is the nearest of the major galaxies and is a mere 169,000 light-years away. The maximum magnitude was 2.3, so that the supernova – 1987A – was a conspicuous naked-eye object for some weeks. Surprisingly, the progenitor star – Sanduleak $-69^{\circ}202$ – was not a red supergiant, but a blue one, and the peak luminosity was only 250 million times that of the Sun, which by supernova standards is low. It seems that the progenitor, about 20 million years old and 20 times as massive as the Sun, was previously a red supergiant; it shed its outer layers and became blue not long before the outburst happened. The ejected material spread out at 10,000 kilometres (over 6000 miles) per second, subsequently lighting up clouds of material lying between the supernova and ourselves. As yet no pulsar has been detected, but if one exists – as is very likely – it should become evident when the main debris has cleared. European astronomers lament the fact that the supernova was so far south in the sky, but at least it has been available to the Hubble Space Telescope, which has taken remarkable pictures of it.

We cannot tell when the next supernova will appear in our Galaxy; it may be tomorrow, or it may not be for many centuries. Astronomers hope that it will be soon, but at least we have learned a great deal from Supernova 1987A in the Large Magellanic Cloud.

Black Holes

In many ways, stellar evolution is now reasonably well understood. We know how stars are born, and how they create their energy; we know how they die – some with a whimper, others with a very pronounced bang. But when we come to consider stars of really enormous mass, we have to admit that there are still some details about which we are far from clear.

Consider a star which is too massive even to explode as a supernova. When its energy runs out, gravity will take over, and it will start to collapse. The process is remarkably rapid; there is no outburst – the star simply goes on becoming smaller and smaller, denser and denser. As it does so, the escape velocity rises, and there comes a time when the escape velocity reaches 300,000 kilometres (186,000 miles) per second. This is the speed of light, so that not even light can escape from the shrunken star – and if light cannot do so, then certainly nothing else can, because light is as fast as anything in the universe. The old star has surrounded itself with a 'forbidden area' from which absolutely nothing can escape. It has created a black hole.

For obvious reasons, we cannot see a black hole – it emits no radiation at all. Therefore, our only hope of locating such an object is by detecting its effects upon something which we can see. Probably the best candidate to date is Cygnus X-1, so called because it is an X-ray source; it lies near the star Eta Cygni (Map 8). The system consists of a B-type supergiant, HDE 226868, which is of the ninth magnitude. It seems to have about 30 times the mass of the Sun, with a diameter of perhaps 18 million kilometres (11.25 million miles); it is associated with an invisible secondary with 14 times the Sun's mass. The orbital period is 5.6 days as we can tell from the behaviour of the supergiant; the distance from us is 5000 light-years. What seems to be happening is that the black hole is pulling material away from the supergiant, and swallowing it up. Before this material disappears, it is whirled around the supergiant, and is so intensely heated that it gives off the X-ray radiation which we can pick up. Other interpretations are possible, of course, but the black hole picture does seem to be the most plausible, and there are other rather similar cases.

The size of a black hole depends upon the mass of the collapsed star. The critical radius of a non-rotating black hole is called the Schwarzschild radius, after the German astronomer who investigated the problem mathematically as long ago as 1916; the boundary around the collapsed star having this radius is termed the 'event horizon'. Once material passes over the event horizon, it is forever cut off from the rest of the universe. For a body the mass of the Sun, the Schwarzschild radius would be about 3 kilometres (1.9 miles); for a body the mass of the Earth the value would be less than a single centimetre (less than half an inch).

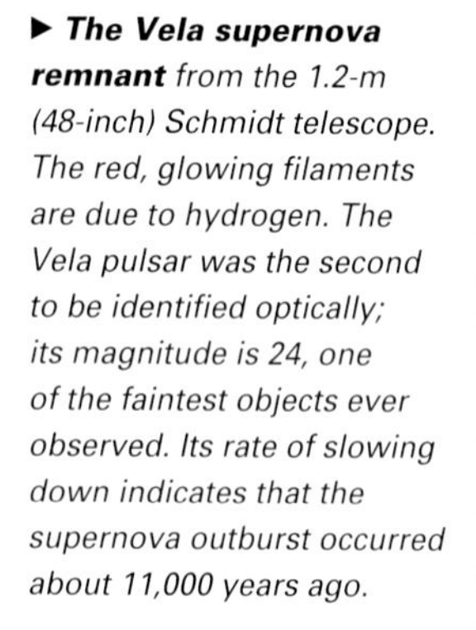

► ***The Vela supernova remnant*** *from the 1.2-m (48-inch) Schmidt telescope. The red, glowing filaments are due to hydrogen. The Vela pulsar was the second to be identified optically; its magnitude is 24, one of the faintest objects ever observed. Its rate of slowing down indicates that the supernova outburst occurred about 11,000 years ago.*

It has been suggested that there may be massive black holes in the centres of very active galaxies, such as Seyfert galaxies, and also inside quasars, which are exceptionally remote and super-luminous. At least the presence of black holes of this kind would account for the immense amount of energy involved. In 1994, research with the Hubble Space Telescope indicated that there is indeed such a black hole at the core of the galaxy M87.

Assuming that black holes really exist, can we decide what happens inside them? The answer is 'no', because beyond the event horizon all the ordinary laws of science break down. Whether the collapsed star crushes itself out of existence altogether is a matter for debate.

All sorts of exotic theories have been proposed to explain what happens inside black holes: for instance, is it possible that material vanishing into a black hole can reappear in a 'white hole' elsewhere in the galaxy, or in a different universe altogether? And if so, can there be any connection between them? Is it possible that over the ages a black hole may lose energy, and slowly evaporate? Can there be very small black holes wandering around, which we cannot detect? Speculation is endless, but it does seem that ideas of this sort must be treated with considerable reserve.

At least we have the satisfaction of knowing that our Sun will never produce a black hole. It is not nearly massive enough, and it will end its career much more gently, passing through the white dwarf stage and coming to its final state as a cold, dead globe.

It has been suggested that many galaxies, and also many globular clusters, may have central black holes. This is a possibility, but at the moment proof is lacking, and we cannot even be sure that there is a black hole at the centre of our own Galaxy.

▲ ***Impression of a black hole*** *(Paul Doherty). Material can be drawn into the black hole, but nothing – absolutely nothing – can escape. Therefore, a black hole can only be detected by its effects upon objects which we can record.*

◄ ***NGC 7742,*** *a spiral galaxy imaged in 1998 with the Hubble Space Telescope. This is a Seyfert active galaxy, probably powered by a black hole at its core. The core of NGC 7742 is the large yellow 'yolk' in the centre of the image. The thick, lumpy ring round this core is an area of active star birth. The ring is about 3000 light-years from the core. Tightly wound spiral arms are visible. Surrounding the inner ring is a wispy band of material, probably the remains of a once very active stellar breeding area.*

Stellar Clusters

▲ ***The Pleiades Cluster**. The 'Seven Sisters' in Taurus; the most famous of all open clusters. Most of the leading Pleiads are hot and bluish white, indicating that the cluster is relatively young; there is also associated nebulosity, not difficult to photograph but very hard to see visually. The cluster is not more than 50 million years old.*

Star clusters are among the most beautiful objects in the sky. Several are easily visible with the naked eye, notably the Pleiades and the Hyades in Taurus, Praesepe in Cancer, and the lovely Jewel Box in the Southern Cross; many are within the range of binoculars or small telescopes.

In 1781, the French astronomer Charles Messier compiled a list of more than a hundred star clusters and nebulae – not because he was interested in them, but because he kept on confusing them with comets, in which he was very interested indeed. Ironically, it is by his catalogue that Messier is now best remembered, and we still use the numbers which he gave; thus Praesepe is M44, while the Pleiades cluster is M45. Also in use are the NGC or New General Catalogue numbers, given in a catalogue by J. L. E. Dreyer in 1888; thus Praesepe is NGC2632. The Caldwell Catalogue (C) which I compiled in 1996 seems to be coming into general use. It includes no Messier objects.

Star clusters are of two types, open and globular. Open or loose clusters may contain anything from a few dozen to a few hundred stars, and have no definite structure; they cannot persist indefinitely, as over a sufficient period of time they will be disrupted by non-cluster stars and will lose their identity. The stars in a cluster are of the same age and were formed from the same interstellar cloud though their differing initial masses mean that they have

SELECTED STELLAR CLUSTERS

M	C	NGC	Name	Constellation	Map	Remarks
2	–	7089		Aquarius	14	Globular; near α and β Aquarii.
3	–	5272		Canes Venatici	1	Globular; easy in binoculars.
4	–	6121		Scorpius	11	Globular cluster near Antares.
5	–	5904		Serpens	10	Fine bright globular.
6	–	6405	Butterfly	Scorpius	11	Naked-eye open cluster.
7	–	6475		Scorpius	11	Fine naked-eye open cluster.
11	–	6705	Wild Duck	Scutum	8	Fan-shaped; fine open cluster.
13	–	6205		Hercules	9	Brightest northern globular.
15	–	7078		Pegasus	13	Fine bright globular.
19	–	6273		Sagittarius	11	Elongated globular.
22	–	6656		Sagittarius	11	Fine globular near λ Sagittarii.
23	–	6494		Sagittarius	11	Bright open cluster near μ.
34	–	1039		Perseus	12	Bright open cluster.
35	–	2168		Gemini	17	Naked-eye open cluster; fine.
36	–	1960		Auriga	18	Bright open cluster.
37	–	2099		Auriga	18	Bright, rich open cluster.
38	–	1912		Auriga	18	Fairly bright open cluster.
41	–	2287		Canis Major	16	Naked-eye open cluster.
44	–	2632	Praesepe	Cancer	5	Famous bright open cluster.
45	–	–	Pleiades	Taurus	17	Brightest open cluster.
46	–	2437		Puppis	19	Open cluster; rich; bright.
47	–	2422		Puppis	19	Fine rich open cluster.
48	–	2548		Hydra	7	Open cluster; not brilliant.
53	–	5024		Coma Berenices	4	Globular, near α Comae.
54	–	6715		Sagittarius	11	Small, bright globular.
62	–	6266		Ophiuchus	10	Small, bright globular.
67	–	2682		Cancer	5	Old open cluster; bright, easy.
79	–	1904		Lepus	16	Small bright globular.
92	–	6341		Hercules	9	Large, bright globular.
93	–	2447		Puppis	19	Bright globular.
–	86	6397		Ara	20	Globular; not difficult to find.
–	102	IC 2602	θ Carinae	Carina	19	Fine open cluster, round θ.
–	96	2516		Carina	19	Fine open cluster, near ε.
–	80	5139	ω Centauri	Centaurus	20	Finest of all globulars.
–	97	3766		Centaurus	20	Open cluster, near λ.
–	78	6541		Corona Australis	11	Globular; binocular object.
–	94	4755	Jewel Box	Crux Australis	20	Fine open cluster, round κ.
–	89	6087	S Normae	Norma	20	Open cluster; binocular object.
–	93	6752		Pavo	21	Bright globular.
–	14	869/884	Sword-Handle	Perseus	12	Naked-eye double open cluster.
–	41	–	Hyades	Taurus	17	Round Aldebaran.
–	95	6025		Triangulum Australe	20	Bright globular, near β.
–	95	IC 2391	o Velorum	Vela	19	Naked-eye open cluster, round o.
–	–	2547		Vela	19	Naked-eye open cluster, near κ.

evolved at different rates. Globular clusters are huge symmetrical systems containing up to a million stars.

The most famous open clusters are the Pleiades and the Hyades, both in Taurus (Map 17). The Pleiades are very conspicuous, and have been known since early times (they are mentioned in the *Odyssey* and in the Bible), and on a clear night anyone with normal eyesight can make out at least seven individual stars – hence the popular nickname of the Seven Sisters. Keen-eyed people can see more (the record is said to be 19), and binoculars show many more, while the total membership of the cluster is around 400. Alcyone or Eta Tauri, the brightest member of the cluster, is of the third magnitude. The Hyades are more scattered, and are overpowered by the brilliant orange light of Aldebaran, which is not a true member of the cluster at all, but simply happens to lie about midway between the Hyades and ourselves. The Hyades were not listed by Messier, presumably because there is not the slightest chance of confusing them with a comet.

Another open cluster easily visible with the naked eye is Praesepe in Cancer (Map 5), nicknamed the Beehive or the Manger; binoculars give an excellent view. It is much older than the Pleiades cluster. Also in Cancer we find M67, which is on the fringe of naked-eye visibility; it is probably the oldest known cluster of its type, and is well away from the main plane of the Galaxy, so that it moves in a sparsely-populated region and is not badly disrupted by the gravitational pull of non-cluster stars. In Perseus there is the double cluster of the Sword-Handle, and in the far south there is the Jewel Box, round Kappa Crucis, which has stars of contrasting colours – including one prominent red giant. Telescopic clusters are common enough and dozens are with the range of a small telescope.

Globular clusters lie round the edge of the main Galaxy; over 100 are known, but all are very remote. Messier listed 28 of them. The two brightest are so far south that they never rise over Europe. Omega Centauri (Map 20) is truly magnificent and is prominent even though it is about 17,000 light-years away. The condensed core is about 100 light-years across, and in it the stars are so closely packed that they are not easy to see individually. 47 Tucanae, also in the far south (Map 21), rivals Omega Centauri; by sheer chance it is almost silhouetted against the Small Cloud of Magellan, but the Small Cloud is an external system far beyond our Galaxy, while 47 Tucanae is about the same distance from us as Omega Centauri. The brightest globular in the northern sky is M13 in Hercules (Map 9), which is just visible with the naked eye.

Globular clusters are very old, so that their leading stars are red giants or supergiants; there is virtually no nebulosity left in them, so that star formation has ceased. They are rich in short-period variables, and this is how their distances were first measured, by Harlow Shapley in 1918. By observing the ways in which the stars behaved, he could find their real luminosities, and hence the distances of the globular clusters in which they lay. Shapley also found that the globulars are not distributed evenly all over the sky; there are more in the south than in the north, particularly towards the constellation of Sagittarius. This is because the Sun lies well away from the centre of the Galaxy, so that we are having what may be called a lop-sided view.

Surprisingly, some hot blue giants are also found in globular clusters and are known as blue stragglers. Logically they ought not to be there, because high-mass stars of this age should long since have left the Main Sequence. What seems to happen is that because stars near the core of the cluster are so close together, relatively speaking, they may 'capture' each other and form binary systems. The less massive member of the new pair will then draw material away from the more evolved, less dense companion, and will heat up, becoming blue again; in a direct collision, two stars will merge. In this case the resulting stars would be of greater than average mass, and would tend to collect near the centre of the cluster, which is precisely what we find.

If there are any inhabited planets moving round stars near the core of a globular cluster, local astronomers will have a very curious sort of sky. There will be many stars brilliant enough to cast shadows; moreover, many of these stars will be red. An astronomer there will be able to examine many stars from relatively close range, but will be unable to learn a great deal about the outer universe.

▲ ***NGC 1818,*** *star cluster in the Large Magellanic Cloud, 170,000 light-years away. The cluster is young – age about 40 million years – and is the site of vigorous star formation. The cluster contains over 20,000 stars. The circled star is a young White Dwarf, which has only recently formed from a Red Giant. The progenitor star was between seven and eight times as massive as the Sun. Photo taken in December 1995 with the Hubble Space Telescope. [Credit: Rebecca Elson and Richard Swird (Cambridge and NASA); original WFPC2 courtesy of J. Westphal (Caltech)]*

Nebulae

When Messier published his catalogue, in 1781, he included nebulae of two types – those which looked as though they were gaseous, and those which gave every impression of being made up of stars. William Herschel was among the first to recognize a definite difference between the two classes, and in 1791 he said of the Orion Nebula: 'Our judgement, I venture to say, will be that the nebulosity is not of a starry nature.'

Proof came in 1864, when Sir William Huggins, the pioneer English astronomical spectroscopist, found that the spectra of bright nebulae were of the emission type, while the starry objects, such as M31 in Andromeda, showed the familiar absorption lines. But there are also two other classes of objects which have to be considered. In particular M1, the first entry in Messier's list, was proved to be a supernova remnant – the wreck of the star seen by the Chinese in 1054; its nickname of the Crab was bestowed on it by the Earl of Rosse when he looked at it with his great 183-centimetre (72-inch) telescope in the mid-19th century. Another supernova remnant is the Gum Nebula in Vela (Map 19), named after the Australian astronomer Colin Gum; in this case the supernova blazed forth in prehistoric times, and must have been exceptionally brilliant back then.

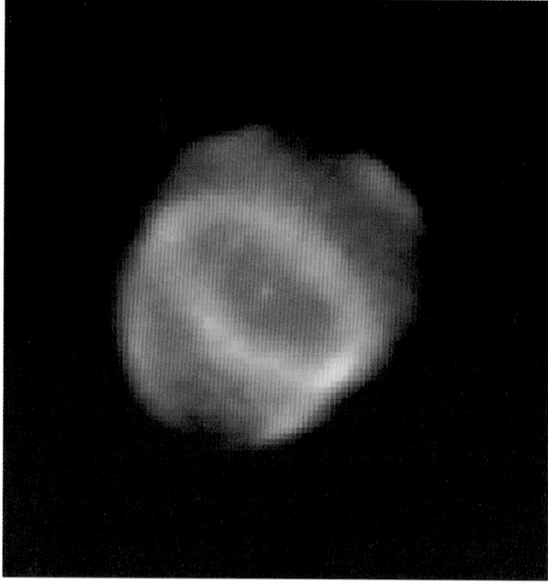

▶ ***Hen 1357**, as imaged by the Hubble Space Telescope. It is about 18,000 light-years away, and lies in Ara (it was the 1357th object in a list of unusual stars compiled by Karl Henize). Previous ground-based observations indicate that over the past few decades Hen 1357 has changed from being an ordinary hot star to an object with the characteristics of a planetary nebula. Only the HST can show it in detail.*

Planetary nebulae have nothing to do with planets, and are not true nebulae; they are old, highly evolved stars which have thrown off their outer layers. The discarded shells shine because of the ultra-violet radiation emitted by the central star, which is extremely hot (with a surface temperature which may reach 400,000 degrees C) and is well on its way to becoming a white dwarf. All planetary nebulae are expanding, and on the cosmic timescale the planetary nebula stage is very brief. The best-known member of the class is M57, the Ring Nebula in Lyra (Map 8), which is easy to locate between the naked-eye stars Beta and Gamma Lyrae; telescopically it looks like a tiny, luminous cycle tyre, with a dim central star (although very recent research suggests that it may really be in the form of a double lobe rather than a ring). Other planetary nebulae are less regular; M27, the Dumbbell Nebula in Vulpecula (Map 8), earns its nickname, while the rather faint M97, in Ursa Major (Map 1), is called the Owl because the positions of two embedded stars do give a slight impression of the eyes in an owl's face.

True nebulae consist mainly of hydrogen, together with what may be termed 'dust'. They shine because of stars in or very near them. Sometimes their light is due to pure reflection, as with the nebulosity in the Pleiades, but in other cases very hot stars make the nebulosity emit a certain amount of luminosity on its own account, making up what are termed HII regions. Such is the Great Nebula M42, in Orion's Sword (Map 16), where the illuminating

▲ ***M17, the Omega Nebula in Sagittarius**: John Fletcher, 25-cm (10-inch) reflector. The Nebula is almost 6000 light-years away, but is an easy telescopic object. It has also been nicknamed the Swan or Horseshoe Nebula.*

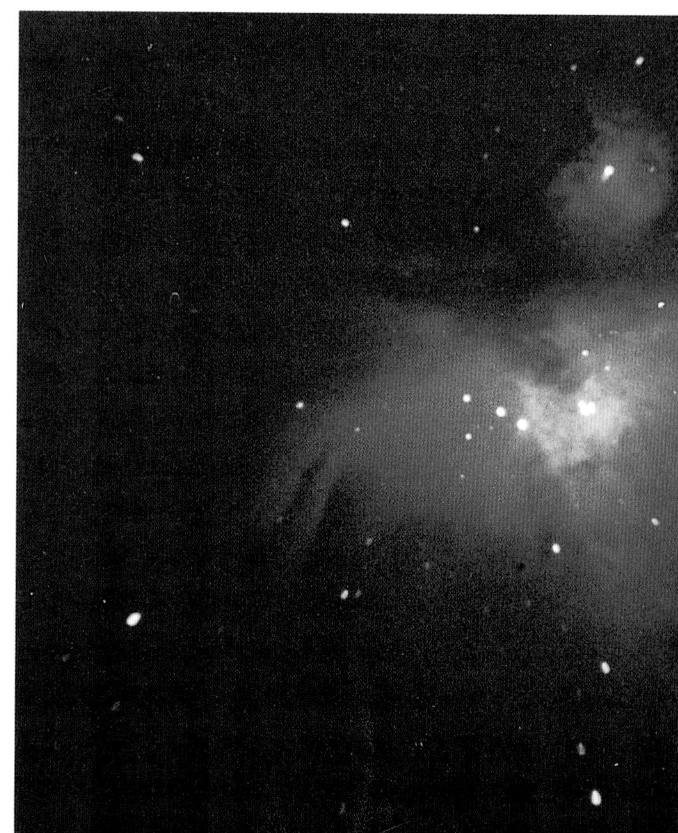

▶ ***M42, the Orion Nebula**: Commander H. R. Hatfield, 30-cm (12-inch) reflector. The Trapezium (Theta Orionis) appears near the centre of the picture, and the bright and dark nebulosity is well shown. Exposure, 10 minutes.*

stars are the members of the multiple Theta Orionis, the Trapezium. The Orion Nebula is about 30 light-years across, and is 1500 light-years away; if one could take a 2.5-centimetre (1-inch) diameter core sample right through it, the total weight of material collected would just about counter-balance a small coin (about 3.5 grams). Yet the Orion Nebula is a stellar birthplace, where fresh stars are being formed from the nebular material. It contains very young T Tauri-type stars, which have not yet reached the Main Sequence and are varying irregularly; there are also immensely powerful stars which we can never see, but which we can detect because their infra-red radiation is not blocked by dust. Such is the Becklin–Neugebauer Object (BN), which is highly luminous, but which will not last for long enough to 'bore a hole' in the nebulosity so that its light could escape. In fact, M42 is only a very small portion of a huge molecular cloud which covers almost the whole of Orion.

Other nebulae are within the range of small telescopes. For instance, in Sagittarius we have the Lagoon Nebula, which is easy to see with binoculars, and the Trifid Nebula, which shows dark lanes of obscuring material (Map 11). The North America Nebula, in Cygnus (Map 8) really does give the impression of the shape of the North American continent. It is dimly visible with the naked eye in the guise of a slightly brighter portion of the Milky Way, and powerful binoculars bring out its shape. It is nearly 50 light-years in diameter; much of its illumination seems to be due to Deneb, which is one of our cosmic searchlights and is at least 70,000 times as luminous as the Sun. Some nebulae are colossal; the Tarantula Nebula round 30 Doradûs, in the Large Cloud of Magellan, would cast shadows if it were as close to us as M42 rather than being a full 169,000 light-years away. Other nebulae are associated with variable stars, so that their aspect changes; such are the nebulae associated with T Tauri, R Monocerotis and R Coronae Australis.

If nebulosity is not illuminated by a suitable star it will not shine, and will be detectable only because it contains enough dust to blot out the light of objects beyond. (It is not 'a hole in the heavens', as William Herschel once suggested.) The best example is the Coal Sack in the Southern Cross (Map 20), near Alpha and Beta Crucis, which produces a starless area easily detectable with the naked eye. Other dark nebulae are smaller, such as the Horse's Head near Zeta Orionis (Map 16), and there are dark rifts in the Milky Way, notably in Cygnus.

There is no difference between a dark nebula and a bright one, except for the lack of illumination. For all we know, there may be a suitable star on the far side of the Coal Sack, so that if we could see the Sack from a different vantage point it would appear bright.

▼ ***The 'Black Cloud'**: dark nebula Barnard 68. It appears as an opaque and sharply defined mass against a rich star background field. No stars are seen in front of B68, showing that the dark nebula is close – around 500 light-years away, in Ophiuchus. This three-colour composite was obtained with Antu, the first component of the VLT, in 1999. [Reproduced by kind permission of the European Southern Observatory.]*

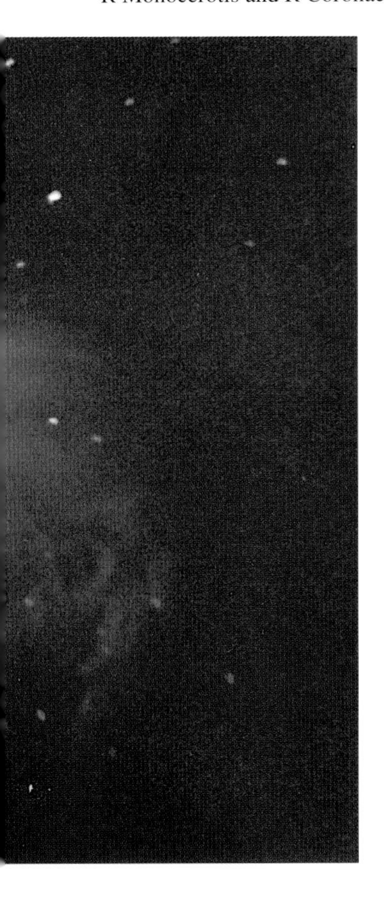

SELECTED NEBULAE

M	C	NGC	Name	Constellation	Map	Remarks
1	–	1952	Crab Nebula	Taurus	17	Supernova remnant.
8	–	6253	Lagoon Nebula	Sagittarius	11	Easy in binoculars.
16	–	6611	Eagle Nebula	Serpens	10	Nebula and cluster.
17	–	6618	Omega Nebula	Sagittarius	11	Omega or Horseshoe.
20	–	6514	Trifid Nebula	Sagittarius	11	Nebula with dark lanes.
27	–	6853	Dumbbell Nebula	Vulpecula	8	Bright planetary.
42	–	1976	Sword of Orion	Orion	16	Great Nebula.
57	–	6720	Ring Nebula	Lyra	8	Bright planetary.
97	–	3587	Owl Nebula	Ursa Major	1	Planetary. Elusive!
–	55	7009	Saturn Nebula	Aquarius	14	Planetary.
–	63	7293	Helix Nebula	Aquarius	14	Bright planetary.
–	31	IC405	Flaming Star Nebula	Auriga	18	Round AE Aurigae.
–	92	3372	Keyhole Nebula	Carina	19	Round Eta Carinae.
–	11	7635	Bubble Nebula	Cassiopeia	3	Faint.
–	68	6729	R Coronae Aust. Nebula	Corona Australis	11	Variable nebula.
–	27	6888	Crescent Nebula	Cygnus	8	Not bright.
–	33/4	6960/92	Veil Nebula	Cygnus	8	Supernova remnant.
–	20	7000	North America Nebula	Cygnus	8	Binocular object.
–	103	2070	Tarantula Nebula	Dorado	22	In Large Magellanic Cloud round 30 Doradûs.
–	39	2392	Eskimo Nebula	Gemini	17	Planetary; faint.
–	49	2237/9	Rosette Nebula	Monoceros	16	Surrounds cluster NGC2244.
–	–	2261	R Monocerotis Nebula	Monoceros	16	Variable. Round R.
–	–	2264	Cone Nebula	Monoceros	16	Variable. Round S.
–	–	1499	California Nebula	Perseus	12	Large but not bright.
–	69	6302	Bug Nebula	Scorpius	11	Planetary; faint.
–	46	1554/5	Hind's Variable Nebula	Taurus	17	Variable; round T Tauri.
–	–	–	Gum Nebula	Vela	19	Supernova remnant.
–	99	–	Coal Sack	Crux	20	Dark nebula.

Hubble Views of Nebulae

The Helix Nebula in Aquarius (NGC7293; C63) is a planetary nebula, 450 light-years away. It is the largest and brightest of the planetary nebulae, and is a very easy telescopic object (see Map 14). In August 1994 it was imaged from the Hubble Space Telescope, and the picture showed the results of the collision of two gases near the dying star. These tadpole-like objects in the upper right-hand corner of the picture have been nicknamed 'cometary knots' because their glowing heads and gossamer tails resemble comets. Thousands of these features were recorded by the Hubble Space Telescope.

Each gaseous head is at least twice the size of the entire Solar System; each tail stretches for around 160,000 million kilometres (100,000 million miles). The most visible gaseous fragments lie along the inner edge of the star's ring, forming a radial pattern round the star like the spokes on a cycle-wheel. Apparently the dying star sends out hot gas from its surface, and this collides with the cooler gas which it had ejected 10,000 years before. The crash fragments the smooth cloud surrounding the star into smaller, denser droplets.

Planetary nebulae are not unique to our Galaxy; other galaxies also have them – but obviously they are not easy to see over intergalactic distances, and this is where the power of the Hubble Space Telescope shows itself. Moreover, it alone can show the fine details in the planetary nebulae of our own Galaxy, such as the Helix. [Credit: Robert O'Dell, Kerry P. Handron; Hubble WFPC-2]

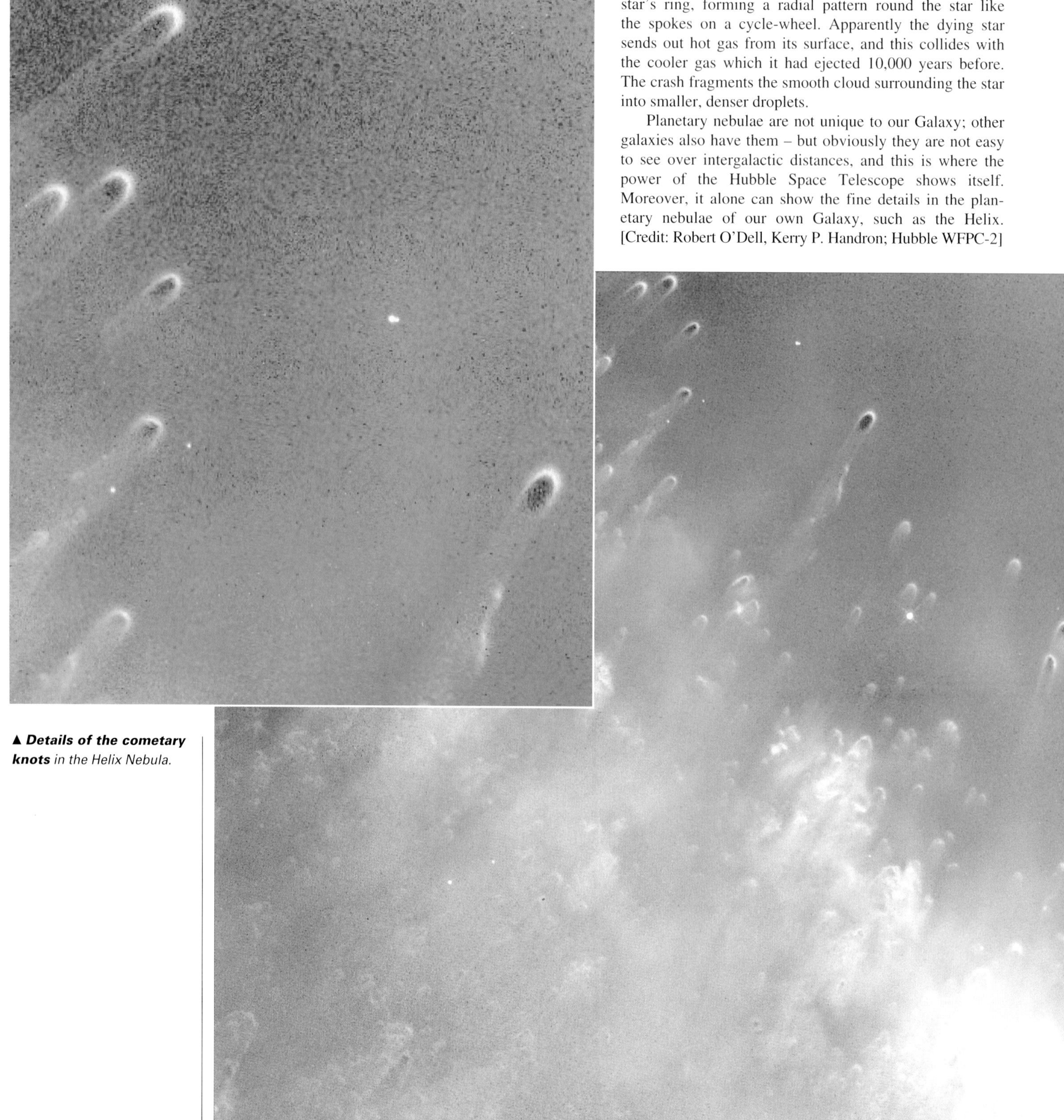

▲ ***Details of the cometary knots*** *in the Helix Nebula.*

▶ ***Helix Nebula****: overall view of the cometary knots.*

◀ ***The Bubble Nebula, NGC 7635**, from the Hubble Space Telescope. This image reveals an expanding shell of glowing gas surrounding a hot, massive star in our Milky Way Galaxy. This shell is being shaped by strong stellar winds of material and radiation produced by the bright star at the left, which is 10 to 20 times more massive than our Sun. These fierce winds are sculpting the surrounding material – composed of gas and dust – into the curve-shaped bubble. Astronomers have dubbed it the Bubble Nebula (NGC 7635). The nebula is 10 light-years across, more than twice the distance from Earth to the nearest star. Only part of the bubble is visible in this image. The glowing gas in the lower right-hand corner is a dense region of material that is getting blasted by radiation from the Bubble Nebula's massive star. The radiation is eating into the gas, creating finger-like features. This interaction also heats up the gas, causing it to glow. [Credit: Hubble Heritage Team (AURA/STScI/NASA)]*

▶ ***The Lagoon Nebula, M8**, 5000 light-years away in Sagittarius, imaged from the Hubble Space Telescope. Nebulae of this kind are stellar birthplaces, where new stars are being born from dusty molecular clouds. The central hot star, O Herschel 36 (upper left), is the main source of ionizing radiation for the brightest region of the nebula. Analogous to the phenomena of terrestrial tornadoes, the large temperature difference between the hot surface and the cold interior of the clouds may produce a strong horizontal shear to twist the clouds into their tornado-like appearance.*

THE UNIVERSE

◀ ***The Large Cloud of Magellan***. *This lies in the far south of the sky, and is a prominent naked-eye object. It is a galaxy 169,000 light-years away, associated with our own Galaxy; it was formerly classed as irregular, but shows clear traces of spiral structure. From Earth it is much the brightest of the external galaxies and contains objects of all types.*

The Structure of the Universe

Look up into the sky on any dark, clear night, and you will see the glorious band of the Milky Way, stretching from one horizon to the other. It must have been known since the dawn of human history, and there are many legends about it, but it was not until 1610 that Galileo, using his primitive telescope, found that it is made up of stars – so many of them that to count each one would be impossible. They look so close together that they seem in imminent danger of colliding, but, as so often in astronomy, appearances are deceptive; the stars in the Milky Way are no more crowded than in other parts of space. We are dealing with a line of sight effect, because the star system or Galaxy in which we live is flattened. Its shape has been likened to that of a double-convex lens or, less romantically, two fried eggs clapped together back to back. But does it make up the whole of the universe?

When Messier drew up his catalogue of nebulous objects, he included nebulae of two different types; those which were fairly obviously gaseous (such as M42 in Orion's Sword) and those which were starry (such as M31 in Andromeda, which is dimly visible with the naked eye). In 1845 the Earl of Rosse, using his great 183-cm (72-inch) reflector at Birr Castle in Ireland, found that many of the starry nebulae are spiral in form, so that they look like Catherine wheels, and it was suggested that they were outer systems ranking as galaxies in their own right (in fact William Herschel had considered this possibility much earlier). The main problem was that whatever their nature, the spirals were too far away to show measurable parallax shifts, so that their distances were very hard even to estimate. As recently as 1920 Harlow Shapley, who had been the first man to make a good estimate of the size of our Galaxy, was still maintaining that the spirals were minor and relatively unimportant features.

It was left to Edwin Hubble to provide an answer. In 1923 he used the Hooker telescope at Mount Wilson, then much the most powerful in the world, to detect Cepheid variables in some of the spirals, including M31. He measured their periods, and worked out their distances. The results were quite clear-cut: the Cepheids, and hence the systems in which they lay, were much too remote to be members of our Galaxy, so that they could only be external systems.

Hubble's first estimate of the distance of M31, the nearest of the large spirals, was 900,000 light-years, later reduced to 750,000 light-years. This later proved to be an underestimate. In 1952 Walter Baade, using the then-new Palomar reflector, showed that there are two types of Cepheids, and that one type is much more luminous than the other; the variables used by Hubble were twice as powerful as he had thought, and therefore much more distant. We now know that M31 is over two million light-years away, though even so it is one of the very closest of the outer galaxies.

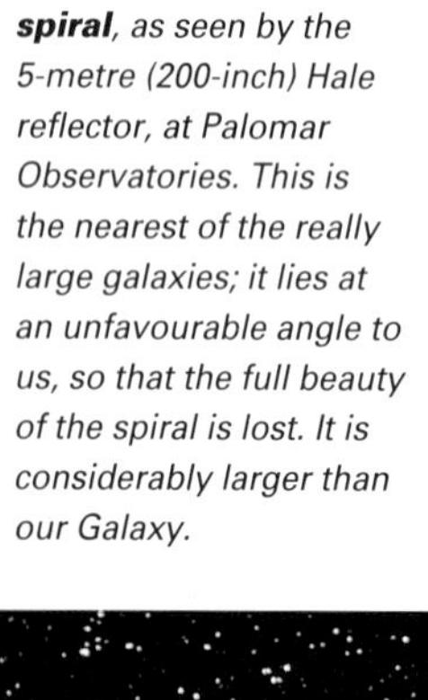

▼ ***M31, the Andromeda spiral**, as seen by the 5-metre (200-inch) Hale reflector, at Palomar Observatories. This is the nearest of the really large galaxies; it lies at an unfavourable angle to us, so that the full beauty of the spiral is lost. It is considerably larger than our Galaxy.*

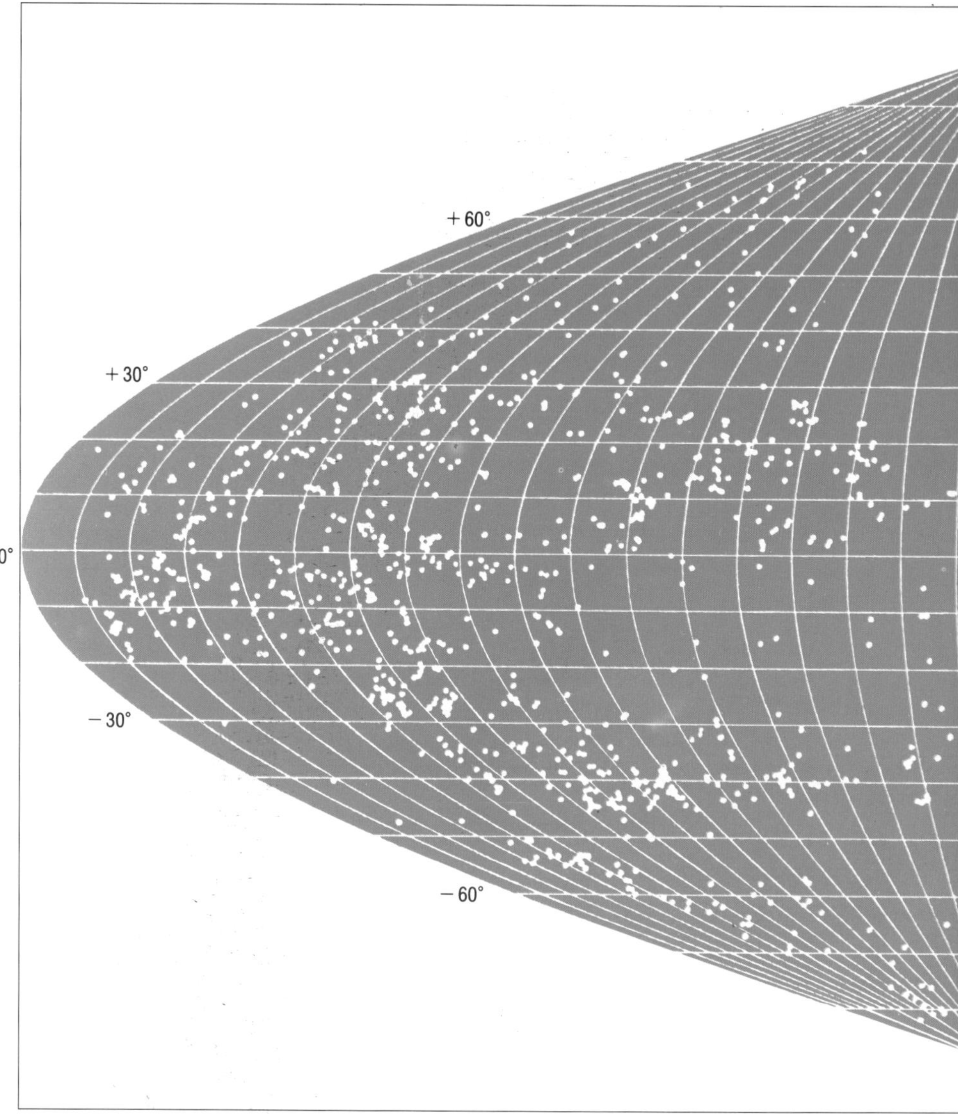

▶ ***This map of the sky** in supergalactic coordinates shows the distribution of over 4000 galaxies with an apparent diameter greater than one arc minute. The concentration of galaxies along the supergalactic equator, particularly in the north (right half), is evident, as is the clustering of galaxies on many different scales, from small groups, such as the Local Group, to large clusters, such as the Virgo Cluster near longitude 100–110°, slightly south of the equator.*

During his work with the Mount Wilson reflector, Hubble also made careful studies of the combined spectra of the galaxies. These spectra are the result of the combined spectra of millions of stars, and are bound to be something of a jumble, but the main absorption lines can be made out, and their Doppler shifts can be measured. Hubble confirmed earlier work, at the Lowell Observatory in Arizona, showing that all galaxies apart from a few which are very close to us (now known to make up what we call the Local Group) showed red shifts, indicating that they are moving away from us. Moreover, the further away they are, the faster they are receding. The entire universe is expanding. This does not mean that we are in a privileged position; every group of galaxies is racing away from every other group.

By now we can observe systems which are thousands of millions of light-years away, so that we are seeing them as they used to be thousands of millions of years ago – long before the Earth or the Sun existed. Once we look beyond the Solar System, our view of the universe is bound to be very out of date.

People used to believe that the Earth was all-important, and lay in the exact centre of the universe with everything else moving round it. We now know better. The Earth, the Sun, even the Galaxy are very insignificant in the universe as a whole. Indeed, the more we find out, the less important we seem to be.

▲ ***The Coma cluster of galaxies**: Hubble Space Telescope, 4 March 1994. The Coma cluster contains 1000 large galaxies and thousands of smaller systems; the mean distance from us is 300 million light-years. The largest galaxy in the cluster, NGC4881, has a diameter of 300,000 light-years, three times that of our Galaxy.*

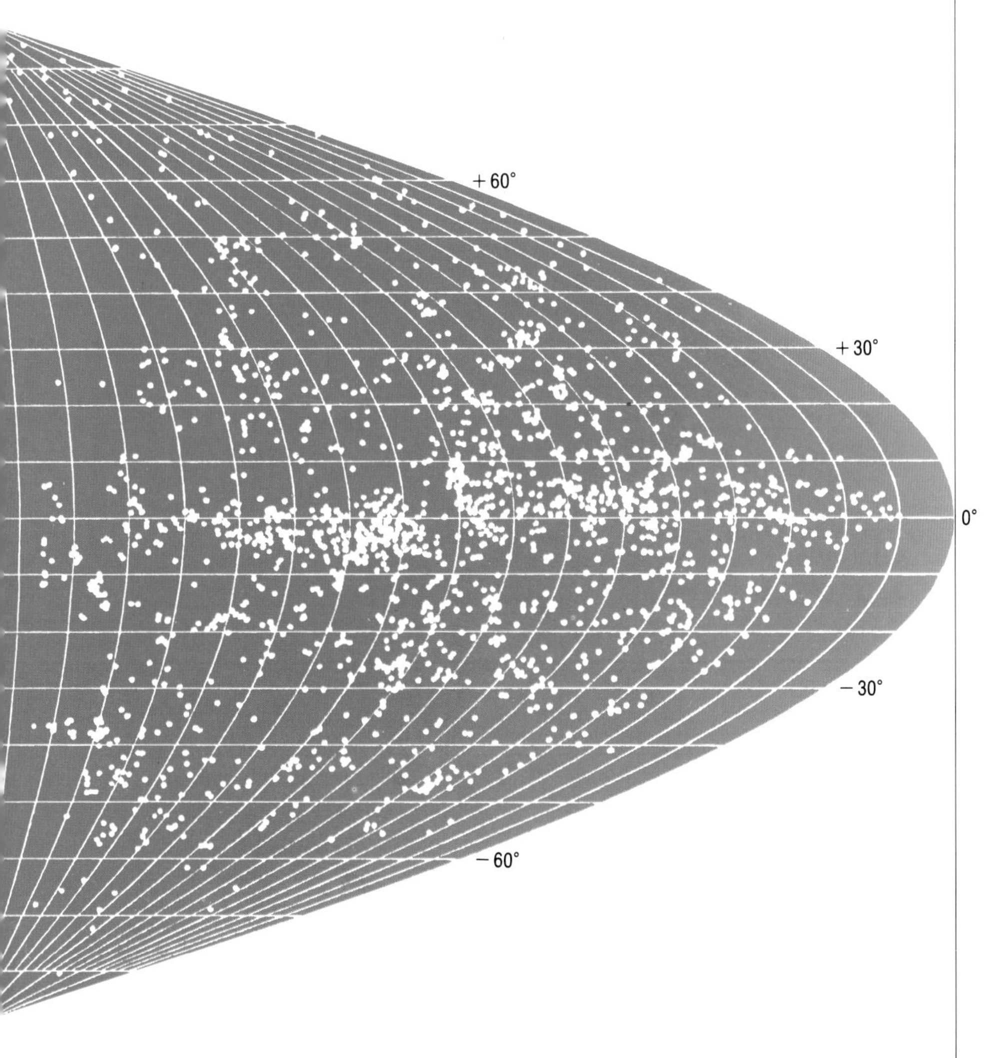

▼ ***The spiral galaxy M100**, in the Virgo cluster: Hubble Space Telescope, 31 December 1993. The pink blobs are huge clouds of glowing hydrogen gas. [Credit: Photo ESA/NASA]*

Our Galaxy

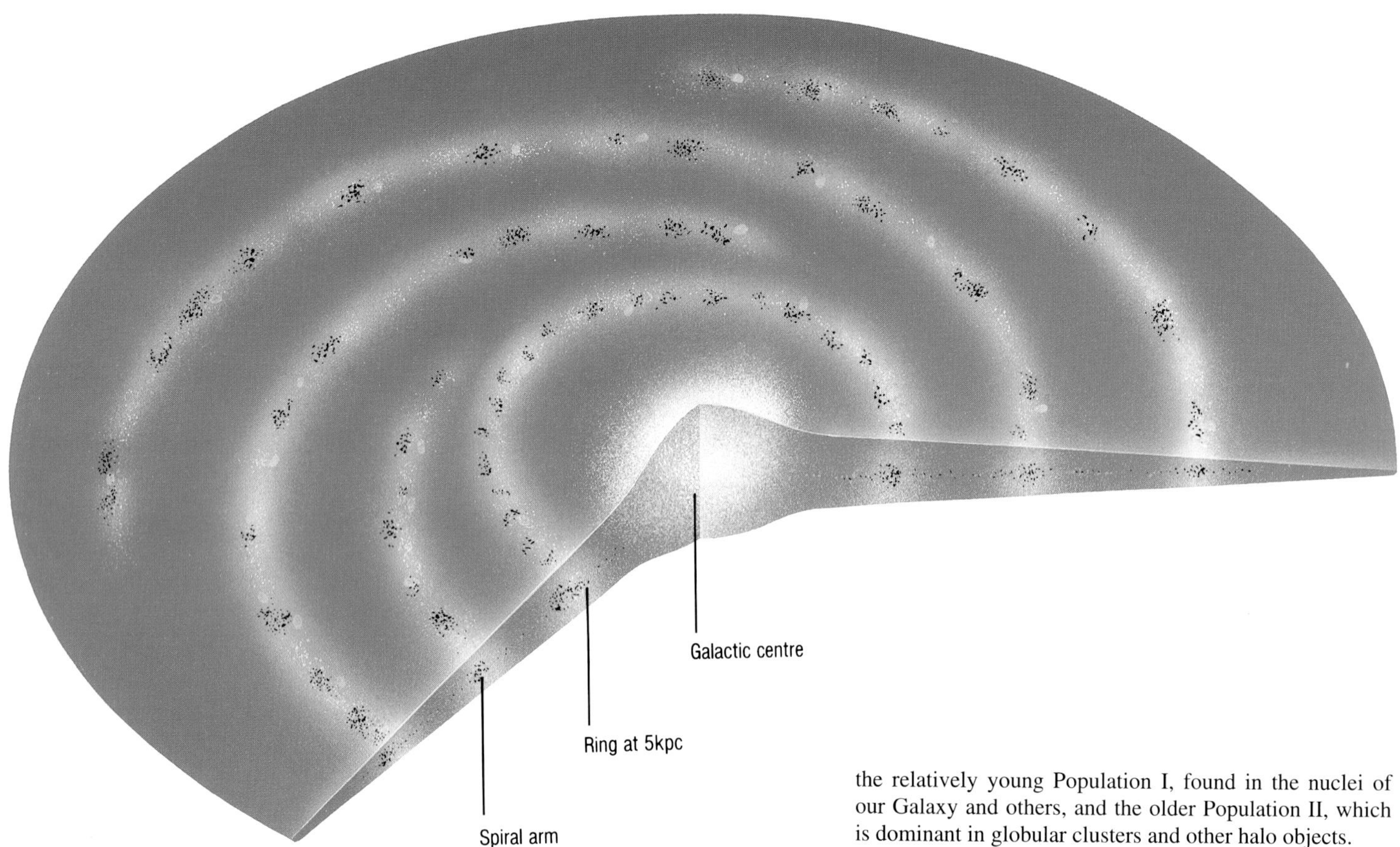

▲ ***In our Galaxy**, neutral hydrogen (in blue) is aggregated mostly along the four large spiral arms where also HII regions (in red) and massive molecular clouds (in black) are clustered. The galactic centre contains numerous expanding regions of ionized hydrogen and giant molecular complexes. It is surrounded by a huge ring of radius about 5 kiloparsecs where a great quantity of atomic and molecular hydrogen is concentrated.*

The main problem about trying to find out the shape of the Galaxy is that we live inside it; the situation is rather like that of a man who is standing in Piccadilly Circus and trying to work out the shape of London. Originally – and quite naturally – most people assumed that the Sun, with its planets, must lie near the centre of the Galaxy; for example William Herschel found that star numbers are much the same all along the Milky Way, though admittedly some parts of it are richer than others.

The first really reliable clue came from radio astronomy, during the 1940s. It was known that there is a great deal of thinly-spread matter between the stars, and it was reasonable to assume that much of this must be hydrogen, which is by far the most plentiful of all the elements. In 1944, H. C. van de Hulst, in Holland, predicted that clouds of cold hydrogen spread through the Galaxy should emit radio waves at one special wavelength: 21.1 centimetres. He proved to be right. The positions and the velocities of the hydrogen clouds were measured, and indicated a spiral structure – which was no surprise, inasmuch as many of the other galaxies are also spiral in form.

By now we are in a position to draw up what we believe to be a reliable picture of the shape and structure of the Galaxy. It is about 100,000 light-years from one end to the other (some authorities believe this to be something of an overestimate), with a central bulge about 10,000 light-years across. The Sun lies between 25,000 and 30,000 light-years from the galactic centre, not far from the main plane and near the edge of a spiral arm. Beyond the main system there is the galactic halo, which is more or less spherical, and contains objects which are very old, such as globular clusters and highly-evolved stars. There are in fact two distinct 'stellar populations'; the relatively young Population I, found in the nuclei of our Galaxy and others, and the older Population II, which is dominant in globular clusters and other halo objects.

We cannot see through to the centre of the Galaxy, because there is too much obscuring material in the way, but we know where it is; it lies beyond the Sagittarius star clouds. It is a decidedly mysterious region, but once again radio astronomy has come to our help, because radio waves are not blocked out in the same way as light. Using the best modern equipment – notably the VLA or Very Large Array radio telescopes in the San Agustin desert of New Mexico – it has been possible to pinpoint a small, compact, very powerful radio source which is called Sagittarius A* (pronounced Sagittarius A-star). It lies in the right position, and probably marks the exact centre of the Galaxy, though we cannot be sure. Its nature is uncertain; a massive black hole is one possibility. Near it there are swirling gas-clouds and groups of highly luminous stars.

We know that the Galaxy is rotating round its centre, and that our Sun takes about 225 million years to complete one circuit. Yet the general rotation does not follow the expected pattern. Kepler's Laws show that in the Solar System, bodies moving close to the centre (in this case the Sun) move quicker than bodies which are further out, so that, for instance, Mercury moves at a greater rate than the Earth, while the Earth moves faster than Mars. In the Galaxy, this sort of situation does not arise, and the speeds are actually greater near the edge of the disk. The only explanation is that the main mass of the Galaxy is not concentrated near the centre at all, and there must be a tremendous amount of material further out. We cannot see it, and we do not know what it is – all we can say for certain is that it exists. the 'missing mass' problem is one of the most puzzling in modern astronomy.

Nowadays the term 'Milky Way' is restricted to describing the luminous band in the sky, though it is true that we still often refer to the Milky Way Galaxy. Sweeping along it with binoculars or a wide-field telescope is fascinating, and it is not always easy to remember that each tiny speck of light is a true sun.

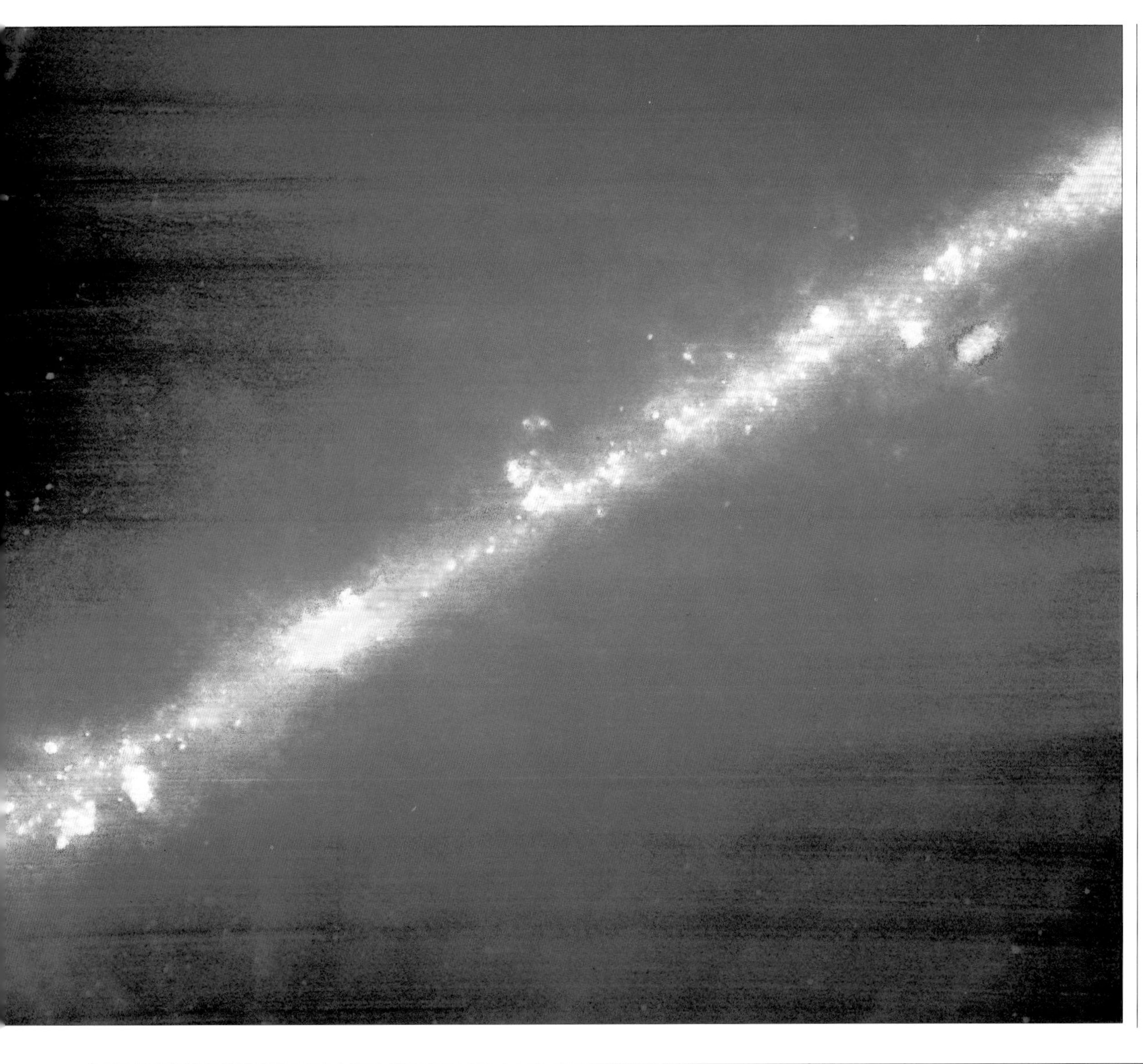

◀ ***The centre of our Galaxy*** *as seen from observations made by the Infra-Red Astronomical Satellite (IRAS). The infra-red telescope carried by IRAS sees through the dust and gas that obscures stars and other objects when viewed by optical telescopes. The bulge in the band is the centre of the Galaxy. The yellow and green knots and blobs scattered along the band are giant clouds of interstellar gas and dust heated by nearby stars. Some are warmed by newly formed stars in the surrounding cloud, and some are heated by nearby massive, hot, blue stars tens of thousands of times brighter than our Sun.*

▼ ***Map of the Milky Way****, by Martin and Tatiana Teskūla (Lund Observatory, Sweden). The coordinates refer to galactic latitude and longitude, measured from the galactic plane and the celestial equator at R.A. 18h 40m. The north galactic pole lies in Coma, the south pole in Sculptor.*

The Local Group of Galaxies

▶ ***The Local Group** is the small band of galaxies to which our Galaxy, the Milky Way, belongs. It also contains the Great Nebula in Andromeda and the Magellanic Clouds.*

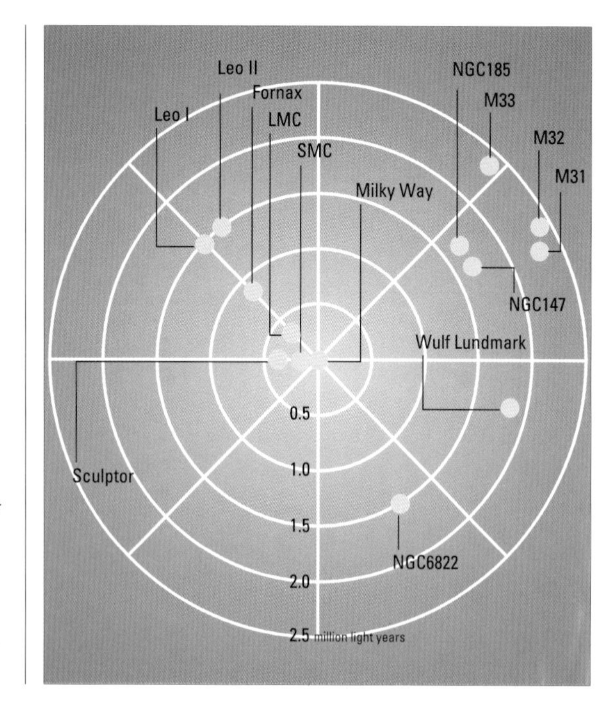

▼ ***The Small Cloud of Magellan** is a system about one-sixth the size of our Galaxy. It is comparatively close, at about 190,000 light-years, and is a prominent naked-eye object in the far south of the sky.*

SELECTED MEMBERS OF THE LOCAL GROUP

Name	Type	Absolute Mag.	Distance 1000 l.y.	Diameter 1000 l.y.
The Galaxy	Spiral	−20.5	–	100
Large Cloud of Magellan	Barred spiral	−18.5	169	30
Small Cloud of Magellan	Barred spiral	−16.8	190	16
Ursa Minor dwarf	Dwarf elliptical	−8.8	250	2
Draco dwarf	Dwarf elliptical	−8.6	250	3
Sculptor dwarf	Dwarf elliptical	−11.7	280	5
Fornax dwarf	Dwarf elliptical	−13.6	420	7
Leo I dwarf	Dwarf elliptical	−11.0	750	2
Leo II dwarf	Dwarf elliptical	−9.4	750	3
NGC6822 (Barnard's Galaxy)	Irregular	−15.7	1700	5
M31 (Andromeda Spiral)	Spiral	−21.1	2200	130
M32	Elliptical	−16.4	2200	12
NGC205	Elliptical	−16.4	2200	8
NGC147	Dwarf elliptical	−14.9	2200	2
NGC1613	Irregular	−14.8	2400	8
M33 (Triangulum Spiral)	Spiral	−18.9	2900	52
Maffei 1	Elliptical	−20	3300	100

The only galaxies which are not moving away from us are the members of what is termed the Local Group. This is a stable collection of more than two dozen systems, of which the largest are the Andromeda Spiral, our Galaxy, the heavily-obscured Maffei 1 and the Triangulum Spiral. Next in order of size come the two Clouds of Magellan, which are satellites of our Galaxy, and M32 and NGC205, which are satellites of the Andromeda Spiral. Most of the rest are dwarfs, some of which are not much more populous than globular clusters and are much less symmetrical and well defined.

The Magellanic Clouds are much the brightest galaxies as seen with the naked eye; they are less than 200,000 light-years away, and cannot be overlooked. Northern observers never cease to regret that the Clouds are so far south in the sky – and in fact this is one of the reasons why most of the large new telescopes have been set up in the southern hemisphere. Both Clouds show vague indications of spiral structure, though the forms are not well marked, and there is nothing of the Catherine-wheel appearance of the classic spirals; it has even been suggested that the Small Cloud is a double system, more or less end-on to us. The Clouds are linked in as much as they form a sort of binary pair, orbiting each other as they travel round our Galaxy. They are joined by a bridge of hydrogen gas, and there is also the 'Magellanic Stream', 300,000 light-years long, reaching over to our Galaxy.

The Clouds are particularly important because they contain objects of all kinds; giant and dwarf stars, doubles and multiples, novae, open and globular clusters, and

◄▼ ***Nearby spirals.*** *M31 in Andromeda (below) and M33 in Triangulum (left), both photographed by John Fletcher with his 25-cm (10-inch) reflector. Both are members of the Local Group; the Triangulum Spiral is more distant, but lies at a more favourable angle.*

gaseous and planetary nebulae. There has even been one recent supernova, 1987A, in the Large Cloud. Telescopically they are magnificent; look, for example, at the Tarantula Nebula, 30 Doradûs, in the Large Cloud, beside which the much-vaunted Orion Nebula seems very puny indeed.

M31, the Andromeda Spiral, is the senior member of the Local Group, and is considerably larger and more luminous than our Galaxy. It too contains objects of all kinds, and there has even been one supernova, S Andromedae of 1885, which reached the fringe of naked-eye visibility. Unfortunately its true nature was not appreciated at the time, and it was not then generally believed that M31 was an independent galaxy.

M31 is a typical spiral, but it lies at a narrow angle to us, and its full beauty is lost. It is frankly rather a disappointing sight in a small telescope (or even a large one), and photography is needed to bring out its details, together with its halo and its 300 globular clusters. The present accepted value for its distance – 2.2 million light-years – may have to be revised slightly upwards if, as now seems possible, the Cepheids are rather more luminous than has been thought. At the moment M31 is actually approaching us, though this will not continue indefinitely and there is no fear of an eventual collision. The two main satellites, M32 and NGC 205, are easy telescopic objects; both are elliptical.

M33, the Triangulum Spiral, is often nicknamed the Pinwheel. It is very close to naked-eye visibility, and is not difficult in binoculars, though users of small telescopes often find it elusive because of its low surface brightness. It is a looser spiral than M31, but lies at a more favourable angle; it too contains objects of all kinds. Its diameter is about half that of our Galaxy. Unlike M31, it does not seem to have been known in ancient times, and in fact it was discovered in 1764 by Messier himself.

Most of the remaining members of the Local Group are dwarfs whose faintness makes them rather hard to identify, particularly if they lie almost behind bright foreground stars of our Galaxy – as with Leo I and Leo II, which are very close to Regulus. Incidentally, no dwarf spiral has ever been found, and it is not likely that any dwarf spirals exist.

Finally there is Maffei 1, discovered by the Italian astronomer Paolo Maffei in 1968. It is probably a giant elliptical, though it has been suggested that there are signs of spirality; it lies in Cassiopeia, not far from the plane of the Milky Way, and is so heavily obscured that we know little about it. A second galaxy discovered by Maffei at the same time was once thought to be a Local Group member, but is now thought to be a spiral at a distance of around 15 million light-years.

In 1994, a dwarf galaxy in Sagittarius was found to be only 80,000 light-years away – the closest yet discovered. These may well be isolated stars and clusters in the space between the galaxies. One globular cluster, NGC5694, seems to be moving on a path which will eventually lead to its escaping from our Galaxy altogether, in which case it will qualify as what can be termed an intergalactic tramp.

The Outer Galaxies

Beyond the Local Group we come to other clusters of galaxies, millions of light-years away. Not surprisingly, it was Edwin Hubble who devised the first really useful system of classification. The diagram which he produced is often called the Tuning Fork, for obvious reasons.

There are three main classes of systems:

1. **Spirals**, from Sa (large nucleus, tightly-wound spiral arms) to Sc (small nucleus, loosely-wound arms). Our own Galaxy is of type Sb, while M51 in Canes Venatici (the Hunting Dogs: Map 1) is Type Sa, and M33, the Pinwheel in Triangulum, is a typical Sc galaxy.

2. **Barred spirals**, where the arms issue from the ends of a sort of bar through the nucleus; they range from SBa through to SBc in order of increasing looseness. It seems that stars in a large rotation disk sometimes 'pile up' in a bar-like structure of this sort, but it does not last for very long on the cosmic scale, which is presumably why SB galaxies are much less common than ordinary spirals.

3. **Ellipticals**, from E0 (virtually spherical) through to E7 (highly flattened). Unlike the spirals, they have little interstellar material left, so that they are more highly evolved, and star formation in them has practically ceased. Giant ellipticals are much more massive than any spirals; for example, M87 in the Virgo cluster of galaxies (Map 6) is of type E0 and is far more massive than our Galaxy or even the Andromeda Spiral. On the other hand many dwarf ellipticals, such as the minor members of our Local Group, are very sparse indeed. It is not always easy to decide upon the type of an elliptical; for example a flattened system, which really should be E7, may be end-on to us and will appear round, so we wrongly class it as E0.

4. **Irregulars**. These are less common than might be thought; they have no definite outline. M82 in Ursa Major (Map 1) is a good example. The Magellanic Clouds were formerly classed as irregular, but it now seems that the Large Cloud, at least, shows definite signs of some faint spiral structure.

The first spiral to be recognized as such was M51, the Whirlpool – by Lord Rosse, who looked at it in 1845 with his giant home-made telescope at Birr Castle in Ireland. It is 37 million light-years away, and is not hard to find, near Alkaid in the tail of the Great Bear. As with all spirals, it is rotating; the arms are trailing – as with a spinning Catherine-wheel. Apparently the arms of a spiral are due to pressure waves which sweep round the system at a rate different from that of the individual stars. The added pressure in these waves triggers off star formation; the most massive stars evolve quickly and explode as supernovae, while the pressure waves sweep on and leave the original spiral arms to disperse. If this is correct, it follows that no particular spiral arm can be a permanent feature.

Recent photographs of the Whirlpool taken with

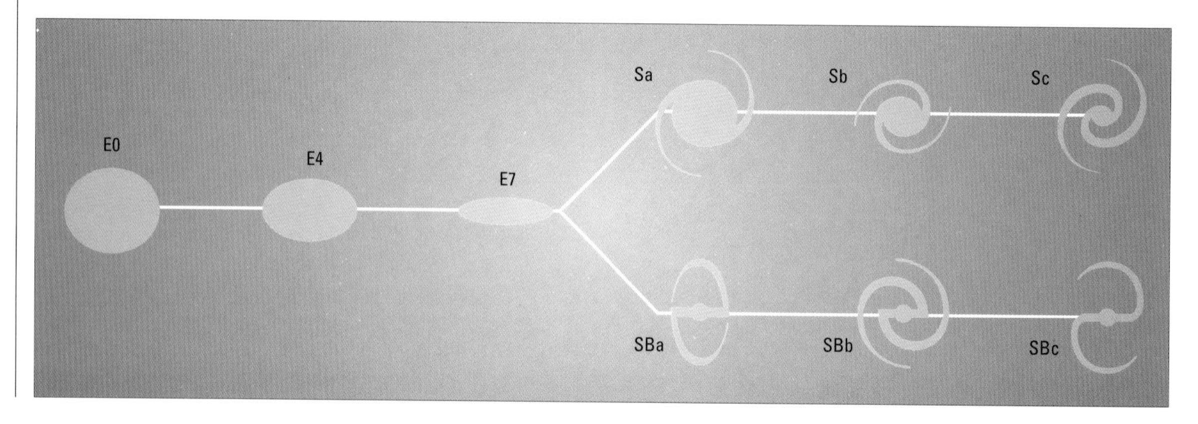

▸ ***Hubble classification of galaxies***. *There are elliptical galaxies (E0 to E7), spirals (Sa, Sb and Sc) as well as irregular systems which are not shown here. There are many refinements; for instance, Seyfert galaxies (many of which are radio sources) have very bright, condensed nuclei.*

▲ ***Type E0***. *M87 in Virgo, magnitude 9.2, distance 41 million light-years. It is a powerful source of radio emission and seems to contain a massive black hole at its core.*

▲ ***Type E4***. *Dwarf galaxy NGC147 in Cassiopeia, magnitude 12.1. Typical of the relatively small systems, and made up entirely of Population II stars, so that there are no very luminous main sequence stars, and star formation has ceased.*

▲ ***Type E6***. *NGC205, photographed in red light. Its system appreciably more elongated than with NGC174, it is the smaller companion of the Andromeda galaxy and is made up of Population II objects.*

the Hubble Space Telescope show a dark X-structure silhouetted across the nucleus, due to absorption by dust. It has been suggested that it may indicate the presence of a black hole at least a million times as massive as the Sun, and in fact it is widely believed that most active galaxies are powered by black holes deep inside them, though on this point there is still disagreement between astronomers.

Certainly there are some galaxies which are very energetic at radio wavelengths; such is NGC5128 in Centaurus (Map 20), also known as Centaurus A, which is crossed by a broad dust lane giving it a most remarkable appearance. There are also Seyfert galaxies, named after Carl Seyfert, who first drew attention to them in 1942. Here we have small, very brilliant nuclei and inconspicuous spiral arms; all seem to be highly active, and most of them, such as the giant M87 in the Virgo cluster, are strong radio sources. Other galaxies emit most of their energy in the infra-red. There are also galaxies with low surface brightness, so that they are difficult to detect even though they may be very massive indeed.

We must admit that our knowledge of the evolution of galaxies is not nearly so complete as we would like. It is tempting to suggest that a spiral may evolve into an elliptical, or vice versa, but the situation does not seem to be nearly as straightforward as this. Because giant ellipticals are so massive, it has been suggested that they may have been formed by the merging of two spirals, but here too opinions differ. At least we have observational evidence that collisions between members of a group of galaxies do occur. The Cartwheel Galaxy A0035, which is 500 million light-years away, is a splendid example of this. It is made up of a circular 'rim' 170,000 light-years across, inside which lie the 'hub' and 'spokes' marked by old red giants and supergiants; apparently the Cartwheel was once a normal spiral, but about 200 million years ago a smaller galaxy passed through it, leading to the formation of very massive stars in the 'rim' region. The invading galaxy can still be seen. Then there are galaxies with double nuclei, probably indicative of cosmic cannibalism, and even the Andromeda Spiral seems to have a double centre, due perhaps to a smaller system swallowed up long ago.

One profitable line of research open to professionals and non-professionals alike is the search for supernovae in external galaxies. Because these outbursts are so colossal they can be seen across vast distances, and are very useful as 'standard candles', because it is reasonable to assume that a supernova in a remote system will be of approximately the same luminosity as a supernova in our own Galaxy. It is important to study a supernova as soon as possible after the flare-up, and amateur 'hunters' have a fine record in this field. One never knows when a normally placid-looking galaxy may be transformed by the dramatic outburst of a brilliant newcomer.

▲ ***Type Sa***. *NGC7217, spiral galaxy in Pegasus. The nucleus is well defined and the arms are symmetrical and tightly wound.*

▲ ***Type SB***. *M81 (NGC3031) in Ursa Major. Seen at a narrower angle than NGC7217, its arms are 'looser'; magnitude is 7.9.*

▲ ***Type Sc***. *M33 (NGC598), the most distant member of the local group, has a less-defined nucleus, and the spiral arms are not so clear.*

▲ ***Type SBa***. *NGC3504 in Leo Minor. The 'bar' through the centre of the system is noticeable.*

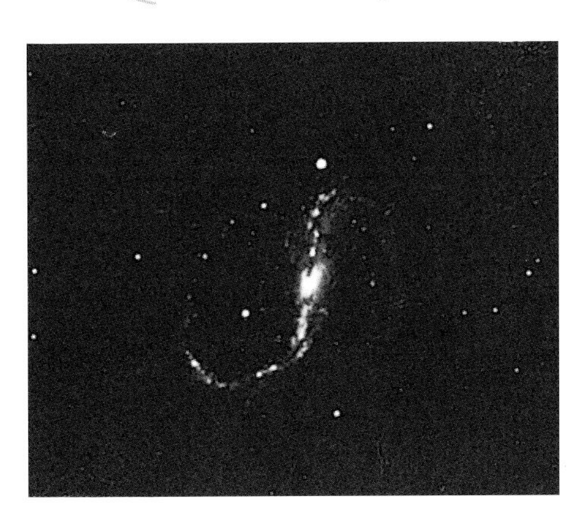

▲ ***Type SBb***. *NGC7479 in Pegasus, magnitude 11.6. The bar formation is much more pronounced.*

▲ ***Type SBc***. *Galaxy in the Hercules cluster. Here the bar formation is dominant, and the arms are secondary.*

Quasars

We have seen that once we move beyond the Local Group, all the galaxies are racing away from us at ever-increasing speeds. There is a definite link between distance and recessional velocity, so that once we know the velocity – which is given by the red shift in the spectral lines – we can work out the distance. The most remote 'normal' galaxies so far found lie at least 10,000 million light-years from us, but we know of objects which are even further away. These are the quasars.

The quasar story began in the early 1960s. By that time there had been several catalogues of radio sources in the sky, several of which had been carried out at Cambridge – but in general the radio emitters did not correspond with visible objects, and in those far-off days radio telescopes were not capable of giving really accurate positions. One source was known as 3C-273, or the 273rd object in the third Cambridge catalogue of radio sources. For once Nature came to the astronomers' assistance. 3C-273 lies in a part of the sky where it can be hidden or occulted by the Moon, and this happened on 5 August 1962. At the Parkes radio astronomy observatory in New South Wales, observers timed the exact moment when the radio emissions were cut off. Since the position of the Moon was known, the position of the radio source could be found. It proved to be an ordinary bluish star.

The results were sent to the Palomar Observatory in California, where Maarten Schmidt used the great Hale reflector to take an optical spectrum of the source. The result was startling. 3C-273 was not a star at all, but something much more dramatic. The red shift of the spectral lines indicated a distance of 3,000 million light-years, and it followed that the total luminosity was much greater than that of an average galaxy, even though the appearance was exactly like that of a star. Other similar discoveries followed, and it became clear that we were dealing with objects of entirely new type. At first they were called QSOs (Quasi-Stellar Radio Sources), but it then emerged that by no means all QSOs are strong radio emitters, and today the objects are always referred to as quasars.

Because the quasars are so powerful, they can be seen across distances even greater than for normal galaxies. According to the best estimate, we can now reach out to at least 13,000 million light-years, so that we are seeing these quasars as they used to be when the universe was young. There are none anywhere near the Local Group, and it may be that no quasars have been formed since the comparatively early history of the universe as we know it.

Quasars are now known to be the cores of very active galaxies, and it seems virtually certain that they are powered by massive central black holes. In many cases it is now possible to see the companion galaxies. Very recent research seems to suggest that quasars are born in environments where two galaxies are violently interacting or even colliding. It is possible that a quasar may remain active for only a limted period on the timescale of the universe, and many large galaxies may go through a 'quasar stage' which does not last for very long.

There are also the BL Lacertae objects, named after the first-discovered member of the class (which was originally taken for an ordinary variable star, and given a variable-star designation). Probably a BL Lac is simply a quasar which we see at a narrow angle, perhaps by looking straight down one of the jets.

The remoteness of the quasars means that they can be used to study interstellar and intergalactic material. A quasar's light will have to pass through this material before reaching us, and the material will leave its imprint on the quasar spectrum; we can tell which lines are due to the quasar and which are not, because the non-quasar lines will not share in the overall red shift. Also, we are becoming increasingly aware of what is termed the 'gravitational lensing effect'. If the light from a remote object passes near a massive object en route, the light will be 'bent', and the result may be that several images will be formed of the object in the background; if the alignment is not perfect there will still be detectable effects. A good example is G2237+0305. Here we have a galaxy 400 million light-years away, behind which is a quasar lying at 8000 million light-years. The light from the quasar is split, producing four images surrounding the image of the lensing galaxy. This is often termed the Einstein Cross, because it was Albert Einstein, in his theory of relativity, who first predicted that such effects could occur.

And yet there is some reason for disquiet. Objects at equal distances from us must have the same red shift – assuming that the shifts themselves are pure Doppler effects. Halton Arp, formerly at the Mount Wilson Observatory and now working in Germany, has found that there are pairs, and groups of objects (quasar/quasar, quasar/galaxy, galaxy/galaxy) which are connected by visible 'bridges' and must therefore be associated, but which have completely different red shifts. If so, then the red shifts are not pure Doppler effects; there is an important non-velocity component as well, so that all our measurements of distance beyond our immediate neighbourhood are unreliable. Arp goes so far as to suggest that quasars are minor features ejected from comparatively nearby galaxies.

Arp is certainly not alone in his views; he is strongly supported by Dr Geoffrey Burbridge, Sir Fred Hoyle and others. While at present this is very much a minority view, it has to be taken very seriously indeed. If it proves to be correct, then many of our cherished ideas will have to be abandoned. It would, indeed, result in a revolution in thought more radical than any since the 1920s, when Hubble and Humason first showed that the 'spiral nebulae' are galaxies in their own right. Time will tell.

► ***Quasar 3C-273.*** *This was the first quasar to be identified, and is also the brightest; its magnitude is 12.8. It lies in Virgo. No other quasar is brighter than magnitude 16.*

▶ ***Quasars and companion galaxies****. The image on the right reveals the galaxy associated with the luminous quasar PKS 2349, which is 1500 million light-years from Earth. The image has enabled astronomers to peer closer into the galaxy's nucleus. Only 11,000 light-years separate the quasar and the companion galaxy (located just above the quasar). This galaxy is similar in size and brightness to the Large Magellanic Cloud galaxy near our Milky Way. The galaxy is closer to the quasar's centre than our Sun is to the centre of our Galaxy. Drawn together by strong gravitational forces, the galaxy will eventually fall into the quasar's engine, the black hole. The black hole will gobble up this companion galaxy in no more than 10 million years.*

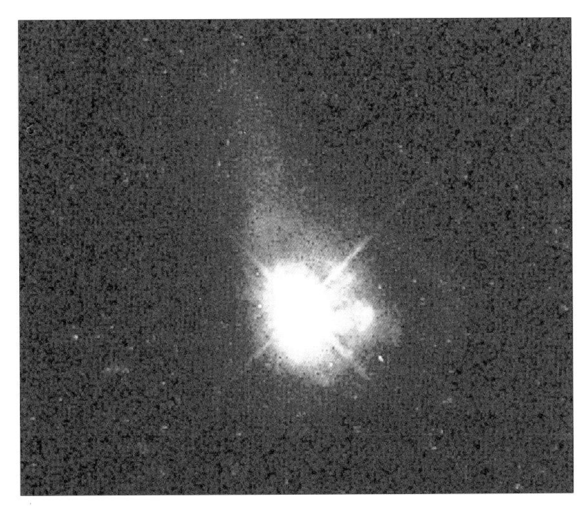

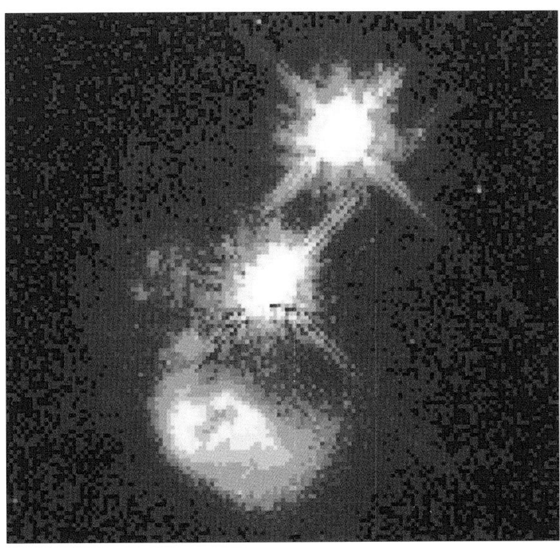

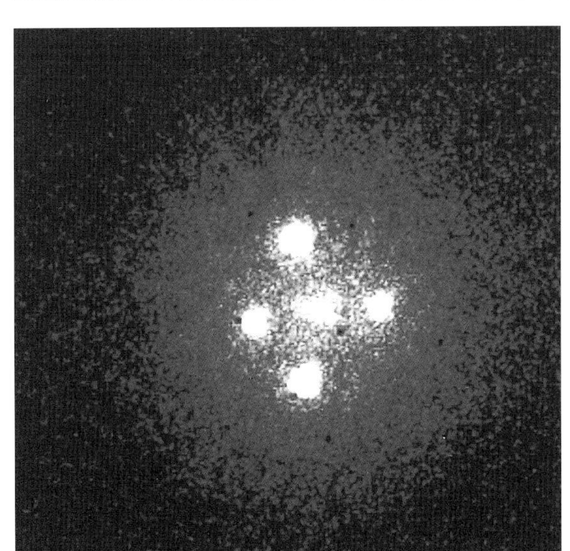

◀ ▼ ***Gravitational lens effect****. The light from a distant quasar passes by an intervening, high-mass galaxy, with the result that the galaxy acts as a 'lens', and produces multiple images of the quasar. In this case the quasar is almost directly behind the galaxy, so the effect is symmetrical; the resulting picture, imaged by the Hubble Space Telescope, is known as the Einstein Cross, since the effect was predicted by Albert Einstein's theory of relativity.*

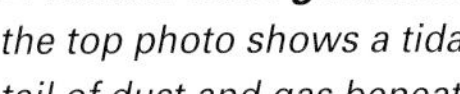

▲ ***Quasar host galaxies****: the top photo shows a tidal tail of dust and gas beneath quasar 0316-346 (2200 million light-years from Earth). The peculiarly-shaped tail suggests that the host galaxy may have interacted with a passing galaxy. The bottom photo shows evidence of a catastrophic collision between two galaxies. The debris from this collision may be fuelling quasar IRAS 04505-2958 (3000 million light-years from Earth). Astronomers believe that a galaxy plunged vertically through the plane of a spiral galaxy, ripping out its core and leaving the spiral ring (at the bottom of the image). The core lies in front of the quasar, the bright object in the centre of the image. Surrounding the core are star-forming regions. The distance between the quasar and spiral ring is 15,000 light-years.*

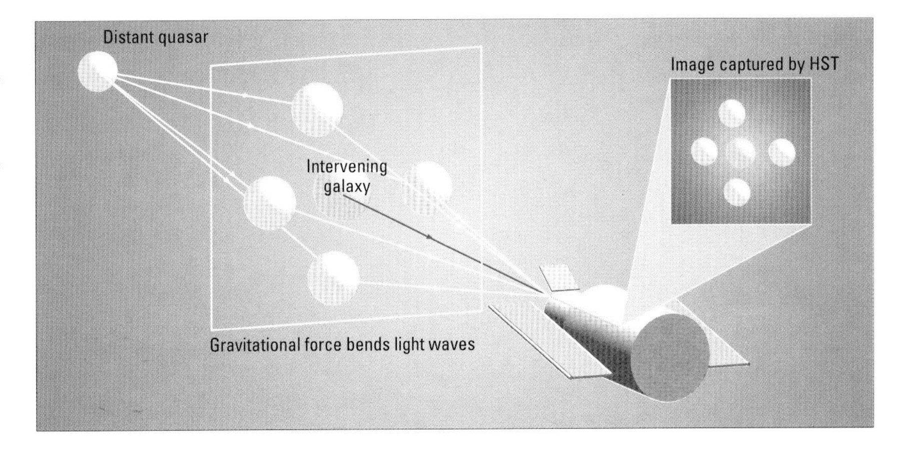

The Expanding Universe

Before we can make any attempt to trace the history of the universe, we must look carefully at the situation we find today. As we have seen, each group of galaxies is receding from each other group, so that the entire universe is expanding. We are in no special position, and the only reasonable analogy – not a good one, admittedly – is to picture what happens when spots of paint are put on to a balloon, and the balloon is then blown up. Each paint spot will move away from all the others, because the balloon is expanding. Similarly, the universe is expanding and carrying all the groups of galaxies along as it does so.

For example, we may say that one particular galaxy is receding at a rate of 2000 kilometres (1250 miles) per second. Anyone living on a planet in that system would maintain that it is the Milky Way Galaxy which is moving away at 2000 kilometres per second; there is no absolute standard of reference.

The distribution of galaxies is not random. They tend to congregate in groups or clusters, and our own Local Group is very far from being exceptional; for instance the Virgo Cluster, at a distance of around 50 million light-years, contains thousands of members, some of which (such as M87, the giant elliptical) are much more massive than our Galaxy. Moreover, there is a definite large-scale structure; there are vast sheets of galaxies – such as the so-called Great Wall, which is 300 million light-years long and joins two populous clusters of galaxies, those in Coma and in Hercules. It seems that the overall pattern of the distribution of the galaxies is cellular, with vast 'voids' containing few or no systems.

If the speed of recession increases with distance, there must come a time when an object will be receding at the full speed of light. We will then be unable to see it, and we will have reached the boundary of the observable universe, though not necessarily of the universe itself. It has generally been assumed that this limit must be somewhere between 12,000 million and 15,000 million light-years, but we have not yet been able to penetrate to such a distance, though the remotest objects known to us are moving away at well over 90 per cent of the speed of light.

Much depends upon what is termed the Hubble Constant, which is a measure of the increase of recessional velocity with distance. Measurements made in 1999, with the Hubble Space Telescope, give a value of 70 kilometres (45 miles) per second per megaparsec (a parsec, as we have already noted, is equal to 3.25 light-years, and a megaparsec is one million times this figure). At one time there was a curious situation – it seemed that the universe might be much younger than had been believed (no more than 10,000 million years) and this would indicate that some known stars are older than the universe itself, which is clearly absurd. However, the Hubble team observed galaxies out to 64 million light-years, and identified 800 Cepheid variables, so that the result seems to be much more reliable than any previous estimate.

This may be the moment to mention what is called Olbers' Paradox, named after the last-century German astronomer Heinrich Olbers, who drew attention to it (though in fact he was not the first to describe it). Olbers asked, 'Why is it dark at night?' If the universe is infinite, then sooner or later we will see a star in whichever direction we look and the whole sky ought to be bright. This is not true, partly because the light from very remote objects is so red shifted that much of it is shifted out of the visible range, and partly because we are now sure that the observable universe is not infinite. Unless our current theories are fundamentally flawed, the limit cannot be more than 20,000 million light-years, and is probably rather less.

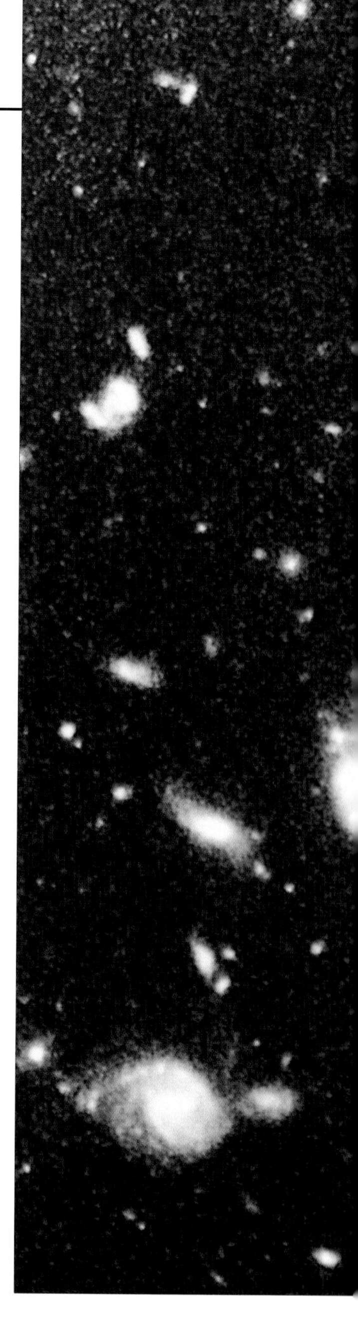

► ***The Hubble deep field picture**: this is the deepest optical image ever obtained. It was taken with the Hubble Space Telescope in 1995–6; images were obtained with four different filters (ultra-violet, blue, orange and infra-red) and then combined. Only about a dozen stars are shown – the brightest is of magnitude 20 – but there are 1500 galaxies of all kinds. The field covers a sky area of 2.5 arc minutes. At this density, the entire sky would contain 50,000 million galaxies, and we are seeing regions as they used to be when the universe was only one-fifth of its present age. The region of the deep field photograph lies in Ursa Major, chosen specifically because it seemed to contain no notable objects.*

But how big is the entire universe as opposed to the observable part of it? If the universe is finite, we are entitled to ask what is outside it; and to say 'nothing' is to beg the question, because 'nothing' is simply space. But if the universe is infinite, we have to visualize something which goes on for ever, and our brains are unequal to the task. All we can really do is say that the universe may be 'infinite, but unbounded'. An ant crawling round a school-room globe will be able to continue indefinitely while covering a limited range; this is a poor analogy, but it is not easy to think of anything better.

New information has come from Hipparcos, the astrometric satellite, which was launched in 1989 to provide a new, much more accurate catalogue of the stars within around 200 light-years of the Sun. The catalogue, finally issued in 1997, provides new data about the luminosities and proper motions of the stars, and this extra knowledge can be extended to further parts of the Galaxy. It has even been suggested that the observable universe may be around 10 per cent larger than has been believed, but final analyses have yet to be made.

Certainly the Hipparcos catalogue will be invaluable as a standard reference for centuries to come. An even more ambitious catalogue, Gaia, is tentatively planned for the early 21st century.

▶ ***Young galaxy survey***. *Embedded in this Hubble Space Telescope image of nearby and distant galaxies are 18 young galaxies or galactic building blocks, each containing dust, gas and a few thousand million stars. Each of the objects is 11,000 million light-years from Earth and much smaller than today's galaxies. This picture is a true-colour image made from separate exposures taken in blue, green and infra-red light with the Wide Field Planetary Camera 2. It required 48 orbits around the Earth to make the observations. The green and red exposures were taken in June 1994; the blue exposures, as well as 15 orbits of the redshifted hydrogen line, were taken in June 1995.*

The Early Universe

Every culture has its own creation myths, and there is of course the account in Genesis which Biblical fundamentalists still take quite literally. But when we come to consider the origin of the universe from a scientific point of view, we are faced with immediate difficulties. The first question to be answered is straightforward enough: 'Did the universe begin at a definite moment, probably about 15,000 million years ago, or has it always existed?' Neither concept is at all easy to grasp.

The idea of a sudden creation in a 'Big Bang' was challenged in 1947 by a group of astronomers at Cambridge, who worked out what came to be called the continuous-creation or 'steady-state' theory. In this picture, the universe had no beginning, and will never come to an end; there is an infinite past and an infinite future. Stars and galaxies have limited lifetimes, but as old galaxies die, or recede beyond the boundary of the observable universe, they are replaced by new ones, formed from material which is spontaneously created out of nothingness in the form of hydrogen atoms. It follows that if we could look forward in time by, say, ten million million years, the numbers of galaxies we would see would be much the same as at present – but they would not be the same galaxies.

The rate at which new hydrogen atoms were created would be so low that it would be quite undetectable, but there were other tests which could be made. If we could invent a time machine and project ourselves back into the remote past, we would be able to see whether the universe looked the same then as it does now. Time machines belong to science fiction, but when we observe very remote galaxies and quasars we are in effect looking back in time, because we see them as they used to be thousands of millions of years ago. Careful studies showed that conditions in those far regions are not identical with those closer to us, so that the universe is not in a steady state.

More definite proof came in 1965. If the universe began with a Big Bang, it would have been incredibly hot. It would then cool down, and calculations indicated that by now the overall temperature should have dropped to three degrees above absolute zero (absolute zero being the coldest temperature that there can possibly be – approximately -273 degrees C). We should therefore be able to detect weak 'background radiation', coming in from all directions all the time, which would represent the remnant of the Big Bang; and in the United States Arno Penzias and Robert Wilson, using a special type of radio telescope, actually detected this background radiation. Theory and observation dovetailed perfectly, and by now the steady-state picture has been abandoned by almost all astronomers. We are back with the Big Bang.

We have to realize that space, time and matter all came into existence simultaneously; this was the start of 'time' and we cannot speculate as to what happened before that, because there was no 'before'. We can work back to 10^{-43} of a second after the Big Bang, but before that all our ordinary laws of physics break down, and we have to confess that our ignorance is complete. (10^{-43} is a convenient way of expressing a very small quantity; it is equivalent to a decimal point followed by 42 zeros and then a 1.)

In any case, we are not really talking about the origin of the universe at all; we are discussing its evolution, which is by no means the same thing. We are rather in the situation of a visitor from Outer Space who arrives in London, spends an hour in Oxford Street, and surveys the passers-by. He will see babies, boys, youths and men; if he is intelligent he will realize that a baby turns into a boy while a boy develops into a man, and he will be able to work out the developmental sequence of a human being. But unless someone has acquainted him with the facts of life, he will have no idea how the baby can have appeared – and so far as the origin of the universe is concerned, our 'baby' is the Big Bang.

If we could go back to the very earliest moment which we consider, 10^{-43} of a second after the Big Bang, we would find an incredibly high temperature of perhaps 10^{32} degrees C. This is so hot that no atoms could possibly form. Various forces were in operation, and when these began to separate there was a period of rapid inflation, when the universe expanded very rapidly. This lasted from about 10^{-36} to 10^{-32} second after the Big Bang, and stopped when the various forces had become fully separated. Since then the rate of expansion has been much slower.

At the end of the inflationary period, the universe was filled with radiation. There were also fundamental particles which we call quarks and antiquarks, which were the exact opposites of each other, so that if they collided they annihilated each other. Had they been equal in number, all of them would have been wiped out, and there would have been no universe as we know it, but in fact there were slightly more quarks than antiquarks, and eventually the surplus quarks combined to form matter of the kind we can understand.

At 10^{-5} second (or one ten-thousandth of a second) after the Big Bang protons and neutrons started to form, and after about 100 seconds, when the temperature had dropped to 1000 million degrees C, these protons and neutrons began to combine to form the nuclei of the lightest elements, hydrogen and helium. Theory predicts that there should have been about ten hydrogen nuclei for every nucleus of helium, and this is still the ratio today, which is yet another argument in favour of the Big Bang picture.

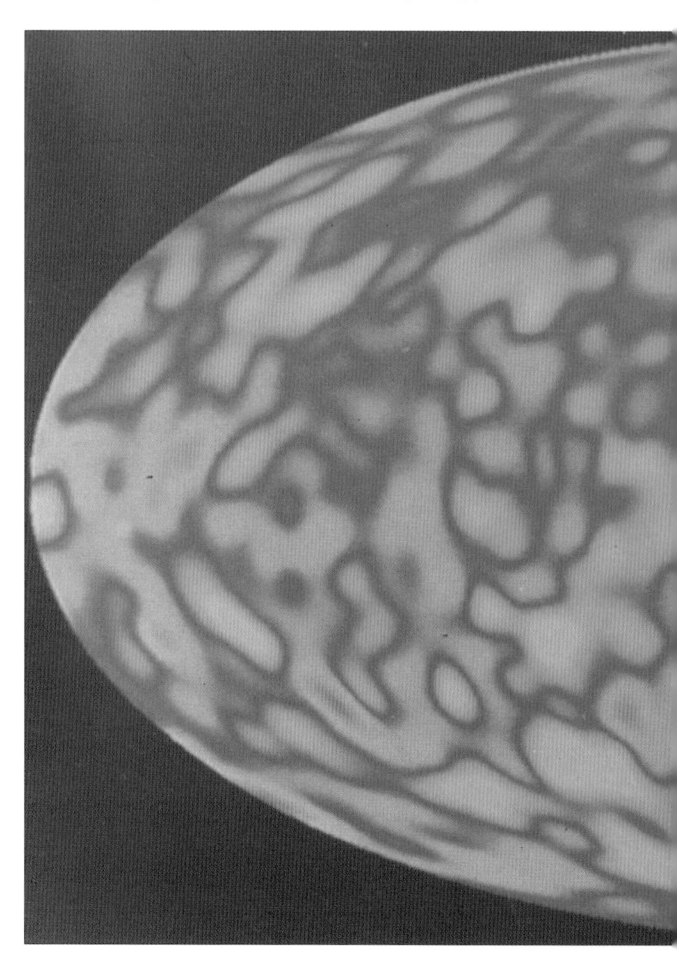

At this stage space was filled with a mish-mash of electrons and atomic nuclei, and it was opaque to radiation; a packet or 'photon' could not go far without colliding with an electron and being blocked. But when the temperature had fallen still further, to between 4000–3000 degrees C, the whole situation changed. By about 300,000 years after the Big Bang, most of the electrons had been captured by protons to make up complete atoms, so that the radiation was no longer blocked and could travel freely across the growing universe. Over a thousand million years after this 'decoupling', galaxies began to form; stars were born, and massive stars built up heavy elements inside them, subsequently exploding as supernovae and spewing their heavy-element enriched material into space to be used to form new stars.

One objection to this whole picture was that the background radiation coming in towards us appeared to be exactly the same from all directions. It followed that the expansion of the universe following the inflationary period would have had to be absolutely smooth – but in that case how could galaxies form? There would have to be some irregularities in the background radiation, but for a long time no trace could be found. Finally, in 1993, tiny 'ripples' were discovered.

The next requirement is to decide whether the present expansion will or will not continue indefinitely, and here everything depends upon the average density of the material spread through the universe. If the density is above a certain critical value, roughly equivalent to one hydrogen atom per cubic metre, the galaxies will not escape from each other; they will stop, turn back and come together again in what may be termed a 'Big Crunch'. If the density is below this critical value, the expansion will go on until all the groups of galaxies have lost contact with each other.

When we look at the material we can see – galaxies, stars, planets and everything else – it is quite obvious that there is not nearly enough to pull the galaxies back. Yet the ways in which the galaxies rotate, and the ways in which they move with respect to each other, indicate that there is a vast quantity of material which we cannot see at all. In the universe as a whole, this 'missing mass' may indeed make up more than 90 per cent of the total. It may be locked up in black holes; it may be accounted for by swarms of low-mass stars which are too dim to be detected; it may be that neutrinos, which are so plentiful in the universe, have a tiny amount of mass instead of none at all; it may be that the invisible material is so utterly unlike ordinary matter that we might not be able to recognize it. At present we have to admit that we simply do not know.

Assume that the overall density is enough to stop the expansion of the universe. After perhaps 40,000 million years following the Big Bang the red shifts will change to blue shifts as the galaxies begin to rush together again at ever-increasing speeds. Between ten and a hundred million years before the Big Crunch, stars will dissolve and the whole of space will become bright; ten minutes before the Crunch, and atomic nuclei will disintegrate into protons and electrons; with one-tenth of a second to go, these will in turn dissolve into quarks – and then will come the crisis. It is possible that the Crunch will be followed by a new Big Bang, and the cycle will begin all over again, though it is just as likely that the universe will destroy itself. If, on the other hand, the density is below the critical value, all that will happen is that the groups of galaxies will go out of contact with each other; their stars will die, and we will end with a dead, radiation-filled universe.

Whether either of these scenarios is right, or both are wrong, remains to be seen.

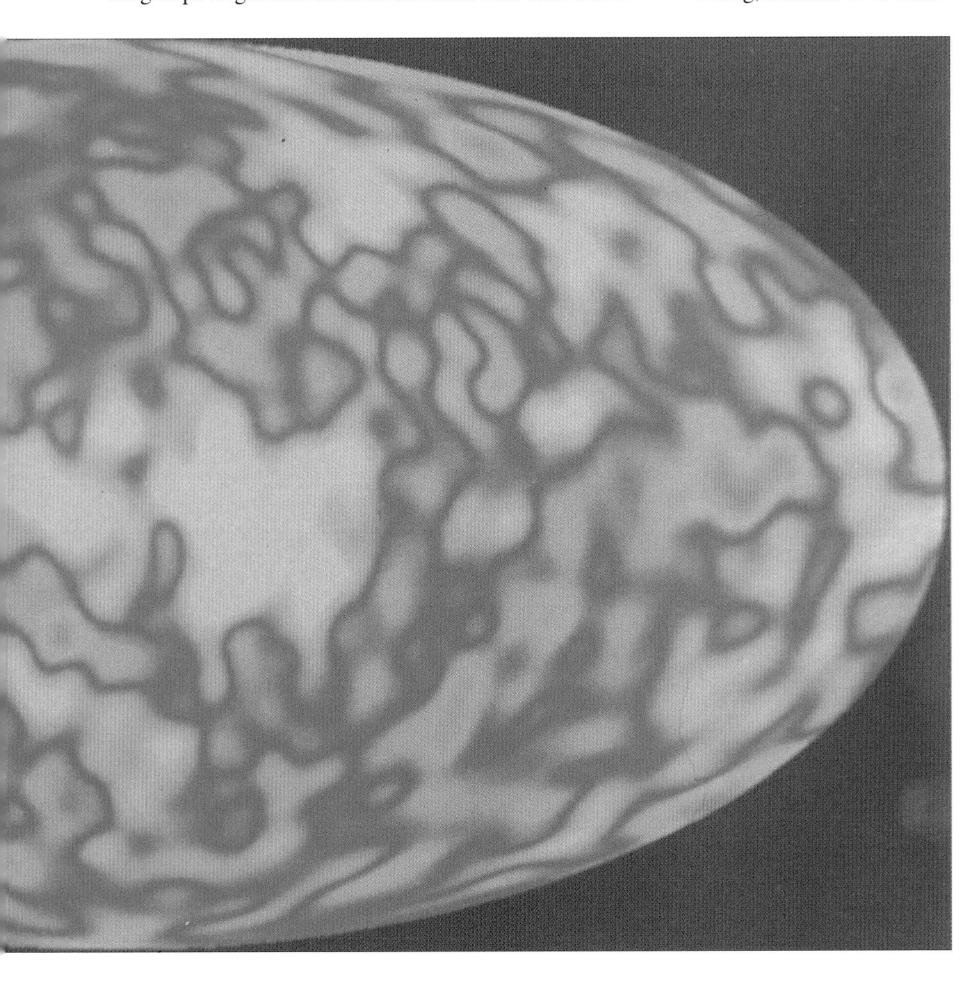

*◄ **COBE**, the specially designed Cosmic Background Explorer, found in 1993 what Big Bang theorists had long been looking for – tiny variations in the temperature of the background radiation, indicating 'ripples'. Here the temperature variations have been colour-coded. These ripples show that early universe was not smooth and uniform, so that matter could concentrate, forming stars and galaxies.*

Life in the Universe

Of all the problems facing mankind, perhaps the most intriguing is: Can there be life on other worlds? Are we alone in the universe, or is life likely to be widespread?

Let us admit at once that we do not yet have the slightest evidence of the existence of life anywhere except on the Earth. Moreover, we must confine ourselves to discussing life of the kind we can understand. All our science tells us that life must be based upon carbon; if this is wrong, then the rest of our science is wrong too, which does not seem very likely. Rather reluctantly, we must reject the weird and wonderful beings so beloved of science-fiction writers, and which are usually classed as BEMs or Bug-Eyed Monsters. Life-forms on other worlds need not necessarily look like us, but they will be made up of the same ingredients – and after all, there is not much outward resemblance between a man, a cat and an earwig.

It seems that there are three main questions to be answered. (1) Do other stars have planet-families of their own, and if so how can we detect them? (2) Assuming that other planets exist, are they likely to support life? And (3) What chances have we of establishing communications with civilizations in other solar systems?

There are 100,000 million stars in our Galaxy, many of which are very like the Sun; we can see 1000 million galaxies – and it does not seem reasonable to believe that in all this host, our Sun alone is attended by a system of planets. Proof is not easy to obtain, but there are several lines of research. Nearby stars have appreciable proper motions, and irregularities in these motions might indicate the presence of a planet or planets; several of our nearer stellar neighbours seem to show evidence of this. More significantly, there are some stars – such as Vega, Fomalhaut and Beta Pictoris – which have been found to be associated with cool, possibly planet-forming material which shows up at infra-red wavelengths; in the case of Beta Pictoris the material has even been photographed. In 1994 Hubble detected pancake-shaped disks of matter around new-born stars in the Orion nebula. This is not to say that planets exist in these locations, but it is certainly a strong possibility.

Attempts have been made to track down planets of other stars by their gravitational effects upon the stars concerned. An orbiting planet could, in theory, make the central star 'wobble' slightly, and this could be detectable by changes in the spectrum – once more we come back to the Doppler effect. There have been claims of planets orbiting pulsars, though it is very difficult to see how such planets could have been formed.

In October 1995, B. Mayor and B. Queloz (Geneva Observatory) stated that they had identified a massive planet orbiting the star 51 Pegasi. Since then, similar claims have been made for other stars. There no longer seems any doubt that planetary systems are common, and in 1999 the spectrum of a planetary body orbiting the star Tau Boötis was actually observed.

The second question is even more difficult to answer, mainly because we are unsure of the origin of life even on Earth. (Suggestions that life did not originate here, but was brought to Earth by way of a comet of a meteorite, seem to raise more problems than they solve.) All we can really say is that if we could locate a planet similar to the Earth, moving round a star similar to the Sun, it would be reasonable to expect life not unlike ours.

So far as communication is concerned, we must concede that in our present state of technology interstellar travel is impossible; even if we could travel at the speed of light it would take a spacecraft years to reach even the nearest star. When we consider 'exotic' forms of travel – teleportation, thought-travel and the like – we are back in the realms of science fiction. It may happen one day, but at the moment we cannot even begin to speculate as to how it might be done.

Therefore, the only hope is to use radio, and various attempts have already been made. The first dates back to 1960, when the powerful telescope at Green Bank, in West Virginia, was used to 'listen out' for signals rhythmical enough to be interpreted as artificial. The wavelength selected was 21.1 centimetres, because emissions at this wavelength are emitted by the clouds of cold hydrogen spread through the Galaxy and radio astronomers anywhere would presumably be on watch. The two stars singled out for special attention were Tau Ceti and Epsilon Eridani, which are the nearest stars which are sufficiently like the Sun to be regarded as possible centres of planetary systems. The experiment – officially known as Project Ozma, but more generally as Project Little Green Men – produced nothing positive, but further surveys have been made since, and the International Astronomical Union has set up a special Commission to concentrate upon SETI, the Search for Extra-Terrestrial Intelligence. At the General Assembly in 1991 it even published a Declaration giving instructions as to the procedure to be followed in the event of an alien contact.

Of course, there is the time-delay factor. Send out a message to, say, Tau Ceti in 2000 and it will reach its destination in 2012; if some obliging operator on a Tau Cetian planet hears it and replies immediately, we would expect an answer in 2024. This means a delay of 22 years, which makes quick-fire repartee difficult. However, no doubt mathematical codes could be devised, because mathematics is universal, and we did not invent it; we merely discovered it. The real significance would be in establishing that ETI does exist. The effect upon our thinking – scientific, religious, political – would indeed be profound.

It has been argued that we really are alone, and that there are no other living things anywhere in the universe. On the other hand it has also been argued that there may be civilizations in all parts of stages of development. It is also possible that there are planets upon which the inhabitants have wiped themselves out in war – as we are ourselves in danger of doing; we have the ability to turn the whole of the Earth into a barren, radioactive waste, and our technology has far outstripped our actual intelligence.

The search goes on; our radio telescopes are used to listen out, and even to send messages in the hope that someone, somewhere, will hear them. The chances of success may be slight, and it is a measure of our changing attitudes that experiments such as SETI are considered worth carrying out at all.

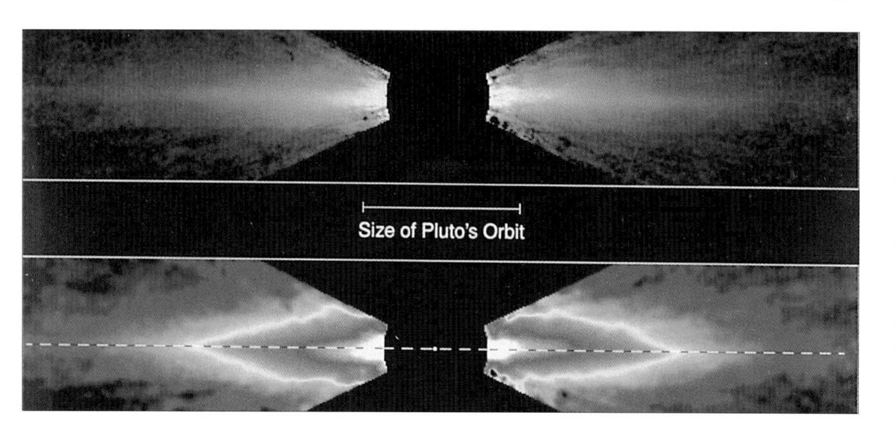

▼ ***The Beta Pictoris disk**, imaged with the Hubble Space Telescope, 2 January 1996.*

SELECTED TARGET STARS

All these stars are within 30 light-years of the Sun, and may be regarded as possible targets for future SETI investigations.

Star	Spectrum	Apparent mag.	Luminosity, Sun=1	Distance l.y.
ε Eridani	K0	3.8	0.3	10.7
ε Indi	K5	4.7	0.1	11.2
τ Ceti	K0	3.5	0.35	11.9
ϱ Ophiuchi	K0	4.0	0.35	17
δ Pavonis	G5	3.6	1.0	18
σ Draconis	K0	4.7	0.3	19 Alrakis
χ Draconis	F7	3.6	2.0	19
β Hydri	G1	2.8	2.3	26
α Piscis Australis Fomalhaut	A3	1.2	13	22
ξ Boötis	G8	4.6	0.5	22
ζ Tucanae	G0	4.2	0.8	23
π^3 Orionis	F6	3.2	2.3	25
α Lyrae	A0	0.0	52	26 Vega
61 Virginis	G8	4.7	0.6	27
μ Herculis	G5	3.4	2.2	26
γ Leporis	F8	3.8	2.0	26
β Comae	G0	4.3	1.2	27
β Canum Venaticorum	G0	4.3	1.3	30 Chara

DECLARATION OF PRINCIPLES

Declaration of principles concerning activities following the detection of extra-terrestrial intelligence. Passed at the General Assembly of the International Astronomical Union at Buenos Aires, Argentina, July 1991.

1. Any individual or institute believing that any sign of ETI has been detected should seek verification and confirmation before taking further action.

2. Before making any such announcement, the discoverer should promptly notify all other observers or organizations which are parties to this Declaration.No public announcement should be made until the credibility of the report has been established. The discoverer should then inform his national authorities.

3. After concluding that the discovery is credible, the discoverer should inform the Central Bureau or Astronomical Telegrams of the International Astronomical Union, and also the Secretary-General of the United Nations. Other organizations to be notified should include the Institute of Space Law, the International Telecommunication Union, and a Commission of the International Astronomical Union.

4. A confirmed detection of ETI should be disseminated promptly, openly and widely through the mass media.

5. All data necessary for confirmation of detection should be made available to the international scientific community.

6. All data relating to the discovery should be recorded, and stored permanently in a form which will make it available for further analysis.

7. If the evidence of detection is in the form of electromagnetic signals, the parties to this Declaration should seek international agreement to protect the appropriate frequencies. Immediate notice should be sent to the Secretary-General of the International Telecommunication Union in Geneva.

8. No response to a signal or other evidence of ETI should be sent until appropriate international consultations have taken place.

9. The SETI (Search for Extra-Terrestrial Intelligence) Committee of the International Academy of Astronautics, in co-ordination with Commission 51 of the International Astronomical Union, will conduct a continuing review of all procedures relating to the detection of ETI and the subsequent handling of the data.

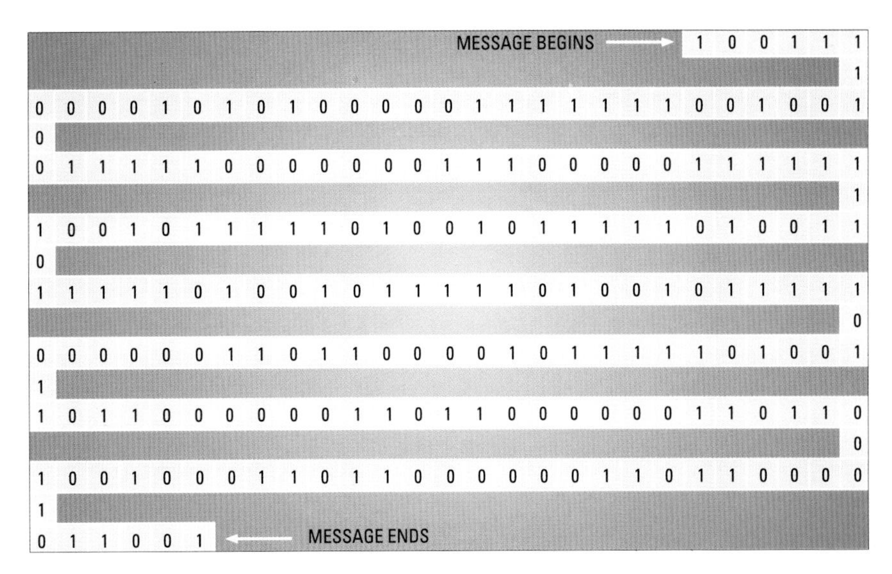

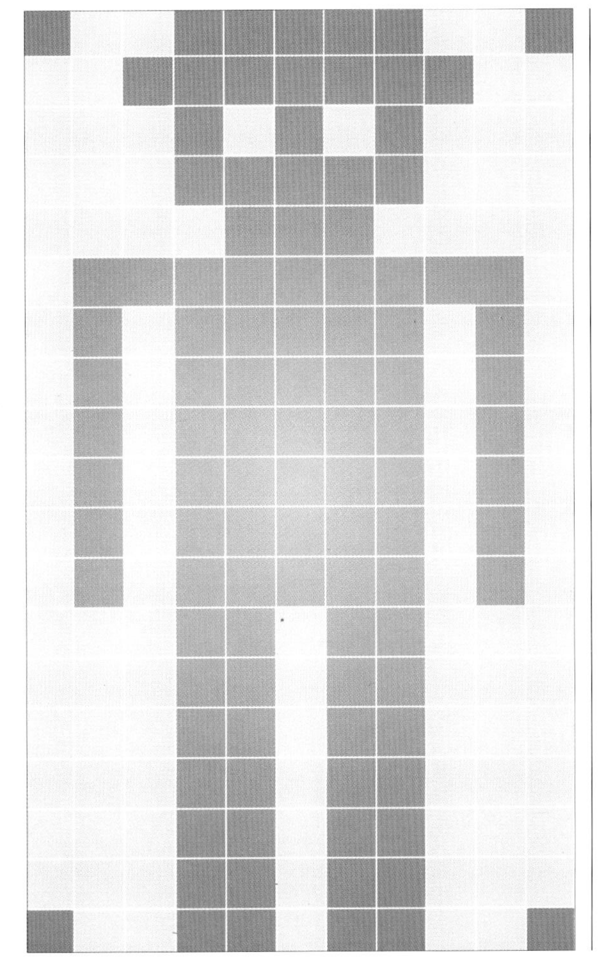

▲◀ *Interstellar code.* *Signals of two definite types – one positive, one negative – are transmitted. If the positive signals are taken as black and the negative are white and arranged in a grid, a pattern emerges. Here 209 signals are sent (top), of which the only factors are 19 and 11 (19 × 11 = 209). If the grid is 19 wide, the pattern is meaningless (above). But if it is 11 wide, a figure emerges (left).*

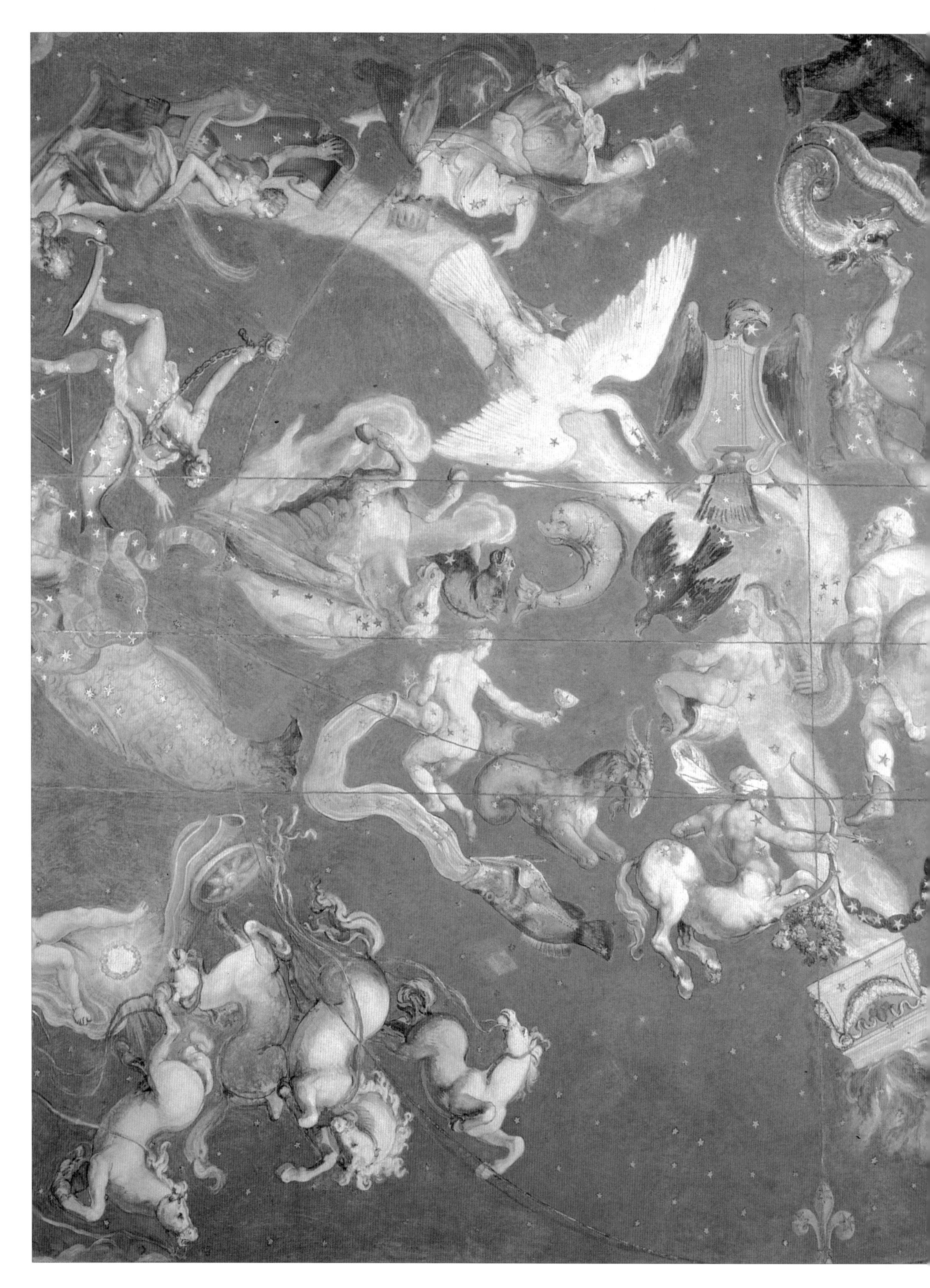

Star Maps

◀ ***The constellations**, as depicted by de Vecchi and da Reggio on the ceiling of the Sala del Mappamondo of the Palazzo Farnese.*

Whole Sky Maps

***▼ Turn the map** for your hemisphere so that the current month is at the bottom. The map will then show the constellations on view at approximately 11 pm GMT (facing south in the northern hemisphere and north in the southern hemisphere). Rotate the map clockwise 15° for each hour before 11 pm; anticlockwise for each hour after 11 pm.*

The origin of the constellation patterns is not known with any certainty. The ancient Chinese and Egyptians drew up fanciful sky maps (two of the Egyptian constellations, for example, were the Cat and the Hippopotamus), and so probably did the Cretans. The pattern followed today is based on that of the ancient Greeks and all of the 48 constellations given by Ptolemy in his book the *Almagest*, written about AD 150, are still in use.

Ptolemy's list contains most of the important constellations visible from the latitude of Alexandria. Among them are the two Bears, Cygnus, Hercules, Hydra and Aquila, as well as the 12 Zodiacal groups. There are also some small, obscure constellations, such as Equuleus (the Foal) and Sagitta (the Arrow), which are surprisingly faint and, one would have thought, too ill-defined to be included in the original 48.

It has been said that the sky is a mythological picture book, and certainly most of the famous old stories are commemorated there. All the characters of the Perseus tale are to be seen – including the sea monster, although nowadays it is better known as Cetus, a harmless whale! Orion, the Hunter, sinks below the horizon as his killer, the Scorpion, rises; Hercules lies in the north, together with his victim the Nemaean lion (Leo). The largest of the constellations, Argo Navis – the ship which carried Jason and his companions in quest of the Golden Fleece – has

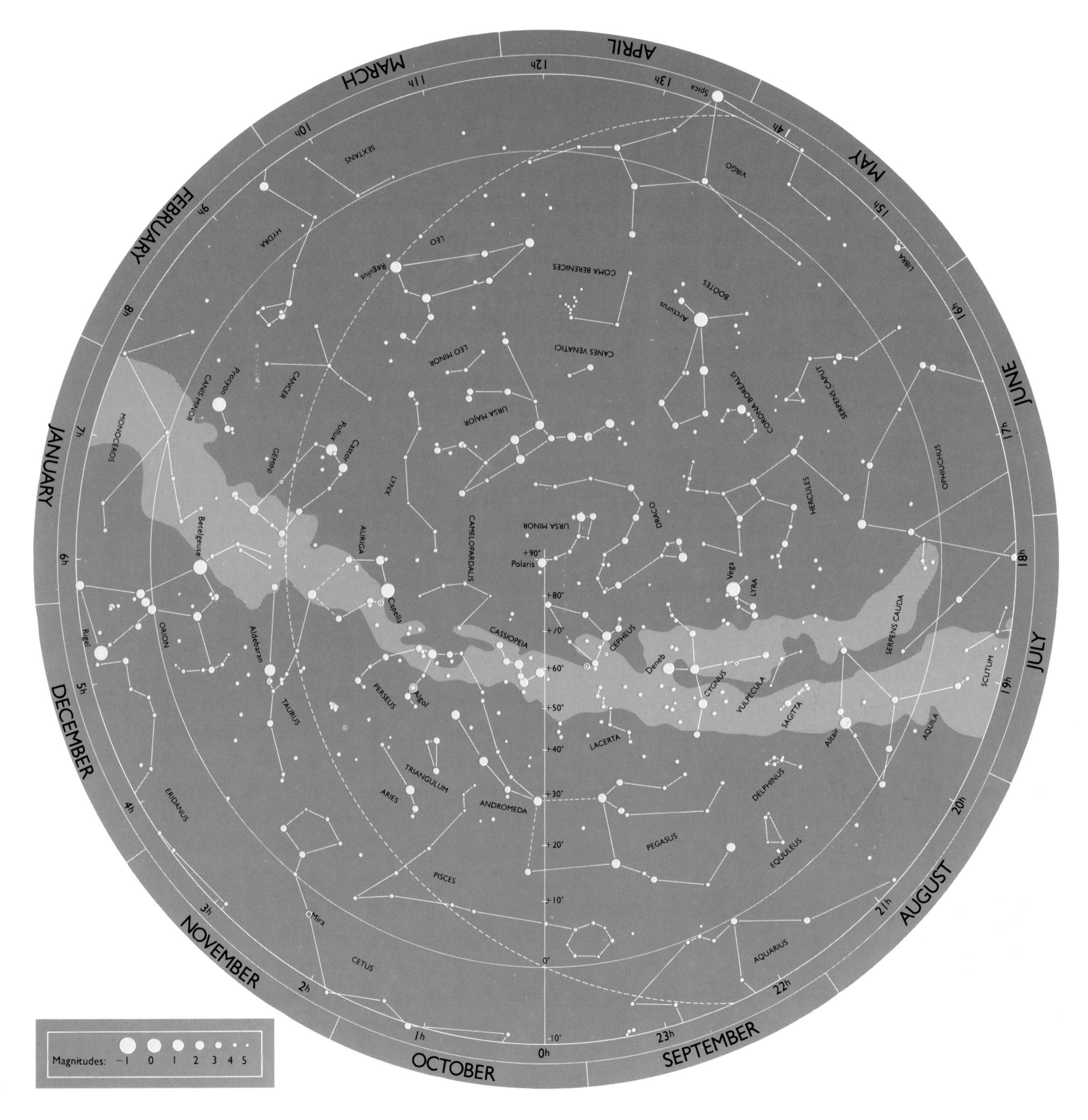

been unceremoniously cut up into its keel (Carina), poop (Puppis) and sails (Vela), because the original constellation was thought to be too unwieldy.

Ptolemy's constellations did not cover the entire sky. There were gaps between them, and inevitably these were filled. Later astronomers added new constellations, sometimes modifying the original boundaries. Later still, the stars of the far south had to be divided into constellations, and some of the names have a very modern flavour. The Telescope, the Microscope and the Air-pump are three of the more recent groups. Even the Southern Cross, Crux Australis, is a 17th-century constellation. It was formed by Royer in 1679, and so has no great claim to antiquity.

Many additional constellations have been proposed from time to time, but these have not been adopted, although one of the rejected groups – Quadrans, the Quadrant – is remembered in the name of the annual Quadrantid meteor shower.

The 19th-century astronomer Sir John Herschel said that the patterns of the constellations had been drawn up to be as inconvenient as possible. In 1933, modified constellation boundaries were laid down by the International Astronomical Union. There have been occasional attempts to revise the entire nomenclature, but it is unlikely any radical change will now be made. The present-day constellations have been accepted for too long to be altered.

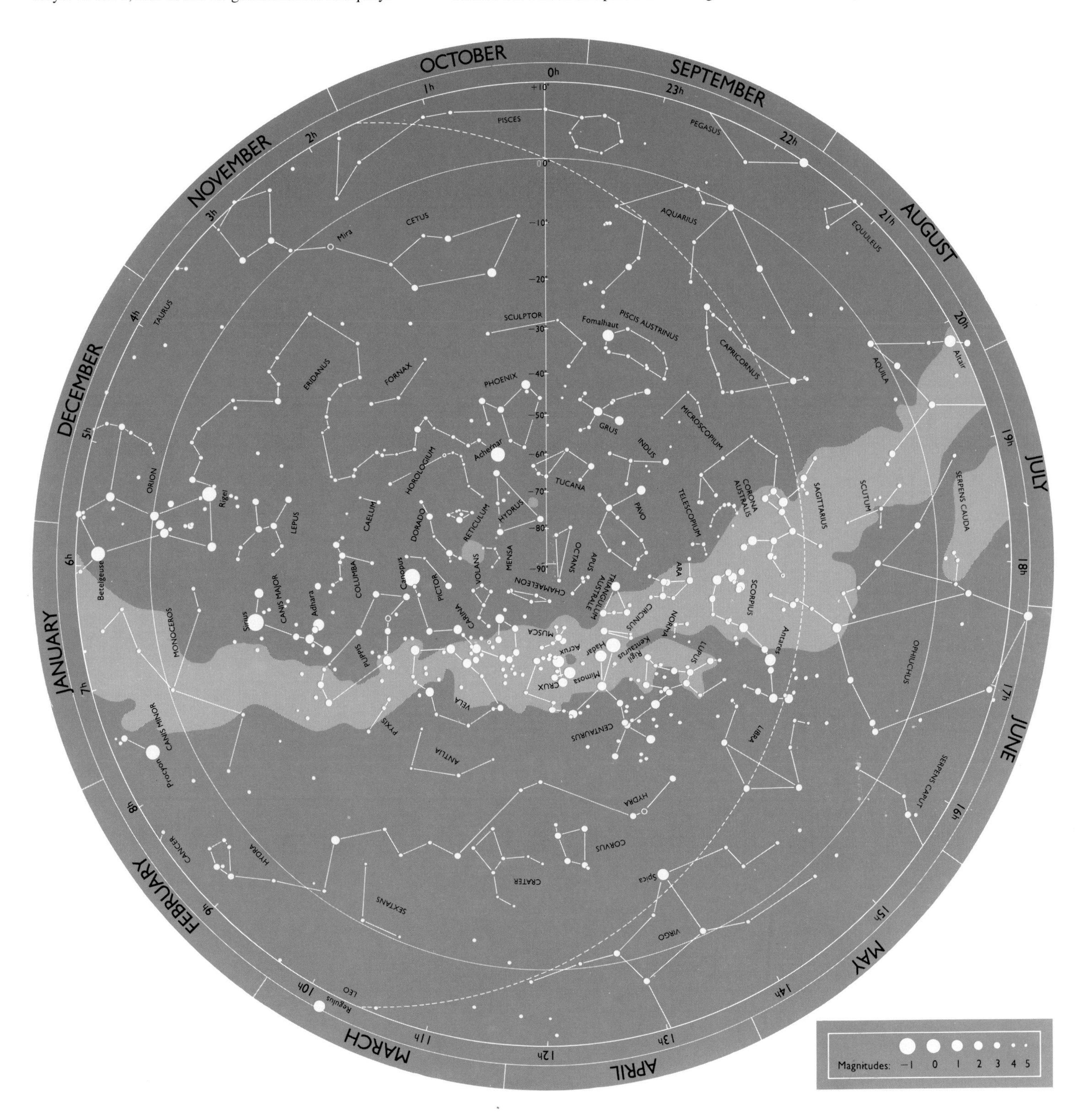

Seasonal Charts: North

▶ ***Latitudes*** *of the major cites of the northern hemisphere. For the observer, all the stars of the northern sky are visible in the course of a year, but he can see only a limited distance south of the equator. If his latitude is x°N, the most southerly point he can see in the sky is 90 – x°S. Thus, for example, to an observer at latitude 50°N, only the sky north of 90 – 50 (or 40°S) is ever visible.*

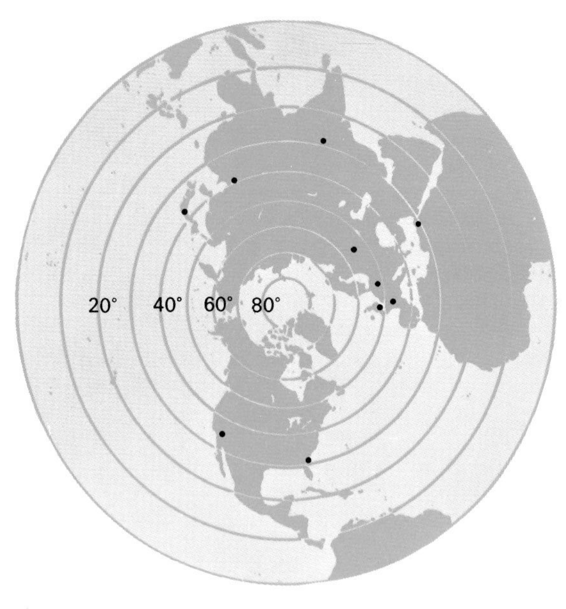

The charts given on this page are suitable for observers who live in the northern hemisphere, between latitudes 50 degrees and 30 degrees north. The horizon is given by the latitude marks near the bottom of the charts. Thus, for an observer who lives at 30 degrees north, the northern horizon in the first map will pass just above Deneb, which will be visible.

A star rises earlier, on average, by two hours a month; thus the chart for 2000 hours on 1 January will be valid for 1800 hours on 1 February and 2200 hours on 1 December.

The limiting visibility of a star for an observer at any latitude can be worked out from its declination. To an observer in the northern hemisphere, a star is at its lowest point in the sky when it is due north; a star which is 'below' the pole by the amount of one's latitude will touch the horizon when at its lowest point. If it is closer to the pole than that it will be circumpolar. From latitude 51 degrees north, for example, a star is circumpolar if its declination is 90 – 51 or 31 degrees north, or greater. Thus Capella, dec. +45 degrees 57 minutes, is circum-

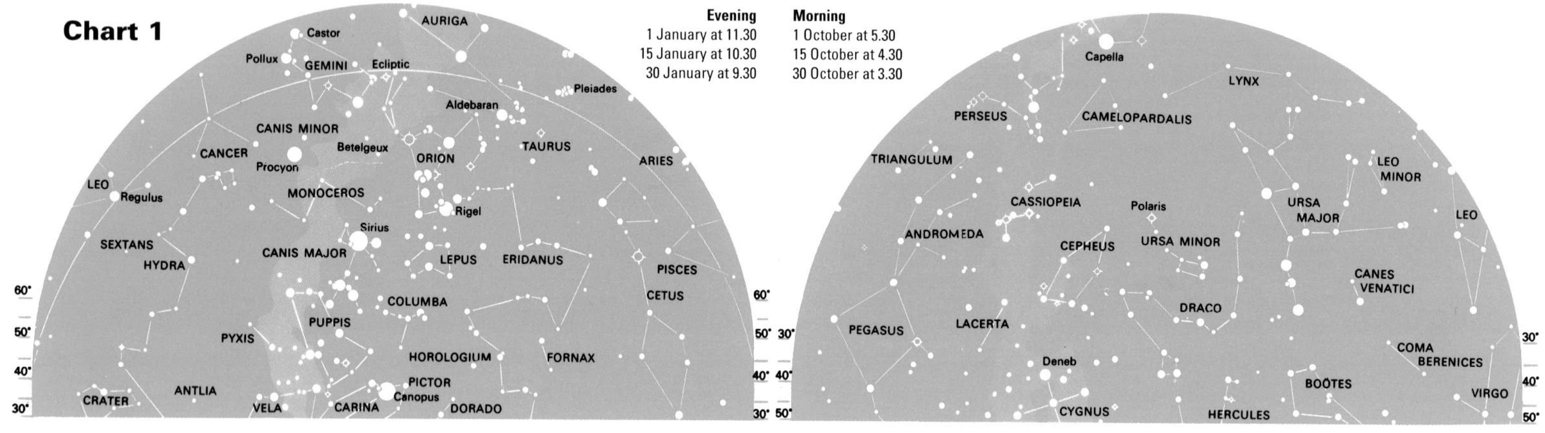

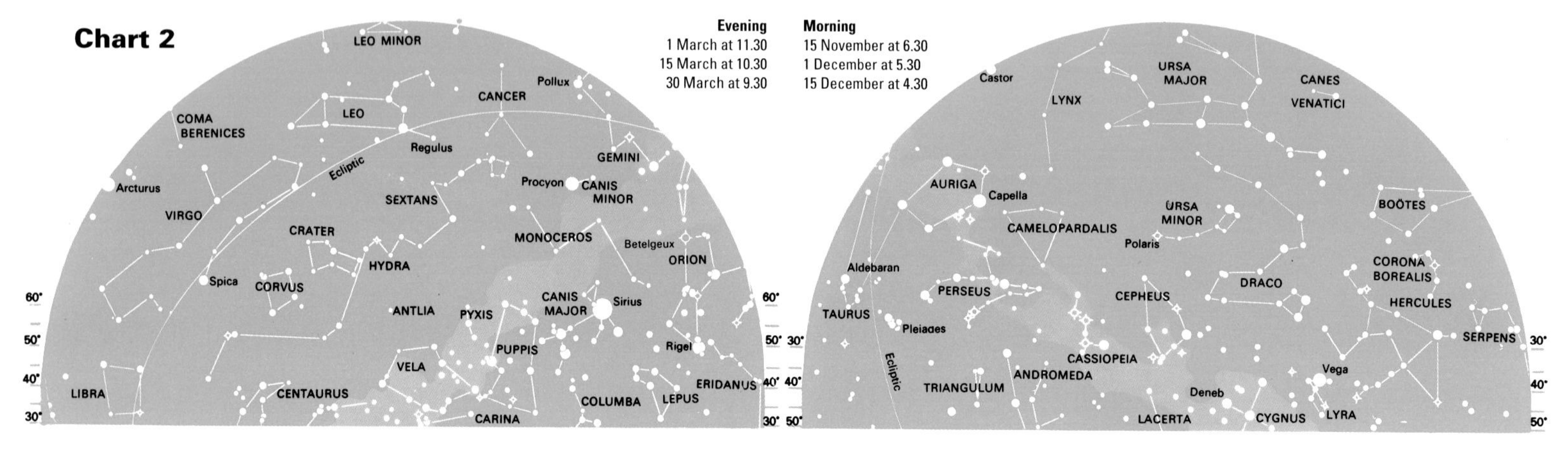

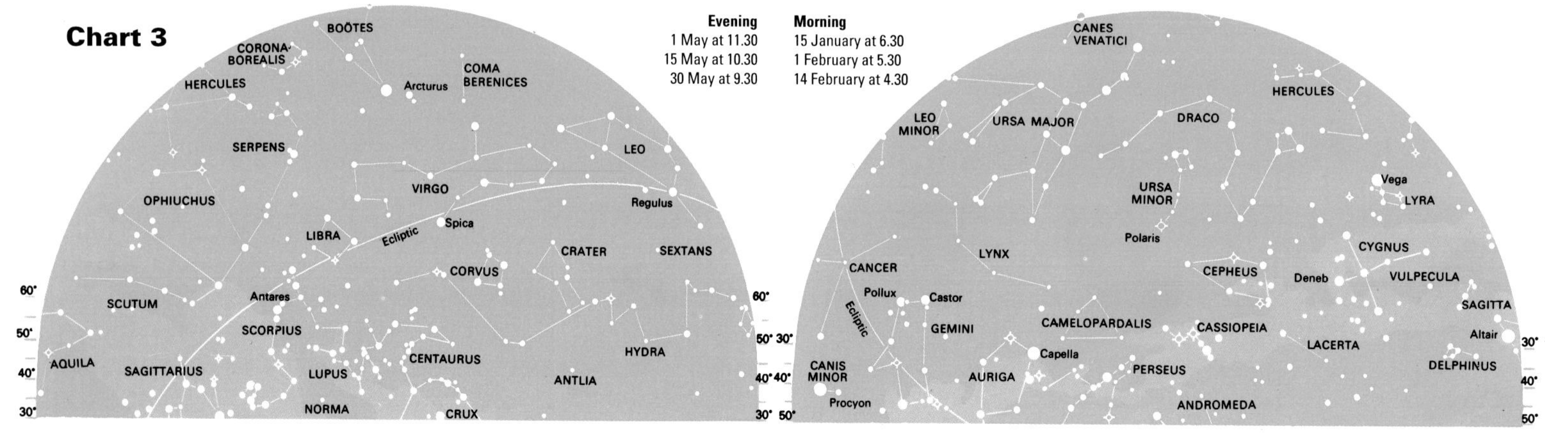

polar to an observer in London, Cologne or Calgary. A minor allowance must be made for atmospheric refraction.

Similarly, to an observer at latitude 51 degrees north, a star with a declination south of −39 degrees will never rise. Canopus lies at declination −52 degrees 42 minutes; therefore it is invisible from London, but can be seen from any latitude south of 37 degrees 20 minutes north, again neglecting the effects of refraction.

The charts given here show the northern (right) and the southern (left) aspects of the sky from the viewpoint of an observer in northern latitudes. They are self-explanatory; the descriptions given below apply in each case to the late evening, but more accurate calculations can be made by consulting the notes at the side of each chart.

Chart 1*. In winter, the southern aspect is dominated by Orion and its retinue. Capella is almost at the zenith or overhead point, and Sirius is at its best. Observers in Britain can see part of Puppis, but Canopus is too far south to be seen from any part of Europe. The sickle of Leo is very prominent in the east; Ursa Major is to the north-east, while Vega is at its lowest in the north. It is circumpolar from London but not from New York, and is not on the first chart.*

Chart 2*. In spring, Orion is still above the horizon until past midnight; Leo is high up, with Virgo to the east. Capella is descending in the north-west, Vega is rising in the north-east; these two stars are so nearly equal in apparent magnitude (0.1 and 0.0) that, in general, whichever is the higher in the sky will also seem the brighter. In the west, Aldebaran and the Pleiades are still visible.*

Charts 3–6*. In early summer (Chart 3), Orion has set and, to British observers, the southern aspect is relatively barren, but observers in more southerly latitudes can see Centaurus and its neighbours. During summer evenings (Chart 4), Vega is at the zenith and Capella low in the north; Antares is at its highest in the south. By early autumn (Chart 5), Aldebaran and the cluster of the Pleiades have reappeared, and the Square of Pegasus is conspicuous in the south, with Fomalhaut well placed. And by early winter (Chart 6), Orion is back in view, with Ursa Major lying low in the northern sky.*

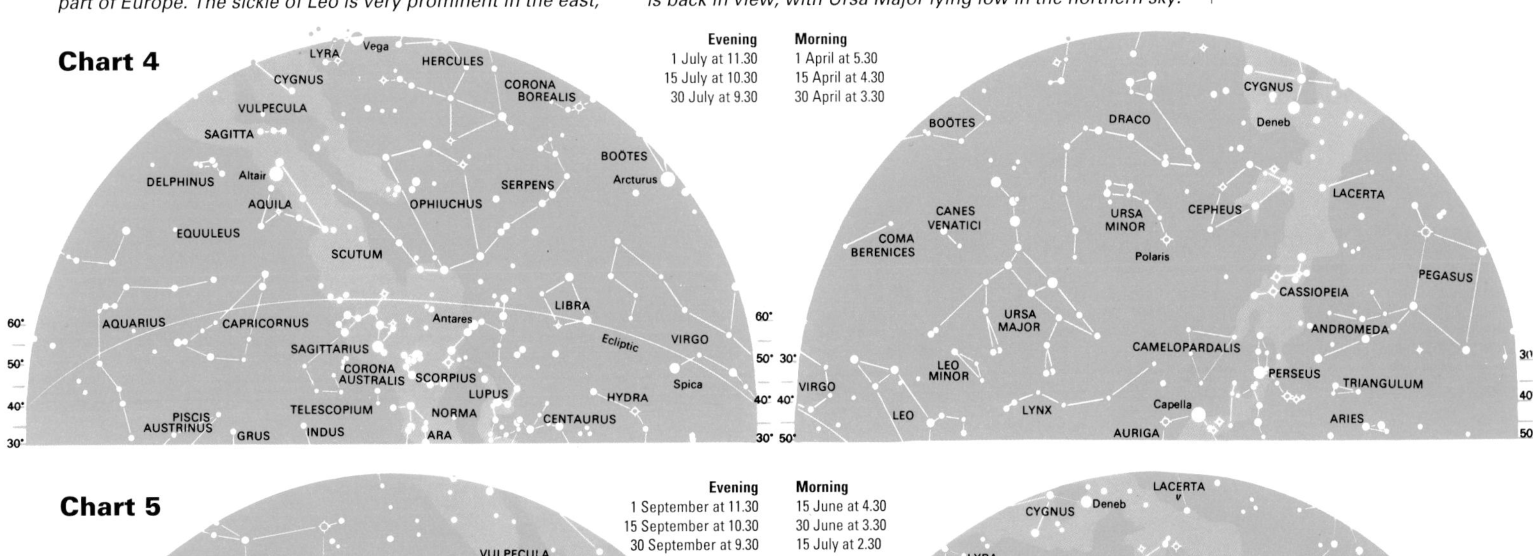

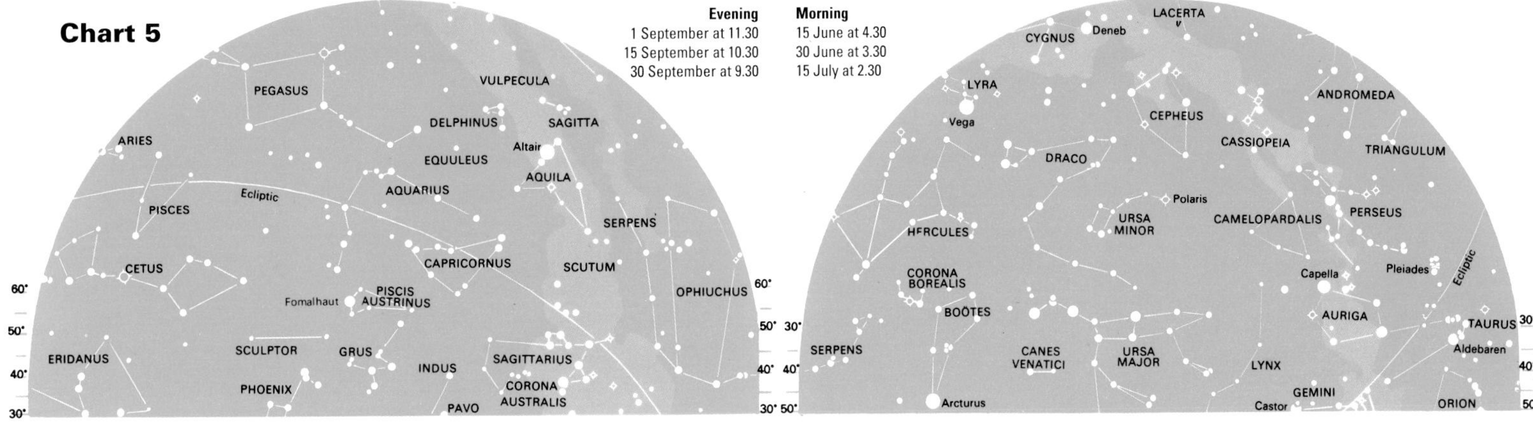

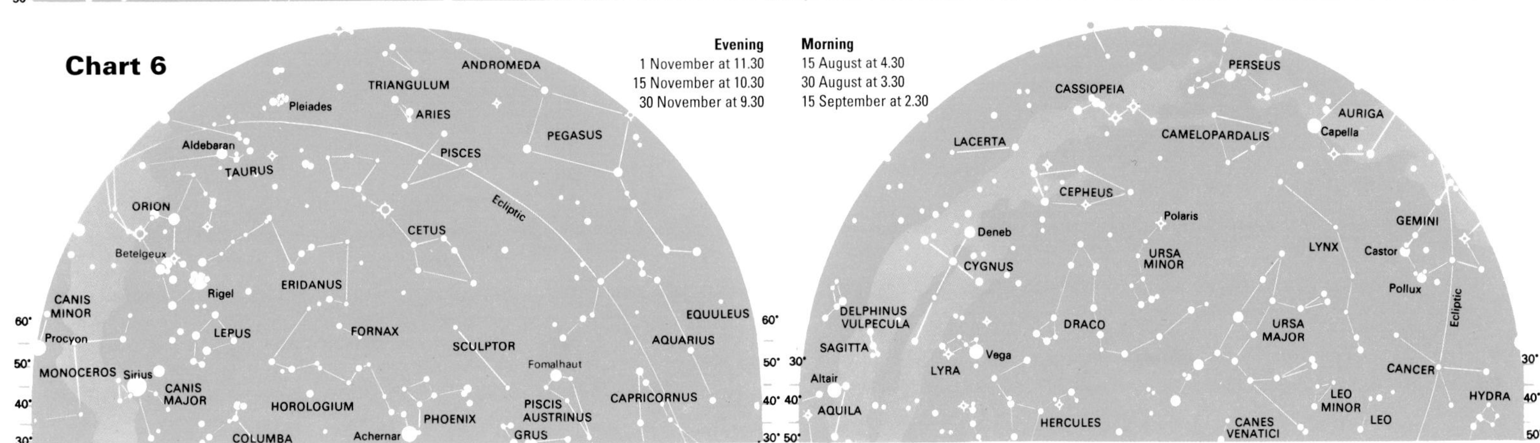

Seasonal Charts: South

Generally, the stars in the South Polar area of the sky are brighter than those of the far north, even though the actual Pole lies in a barren region, and there is no pattern of stars so distinctive as the Great Bear – apart from the Southern Cross, which covers a much smaller area. Canopus, the brightest star in the sky apart from Sirius, has a declination of some −53 degrees, and is not visible from Europe, but rises well above the horizon from Mexico, and from Australia and New Zealand it is visible for much of the year. In the far south, too, there are the Clouds of Magellan. They are prominent naked-eye objects, and the Large Cloud can be seen without optical aid even under conditions of full moonlight.

An observer at one of the Earth's poles would see one hemisphere of the sky only, and all the visible stars would be circumpolar. It is not even strictly correct to say that Orion is visible from the entire surface of the Earth. An observer at the South Pole would never see Betelgeux, whose declination is +7 degrees. From latitudes above −83 degrees (90 − 7) Betelgeux would never rise.

These charts may be used for almost all the densely populated regions of the southern hemisphere which lie between 15 and 35 degrees south. The northern view is given in the left chart, the southern in the right.

***Chart 1**. In January, the two most brilliant stars, Sirius and Canopus, are high up. Sirius seems appreciably the brighter of the two (magnitude −1.5 as against −0.8), but its eminence is due to its closeness rather than its real luminosity. It is an A-type Main Sequence star, only 26 times as luminous as the Sun; Canopus is an F-type supergiant, whose luminosity may be 200,000 times that of the Sun, according to one estimate, though both its distance and its luminosity are uncertain and estimates vary widely. Lower down, the Southern Cross is a prominent feature, and the brilliant pair of stars Alpha and Beta Centauri are also found in the same area. In the north, Capella is well above the horizon; Orion is not far from the zenith, and if the sky is clear a few stars of Ursa Major may be seen low over the northern horizon.*

***Chart 2**. In March, Canopus is descending in the south-west,*

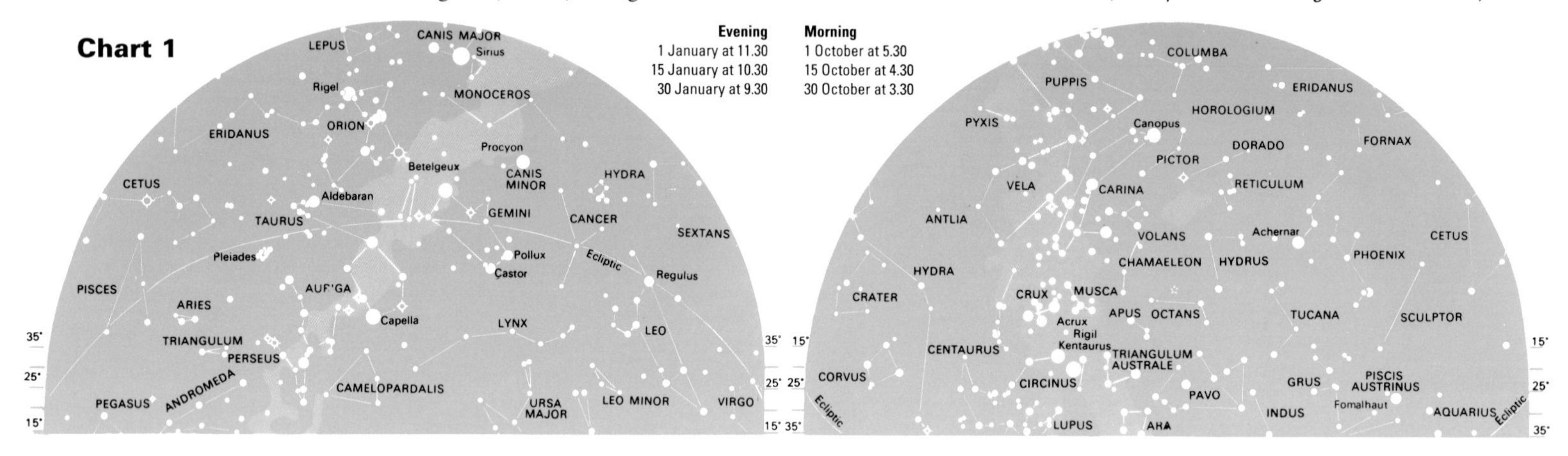

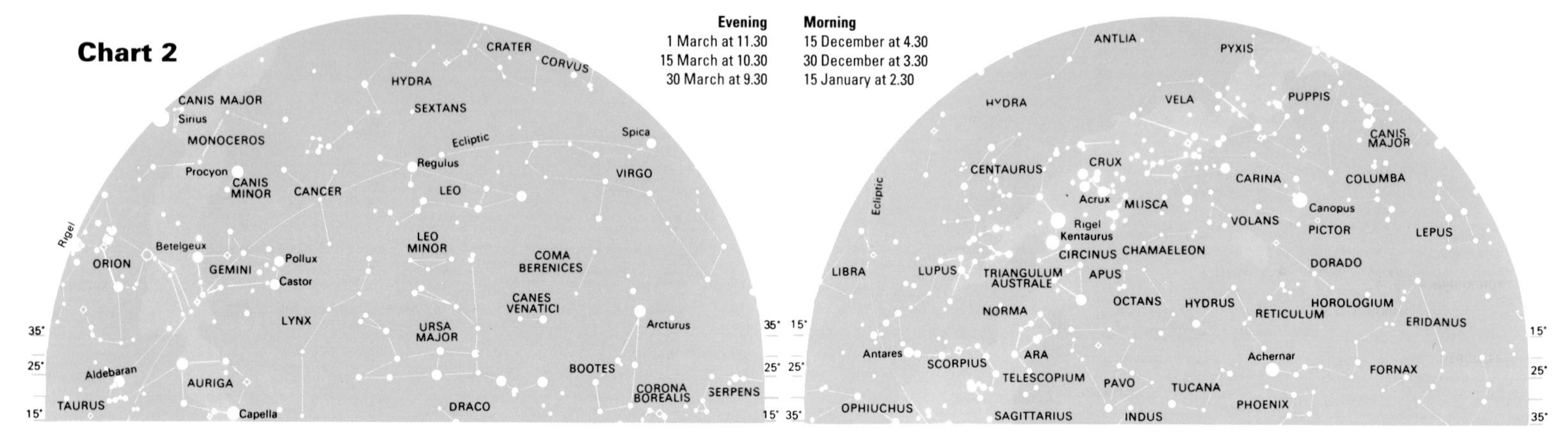

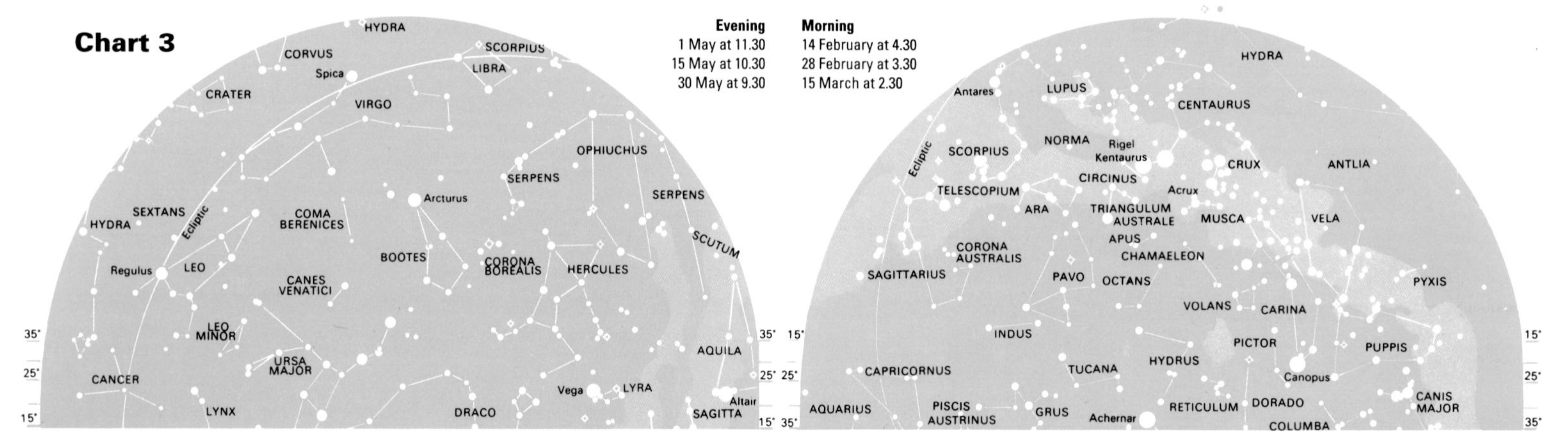

and Crux rising to its greatest altitude; the south-east is dominated by the brilliant groups of Scorpius and Centaurus. (Scorpius is a magnificent constellation. Its leading star, Antares, is well visible from Europe, but the 'tail' is too far south to be seen properly.) To the north, the Great Bear is seen; Orion is descending in the west.

Charts 3–4*. The May aspect (Chart 3) shows Alpha and Beta Centauri very high up, and Canopus in the south-west; Sirius and Orion have set, but Scorpius is brilliant in the south-east. In the north, Arcturus is prominent, with Spica in Virgo near the zenith. By July (Chart 4) Vega, Altair and Deneb are all conspicuous in the north. Arcturus is still high above the north-west horizon. Antares is not far from its zenith.*

Charts 5–6*. The September view (Chart 5) shows Pegasus in the north, and the 'W' of Cassiopeia is above the horizon. The Southern Cross is almost at its lowest. By November (Chart 6) Sirius and Canopus are back in view; Alpha and Beta Centauri graze the horizon, and the region of the zenith is occupied by large, comparatively barren groups such as Cetus and Eridanus.*

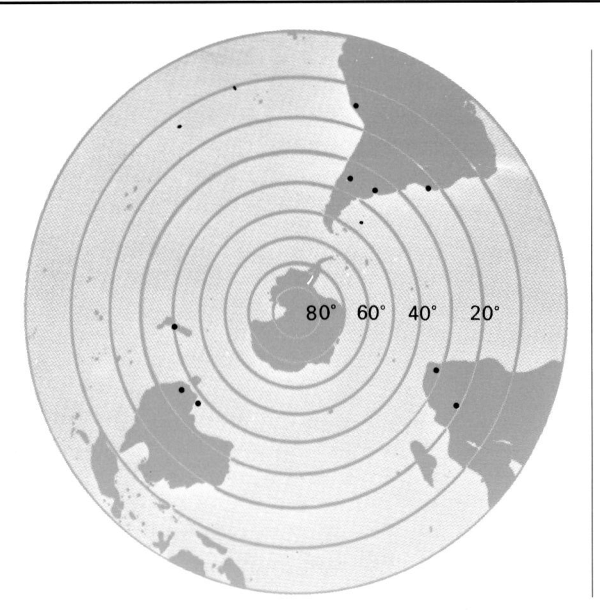

◀ ***For the observer*** *in the southern hemisphere all the stars of the southern sky are visible in the course of a year, but he can only see a limited distance north of the celestial equator. If his latitude is x°S, the most northerly point he can see is 90 – x°N. Thus, for example, to an observer at latitude 50°S only the sky south of 90 – 50 (or 40°N) is ever visible.*

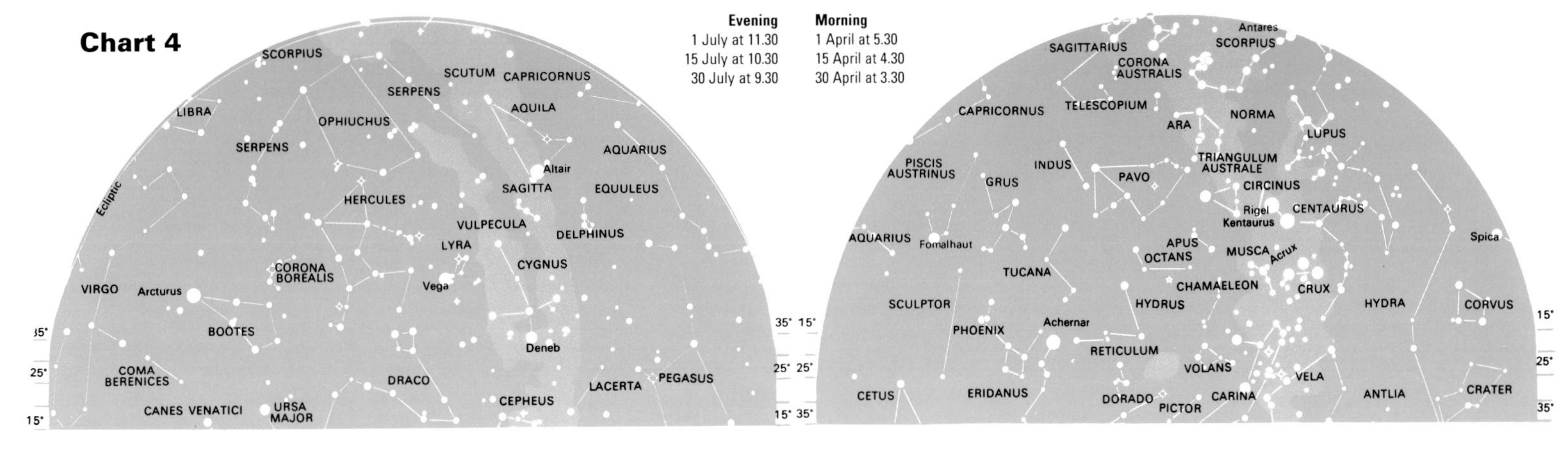

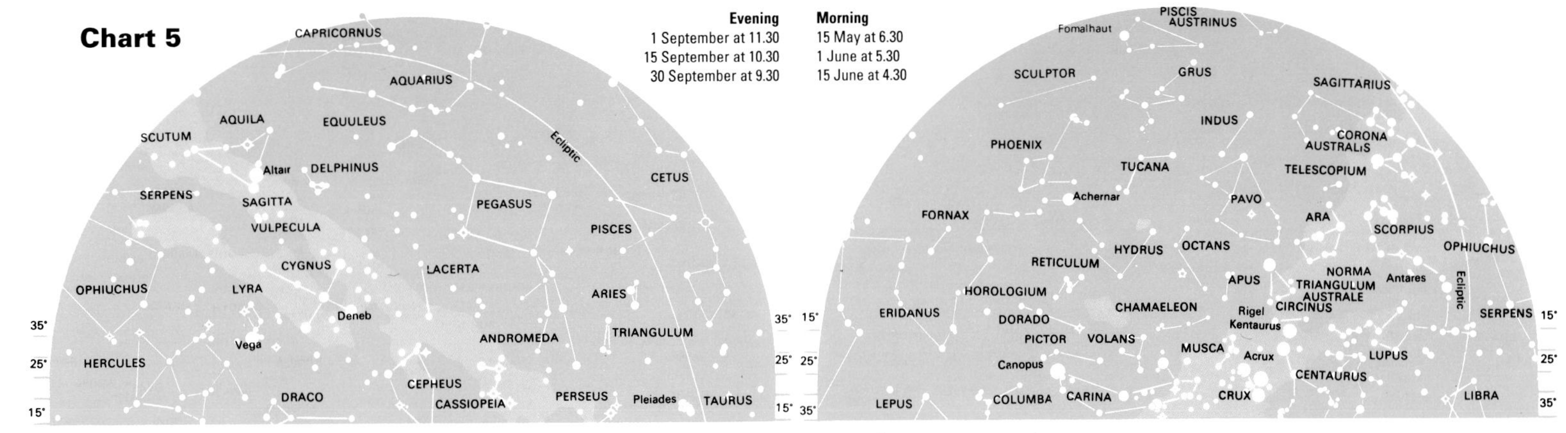

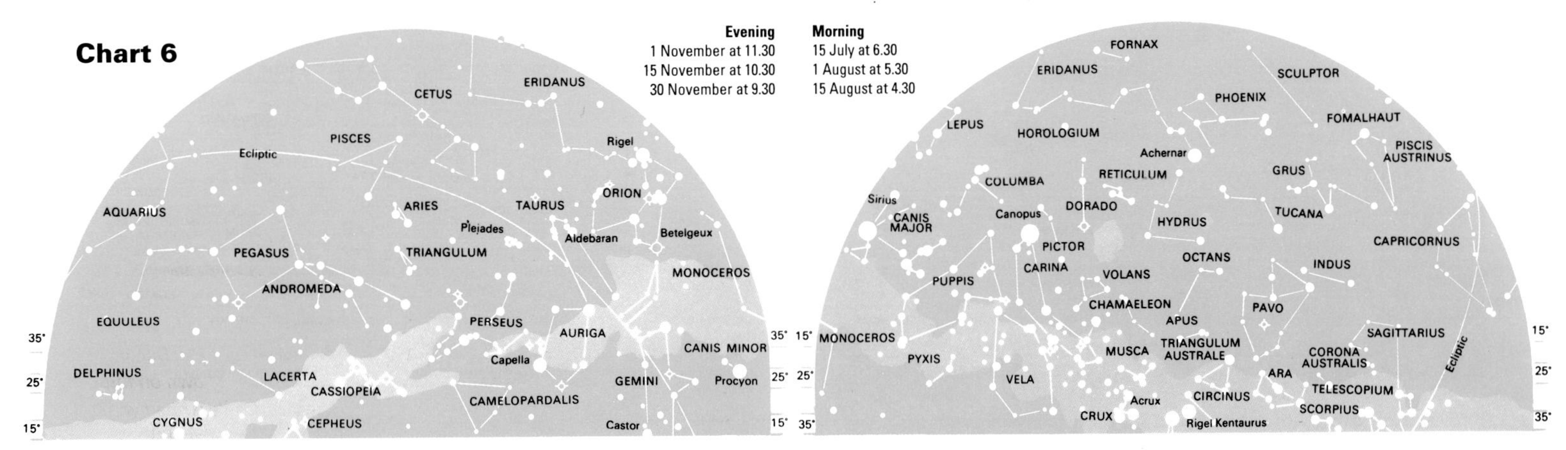

Ursa Major, Canes Venatici, Leo Minor

Ursa Major. There can be few people who cannot recognize the Plough – alternatively nicknamed King Charles' Wain or, in America, the Big Dipper. The seven main stars make up an unmistakable pattern; but in fact only five of them share a common motion in space and presumably have a common origin; the remaining two – Dubhe and Alkaid – are moving through space in the opposite direction, so that after a sufficient length of time the plough-shape will become distorted. Of the seven, six are hot and white, but Dubhe is obviously orange; the colour is detectable with the naked eye, and binoculars bring it out well.

It is interesting that Megrez (δ Ursae Majoris) is about a magnitude fainter than the rest. In 1603 Bayer, who drew up a famous star catalogue and gave the stars their Greek letters, gave its magnitude as 2; but earlier cataloguers ranked it as 3, and there has probably been no real change. It is 65 light-years away, and 17 times as luminous as the Sun.

Of course the most celebrated star in the Great Bear is ζ (Mizar) with its naked-eye companion Alcor. Strangely, the Arabs of a thousand years ago regarded Alcor as a test of keen eyesight, but today anyone with average eyes can see it when the sky is reasonably dark and clear. A small telescope will show that Mizar itself is double, but the separation (14.4 seconds of arc) is too small for the two stars to be seen separately with the naked eye, or even binoculars. Between Alcor and the two Mizars is a fainter star which was named Sidus Ludovicianum in 1723 by courtiers of Emperor Ludwig V, who believed that it had appeared suddenly. Ludwig's Star can be seen with powerful binoculars, and it has been suggested that it might have been the 'test' referred to by the Arabs, but certainly it would have been a very severe one – even if the star is slightly variable.

Outside the Plough pattern is a triangle of fainter stars: ψ, λ and μ. The two latter stars are in the same binocular field, and make a good colour contrast. λ is white, while μ, with its M-type spectrum, is very red.

ξ Ursae Majoris, close to ν, was one of the first binary stars to have its orbit computed. The components are equal at magnitude 4.8, and the period is 59.8 years; but the separation is currently only about 1 arc second, so that a very small telescope will not split the pair. (Generally speaking, a 3-inch or 7.6-centimetre refractor will be able to divide pairs down to a separation of about 1.8 arc seconds, assuming that the components are more or less equal and are not too faint.) There are not many notable variables in Ursa Major, but the red semi-regular Z, easy to find because of its closeness to Megrez, is a favourite test subject for newcomers to variable-star work.

There are four Messier objects in the constellation. One of these, M97, is the famous Owl planetary nebula. It was discovered by Pierre Méchain in 1781, who recorded it as being 'difficult to see', and certainly it can be elusive; the two embedded stars which give it its owlish appearance are no brighter than magnitude 14, and the whole nebula is faint. It lies not far from β or Merak, and it can be seen with a 7.6-centimetre (3-inch) telescope when the sky is dark and clear. M81 and M82 are within binocular range, not far from 24; M81 is a spiral, while M82 is a peculiar system which is a strong radio source. Each is about 8.5 million light-years away, and they are associated with each other. The other Messier object, M101, was also discovered by Méchain; it forms an equilateral triangle with Mizar and Alkaid (ζ and η Ursae Majoris), and is a loose spiral whose surface brightness is rather low. It is face-on to us, and photographs can often show it beautifully.

Though all the main stars of Ursa Major are well below the first magnitude, their proper names are often used. There are, incidentally, two alternatives; η may be called Benetnasch as well as Alkaid, while γ is also known as Phekda or Phecda.

Magnitudes
−1
0
1
2
3
4
5
Variable star
Galaxy
Planetary nebula
Gaseous nebula
Globular cluster
Open cluster

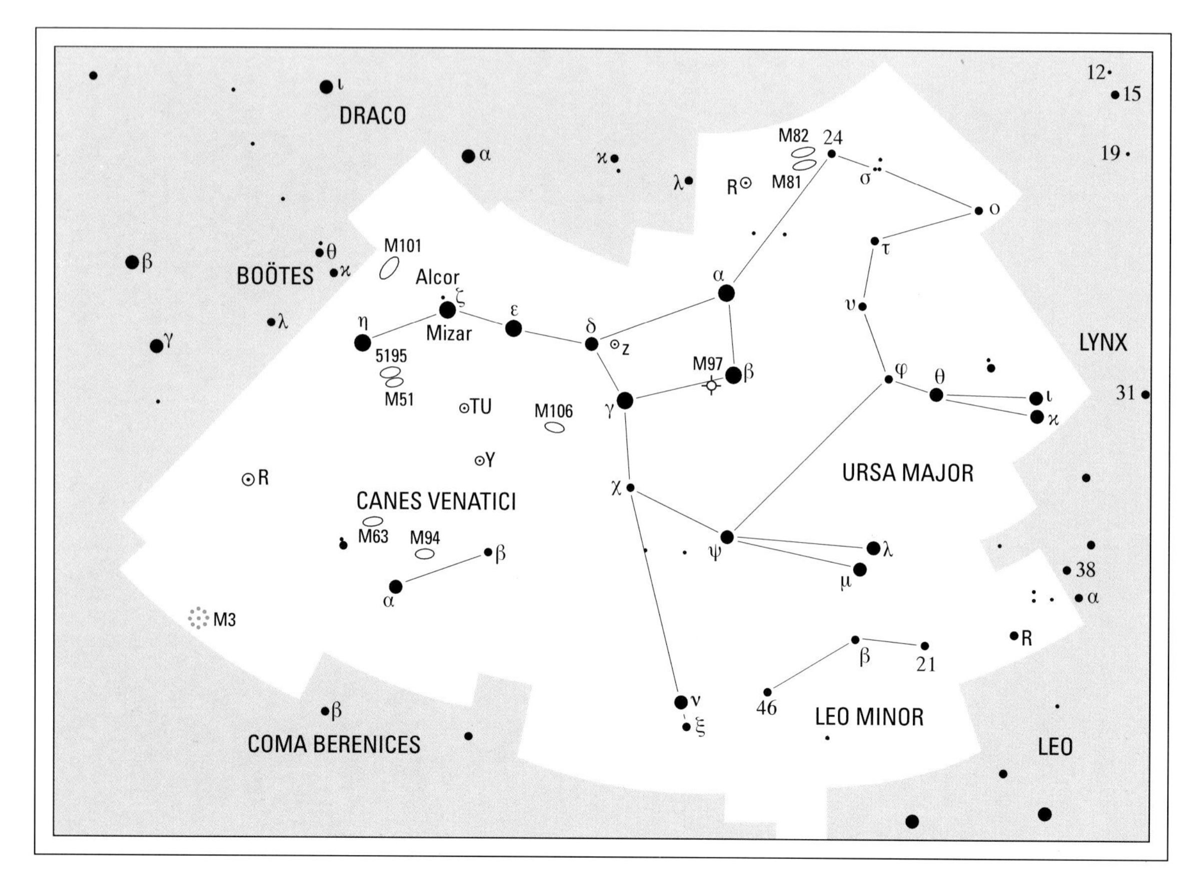

◀ ***Ursa Major** is the most famous of all the northern constellations, and can be used as a guide to find many of the less prominent groups. The 'Plough' is only part of the entire constellation, but its seven main stars cannot be mistaken; they are circumpolar over the British Isles and parts of Europe and North America. They are always low over South Africa and Australia, but only from parts of New Zealand is the 'Plough' completely lost. Canes Venatici and Leo Minor adjoin Ursa Major; Lynx is also shown on this map, but is described with Auriga (Star Map 18).*

Canes Venatici, the Hunting Dogs – Asterion and Chara – were added to the sky by Hevelius in 1690; they are held by the herdsman Boötes – possibly to stop them from chasing the Bears round the celestial pole. The only bright star, α^2, was named Cor Caroli (Charles' Heart) by the second Astronomer Royal, Edmond Halley, in honour of King Charles I of England. The star shows interesting periodical changes in its spectrum, due probably to variations in its magnetic field. It is 65 light-years away, and 80 times as luminous as the Sun. Its companion, of magnitude 5.5, lies at a separation of over 19 seconds of arc, so that this is a very easy pair.

The semi-regular variable Y Canum Venaticorum lies about midway between Mizar and β Canum Venaticorum. It is one of the reddest stars known, and has been named La Superba; at maximum it is visible with the naked eye, but binoculars are needed to bring out its vivid colour.

M51, the Whirlpool, lies near the border of Canes Venatici, less than four degrees from Alkaid in the Great Bear. It was discovered by Messier himself in 1773, and is the perfect example of a face-on spiral; its distance is around 37 million light-years. It was the first spiral to be seen as such, by Lord Rosse in 1845.

Though a difficult binocular object, a modest telescope – say a 30-centimetre (12-inch) – is adequate to show its form; it is linked with its companion, NGC5195. M94, not far from Cor Caroli, is also a face-on spiral, and though it is small it is not difficult to find, because its nucleus is bright and distinct.

The other Messier spirals in Canes Venatici, M63 and M106, are less striking. M63 is also a spiral, but the arms are much less obvious. And M106 – added later to the Messier catalogue – has one arm which is within range of a 25-centimetre reflector.

M3 is one of the most splendid globular clusters in the sky. It lies almost midway between Cor Caroli and Arcturus, near the fainter star Beta Comae (magnitude 4.6) and is easy to find with binoculars, while it can be partly resolved into stars with a telescope of more than 8-centimetre aperture. Like all globular clusters, it is a very long way away – over 48,000 light-years – and is particularly rich in RR Lyrae variables. The total mass has been given as around 245,000 times that of the Sun. Not surprisingly, it is a favourite target for amateur astrophotographers. The integrated magnitude is about 6.4, so that it is not very far below naked-eye visibility; Messier discovered it in the year 1764.

Leo Minor is a small constellation with very dubious claims to a separate identity; it was first shown by Hevelius, on his maps of 1690. The system of allotting Greek letters has gone badly wrong here, and the only star so honoured is β, which is not even the brightest star in the group. The leader is 46, which has been given a separate name: Praecipua. It is in the same binocular field as ν and ξ Ursae Majoris, and can be identified by its decidedly orange colour. The only object of any interest is the Mira-type variable R Leonis Minoris, which can reach magnitude 6.3 at maximum, but sinks to below 13 when at its faintest.

Hevelius had a habit of creating new constellations. Some of these have survived; as well as Leo Minor, there are Camelopardalis, Canes Venatici, Lacerta, Lynx, Scutum, Monoceros, Sextans and Vulpecula, while others, such as Triangulum Minor (the Little Triangle) and Cerberus (Pluto's three-headed dog), have now been rejected. The constellation Leo Minor – which has no mythological significance – was formerly included in Ursa Major, and logically should probably have remained there.

URSA MAJOR

BRIGHTEST STARS

No.	Star	R.A. h m s	Dec. ° ′ ″	Mag.	Spectrum	Proper name
77	ε	12 54 02	+55 57 35	1.77	A0	Alioth
50	α	11 03 44	+61 45 03	1.79	K0	Dubhe
85	η	13 47 32	+49 18 48	1.86	B3	Alkaid
79	ζ	13 23 56	+54 55 31	2.09	A0	Mizar
48	β	11 01 50	+56 22 56	2.37	A1	Merak
64	γ	11 53 50	+53 41 41	2.44	A0	Phad
52	ψ	11 09 40	+44 29 54	3.01	K1	
34	μ	10 22 20	+41 29 58	3.05	M0	Tania Australis
9	ι	08 59 12	+48 02 29	3.14	A7	Talita
25	θ	09 32 51	+51 40 38	3.17	F6	
69	δ	12 15 25	+57 01 57	3.31	A3	Megrez
1	ο	08 30 16	+60 43 05	3.36	G4	Muscida
33	λ	10 17 06	+42 54 52	3.45	A2	Tania Borealis
54	ν	11 18 29	+33 05 39	3.48	K3	Alula Borealis

Also above mag. 4.3: κ (A1 Kaprah) (3.60), h (3.67), χ (Alkafzah) (3.71), ξ (3.79), 10 (4.01).

VARIABLES

Star	R.A. h m	Dec. ° ′	Range (mags)	Type	Period (d)	Spectrum
R	10 44.6	+68 47	6.7–13.4	Mira	302	M
Z	11 56.5	+57 52	6.8–9.1	Semi-reg.	196	M

DOUBLES

Star	R.A. h m	Dec. ° ′	P.A. °	Sep. ″	Mags	
ν	11 18.5	+33 06	147	7.2	3.5, 9.9	
ζ	13 23.9	+54 56	AB 152	14.4	2.3, 4.0	Mizar/Alcor
			AC 071	708.7	2.1, 4.0	

CLUSTERS AND NEBULAE

M	NGC	R.A. h m	Dec. ° ′	Mag.	Dimensions ′	Type
81	3031	09 55.6	+69 04	6.9	25.7 × 14.1	Sb galaxy
82	3034	09 55.8	+69 41	8.4	11.2 × 4.6	Peculiar galaxy
97	3587	11 14.8	+55 01	12	194″	Planetary (Owl Nebula)
101	5457	14 03.2	+54 21	7.7	26.9 × 26.3	Sc galaxy

CANES VENATICI

BRIGHTEST STARS

No.	Star	R.A. h m s	Dec. ° ′ ″	Mag.	Spectrum	Proper name
12	α^2	12 56 02	+38 19 06	2.90	A0p	Cor Caroli

Also above mag. 4.3: β (Chara) (4.26).

VARIABLES

Star	R.A. h m	Dec. ° ′	Range (mags)	Type	Period (d)	Spectrum
R	13 49.0	+39 33	6.5–12.9	Mira	329	M
TU	12 54.9	+47 12	5.6–6.6	Semi-reg.	50	M
Y	12 45.1	+45 26	4.8–6.6	Semi-reg.	157	N

DOUBLE

Star	R.A. h m	Dec. ° ′	P.A. °	Sep. ″	Mags
α^2	12 56.0	+38 19	22.9	19.4	2.9,5.5

CLUSTERS AND NEBULAE

M	NGC	R.A. h m	Dec. ° ′	Mag.	Dimensions ′	Type
3	5272	13 42.2	+28 23	6.4	16.2	Globular cluster
51	5195	13 29.9	+47 12	8.4	11.0 × 7.8	Sc galaxy (Whirlpool)
63	5055	13 15.8	+42 02	8.6	12.3 × 7.6	Sb galaxy
94	4736	12 50.9	+41 07	8.2	11.0 × 9.1	Sb galaxy
106	4258	12 19.0	+47 18	8.3	18.2 × 7.9	Sb galaxy
	5195	13 30.0	+47 16	9.6	5.4 × 4.3	Companion to M51

LEO MINOR

The brightest star is 46 (Præcipua), R.A. 10h 53m, dec. +34° 13′, mag. 3.83. Also above mag. 4.3: β (4.21).

VARIABLE

Star	R.A. h m	Dec. ° ′	Range (mags)	Type	Period (d)	Spectrum
R	09 45.6	+34 31	6.3–13.2	Mira	372	M

Ursa Minor, Draco

Magnitudes
−1
0
1
2
3
4
5
Variable star
Galaxy
Planetary nebula
Gaseous nebula
Globular cluster
Open cluster

Ursa Minor, the Little Bear is notable chiefly because it contains the north celestial pole, now marked within one degree by the second-magnitude star α (Polaris). At present it is moving even closer to the pole, and will be at its nearest (within 28 minutes 31 seconds) in the year 2102. Navigators have found it very useful indeed, because to find one's latitude on the surface of the Earth all that has to be done is to measure the height of Polaris above the horizon and then make a minor correction. (Southern-hemisphere navigators are not so lucky; their pole star, σ Octantis, is very faint indeed.) As a matter of interest, the actual pole lies almost along a line connecting Polaris with Alkaid in the tail of the Great Bear.

Polaris itself was known to the early Greeks as 'Phoenice', and another name for it, current during the 16th and 17th centuries, was Cynosura. It is of spectral type F8, so that in theory it should look slightly yellowish, but most observers will certainly call it white. The ninth-magnitude companion, lying at a distance of over 18 seconds of arc, is by no means a difficult object; it was discovered in 1780 by William Herschel, and is said to have been glimpsed with a 5-centimetre (2-inch) telescope, though at least a 7.6-centimetre (3-inch) instrument is needed to show it clearly. Polaris lies at a distance of 680 light-years. It is a powerful star, about 6000 times as luminous as the Sun.

The only other reasonably bright star in Ursa Minor is β (Kocab), which is very different to Polaris; it is of type K, and its orange colour is evident even with the naked eye. It is 29 light-years from us, and equal to 95 Suns. Kocab and its neighbour γ (Perkad Major) are often called 'the Guardians of the Pole'. The rest of the Little Bear pattern is very dim, and any mist or moonlight will drown it. Neither are there any other objects of immediate interest.

Draco, the Dragon, is a large constellation, covering more than 1000 square degrees of the sky, but it contains no really bright stars. It is not difficult to trace. Beginning more or less between the Pointers and Polaris, it winds its way around Ursa Minor, extending up to Cepheus and then towards Lyra; the 'head', not far from Vega, is the most prominent part of the constellation, and is made up of γ (Eltamin), β, ν and ξ. ν is a particularly wide, easy double, with equal components; really keen-sighted people claim to be able to split it with the naked eye, and certainly it is very evident with binoculars. The two are genuinely associated, and share a common motion through space, but the real separation between them is of the order of 350,000 million kilometres. Each component is about 11 times as luminous as the Sun.

Eltamin is an ordinary orange star, 100 light-years away and 107 times as luminous as the Sun, but it has a place in scientific history because of observations made of it in 1725–6 by James Bradley, later to become Astronomer Royal. Bradley was attempting to measure stellar parallaxes, and Eltamin was a suitable target because it passed directly over Kew, in Outer London, where Bradley had his observatory. He found that there was indeed a displacement, but was too large to be put down to parallax – and this led him on to the discovery of the aberration of light, which is an apparent displacement of a stationary object when observed from a moving one.

ε Draconis, close to the rather brighter δ, is an easy double. The primary was once suspected of being variable between magnitudes 3¾ and 4¾, but this has not been confirmed. The spectral type is G8. σ Draconis or Alrakis, magnitude 4.68, is one of the closest of the naked-eye stars; its distance from us is less than 19 light-years. It is a K-type dwarf, much less luminous than the Sun.

α Draconis (Thuban) was the north pole star at the time when the Pyramids were built. Since then the pole has shifted out of Draco into Ursa Minor; in the future it will migrate through Cepheus and Cygnus, reaching Lyra in 12,000 years from now – though Vega will never be as

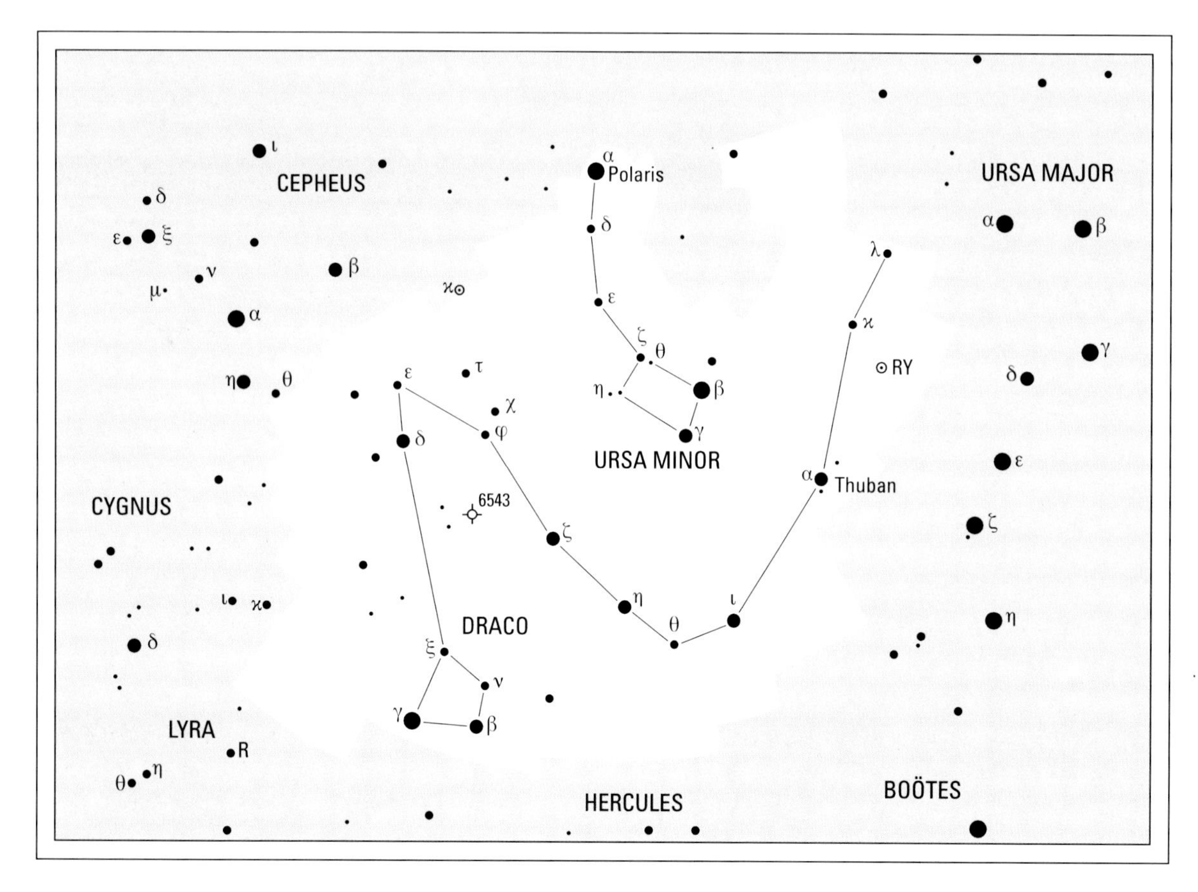

◀ ***The north celestial pole*** *is marked within one degree by Polaris in Ursa Minor. All the constellations shown here are circumpolar from Britain and much of Europe and North America. Polaris can be identified by using the 'Pointers', Merak and Dubhe, as guides; Draco sprawls from the region near the Pointers almost as far as Vega. Lyra is shown here, but described in Star Map 8.*

close to the pole as Polaris is at present. The pole will then pass through Hercules, returning to Draco and again passing close to Thuban.

Though Thuban has been given the Greek letter α, it is not the brightest star in the constellation; it is well over a magnitude fainter than γ. William Herschel believed that it had faded in historic times, and certainly both Tycho Brahe and Bayer ranked Thuban as of the second magnitude but, all in all, it is not likely that there has been any real change. The distance is 230 light-years, and the luminosity 150 times that of the Sun.

The most interesting nebular object in Draco is NGC6543, which lies almost midway between δ and ζ. It is a small but fairly bright planetary nebula, with a central star of magnitude 9.6. With a small telescope it has been described as looking like 'a luminous disk, resembling a star out of focus', and many observers have claimed that it shows a bluish colour. It was the first nebular object to be examined spectroscopically – by William Huggins in 1864. At once Huggins saw that the spectrum was of the emission type, so that it could not possibly be made up of stars. The real diameter is about one-third of a light-year; the central star is particularly hot, with a surface temperature of around 35,000 degrees C. The distance has been given as 3200 light-years.

Draco is one of the original constellations. In mythology it has been said to honour the dragon which guarded the golden apples in the Garden of the Hesperides, though it has also been said to represent the dragon which was killed by the hero Cadmus before the founding of the city of Boeotia.

Draco is also one of the largest of the constellations; it covers 1083 square degrees of the sky, and there are not very many constellations larger than that. And though Draco contains no brilliant stars, it is easy enough to identify. From Britain and similar northern latitudes it is, of course, circumpolar.

URSA MINOR

BRIGHTEST STARS

No.	Star	R.A. h m s	Dec. ° ′ ″	Mag.	Spectrum	Proper name
1	α	02 31 50	+89 15 51	1.99	K0	Polaris
7	β	14 50 42	+74 09 19	2.08	K4	Kocab
13	γ	15 20 44	+71 50 02	3.05	A3	Pherkad Major

Also above mag. 4.3: ε (4.23), 5 (4.25). The other stars of the 'Little Dipper' are ζ (Alifa) (4.32), δ (Yildun) (4.36) and η (Alasco) (4.95).

DOUBLE

Star	R.A. h m	Dec. ° ′	P.A. °	Sep. ″	Mags
α	02 31.8	+89 16	218	18.4	2.0, 9.0

DRACO

BRIGHTEST STARS

No.	Star	R.A. h m s	Dec. ° ′ ″	Mag.	Spectrum	Proper name
33	γ	17 56 36	+51 29 20	2.23	K5	Eltamin
14	η	16 23 59	+61 30 50	2.74	G8	Aldhibain
23	β	17 30 26	+52 18 05	2.79	G2	Alwaid
57	δ	19 12 33	+67 39 41	3.07	G9	Taïs
22	ζ	17 08 47	+65 42 53	3.17	B6	Aldhibah
12	ι	15 24 56	+58 57 58	3.29	K2	Edasich

Also above mag. 4.3: χ (3.57), α (Thuban) (3.65), ξ (Tuza) (3.75), ε (Tyl) (3.83), λ (Giansar) (3.84), κ (3.87), θ (4.01) and φ (4.22)

VARIABLES

Star	R.A. h m	Dec. ° ′	Range (mags)	Type	Period (d)	Spectrum
RY	12 56.4	+66 00	5.6–8.0	Semi-reg.	173	N

DOUBLES

Star	R.A. h m	Dec. ° ′	P.A. °	Sep. ″	Mags	
η	16 24.0	+61 31	142	5.2	2.7, 8.7	
ν	17 32.2	+55 11	312	61.9	4.9, 4.9	Binocular pair
ψ	17 41.9	+72 09	015	30.3	4.9, 6.1	
ε	19 48.2	+70 16	016	3.1	3.8, 7.4	

CLUSTERS AND NEBULAE

M	NGC	R.A. h m	Dec. ° ′	Mag.	Dimensions ″	Type
	6543	7 58.7	+66 38	8.8	18 × 350	Planetary nebula

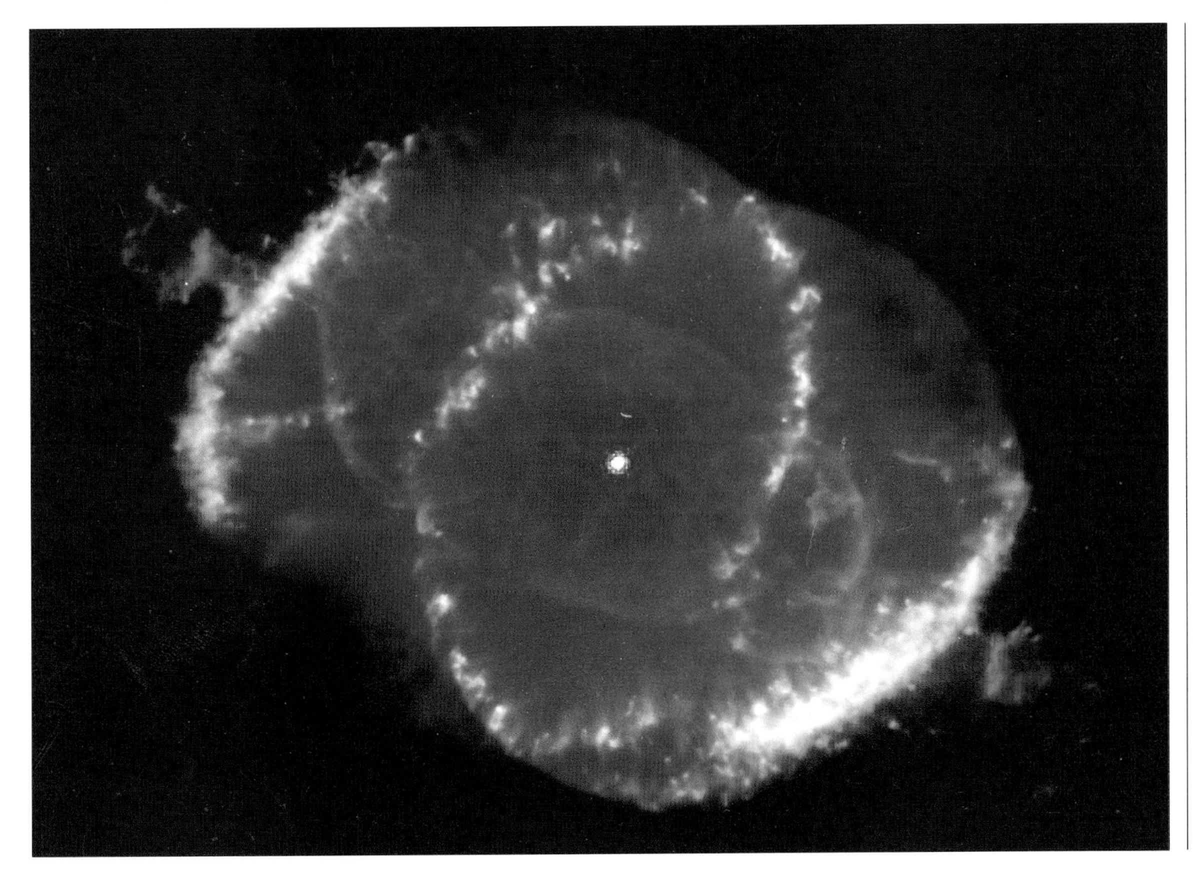

◀ ***NGC 6543, the Cat's Eye Nebula.*** *This is a complex planetary nebula, with intricate structures including concentric gas shells, jets of high-speed gas, and unusual shock-induced knots of gas. The nebula is about 1000 years old, and represents a dying star. It could even be a double-star system. It lies in Draco, and is 3000 light-years away. [Credit: J. P. Harrington and K. J. Borkowski (University of Maryland) and NASA.]*

Cassiopeia, Cepheus, Camelopardalis,

Magnitudes

- −1
- 0
- 1
- 2
- 3
- 4
- 5

Variable star

Galaxy

Planetary nebula

Gaseous nebula

Globular cluster

Open cluster

Cassiopeia. The W shape of the constellation Cassiopeia is unmistakable, and is of special interest because one member of the pattern is variable, while another probably is.

The confirmed variable is γ, with a peculiar spectrum which shows marked variations. No changes in light seem to have been recorded until about 1910, and the magnitude had been given as 2.25. The star then slowly brightened, and there was a rapid increase during late 1936 and early 1937, when the magnitude rose to 1.6. A decline to below magnitude 3 followed by 1940, and then came a slow brightening; ever since the mid-1950s the magnitude has hovered around 2.2, slightly fainter than Polaris and slightly brighter than β Cassiopeiae. There is certainly no period; what apparently happens is that the star throws off shells of material and brightens during the process. A few other stars of the same type are known – Pleione in the Pleiades is a good example – but what are now known as 'GCAS' or Gamma Cassiopeiae variables, are rare. All of them seem to be rapid rotators. There may be a new brightening at any time, so that luminosity rises to about 6000 times that of our own Sun.

α Cassiopeiae (Shedir) is decidedly orange, with a K-type spectrum. It is 120 light-years away, and 190 times as luminous as the Sun. During the last century it was accepted as being variable, with a probable range of between magnitude 2.2 and 2.8; it was even suggested that there might be a rough period of about 80 days. Later observers failed to confirm the changes, and in modern catalogues α is often listed as 'constant', though my own observations between 1933 and the present time indicate that there are slight, random fluctuations between magnitudes 2.1 and 2.4, with a mean of 2.3. Generally speaking, the order of brilliance of the three main members of the W is γ, α, β, but this is not always the case, and watching the slight variations is a good exercise for the naked-eye observer. β itself fluctuates very slightly, but the range is less than 0.04 of a magnitude, so that in estimating γ and α it is safe to take the magnitude of β as 2.27.

ϱ, which lies close to β and midway between σ (magnitude 4.88) and τ (4.87), is unquestionably variable, but nobody is sure of its type. It is an exceptionally luminous supergiant, equal to at least 130,000 Suns, and is 4800 light-years away; for most of the time it hovers around magnitude 4.8, so that σ makes a convenient comparison star (although τ is also at a comparable magnitude, it is suspected of variability, and is better avoided). Occasionally ϱ drops to below the sixth magnitude, though this has not happened now for more than 40 years. The spectrum too is variable, and can range from type F8 to early M. The star is an excellent target for the binocular-user, since when a new minimum occurs early warning will be important.

R Cassiopeiae, a normal Mira star, can reach naked-eye visibility at maximum. The supernova of 1572 flared up near ϰ; the site is now identified by its radio emissions. η Cassiopeiae is a wide, easy double. ι is also easy, and there is another seventh-magnitude companion at a separation of just over 8 seconds of arc.

There are two Messier open clusters in Cassiopeia, neither of which is of special note; indeed M103 is less prominent than its neighbour NGC663 (C10), and it is not easy to see why Messier gave it preference. NGC457 is of more interest. It contains several thousands of stars, and is an easy binocular object. φ Cassiopeiae, magnitude 4.98, lies in its south-eastern edge, and if it is a genuine cluster member – as seems likely – it must have a luminosity well over 200,000 times that of the Sun; the distance is at least 9000 light-years.

The Milky Way crosses Cassiopeia, and the whole constellation is very rich. Here too we find the galaxies Maffei 1 and 2, which are so heavily obscured that they are difficult to see; Maffei 1 is almost certainly a member of the Local Group.

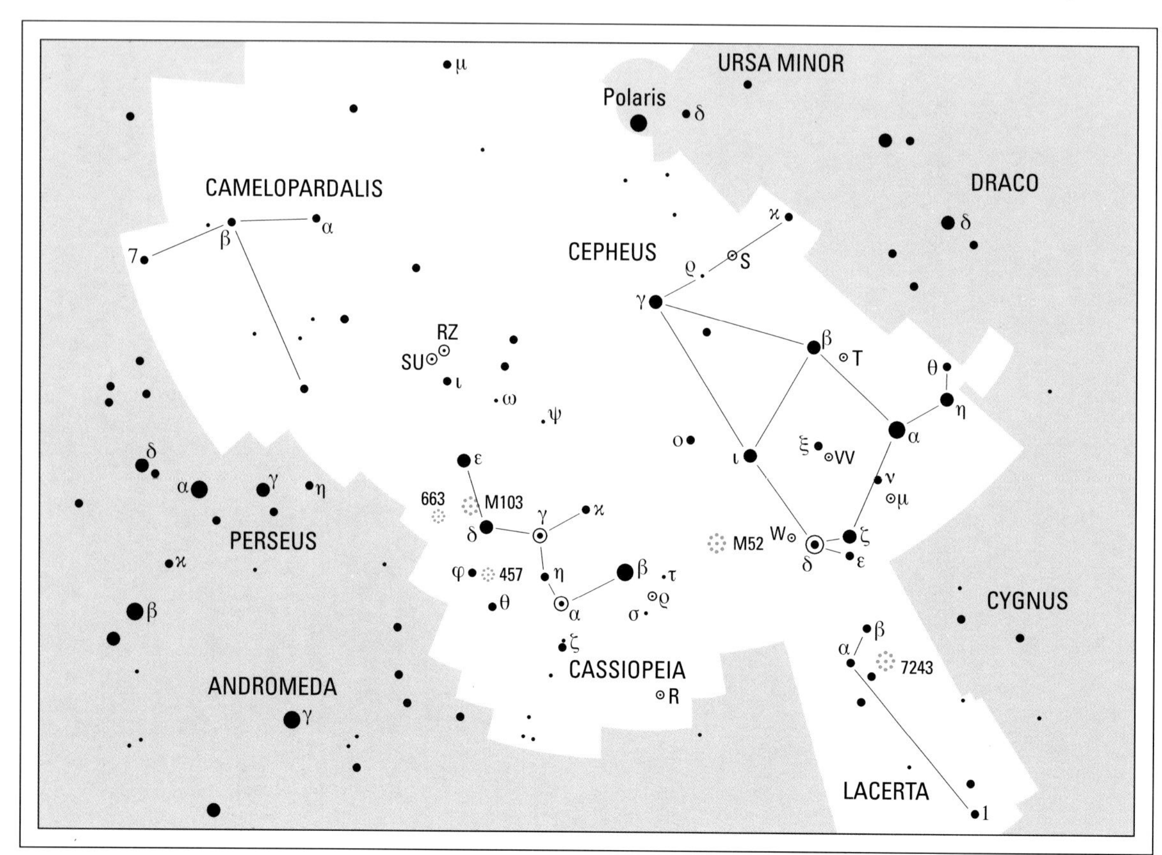

◀ ***Apart from Ursa Major**, Cassiopeia is much the most conspicuous of the far northern constellations. It and Ursa Major lie on opposite sides of the celestial pole, so that when Ursa Major is high up, Cassiopeia is low down, and vice versa – though neither actually sets over any part of the British Isles or the northern United States. Cepheus is much less prominent, and is almost lost from southern countries; Lacerta and Camelopardalis are very obscure.*

Lacerta

Cepheus, the King, is much less prominent than his Queen. α (Alderamin) is of magnitude 2.4, and is 45 light-years away, with a luminosity 14 times that of the Sun. The quadrilateral made up of α, β, ι and ζ is not hard to identify.

The main interest in Cepheus is centred upon three variables stars: δ, μ and VV. δ is the prototype Cepheid, and has given its name to the whole class; its behaviour was explained in the 18th century by the young deaf-mute astronomer John Goodricke. It forms a small triangle with ζ (3.55) and ε (4.19) which make good comparison stars, though δ never becomes as bright as ζ. The 7.5-magnitude companion is an easy telescopic object, and seems to be genuinely associated, since it and the variable share a common motion in space.

μ Cephei is so red that William Herschel nicknamed it the Garnet Star; although the light-level is too low for the colour to be evident with the naked-eye, binoculars bring it out beautifully. The range is between magnitudes 3.4 and 5.1, but the usual value is about 4.3, so that the nearby ν (4.29) makes a convenient comparison. It has been suggested that μ may be of the semi-regular type, but it is difficult to find any real periodicity. The distance has been given as 1500 light-years, in which case the luminosity is more than 50,000 times that of the Sun – making it much more powerful than Betelgeux in Orion, which is of the same type. The Garnet Star is so luminous that if it were as close as, say, Pollux in Gemini, the apparent magnitude would be −7, and it would be conspicuous in the sky even in broad daylight.

VV Cephei, close to ξ (4.29), is a huge eclipsing binary of the Zeta Aurigae type. The system consists of a red supergiant together with a smaller hot blue companion; the range is small – magnitude 4.7 to 5.4 – and the orbital period is 7430 days, or 20.3 years. It is thought that the diameter of the supergiant may be as much as 1600 times that of the Sun, in which case it is one of the largest stars known. The last eclipse occurred in 1996.

The two variables in the constellation are worth mentioning. W, close to δ, is a red semi-regular with a long but uncertain period. Telescope users may care to pick out the Mira variable S, which is one of the reddest of all stars. There are no Messier objects in Cepheus.

Camelopardalis (alternatively known as Camelopardus). This is a very barren far-northern constellation, and was introduced to the sky by Hevelius in 1690. There is little really of much interest here, but it is worth noting that the three brightest stars, α, β and 7, are all very remote and luminous; 7 is well over 50,000 times as powerful as the Sun.

Lacerta. Although Lacerta the Lizard, was one of the original constellations listed by Ptolemy, it is very small and obscure. There is a small 'diamond' of dim stars, of which α is the brightest; to find them, use ζ and ε Cephei as guides. ε Cephei is just in the same binocular field with β Lacertae.

The only object of any note is the open cluster NGC7243 (C16), which forms an equilateral triangle with α and β and is just within binocular range. A bright nova (CP) flared up in Lacerta in 1936, and reached magnitude 1.9, but it faded quickly, and is now below the 15th magnitude.

BL Lacertae, which is too faint to be of interest to the user of a small telescope, was once regarded as an ordinary run-of-the-mill variable star, and was given the appropriate designation, but when its spectrum was examined it was found to be something much more dramatic, and is more akin to a quasar. It has given its name to the whole class of such objects (see page 200), which are conventionally known as 'BL Lacs'.

CASSIOPEIA

BRIGHTEST STARS

No.	Star	R.A.			Dec.			Mag.	Spectrum	Proper name
		h	m	s	°	′	″			
27	γ	00	56	42	+60	43	00	2.2v	B0p	
18	α	00	40	30	+56	32	15	2.2v?	K0	Shedir
11	β	00	09	11	+59	08	59	2.27	F2	Chaph
37	δ	0	25	49	+60	14	07	2.68	A5	Ruchbah
45	ε	01	54	24	+63	40	13	3.38	B3	Segin
24	η	00	49	06	+57	48	58	3.44	G0	Achird

Also above magnitude 4.3: ζ (3.67), ι (3.98), ϰ (4.16); next comes θ (Marfak) (4.33). ϱ is an irregular variable which can at times exceed magnitude 4.3, but is usually nearer 4.8.

VARIABLES

Star	R.A.		Dec.		Range	Type	Period	Spectrum
	h	m	°	′	(mags)		(d)	
ϱ	23	54.4	+58	30	4.1–6.2	?	–	F
γ	00	56.7	+60	43	1.6–3.3	Irregular	–	Bp
α	00	40.5	+56	22	2.1–2.5?	Suspected	–	K
R	23	58.4	+51	24	4.7–13.5	Mira	431	M
SU	02	52.0	+68	53	5.7–6.2	Cepheid	1.95	F
RZ	02	48.9	+69	38	6.2–7.7	Algol	1.19	A

DOUBLES

Star	R.A.		Dec.		P.A.	Sep.	Mags	
	h	m	°	′	°	″		
η	00	49.1	+57	49	315	12.6	3.4, 7.5	Binary, 480y
ι	02	29.1	+67	24	232	2.4	4.9, 6.9	Binary, 840y

CLUSTERS AND NEBULAE

M	NGC	R.A.		Dec.		Mag.	Dimensions	Type
		h	m	°	′		′	
52	7654	23	24.2	+61	35	6.9	13	Open cluster
103	581	01	33.2	+60	42	7.4	6	Open cluster
	663	01	46.0	+61	15	7.1	116	Open cluster
	457	01	19.1	+58	20	6.4	13	Open cluster round φ Cas

CEPHEUS

BRIGHTEST STARS

No.	Star	R.A.			Dec.			Mag.	Spectrum	Proper name
		h	m	s	°	′	″			
5	α	21	18	35	+62	35	08	2.44	A7	Alderamin
35	γ	23	39	21	+77	37	57	3.21	K1	Alrai
8	β	21	28	39	+70	33	39	3.23v	B2	Alphirk
21	ζ	22	10	51	+58	12	05	3.35	K1	
3	η	20	45	17	+61	50	20	3.43	K0	

Also above magnitude 4.3: ι (3.52), ε (4.19), θ (4.22), ν (4.29), ξ (4.29).
The two famous variables can exceed magnitude 4 at maximum, δ and μ.

VARIABLES

Star	R.A.		Dec.		Range	Type	Period	Spectrum
	h	m	°	′	(mags)		(d)	
δ	22	29.2	+58	25	3.5–4.4	Cepheid	5.37	F-G
μ	21	43.5	+58	47	3.4–5.1	Irregular	–	M
T	21	09.5	+68	29	5.2–11.3	Mira	388	M
VV	21	56.7	+63	38	4.8–5.4	Eclipsing	7430	M+B
W	22	36.5	+58	26	7.0–9.2	Semi-regular	Long	K-M
S	21	35.2	+78	37	7.4–12.9	Mira	487	N

DOUBLES

Star	R.A.		Dec.		P.A.	Sep.	Mags	
	h	m	°	′	°	″		
ϰ	20	08.9	+77	43	122	7.4	4.4, 8.4	
β	21	28.7	+70	34	249	13.3	3.2, 7.9	
δ	22	29.2	+58	25	191	41.0	var, 7.5.	
ο	23	18.6	+68	07	220	2.9	4.9, 7.1	Binary, 796y
ξ	22	03.8	+64	38	277	7.7	4.4, 6.5	Binary, 3800y

CAMELOPARDALIS

The brightest star is β; R.A. 05h 03m, 25s 1, dec. +60° 26′ 32″, mag. 4.03. The only other stars above mag. 4.3 are 7 (4.21) and α (4.29).

LACERTA

A small, obscure constellation. The brightest star is α: R.A. 22h 31m 17s.3, dec. +50° 16′ 17″, mag. 3.77. The only other star above mag. 4.3 is 2 (4.13).

CLUSTERS AND NEBULAE

M	NGC	R.A.		Dec.		Mag.	Dimensions	Type
		h	m	°	′		′	
	7243	22	15.3	+49	53	6.4	21	Open cluster; about 40 stars

Boötes, Corona Borealis, Coma Berenices

Boötes. A large and important northern constellation, said to represent a herdsman who invented the plough drawn by two oxen – for which service to mankind he was rewarded with a place in the heavens.

Of course the whole area is dominated by Arcturus, which is the brightest star in the northern hemisphere of the sky and is one of only four with negative magnitudes (the other three are Sirius, Canopus and α Centauri). It is a light orange K-type star, 36 light-years away and with a luminosity 115 times that of the Sun; the diameter is about 30 million kilometres (about 19 million miles). It is too bright to be mistaken, but in case of any doubt it can be located by following through the tail of the Great Bear (Alioth, Mizar, Alkaid).

Arcturus has the exceptionally large proper motion of 2.3 seconds of arc per year, and as long ago as 1718 Edmond Halley found that its position relative to the background stars had shifted appreciably since ancient times. At the moment it is approaching us at the rate of 5 kilometres per second (3 miles per second), but this will not continue indefinitely; in several thousand years' time it will pass by us and start to recede, moving from Boötes into Virgo and dropping below naked-eye visibility in half a million years. It is a Population II star belonging to the galactic halo, so that its orbit is sharply inclined, and it is now cutting through the main plane of the Galaxy.

In 1860 a famous variable star observer, Joseph Baxendell, found a star of magnitude 9.7 in the field of Arcturus, at a P.A. of 250 degrees and a separation of 25 minutes of arc. Within a week it had disappeared, and has never been seen again, though it is still listed in catalogues as T Boötis. It may have been a nova or recurrent nova, and there is always the chance that it will reappear, so that amateur observers make routine checks to see whether it has done so.

ε, the second brightest star in the constellation, is a fine double; the primary is an orange K-type star, while the companion looks rather bluish by contrast. No doubt the two stars have a common origin, but the revolution period must be immensely long. The primary is 200 times more luminous than the Sun, so that it is more powerful than Arcturus, but it is also further away – around 150 light-years. The semi-regular variable W Boötis is in the same binocular field with ε, and is easily recognizable because of its orange-red hue.

ζ is a binary, with almost equal components (magnitudes 4.5 and 4.6) and an orbital period of 123 years, but the separation is never more than 1 second of arc, so that a telescope of at least 13-centimetre (5-inch) aperture is needed to split it. There are no Messier objects in Boötes, and in fact no nebular objects with integrated magnitude as bright as 10.

One constellation which is still remembered, even though it is no longer to be found on our maps, is Quadrans Muralis (the Mural Quadrant), added to the sky by Bode in 1775. The nearest brightish star to the site is β Boötis (Nekkar), magnitude 3.5. Like all the rest of Bode's groups, Quadrans was later rejected, but it so happens that the meteors of the early January shower radiate from there, which is why we call them the Quadrantids. For this reason alone, there might have been some justification for retaining Quadrans.

Corona Borealis is a very small constellation, covering less than 180 square degrees of the sky (as against over 900 square degrees for Boötes), but it contains far more than its fair share of interesting objects. The brightest star, α or Alphekka (also known as Gemma), is of the second magnitude, and is actually an eclipsing binary with an unusually small range; the main component is 50 times as luminous as the Sun, while the fainter member of the pair has just twice the Sun's power. The real separation between the two is less than 30 million kilometres (19 million miles), so that they cannot be seen separately. The distance from the Solar System is some 78 light-years.

Magnitudes
−1
0
1
2
3
4
5
Variable star
Galaxy
Planetary nebula
Gaseous nebula
Globular cluster
Open cluster

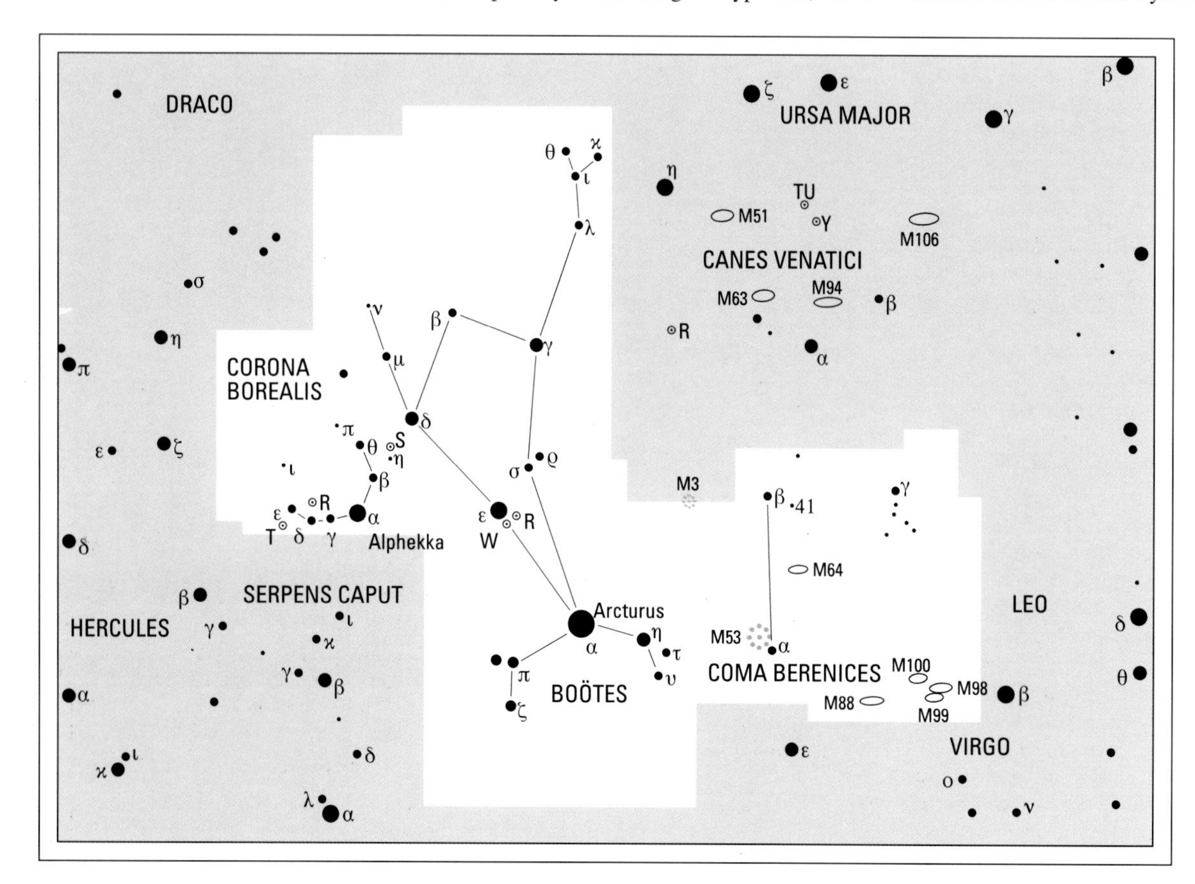

◀ ***The map*** *is dominated by Arcturus, the brightest star in the northern hemisphere of the sky; it is sufficiently close to the celestial equator to be visible from every inhabited country, and is at its best during evenings in northern spring (southern autumn). The Y-formation made up of Arcturus, ε and γ Boötis, and Alphekka (α Coronae) is distinctive. The rejected constellation of Quadrans is now included in Boötes, near β; it is from here that the January meteors radiate, which is why they are known as the Quadrantids.*

η Coronae is a close binary, with an average separation of 1 second of arc and components of magnitudes 5.6 and 5.9; it is a binary with a period of 41.6 years, and is a useful test object for telescopes of around 13-centimetre (5-inch) aperture. There are optical companions at 58 seconds of arc (magnitude 12.5) and 215 seconds of arc (magnitude 10.0). ζ and σ are both easy doubles, while β is a spectroscopic binary, and is also a magnetic variable of the same type as Cor Caroli.

Inside the bowl of the Crown lies the celebrated variable R Coronae, which periodically veils itself behind clouds of soot in its atmosphere. Usually it is on the brink of naked-eye visibility, but it shows sudden, unpredictable drops to minimum. At its faintest it fades below magnitude 15, so that it passes well out of the range of small telescopes; on the other hand there may be long periods when the light remains almost steady, as happened between 1924 and 1934. It is much the brightest member of its class, and of the rest only RY Sagittarii approaches naked-eye visibility.

R Coronae is a splendid target for binocular observers. Generally, binoculars will show two stars in the bowl, R and a star (M) of magnitude 6.6. If you examine the area with a low power and see only one star instead of two, you may be sure that R Coronae has 'taken a dive'.

Outside the bowl, near ε, is the Blaze Star, T Coronae, which is normally of around the tenth magnitude, but has shown two outbursts during the past century and a half; in 1866, when it reached magnitude 2.2 (equal to Alphekka), and again in 1946, when the maximum magnitude was about 3. On neither occasion did it remain a naked-eye object for more than a week. But it is worth keeping a watch on it, though if these sudden outbursts have any periodicity there is not likely to be another until around 2026.

Spectroscopic examination has shown that T Coronae is in fact, a binary, made up of a hot B-type star together with a cool red giant. It is the B-star which is the site of the outbursts, while the red giant seems to be irregularly variable over a range of about a magnitude, causing the much smaller fluctuations observed when the star is at minimum. Other recurrent novae are known, but only the Blaze Star seems to be capable of becoming really prominent. Also in the constellation is S Coronae, a normal Mira variable which rises to the verge of naked-eye visibility when it is at its maximum.

Mythologically, Corona is said to represent a crown given by the wine-god Bacchus to Ariadne, daughter of King Minos of Crete.

Coma Berenices is not an original constellation – it was added to the sky by Tycho Brahe in 1690 – but there is a legend attached to it. When the King of Egypt set out upon a dangerous military expedition, his wife Berenice vowed that if he returned safely she would cut off her lovely hair and place it in the Temple of Venus. The king returned; Berenice kept her promise, and Jupiter placed the shining tresses in the sky.

Coma gives the impression of a vast, dim cluster. It abounds in galaxies, of which five are in Messier's list. Of these, the most notable is M64, which is known as the Black-Eye Galaxy because of a dark region in it north of the centre – though this feature cannot be seen with any telescope below around 25-centimetre (10-inch) aperture. There is also a globular cluster, M53, close to α Comae which is an easy telescopic object. β Comae and its neighbour 41 act as good guides to the globular cluster M3, which lies just across the border of Canes Venatici and is described with Star Map 1.

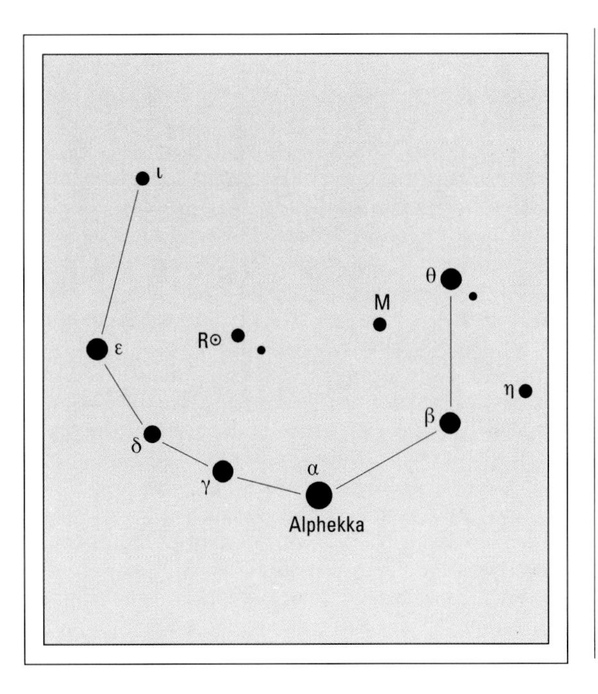

◄ ***R Coronae*** *is a splendid target for binocular observers. It is found within the bowl of Corona Borealis. If only one star is visible there, it is M with a magnitude of 6.6, and R Coronae has taken one of its periodic 'dives'.*

BOÖTES

BRIGHTEST STARS

No.	Star	R.A. h	m	s	Dec. °	′	″	Mag.	Spectrum	Proper name
16	α	14	15	40	+19	10	57	−0.04	K2	Arcturus
36	ε	14	44	59	+27	04	27	2.37	K0	Izar
8	η	13	54	41	+18	23	51	2.68	G0	
27	γ	14	32	05	+38	13	30	3.03	A7	Seginus
49	δ	15	15	30	+33	18	53	3.47	G8	Alkalurops
42	β	15	01	57	+40	23	26	3.50	G8	Nekkar

Also above magnitude 4.3; ρ (3.58), ζ (3.78), θ (4.05), υ (4.06), λ (4.18).

VARIABLES

Star	R.A. h	m	Dec. °	′	Range (mags)	Type	Period (d)	Spectrum
R	14	37.2	+26	44	6.2–13.1	Mira	223	M
W	14	43.4	+26	32	4.7–5.4	Semi-regular	450	M

DOUBLES

Star	R.A. h	m	Dec. °	′	P.A. °	Sep. ″	Mags
κ	14	13.5	+51	47	236	13.4	4.6, 6.6
ι	14	16.2	+51	22	033	38.5	4.9, 7.5
π	14	40.7	+16	25	108	5.6	4.9, 5.8
μ	15	24.5	+37	23	171	108.3	4.3, 7.0
ε	14	45.0	+27	04	339	2.8	2.5, 4.9

CORONA BOREALIS

BRIGHTEST STARS

No.	Star	R.A. h	m	s	Dec. °	′	″	Mag.	Spectrum	Proper name
5	α	15	34	41	+26	42	53	2.23	A0	Alphekka

The 'crown' is made up of α together with ε (4.15), δ (4.63), γ (3.84), β (3.68) and θ (4.14).

VARIABLES

Star	R.A. h	m	Dec. °	′	Range (mags)	Type	Period (d)	Spectrum
R	15	48.6	+28	09	5.7–15	R Coronae	–	F8p
S	15	21.4	+31	22	5.8–14.1	Mira	360	M
T	15	59.5	+25	55	2.0–10.8	Recurrent nova	–	M+Q

DOUBLE

Star	R.A. h	m	Dec. °	′	P.A. °	Sep. ″	Mags	
η	15	23.2	130	17	030	1.0	5.8, 5.9	
ζ	15	39.4	+36	38	305	6.3	5.1, 6.0	
σ	16	14.7	+33	52	234	7.0	5.6, 6.6	Binary, 1000y.

COMA BERENICES

The brightest star in this vast, dim cluster is β; R.A. 13h 11m 52s, dec. +27° 52′ 41″, mag. 4.26. Then come α (Diadem) (4.32) and γ (4.35).

CLUSTERS AND NEBULAE

M	NGC	R.A. h	m	Dec. °	′	Mag.	Dimensions ′	Type
53	5024	13	12.9	+18	10	7.7	12.6	Globular cluster
64	4826	12	56.7	+21	41	8.5	9.3 × 5.4	Sb (Black-Eye) galaxy
88	4501	12	32.0	+14	25	9.5	6.9 × 3.9	SBb galaxy
98	4192	12	13.8	+14	54	10.1	9.5 × 3.2	Sb galaxy
99	4254	12	18.8	+14	25	9.8	5.4 × 4.8	Sc galaxy
100	4321	12	22.9	+15	49	9.4	6.9 × 6.2	Sc galaxy

Leo, Cancer, Sextans

Magnitudes
–1
0
1
2
3
4
5
Variable star
Galaxy
Planetary nebula
Gaseous nebula
Globular cluster
Open cluster

Leo was the mythological Nemaean lion which became one of Hercules' many victims, but in the sky the Lion is much more imposing than his conqueror, and is indeed one of the brightest of the Zodiacal constellations. The celestial equator cuts its southernmost extension, and Regulus, at the end of the Sickle, is so close to the ecliptic that it can be occulted by the Moon and planets – as happened on 7 July 1959, when Venus passed in front of it. On that occasion the fading of Regulus before the actual occultation, when the light was coming to us by way of Venus' atmosphere, provided very useful information about the atmosphere itself (of course, this was well before any successful interplanetary spacecraft had been launched to investigate the atmosphere of Venus more directly).

Regulus is a normal white star, some 85 light-years away and around 130 times as luminous as the Sun. It is a wide and easy double; the companion shares Regulus' motion through space, so that presumably the two have a common origin. The companion is itself a very close double, difficult to resolve partly because of the faintness of the third star and partly because of the glare from the brilliant Regulus.

About 20 minutes of arc north of Regulus is the dwarf galaxy Leo I, a member of the Local Group, about 750,000 light-years from us. It was discovered photographically as long ago as 1950, but even giant telescopes are hard pressed to show it visually, because its surface brightness is so low; it is also one of the smallest and least luminous galaxies known. Even feebler is another member of the Local Group, Leo II, which lies about two degrees north of δ.

There is a minor mystery associated with Denebola or β Leonis. All observers up to and including Bayer, in 1603, ranked it as being of the first magnitude, equal to Regulus, but it is now almost a whole magnitude fainter. Yet it is a perfectly normal Main Sequence star of type A, 39 light-years away and 17 times as luminous as the Sun – not at all the kind of star expected to show a slow, permanent change. It is probable that there has been a mistake in recording or interpretation; all the same, a certain doubt remains, and naked-eye observers may care to check on it to see if there are any detectable fluctuations. The obvious comparison star is γ, which is of virtually the same brightness and can often be seen at the same altitude above the horizon.

γ is a magnificent double, easily split with a very small telescope. The primary is orange, and the G-type companion usually looks slightly yellowish. The main star is 60 times as luminous as the Sun, and the companion is the equal of at least 20 Suns; the distance from us is 91 light-years. Two other stars, some distance away, are not genuinely connected with the bright pair.

Two fainter stars (not on the map), 18 Leonis (magnitude 5.8) and 19 Leonis (6.5), lie near Regulus and are easily identified with binoculars. Forming a group with them is the Mira variable R Leonis, which can reach naked-eye brightness when at maximum and seldom falls below the tenth magnitude. Like most stars of its type, it is very red, and is a suitable target for novice observers, particularly since it is so easy to find.

There are five Messier galaxies in Leo. M65 and M66, which lie more or less between θ and ι Leonis, can be seen with binoculars, and are only 21 minutes of arc apart, so that they are in the same field of a low-power telescope. Both are spiral galaxies; M66 is actually the brighter of the two, though M65 is often regarded as the easier to see. Unfortunately, both are placed at an unfavourable angle to us, so that the full beauty of the spiral forms is lost. They are around 35 million light-years away, and form a true pair. Another pair of spirals, M95 and M96, lies between ϱ and θ Leonis; close by is the elliptical galaxy M105, which is an easy object. Leo contains many additional galaxies, and the whole area is worth sweeping.

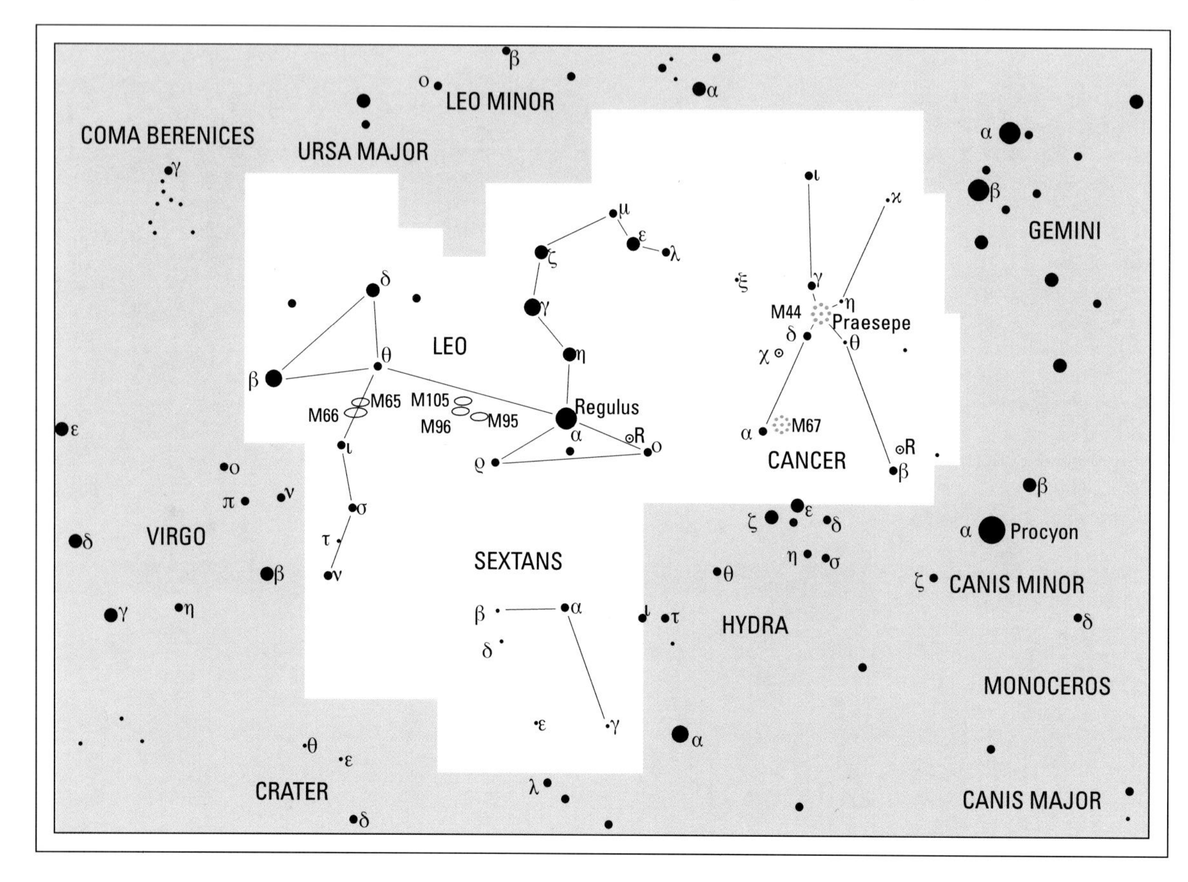

◀ ***Two Zodiacal constellations** are shown here, Leo and Cancer: Leo is large and prominent, Cancer decidedly obscure. Both are at their best during evenings in northern spring (southern autumn). Leo is distinguished by the 'Sickle', of which Regulus is the brightest member, while Cancer contains Praesepe, one of the finest open clusters in the sky. Sextans is very barren and obscure. The equator crosses this map, and actually passes through the southernmost part of the constellation of Leo.*

Before leaving the Lion, it is worth mentioning Wolf 359, which lies at R.A. 15h 54m.1, dec. +07 degrees 20 minutes. Apart from Barnard's Star and the members of the α Centauri group, Wolf 359 is the closest of our stellar neighbours, at a mere 7.6 light-years; even so, its apparent magnitude is only 13.5, so that it is by no means easy to identify. It is one of the feeblest red dwarfs yet to be discovered, and its luminosity is less than 1/60,000 that of our own Sun.

Cancer, the celestial Crab, which according to legend met an untimely fate when Hercules trod upon it, looks a little like a dim and ghostly version of Orion. It is easy to locate, since it lies almost directly between the Twins (Castor and Pollux, α and β Geminorum) and Regulus; it is of course in the Zodiac, and wholly north of the equator. ζ (Tegmine) is, in fact, a triple system; the main pair is easy to resolve, and the brighter component is itself a close binary, with a separation which never exceeds 1.2 seconds of arc.

The semi-regular variable X Cancri, near δ, is worth finding because of its striking red colour. As it never fades below magnitude 7.5, it is always within binocular range, and its colour makes it stand out at once. R Cancri, near β, is a normal Mira variable which can rise to almost magnitude 6 at maximum.

The most interesting objects in Cancer are the open clusters, M44 (Praesepe) and M67. Praesepe is easily visible without optical aid, and has been known since very early times; Hipparchus, in the second century BC, referred to it as 'a little cloud'. It was also familiar to the Chinese, though it is not easy to decide why they gave it the unprepossessing nickname of 'the Exhalation of Piled-up Corpses'. Because it is also known as the Manger, the two stars flanking it, δ and γ Cancri, are called the Asses. Yet another nickname for Praesepe is the Beehive.

Praesepe is about 525 light-years away. It contains no detectable nebulosity, so that star formation there has presumably ceased, and since many of the leading stars are of fairly late spectral type, it may be assumed that the cluster is fairly old. Because Praesepe covers a wide area – the apparent diameter is well over one degree – it is probably best seen with binoculars or else with a very low-power eyepiece. The real diameter is of the order of 10 to 15 light-years, though, as with all open clusters, there is no sharp boundary.

M67 is on the fringe of naked-eye visibility, and is easily found; it lies within two degrees of α (Acubens). It contains at least 200 stars; the French astronomer Camille Flammarion likened it to 'a sheaf of corn'. Its main characteristic is its great age.

Most open clusters lose their identity before very long, cosmically speaking, because they are disrupted by passing field stars, but M67 lies at around 1500 light-years away from the main plane of the Galaxy, so that it moves in a comparatively sparsely-populated region and there is little danger of this happening. Consequently, M67 has retained much of its original structure. It may be considerably older than the Sun. Despite its great distance, it is easy to resolve into stars, and in appearance it is not greatly inferior to Praesepe. The distance from the Earth is around 2700 light-years, and the real diameter is about 11 light-years.

Sextans (originally Sextans Uraniae, Urania's Sextant) is one of the groups formed by Hevelius, but for no obvious reason, since it contains no star brighter than magnitude 4.5 and no objects of immediate interest to the telescope-user, though there are several galaxies with integrated magnitudes of between 9 and 12.

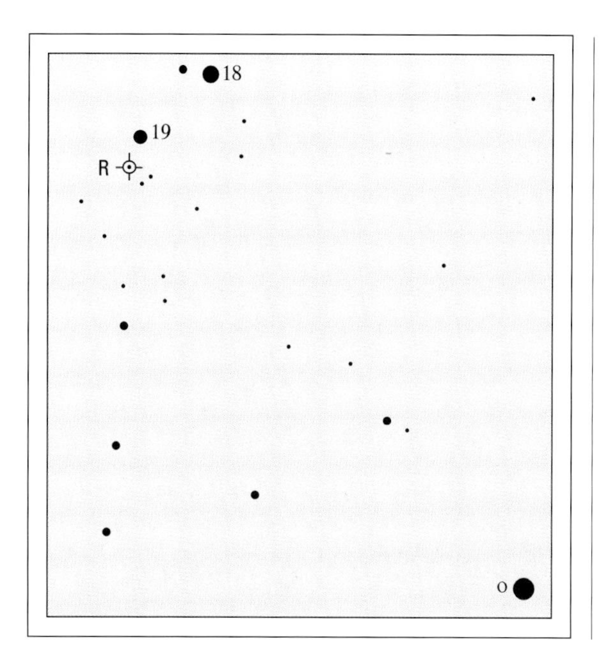

◄ ***R Leonis**. The bright star is ο Leonis, at magnitude 3.8. The comparisons for R Leonis are 18 Leonis, magnitude 5.8, and 19 Leonis, magnitude 6.4.*

LEO

BRIGHTEST STARS

No.	Star	R.A. h	m	s	Dec. °	′	″	Mag.	Spectrum	Proper name
32	α	10	08	22	+11	58	02	1.35	B7	Regulus
41	γ	10	19	58	+19	50	30	1.99	K0+G7	Algieba
94	β	11	49	04	+14	34	19	2.14	A3	Denebola
68	δ	11	14	06	+20	31	26	2.56	A4	Zosma
17	ε	09	45	51	+23	46	27	2.98	G0	Asad Australis
70	θ	11	14	14	+15	25	46	3.34	A2	Chort
36	ζ	10	16	41	+23	25	02	3.44	F0	Adhafera

Also above mag. 4.3: η (3.52), ο (Subra) (3.52), ϱ (3.85), μ (3.88), ι (3.94), σ (4.05) and ν (4.30)

VARIABLE

Star	R.A. h	m	Dec. °	′	Range (mags)	Type	Period (d)	Spectrum
R	09	47.6	+11	25	4.4–11.3	Mira	312	M

DOUBLES

Star	R.A. h	m	Dec. °	′	P.A. °	Sep. ″	Mags	
α	10	08.4	+11	58	307	176.9	1.4, 7.7	
τ	11	27.9	+02	51	176	91.1	4.9, 8.0	
γ	10	20.0	+19	51	AB 124	4.3	2.2, 3.5	Binary, 619y
					AC 291	259.9	9.2	
					AD 302	333.0	9.6	

γ is a fine binary with an orange primary; it is out of binocular range, but a small telescope will split it. The more distant companions are optical.

CLUSTERS AND NEBULAE

M	NGC	R.A. h	m	Dec. °	′	Mag.	Dimensions ′	Type
65	3623	11	18.9	+13	05	9.3	10.0 × 3.3	Sb galaxy
66	3627	11	20.2	+12	59	9.0	8.7 × 4.4	Sb galaxy
95	3351	10	44.0	+11	42	9.7	7.4 × 5.1	SBb galaxy
96	3368	10	46.8	+11	49	9.2	7.1 × 5.1	Sb galaxy
105	3379	10	47.8	+12	35	9.3	4.5 × 4.0	E1 galaxy

CANCER

A dim Zodiacal constellation. The brightest star is β (Altarf); R.A. 08h 16m 30s.9, dec. +09° 11′ 08″, magnitude 3.52. The other stars making up the dim 'pseudo-Orion' pattern are δ (Asellus Australis) (3.84), γ (Asellus Borealis) (4.66), α (Acubens) (4.25), ι (4.02) and χ (5.14).

VARIABLES

Star	R.A. h	m	Dec. °	′	Range (mags)	Type	Period (d)	Spectrum
R	08	16.6	+11	44	6.1–11.8	Mira	362	M
X	08	55.4	+17	14	5.6–7.5	Semi-reg.	195	N

DOUBLE

Star	R.A. h	m	Dec. °	′	P.A. °	Sep. ″	Mags	
ζ (Tegmine)	08	12.2	+17	39	088	5.7	5.0, 6.2	Binary, 1150 y. A is a close binary, and there is a 9.7-mag. third component, P.A. 108°, sep. 288″.

CLUSTERS

M	NGC	R.A. h	m	Dec. °	′	Mag.	Dimensions ′	Type
44	2632	08	40.1	+19	59	3.1	95	Open cluster (Praesepe)
67	2682	08	50.4	+11	49	6.9	30	Open cluster

SEXTANS

The brightest star is α: R.A. 10h 07m 56s.2, dec. −00° 22′ 18″, mag. 4.49. There is nothing here of interest to the user of a small telescope.

Virgo, Libra

Magnitudes

−1

0

1

2

3

4

5

Variable star

Galaxy

Planetary nebula

Gaseous nebula

Globular cluster

Open cluster

Virgo is one of the largest of all the constellations, covering almost 1300 square degrees of the sky, though it has only one star of the first magnitude (Spica) and two more above the third. Mythologically it represents the Goddess of Justice, Astraea, daughter of Jupiter and Themis; the name Spica is said to mean 'the ear of wheat' which the Virgin is holding in her left hand. The main stars of Virgo make up a Y-pattern, with Spica at the base and γ at the junction between the 'stem' and the 'bowl'. The bowl, bounded on the far side by β Leonis, is crowded with galaxies.

Spica can be found by continuing the curve from the Great Bear's tail through Arcturus; if sufficiently prolonged it will reach Spica, which is in any case brilliant enough to be really conspicuous. It is an eclipsing binary, with a very small magnitude range from 0.91 to 1.01; the components are only about 18 million kilometres (11 million miles) apart. Around 80 per cent of the total light comes from the primary, which is more than ten times as massive as the Sun and is itself intrinsically variable, though the fluctuations are very slight indeed. The distance from us is 257 light-years, and the combined luminosity is well over 2000 times that of the Sun. Like Regulus, Spica is so close to the ecliptic that it can at times be occulted by the Moon or a planet; the only other first-magnitude stars similarly placed are Aldebaran and Antares.

The bowl of Virgo is formed by ε, δ, γ, η and β. The last two are much fainter than the rest; early catalogues made them equal to the others, but one must be very wary of placing too much reliance on these old records, and neither star seems to be of the type expected to show long-term changes in brightness. Of the other stars in the main pattern, δ (Minelauva) is a fine red star of type M; Angelo Secchi, the great Italian pioneer of astronomical spectroscopy, nicknamed it Bellissima because of its beautifully-banded spectrum. It is 147 light-years away, and 130 times as luminous as the Sun. ε, named Vindemiatrix or The Grape-Gatherer, is of type G; distance 104 light-years, luminosity 75 times greater than that of the Sun.

γ has three accepted proper names; Arich, Porrima and Postvarta. It is a famous binary, whose components are identical twins; the orbital period is 171.4 years, and a few decades ago the separation was great enough to make Arich one of the most spectacular doubles in the sky. We are now seeing it from a less favourable angle, and by 2007 the star will appear single except in giant telescopes, after which it will start to open out again; the minimum separation will be no more than 0.3 of a second of arc. The orbit is eccentric, and the real separation between the components ranges from 10,500 million kilometres (6520 million miles) to only 450 million kilometres (280 million miles). Arich is relatively near, at 36 light-years.

There are no bright variables in Virgo, but there is one much fainter star, W Virginis, which is worthy of special mention. Its position is R.A. 13h 23m.5, declination −03 degrees 07 minutes, less than four degrees away from ζ, but it is not likely to be of much interest to the user of a small telescope; the magnitude never rises above 9.5, and drops to 10.6 at minimum. The period is 17.3 days. Originally W Virginis was classed as a Cepheid, but it belongs to Population II, and short-period stars of this sort are considerably less luminous than classical Cepheids. It was this which led Edwin Hubble to underestimate the distance of the Andromeda Galaxy; at the time he had no way of knowing that there are two kinds of short-period variables with very different period–luminosity relationships. For a while the less luminous stars were called Type II Cepheids, but they are now known officially as W Virginis stars. The brightest of them, and the only member of the class easily visible with the naked eye, is ϰ Pavonis, described with Star Map 21; unfortunately it is too far south in the sky to be seen from Britain or any part of Europe. W Virginis itself peaks at

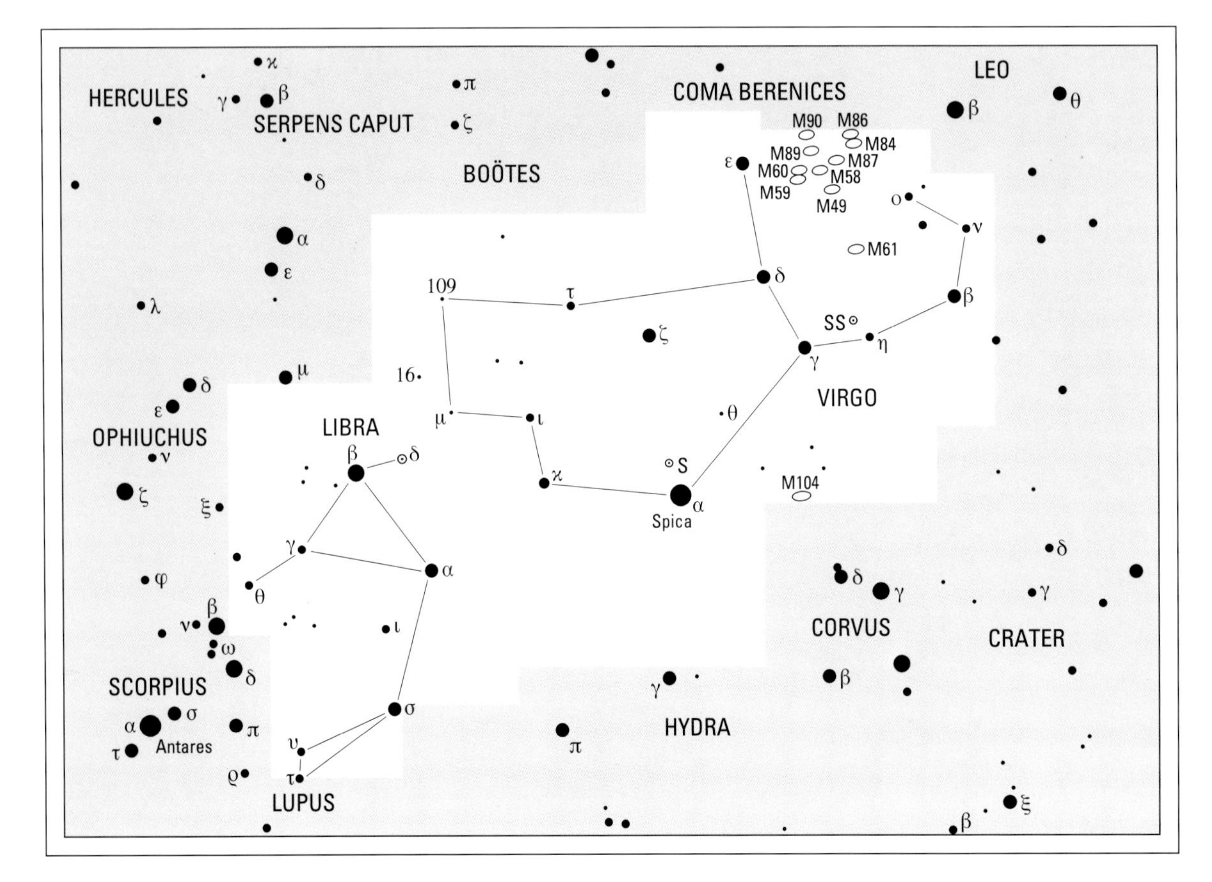

◀ The constellations *in this map are best seen during evenings around April to June. Both Virgo and Libra are in the Zodiac, and Virgo is crossed by the celestial equator; the 'Y' of Virgo is unmistakable, and the 'bowl' of the Y is crowded with rather faint galaxies. Libra adjoins Virgo to the one side and Scorpius to the other. The conspicuous quadrilateral of Corvus is also shown here, but is described in Star Map 7.*

1500 times the luminosity of the Sun, but this is not very much compared with the 6000 Sun-power of δ Cephei.

Of course, the main feature of Virgo is the cluster of galaxies, which spreads into the adjacent constellations of Leo and Coma. The average distance of the cluster members is between 40 and 50 million light-years, and since there are thousands of systems it makes our Local Group seem very puny. In Virgo there are no less than 11 Messier objects, plus many more galaxies with integrated magnitudes of 12 or brighter.

Pride of place must go to the giant elliptical M87, discovered by Messier himself in 1781. A curious jet, several thousands of light-years long, issues from it, and it is attended by many globular clusters – perhaps as many as a thousand. M87 is a very strong radio source, and is known to radio astronomers as Virgo A or as 3C-274; it is also a source of X-rays, and it is clear that tremendous activity is going on there. There is considerable evidence that in the heart of the galaxy there is a super-massive black hole.

M104 is different; it is an Sb spiral distinguished by the dark dust-lane which crosses it and gives it the nickname of the Sombrero Hat. It too is associated with a wealth of globular clusters. With a telescope of more than 30-centimetre (12-inch) aperture the dust-lane is not hard to see, and there is nothing else quite like it, so that it is also a favourite target for astro-photographers.

M49, which makes up an equilateral triangle with δ and ε, is another giant elliptical, strictly comparable with M87 apart from the fact that it is not a strong radio emitter. All the other galaxies in the catalogue are fairly easy to locate, as also are various others not in Messier's list; this is a favourite hunting-ground for observers who concentrate upon searching for supernovae in external systems.

Libra, the Scales or Balance, adjoins Virgo and is one of the least conspicuous of the Zodiacal groups. It is the only constellation of the Zodiac named after an inanimate object, but it was originally the Scorpion's Claws; some early Greek legends link it, rather vaguely, with Mochis, the inventor of weights and measures. Its main stars (α, β, γ and σ) make up a distorted quadrilateral; σ has been filched from Scorpius, and was formerly known as γ Scorpii.

α Librae – Zubenelgenubi, the Southern Claw – makes a very wide pair with 8 Librae, of magnitude 5.2; the separation is so great that the pair can be well seen with binoculars. The brighter member of the pair is a spectroscopic binary, 72 light-years away and 31 times as luminous as the Sun. Of rather more interest is β Librae or Zubenelchemale, the Northern Claw. It is 121 light-years away and 100 Sun-power; the spectral type is B8, and it has often been said to be the only single star with a decidedly greenish hue. T. W. Webb, a famous last-century English observer, referred to its 'beautiful pale green' colour. This is certainly an exaggeration, and most people will call it white, but it is worth examining. There have been suggestions that it is yet another star to have faded in historic times, and there is evidence that Ptolemy ranked it as of the first magnitude, but – as with other similar cases – the evidence is very slender.

There is not much else of interest in Libra, and there are no Messier objects. There is, however, one eclipsing binary of the Algol type. This is δ Librae, which makes up a triangle with 16 Librae (magnitude 4.5) and β; the range is from magnitude 4.8 to 6.1, so that it is never conspicuous, and at minimum sinks to the very limit of naked-eye visibility. With low-power binoculars it is in the same field with β, and this is probably the best way to locate it.

VIRGO

BRIGHTEST STARS

No.	Star	R.A.	Dec.	Mag.	Spectrum	Proper name
		h m s	° ′ ″			
67	α	13 25 11.5	−11 09 41	0.98	B1	Spica
29	γ	12 41 39.5	−01 26 57	2.6	F0+F0	Arich
47	ε	13 02 10.5	+10 57 33	2.83	G9	Vindemiatrix
79	ζ	13 34 41.5	−00 35 46	3.37	A3	Heze
43	δ	12 55 36.1	+03 23 51	3.38	M3	Minelauva

Also above magnitude 4.3: β (Zavijava) (3.61), 109 (3.72), μ (Rijl al Awwa) (3.88), η (Zaniah) (3.89), ν (4.03), ι (Syrma) (4.08), ο (4.12), κ (4.19) and τ (4.26).

DOUBLES

Star	R.A.	Dec.	P.A.	Sep.	Mags
	h m	° ′	°	″	
γ	12 41.7	−01 27	287	3.0	3.5, 3.5
θ	13 09.9	−05 32	343	7.1	4.4, 9.4
(Apami-Atsa)					

CLUSTERS AND NEBULAE

M	NGC	R.A.	Dec.	Mag.	Dimensions	Type
		h m	° ′		′	
49	4472	12 29.8	+08 00	8.4	8.9 × 7.4	E3 galaxy
58	4579	12 37.7	+11 49	9.8	5.4 × 4.4	SB galaxy
59	4621	12 42.0	+11 39	9.8	5.1 × 3.4	E3 galaxy
60	4649	12 43.7	+11 33	8.8	7.2 × 6.2	E1 galaxy
61	4303	12 21.9	+04 28	9.7	6.0 × 5.5	Sc galaxy
84	4374	12 25.1	+12 53	9.3	5.0 × 4.4	E1 galaxy
86	4406	12 26.2	+12 57	9.2	7.4 × 5.5	E3 galaxy
87	4486	12 30.8	+12 24	8.6	7.2 × 6.8	E1 galaxy (Virgo A)
89	4552	12 35.7	+12 33	9.8	9.5 × 4.7	E0 galaxy
90	4569	12 36.8	+13 10	9.5	9.5 × 4.7	Sb galaxy
104	4594	12 40.0	−11 37	8.3	8.9 × 4.1	Sb galaxy (Sombrero Hat)

LIBRA

BRIGHTEST STARS

No.	Star	R.A.	Dec.	Mag.	Spectrum	Proper name
		h m s	° ′ ″			
27	β	15 17 00.3	−09 22 58	2.61	B8	Zubenelchemale
9	α	14 50 52.6	−16 02 30	2.75	A3	Zubenelgenubi
20	σ	15 04 04.1	−25 16 55	3.29	M4	Zubenalgubi

Also above magnitude 4.3: υ (3.58), τ (3.66), γ (Zubenelhakrabi) (3.91), ι (4.15).
σ Librae was formerly included in Scorpius, as γ Scorpii.

VARIABLE

Star	R.A.	Dec.	Range	Type	Period	Spectrum
	h m	° ′	(mags)		(d)	
δ	15 01.1	−08 31	4.9–5.9	Algol	2.33	B

DOUBLE

Star	R.A.	Dec.	P.A.	Sep.	Mags
	h m	° ′	°	″	
α	14 50.9	−16 02	314	231.0	2.8,5.2

▼ ***The constellation Virgo*** *with the planets Mars and Saturn.*

Hydra, Corvus, Crater

Magnitudes
-1
0
1
2
3
4
5
Variable star
Galaxy
Planetary nebula
Gaseous nebula
Globular cluster
Open cluster

Hydra, with an area of 1303 square degrees, is the largest constellation in the sky; it gained that distinction when the old, unwieldy Argo Navis was dismembered. As a matter of casual interest, the only other constellations with areas of more than 1000 square degrees are Virgo (1294), Ursa Major (1280), Cetus (1232), Eridanus (1138), Pegasus (1121), Draco (1083) and Centaurus (1060). At the other end of the scale comes the Southern Cross, with a mere 68 square degrees.

Despite its size, Hydra is very far from being conspicuous, since there is only one star above the third magnitude and only ten above the fourth. The only reasonably well-defined pattern is the 'head', made up of ζ, ε, δ and η; it is easy to find, more or less between Procyon and Regulus, but there is nothing in the least striking about it. Mythologically, Hydra is said to represent the multi-headed monster who became yet another of Hercules' victims, but many lists relegate it to the status of a harmless watersnake.

The only bright star, α (Alphard), is prominent enough. The Twins, Castor and Pollux, point directly to it, but in any case it is readily identifiable simply because it lies in so barren an area; there are no other bright stars in the region, and Alphard has been nicknamed the Solitary One. It has a K-type spectrum, and is obviously reddish. It is 85 light-years away, and 115 times as luminous as the Sun.

During the 1830s Sir John Herschel, son of the discoverer of Uranus, went to the Cape of Good Hope to survey the far-southern stars. On the voyage home he made some observations of Alphard, and concluded that it was decidedly variable. This has never been confirmed, and today the star is regarded as being constant in light; however, it may be worth watching – though it is awkward to estimate with the naked eye because of the lack of suitable comparisons. If there are any fluctuations, they cannot amount to more than a few tenths of a magnitude.

There is, however, one interesting variable in the constellation. This is R Hydrae, close to γ and therefore rather inconveniently low for observers in Britain, Europe and the northern United States. At its maximum it can attain the fourth magnitude, and never falls below 10, so that it is always an easy object; but it is not a typical Mira star, because there seems no doubt that the period has changed during the last couple of centuries. It used to be around 500 days; by the 1930s it had fallen to 425 days, and the latest official value is 390 days, so that we seem to be dealing with a definite and probably permanent change in the star's evolutionary cycle. Observations are of value, because there is no reason to assume that the shortening in period has stopped. U Hydrae, which forms a triangle with ν and μ, is a semi-regular variable which is worth finding because, like virtually all stars of spectral type N, it is intensely red.

ε Hydrae is a multiple system. The two main components are easy to resolve; the primary is an extremely close binary with an orbital period of 15 years, and there is a third and probably a fourth star sharing a common motion in space. β is also double, but is a difficult test for a 15-centimetre (6-inch) telescope. Though given the second Greek letter, β is below the fourth magnitude, and therefore more than a magnitude fainter than γ, ζ or ν.

There are three Messier objects in Hydra. M48 is an open cluster on the edge of the constellation, close to the boundary with Monoceros (in fact the fourth-magnitude star ζ Monocerotis is the best guide to it). It is just visible with the naked eye, but it is not too easy to identify. M68, discovered by Pierre Méchain in 1780, is a globular cluster about 39,000 light-years away, lying almost due south of β Corvi and more or less between γ and β Hydrae. M83, south of γ and near the border between Hydra and Centaurus, is a fine face-on spiral galaxy, about 8.5 million light-years away and therefore not far beyond the Local Group; it is easy to locate with a telescope of 10-centimetre (4-inch) aperture or larger, and is a favourite photographic target, but it is of course best seen from the

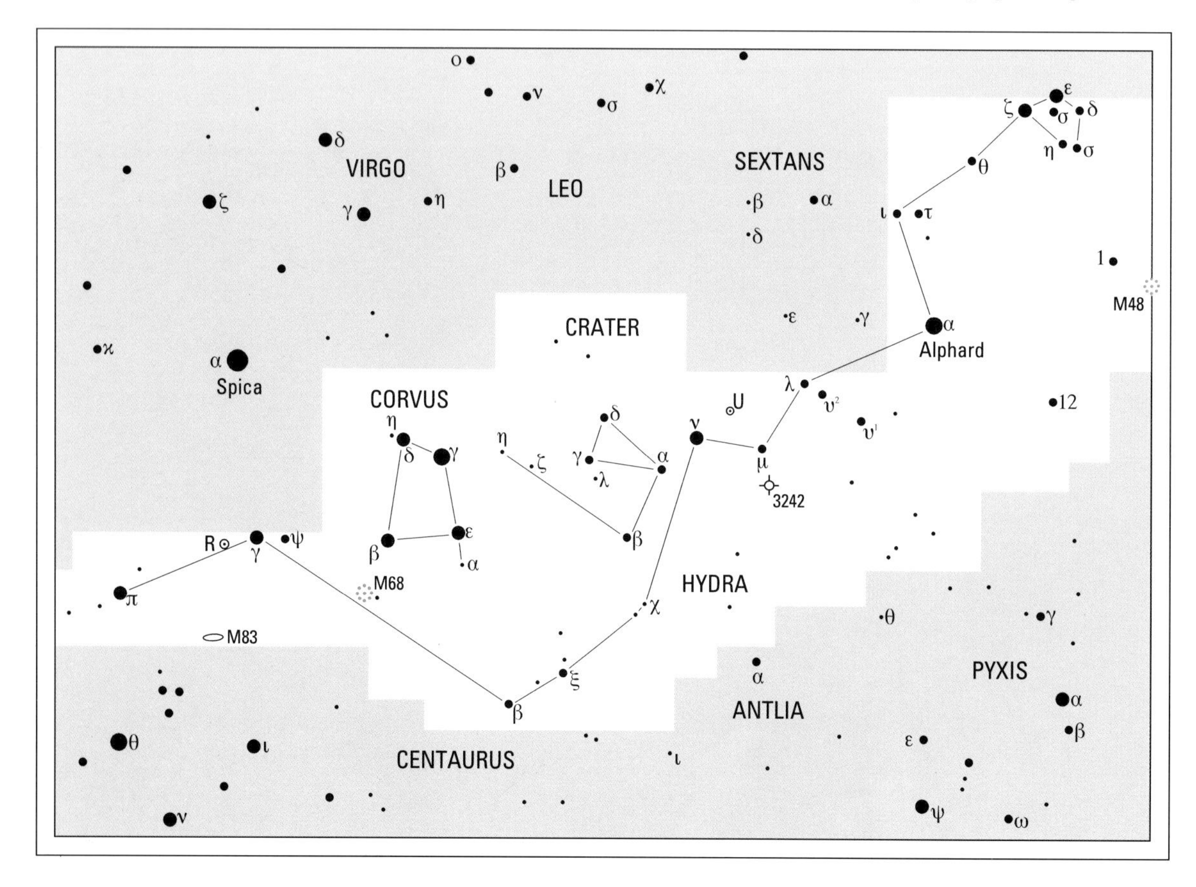

◀ ***This map*** *shows a decidedly barren region. Hydra is the largest of all the constellations, but contains only one fairly bright star, α (Alphard). The 'head' lies near Cancer, the 'tail' extends to the south of Virgo. Corvus is fairly prominent, though none of its stars is as bright as the second magnitude; Crater is very obscure.*

southern hemisphere. Several supernovae have been seen in it during recent years.

There is also a planetary nebula, NGC3242, which has been nicknamed the Ghost of Jupiter. There is a relatively bright oval ring, and a hot 12th-magnitude central star; the whole nebula is said to show a bluish-green colour, and it is certainly well worth locating. It forms a triangle with ν and μ, so that it is well placed for northern-hemisphere observers.

Two now-rejected constellations border Hydra. Noctua, the Night Owl, was sited close to γ (a rather dim globular cluster, NGC5694, lies here), while Felis, the Cat, nestled against the watersnake's body south of λ. Generally speaking, these small and obscure constellations tend to confuse the sky maps, but in some ways it is sad to think that we have said good-bye to the Owl and the Pussycat!

Corvus is one of the original constellations, listed by Ptolemy. According to legend, the god Apollo became enamoured of Coronis, mother of the great doctor Aesculapius, and sent a crow to watch her and report on her behaviour. Despite the fact that the crow's report was decidedly adverse, Apollo rewarded the bird with a place in the sky.

Corvus is easy to identify, because its four main stars, γ, β, δ and ε, all between magnitudes 2.5 and 3, make up a quadrilateral which stands out because there are no other bright stars in the vicinity. It has to be admitted that the constellation is remarkably devoid of interesting objects. Curiously, the star lettered α (Alkhiba) is more than a magnitude fainter than the four which make up the quadrilateral.

Crater, the Cup, said to represent the wine-goblet of Bacchus, is so dim and obscure that it is rather surprising to find that it was one of Ptolemy's original 48 constellations. It is not hard to identify, close to ν Hydrae, but there is nothing of immediate interest in it.

HYDRA

BRIGHTEST STARS

No.	Star	R.A. h m s	Dec. ° ′ ″	Mag.	Spectrum	Proper name
30	α	09 27 35	−08 39 31	1.98	K3	Alphard
46	γ	13 18 55	−23 10 17	3.00	G5	
16	ζ	08 55 24	+05 56 44	3.11	K0	
	ν	10 49 37	−16 11 37	3.11	K2	
49	π	14 06 22	−26 40 56	3.27	K2	
11	ε	08 46 46	+06 25 07	3.38	G0	

Also above magnitude 4.3: ξ (3.54), λ (3.61), υ (4.12), δ (4.16), β (4.28), η (4.30).

VARIABLES

Star	R.A. h m	Dec. ° ′	Range (mags)	Type	Period (d)	Spectrum
R	13 29.7	−23 17	4.0–10.0	Mira	390	M
U	10 37.6	−13 23	4.8–5.8	Semi-reg.	450	N

DOUBLES

Star	R.A. h m	Dec. ° ′	P.A. °	Sep. ″	Mags	
ε	08 46.8	+06 25	281	2.8	3.8, 6.8	A is a close binary
β	11 52.9	−33 54	008	0.9	4.7, 5.5	Difficult test

CLUSTERS AND NEBULAE

M	NGC	R.A. h m	Dec. ° ′	Mag.	Dimensions ′	Type
48	2548	08 13.8	−05 48	5.8	54	Open cluster
68	4590	12 39.5	−26 45	8.2	12	Globular cluster
83	5236	13 37.0	−29 52	8.2	11.3 × 10.2	Sc galaxy
	3242	10 24.8	−18 38	8.6	16″ × 26″	Planetary nebula (Ghost of Jupiter)

CORVUS

BRIGHTEST STARS

No.	Star	R.A. h m s	Dec. ° ′ ″	Mag.	Spectrum	Proper name
4	γ	12 15 48	−17 32 31	2.59	B8	Minkar
9	β	12 34 23	−23 23 48	2.65	G5	Kraz
7	δ	12 29 52	−16 30 55	2.95	B9	Algorel
2	ε	12 10 07	−22 37 11	3.00	K2	

Also above magnitude 4.3: α (Alkhiba) (4.02).

CRATER

A small, dim group. The brightest stars are α (Alkes) and γ, each of magnitude 4.08. Alkes lies at R.A. 10h 59m 46s, dec. −18° 17′ 56″.

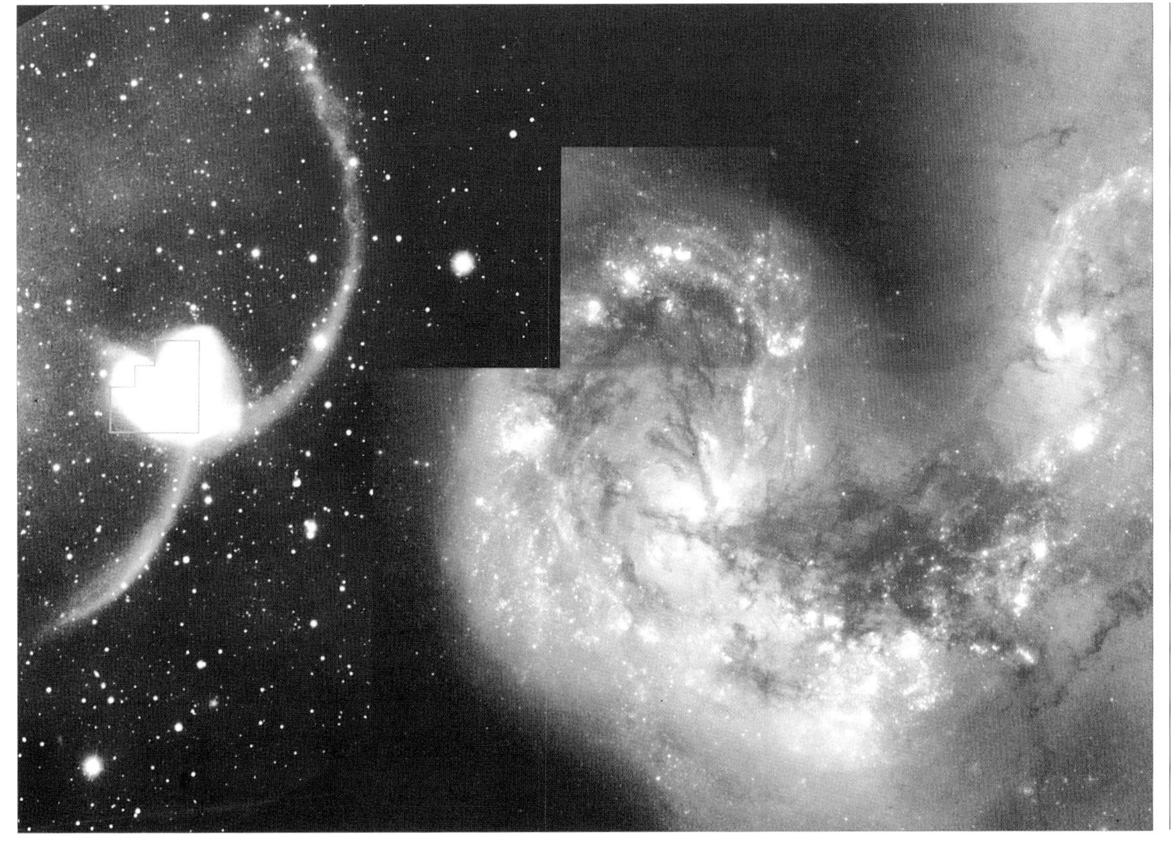

◀ ***The Antennae: colliding galaxies in Corvus, NGC 4038 and 4039.*** *(Left) Ground-based view, showing long tails of luminous matter formed by the gravitational tidal forces of the encounter. Distance 63 million light-years. (Right) Hubble Space Telescope view: the cores of the galaxies are the orange blobs left and right of image centre, criss-crossed by filaments of dark dust. A wide band of chaotic dust, called the overlap region, stretches between the cores of the two galaxies. The sweeping spiral-like patterns, traced by bright blue star clusters, shows the result of energetic star-birth activity which was triggered off by the collision. [Credit: Brad Whitmore (STScI) and NASA.]*

Lyra, Cygnus, Aquila, Scutum, Sagitta,

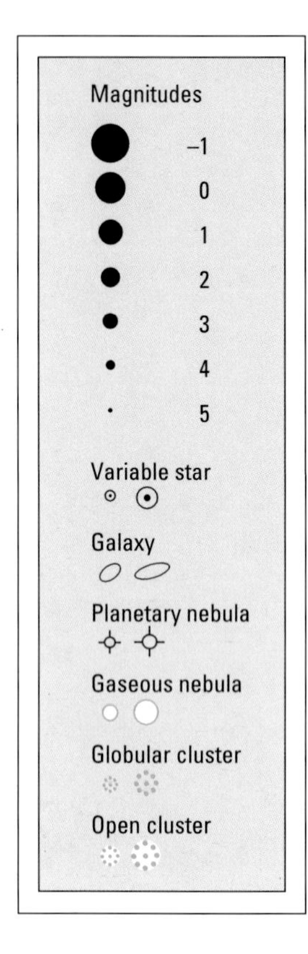

Lyra is a small constellation, but it contains a wealth of interesting objects. α (Vega) is the brightest star in the northern hemisphere of the sky apart from Arcturus, and is distinguished by its steely-blue colour; it is 26 light-years away, and 52 times as luminous as the Sun. During 1983 observations made from IRAS, the Infra-Red Astronomical Satellite, showed that Vega is associated with a cloud of cool material which may be planet-forming, though it would certainly be premature to claim that any planets actually exist there. Vega's tenth-magnitude companion, at a separation of 60 seconds of arc, merely happens to lie in almost the same line of sight; there is no real connection.

β Lyrae (Sheliak) is an eclipsing binary with alternative deep and shallow minima; it is the prototype star of its class. Its variations are very easy to follow, because the neighbouring γ (3.24) makes an ideal comparison star; when β is faint there are other comparison stars in ϰ (4.3), δ (also 4.3), and ζ (4.4). R Lyrae is a semi-regular variable, very red in colour, with a rough period of 46 days; useful comparison stars are η and θ, both of which are listed as magnitude 4.4, though I find θ to be appreciably the brighter of the two.

Close to Vega lies ε Lyrae, a splendid example of a quadruple star. Keen-eyed people can split the two main components, while a 7.6-centimetre (3-inch) telescope is powerful enough to show that each component is again double. It is worth using binoculars to look at the pair consisting of δ^1 and δ^2; here we have a good colour contrast, because the brighter star is an M-type red giant and the fainter member is white. ζ is another wide, easy double.

M57, the Ring, is the most famous of all planetary nebulae, though not actually the brightest. It is extremely easy to find, since it lies between β and γ, and a small telescope will show it. The globular cluster M56 is within binocular range, between γ Lyrae and β Cygni; it is very remote, at a distance of over 45,000 light-years. Mythologically, Lyra represents the Lyre which Apollo gave to the great musician Orpheus.

Cygnus, the Swan, said to represent the bird into which Jupiter once transformed himself while upon a clandestine visit to the Queen of Sparta, is often called the Northern Cross for obvious reasons; the X-pattern is striking. The brightest star, Deneb, is an exceptionally luminous supergiant, at least 70,000 times brighter than the Sun, and 1800 light-years away, so that we now see it as it used to be when Britain was occupied by the Romans. γ Cygni or Sadr, the central star of the X, is of type F8, and equal to 6000 Suns. One member of the pattern, β Cygni or Albireo, is fainter than the rest and also further away from the centre, so that it rather spoils the symmetry; but it compensates for this by being probably the loveliest coloured double in the sky. The primary is golden yellow, the companion vivid blue; the separation is over 34 seconds of arc, so that almost any small telescope will show both stars. It is an easy double; so too is the dim 61 Cygni, which was the first star to have its distance measured.

There are several variable stars of note. χ Cygni is a Mira star, with a period of 407 days and an exceptionally large magnitude range; at maximum it may rise to 3.3, brighter than its neighbour η, but at minimum it sinks to below 14, and since it lies in a rich area it is then none too easy to identify. χ is one of the strongest infra-red sources in the sky. U Cygni (close to the little pair consisting of o^1 and o^2) and R Cygni (in the same telescopic field with θ, magnitude 4.48) are also very red Mira variables.

P Cygni, close to γ, has a curious history. In 1600 it flared up from obscurity to the third magnitude; ever since 1715 it has hovered around magnitude 5. It is very luminous and remote, and is also known to be unstable. It is worth monitoring, because there is always the chance of a new increase in brightness; good comparison stars are 28 Cygni (4.9) and 29 Cygni (5.0)

The Milky Way flows through Cygnus, and there are conspicuous dark rifts, indicating the presence of obscuring dust. There are also various clusters and nebulae. The open cluster M29 is in the same binocular field as P and γ, and though it is sparse it is not hard to identify. M39, near ϱ, is also loose and contains about 30 stars.

NGC7000 is known as the North America Nebula. It is dimly visible with the naked eye in the guise of a slightly brighter portion of the Milky Way, and binoculars show it well as a wide region of diffuse nebulosity; photographs show that its shape really does bear a marked resemblance to that of the North American continent. It is nearly 500

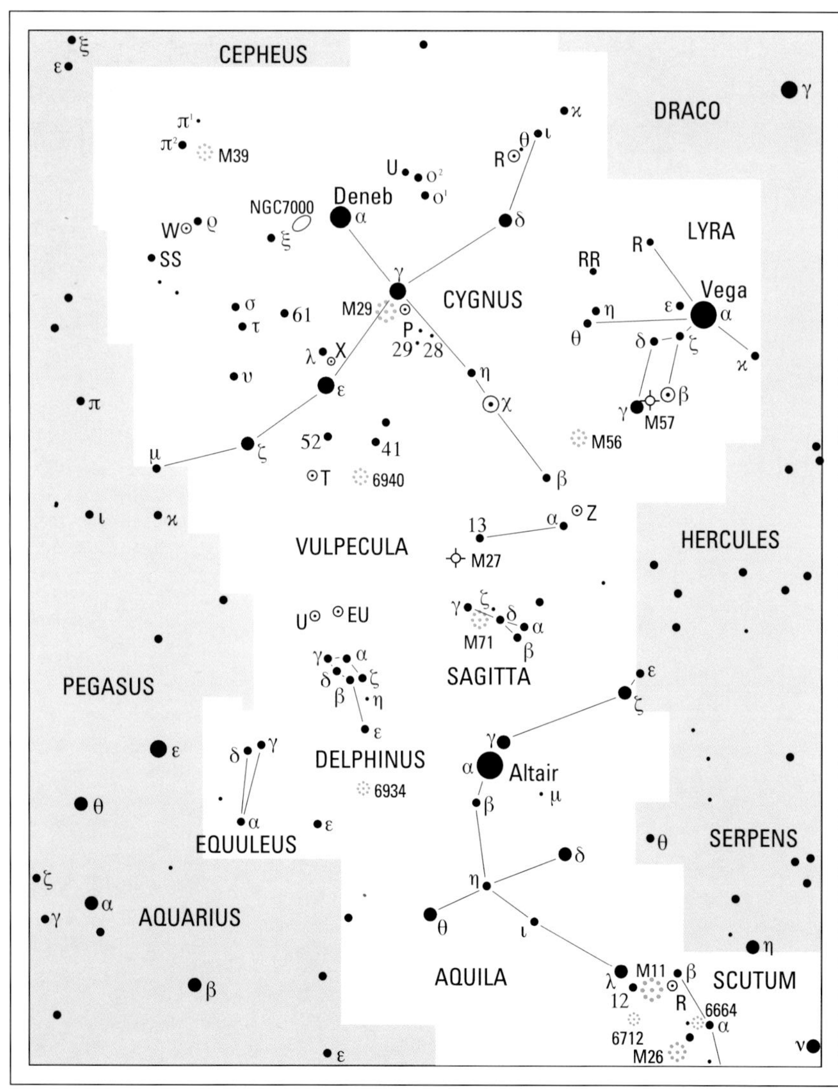

◀ ***This map*** *is dominated by three bright stars: Deneb, Vega and Altair. Because they are so prominent during summer evenings in the northern hemisphere, I once referred to them as 'the Summer Triangle', and the name has come into general use, though it is quite unofficial and is inappropriate in the southern hemisphere. The Milky Way crosses the area, which is very rich. All three stars of the 'Triangle' can be seen from most inhabited countries, though from New Zealand Deneb and Vega are always very low.*

Vulpecula, Delphinus, Equuleus

LYRA

BRIGHTEST STARS

No.	Star	R.A.			Dec.			Mag.	Spectrum	Proper name
		h	m	s	°	′	″			
3	α	18	36	56	+38	47	01	0.03	A0	Vega
14	γ	18	58	56	+32	41	22	3.24	B9	Sulaphat
10	β	18	50	05	+33	21	46	3.3 (max)	B7	Shelik

Also above magnitude 4.3: ε (3.9) (combined magnitude); R (3.9, max).

VARIABLES

Star	R.A.		Dec.		Range	Type	Period	Spectrum
	h	m	°	′	(mags)		(d)	
β	18	50.1	+33	22	3.3–4.3	β Lyrae	12.94	B+A
R	18	55.3	+43	57	3.9–5.0	Semi-reg.	46	M
RR	19	25.5	+42	47	7.1–8.1	RR Lyrae	0.57	A–F

DOUBLES

Star	R.A.		Dec.		P.A.	Sep.	Mags	
	h	m	°	′	°	″		
ε	18	44	+39	40	AB+CD 173	207.7	4.7, 5.1	
					ε^1 = AB 357	2.8	5.0, 5.1	Quadruple star
					ε^2 = CD 094	2.3	5.2, 5.5	
ζ	18	44.8	+37	36	150	43.7	4.3, 5.9	
β	18	50.1	+33	22	149	45.7	var, 8.6	

CLUSTERS AND NEBULAE

M	NGC	R.A.		Dec.		Mag.	Dimensions	Type
		h	m	°	′		′	
56	6779	19	16.6	+30	11	8.2	7.1	Globular cluster
57	6720	18	53.6	+33	02	9.7	70″ × 150″	Planetary nebula (Ring Nebula)

CYGNUS

BRIGHTEST STARS

No.	Star	R.A.			Dec.			Mag.	Spectrum	Proper name
		h	m	s	°	′	″			
50	α	20	41	26	+45	16	49	1.25	A2	Deneb
37	γ	20	22	13	+40	15	24	2.20	F8	Sadr
53	ε	20	46	12	+33	58	13	2.46	K0	Gienah
18	δ	19	44	58	+45	07	51	2.87	A0	
6	β	19	30	43	+27	57	35	3.08	K5	Albireo
64	ζ	21	12	56	+30	13	37	3.20	G8	

Also above magnitude 4.3: ξ (3.72); τ (3.72), κ (3.77), ι (3.79), o^1 (3.79), η (3.89), ν (3.94), o^2 (3.98), 41 (4.01), ρ (4.02), 52 (4.22), σ (4.23), π^2 (4.23). The curious variable P Cygni has been known to reach magnitude 3, but is usually nearer 5. At some maxima the red Mira variable χ Cygni can reach magnitude 3.3, but most maxima are considerably fainter than this.

VARIABLES

Star	R.A.		Dec.		Range	Type	Period	Spectrum
	h	m	°	′	(mags)		(d)	
χ	19	50.6	+32	55	3.3–14.2	Mira	407	S
P	20	17.8	+38	02	3–6	Recurrent nova	–	Pec.
R	19	36.8	+50	12	5.9–14.2	Mira	426	M
U	20	19.6	+47	54	5.9–12.1	Mira	462	N
X	20	43.4	+35	35	5.9–6.9	Cepheid	16.4	F–G
W	21	36.0	+45	22	5.0–7.6	Semi-reg.	126	M
SS	21	42.7	+43	35	8.4–12.4	SS Cyg (U Gem)	±50	A–G

DOUBLES

Star	R.A.		Dec.		P.A.	Sep.	Mags	
	h	m	°	′	°	″		
β	19	30.7	+27	58	054	34.4	3.1, 5.1	Yellow, blue
δ	19	45.0	+45	07	225	2.4	2.9, 6.3	Binary, 828 years
61	21	06.9	+38	45	148	29.9	5.2, 6.0	

CLUSTERS AND NEBULAE

M	NGC	R.A.		Dec.		Mag.	Dimensions	Type
		h	m	°	′		′	
29	6913	20	23.9	+38	32	6.6	7	Open cluster
39	7092	21	32.2	+48	26	4.6	32	Open cluster
	7000	20	58.8	+44	20	6.0	120 × 100	Nebula (North America Nebula)

AQUILA

BRIGHTEST STARS

No.	Star	R.A.			Dec.			Mag.	Spectrum	Proper name
		h	m	s	°	′	″			
53	α	19	50	47	+08	52	06	0.77	A7	Altair
50	γ	19	46	15	+10	36	48	2.72	K3	Tarazed
17	ζ	19	05	24	+13	51	48	2.99	B9	Dheneb
65	θ	20	11	18	−00	49	17	3.23	B9	
30	δ	19	25	30	+03	06	53	3.36	F0	
16	λ	19	06	15	−04	52	57	3.44	B9	Althalimain

Also above magnitude 4.3: β (Alshain) (3.71), ε (4.02), 12 (4.02). The Cepheid variable η rises to 3.5 at maximum.

VARIABLES

Star	R.A.		Dec.		Range	Type	Period	Spectrum
	h	m	°	′	(mags)		(d)	
η	19	52.5	+01	00	3.5–4.4	Cepheid	7.2	F–G
R	19	06.4	+08	14	5.5–12.0	Mira	284	M

SCUTUM

The brightest star is α: R.A. 18h 35m 12s.1, dec. −08° 14′ 39″, mag. 3.85.
Also above magnitude 4.3: β (4.22).

VARIABLE

Star	R.A.		Dec.		Range	Type	Period	Spectrum
	h	m	°	′	(mags)		(d)	
R	18	47.5	−05	42	4.4–8.2	RV Tauri	140	G–K

CLUSTERS AND NEBULAE

M	NGC	R.A.		Dec.		Mag.	Dimensions	Type
		h	m	°	′		′	
11	6705	18	51.1	−06	16	5.8	14	Open cluster (Wild Duck)
26	6694	18	45.2	−09	24	8.0	15	Open cluster
	6664	18	36.7	−08	13	7.8	16	Open cluster (EV Scuti cl.)
	6712	18	53.1	−08	42	8.2	7	Globular cluster

SAGITTA

The only two stars above magnitude 4.3 are γ and δ. The brightest, γ, lies at R.A. 19h 58m 45s.3, dec. +19° 29′ 32″; magnitude 3.47. The magnitude of δ is 3.82.

CLUSTERS AND NEBULAE

M	NGC	R.A.		Dec.		Mag.	Dimensions	Type
		h	m	°	′		′	
71	6838	19	53.8	+18	47	8.3	7.2	Globular cluster

VULPECULA

The brightest star is α: R.A. 19h 28m 42s.2, dec. +24° 39′ 24″, magnitude 4.44.

VARIABLES

Star	R.A.		Dec.		Range	Type	Period	Spectrum
	h	m	°	′	(mags)		(d)	
T	20	51.5	+28	15	5.4–6.1	Cepheid	4.4	F–G
Z	19	21.7	+25	34	7.4–9.2	Algol	2.45	B+A

DOUBLE

Star	R.A.		Dec.		P.A.	Sep.	Mags
	h	m	°	′	°	″	
α - 8	19	28.7	+24	40	028	413.7	4.4, 5.8

CLUSTERS AND NEBULAE

M	NGC	R.A.		Dec.		Mag.	Dimensions	Type
		h	m	°	′		′	
27	6853	19	59.6	+22	43	7.6	350″ × 910″	Planetary (Dumbbell)
	6940	20	34.6	+28	18	6.3	31	Open cluster

DELPHINUS

The brightest star is β: R.A. 20h 37m 32s.8, dec. 14° 35′ 43″, mag. 3.54. Also above magnitude 4.3: α (3.77), γ (3.9 combined magnitude), ε (4.03).

VARIABLES

Star	R.A.		Dec.		Range	Type	Period	Spectrum
	h	m	°	′	(mags)		(d)	
U	20	45.5	+18	05	5.6-8.9	Semi-regular	110	M
EU	20	37.9	+18	16	5.8-6.9	Semi-regular	59	M

DOUBLE

Star	R.A.		Dec.		P.A.	Sep.	Mags
	h	m	°	′	°	″	
γ	20	46.7	+16	07	268	9.6	4.5,5.5

CLUSTERS AND NEBULAE

M	NGC	R.A.		Dec.		Mag.	Dimensions	Type
		h	m	°	′		′	
	6934	20	34.2	+07	24	8.9	5.9	Globular cluster

EQUULEUS

The only star above mag. 4.3 is α (Kitalpha), R.A. 21h 15m 49s.3, dec. 05° 14′ 52″, mag. 3.92.

DOUBLE

Star	R.A.		Dec.		P.A.	Sep.	Mags	
	h	m	°	′	°	″		
ε	20	59.1	+04	18	AB 285	1.0	6.0, 6.3	Binary,101 years
					AB+C 070	10.7	7.1	
					AD 280	74.8	12.4	

Hercules

light-years in diameter, and may owe much of its illumination to Deneb. It lies in the same field as reddish ξ Cygni.

Aquila, the Eagle, commemorates the bird sent by Jupiter to fetch a shepherd boy, Ganymede, who was destined to become the cup-bearer of the gods. Altair, at a distance of 16.6 light-years, is the closest of the first-magnitude stars apart from α Centauri, Sirius and Procyon; it is ten times as luminous as the Sun, and is known to be rotating so rapidly that it must be egg-shaped. It is flanked to either side by two fainter stars, γ (Tarazed) and β (Alshain). γ is an orange K-type star, much more powerful than Altair but also much more remote.

η Aquilae is a Cepheid variable. The range is from magnitude 3.4 to 4.4, so that δ and θ make ideal comparison stars; when η is near minimum, a useful comparison star is ι (4.0).

Scutum is not an original constellation; it was one of Hevelius' inventions, and was originally Scutum Sobieskii, Sobieski's Shield. The variable R Scuti is the brightest member of the RV Tauri class, and is a favourite binocular target; there are alternate deep and shallow minima, with occasional periods of irregularity. Of the two Messier open clusters, much the more striking is the fan-shaped M11, which has been nicknamed the Wild Duck cluster and is a glorious sight in any telescope; it contains hundreds of stars, and is easily identified, being close to λ and 12 Aquilae.

Sagitta, the Arrow – Cupid's Bow – is distinctive; the main arrow pattern is made up of the two bright stars δ and γ, together with α and β (each of magnitude 4.37). There is one Messier object, M71, which was formerly classified as an open cluster, but is now thought to be a globular, though it is much less condensed than other systems of this type. It lies a little less than halfway from γ to δ.

Vulpecula was originally Vulpecula et Anser, the Fox and Goose, but the goose has long since vanished from the maps. The constellation is very dim, but is redeemed by the presence of M27, the Dumbbell, probably the finest of all planetary nebulae. There is no problem in finding it with binoculars; it is close to γ Sagittae, which is the best guide to it. A moderate power will reveal its characteristic shape. Like all planetaries it is expanding; the present diameter is of the order of two and a half light-years.

Delphinus is one of Ptolemy's original constellations. It honours a dolphin which carried the great singer Arion to safety when he had been thrown overboard by the crew of the ship which was carrying him home after winning all the prizes in a competition. Delphinus is a compact little group – unwary observers have been known to confuse it with the Pleiades. Its two leading stars have curious names: α is Svalocin, β is Rotanev. These names were given by one Nicolaus Venator, and the association is obvious enough.

γ Delphini is a wide, easy double, and the two red semi-regular variables U and EU are good binocular objects. Near them an interesting nova, HR Delphini, flared up in 1967 and was discovered by the English amateur G. E. D. Alcock; it reached magnitude 3.7, and remained a naked-eye object for months. Its present magnitude is between 12 and 13, and as this was also the pre-outburst value it is unlikely to fade much further.

Equuleus represents a foal given by Mercury to Castor, one of the Heavenly Twins. It is so small and dim that it is surprising to find it in Ptolemy's original list, but the little triangle made up of α, δ (4.49) and γ (4.69) is not hard to identify, between Delphinus and β Aquarii. ε is a triple star, but otherwise Equuleus contains nothing of immediate interest.

Hercules is a very large constellation; it covers 1225 square degrees, but it is not particularly rich. The best guide to it is Rasalhague, or α Ophiuchi, which is of the second magnitude – bright enough to be prominent – and is also rather isolated. Not far from it is Rasalgethi or α Herculis, which is some way away from the other main stars of the constellation. The main part of Hercules lies inside the triangle bounded by Alphekka in Corona Borealis, Rasalhague and Vega. With its high northern declination, part of it is circumpolar from the latitudes of Britain or the northern United States. Its main features are the red supergiant Rasalgethi, a wide and easy binary (ζ), and two spectacular globular clusters (M13 and M92).

In 1759 William Herschel discovered that Rasalgethi is variable. At that time only four variables had been found – Mira Ceti, Algol in Perseus, χ Cygni and R Hydrae – so that the discovery was regarded as very important. Certainly there is no doubt about the fluctuations; it is said that the extreme range is from magnitude 3.0 to 4.0, though for most of the time the star remains between 3.1 and 3.7. Officially it is classed as a semi-regular with a rough period of 90 to 100 days, but this period is by no means well marked. The variations are slow, but can be followed with the naked eye; suitable comparison stars are ϰ Ophiuchi (3.20), δ Herculis (3.14) and γ Herculis (3.75). β Herculis (2.71), the brightest star in the constellation, is always considerably superior to Rasalgethi.

The distance of Rasalgethi is 218 light-years; the spectral type is M. What makes it so notable is its vast size. It may be even larger than Betelgeux, in which case its diameter exceeds 400 million kilometres (250 million miles). It is relatively cool – the surface temperature is well below 3000 degrees C – and its outer layers, at least, are very rarefied. It is a very powerful emitter of infra-red radiation.

Rasalgethi is also a fine double. The companion is of magnitude 5.3, and since the separation is not much short of 5 seconds of arc a small telescope will resolve the pair. The companion is often described as vivid green, though this is due mainly, if not entirely, to contrast with the redness of the primary. The companion is itself an excessively close binary, with a period of 51.6 days, and there is every reason to believe that both stars are enveloped in a huge, rarefied cloud. Rasalgethi is indeed a remarkable system.

δ Herculis has an eighth-magnitude companion at a separation of 9 seconds of arc (position angle 236 degrees), but this is an optical pair; there is no connection between the two components, and the secondary lies well in the background. δ itself is an ordinary A-type star, 35 times as luminous as the Sun and 91 light-years away.

Of more interest is ζ Herculis, or Rutilicus, which is a fine binary; its duplicity was discovered by William Herschel in 1782. The magnitudes are 2.9 and 3.5; the period is only 34.5 years, so that both separation and position angle change quickly. In 1994, the separation was 1.6 seconds of arc, so that this is a very wide, easy pair. The primary is a G-type subgiant, 31 light-years away and rather more than five times as luminous as the Sun.

68 (u) Herculis is an interesting variable of the β Lyrae type. The secondary minimum takes the magnitude down to 5.0 and the deep minimum to only 5.3, so that the star is always within binocular or even naked-eye range. Both components are B-type giants, so close that they almost touch; as with β Lyrae, each must be pulled out into the shape of an egg. If the distance is around 600 light-years, as seems possible, each star must be well over 100 times as luminous as the Sun.

On 13 December 1934 the English amateur J. P. M. Prentice discovered a bright nova in Hercules, near ι and

HERCULES

BRIGHTEST STARS

No.	Star	R.A. h	m	s	Dec. °	′	″	Mag.	Spectrum	Proper name
27	β	16	30	13	+21	29	22	2.77	G8	Kornephoros
40	ζ	16	41	17	+31	36	10	2.81	G0	Rutilicus
64	α	17	14	39	+14	23	25	3.0 (max)	M5	Rasalgethi
65	δ	17	15	02	+24	50	21	3.14	A3	Sarin
67	π	17	15	03	+36	48	33	3.16	K3	
86	μ	17	46	27	+27	43	15	3.42	G5	

Also above magnitude 4.3: η (3.53), ξ (3.70), γ (3.75), ι (3.80), ο (3.83), 109 (3.84), θ (3.86), τ (3.89), ε (3.92), 110 (4.19), σ (4.20), 95 (4.27).

VARIABLES

Star	R.A. h	m	Dec. °	′	Range (mags)	Type	Period (d)	Spectrum
α	17	14.6	+14	23	3–4	Semi-reg.	±100?	M
30 (g)	16	28.6	+41	53	5.7–7.2	Semi-reg.	70	M
68 (u)	17	17.3	+35	06	4.6–5.3	β Lyrae	2.05	B+B

DOUBLES

Star	R.A. h	m	Dec. °	′	P.A. °	Sep. ″	Mags	
α	17	14.6	+14	23	107	4.7	var, 5.4	Binary, 3600y red, green
ζ	16	41.3	+31	36	089	1.6	2.9, 3.5	Binary, 34.5y
ϱ	17	23.7	+37	09	316	4.1	4.6, 5.6	

CLUSTERS AND NEBULAE

M	NGC	R.A. h	m	Dec. °	′	Mag.	Dimensions ′	Type
13	6205	16	41.7	+36	28	5.9	16.6	Globular cluster
92	6341	17	17.1	+43	08	5.5	11.2	Globular cluster

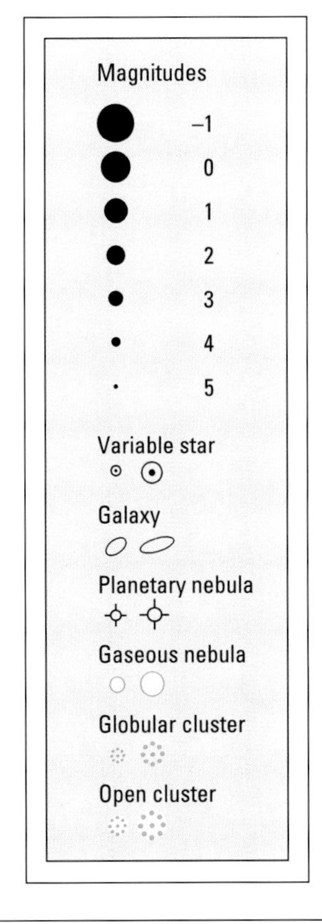

▼ ***The area** enclosed in the quadrilateral formed by imaginary lines joining Arcturus, Vega, Altair and Antares is occupied by three large, dim constellations: Hercules, Ophiuchus and Serpens. The region is best seen during evenings in the northern summer (southern winter), but there are no really distinctive patterns. Although Hercules is so extensive it has no star much brighter than the third magnitude.*

not far from the Dragon's head. It rose to magnitude 1.2, so that it remains the brightest nova to have appeared in the northern hemisphere of the sky since Nova Aquilae 1918. During its decline it was strongly green for a while, and was also unusual inasmuch as it remained a naked-eye object for several months. It has now faded back to around its pre-outburst magnitude of 15, and has been found to be an eclipsing binary with the very short period of 4 hours 39 minutes. Both components are dwarfs – one white, one red.

M13, which lies rather more than halfway between ζ and η, is the brightest globular cluster north of the celestial equator; its only superiors are ω Centauri and 47 Tucanae. M13 is just visible with the naked eye on a clear night, but it is far from obvious, and it is not surprising that it was overlooked until Edmond Halley chanced upon it in 1714 – describing it as 'a little patch, but it shows itself to the naked eye when the sky is serene and the Moon absent'. William Herschel was more enthusiastic about it: 'A most beautiful cluster of stars exceedingly condensed in the middle and very rich. Contains about 14,000 stars.' In fact this is a gross underestimate; half a million would be closer to the truth.

Like all globulars, M13 is a long way away. Its distance has been given as 22,500 light-years and its real diameter perhaps 160 light-years, with a condensed central region 100 light-years across. It lies well away from the main plane of the Milky Way, so that it has not been greatly disturbed by the concentration of mass in the centre of the Galaxy, and is certainly very old indeed. Binoculars give good views of it, and even a small telescope will resolve the outer parts into stars.

The second globular, M92, lies directly between η and ι. It is on the fringe of naked-eye visibility – very keen-eyed observers claim that they can glimpse it – and telescopically it is not much inferior to M3. It is rather further away, at a distance of 37,000 light-years, and in most respects it seems to be similar to M13, though it contains a larger number of variable stars.

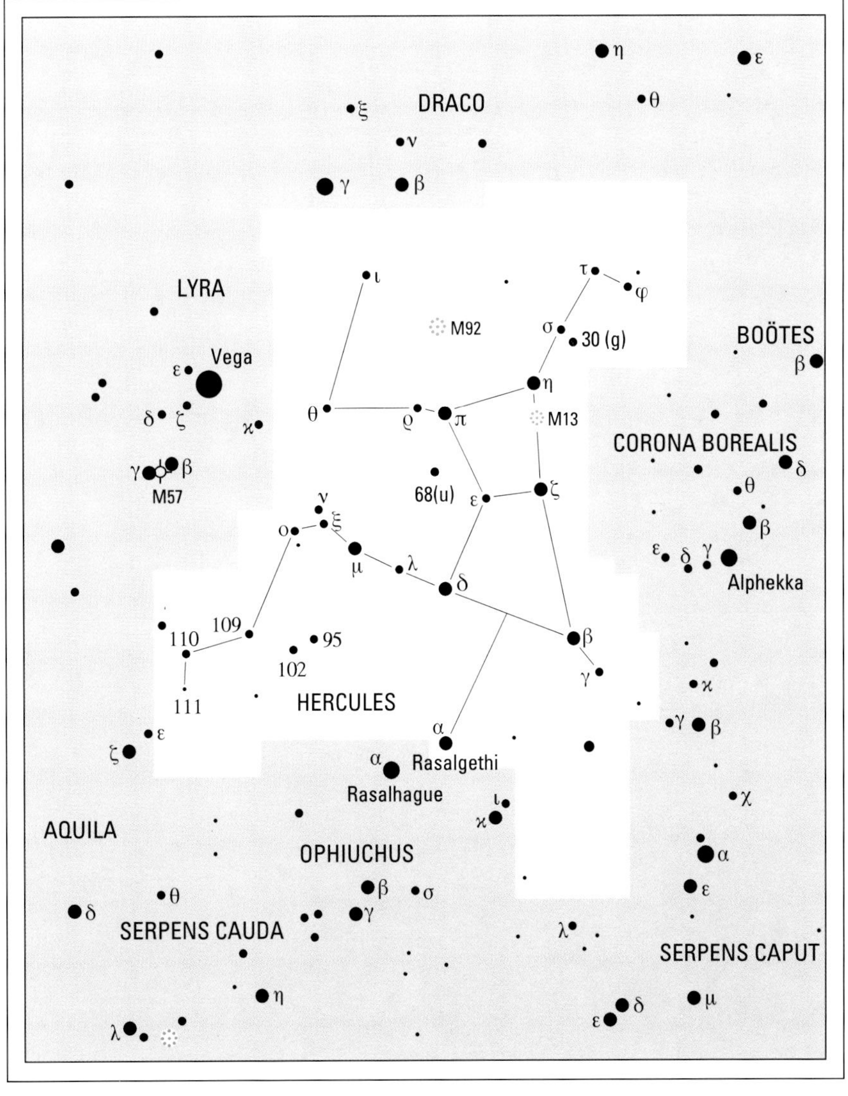

Ophiuchus, Serpens

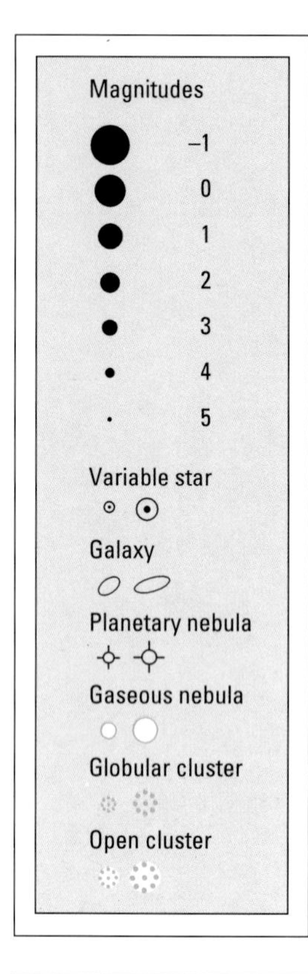

Ophiuchus is another large constellation, covering 948 square degrees of the sky, but it is confusingly intertwined with the two parts of Serpens. It straddles the equator; of its brighter stars ϰ is at declination 9 degrees north, θ almost 25 degrees south. There is no distinctive pattern, and the only star bright enough to be really prominent is α (Rasalhague), which is 62 light-years away and 67 times as luminous as the Sun. Rasalhague is much closer and less powerful than its neighbour Rasalgethi, in Hercules (see Star Map 9), even though it looks a full magnitude the brighter of the two. It is also different in colour; Rasalhague is white, while Rasalgethi (α Hercules) is a red supergiant.

Of the other leaders of Ophiuchus, δ is of type M, and its redness contrasts well with that of the nearby ε, which is only slightly yellowish. The two are not genuine neighbours; δ is 140 light-years away, ε only just over 100. It is worth looking at them with binoculars, because they are in the same field.

On the very edge of the constellation close to σ Scorpii (one of the two stars flanking Antares) is a wide double ϱ Ophiuchi (not shown on map) which lies close to a very rich region which is a favourite photographic target. η is a binary with components which are not very unequal (the primary is no more than half a magnitude brighter than the secondary), but the separation is less than one second of arc, so that it is a good test for telescopes of around 25-centimetre (10-inch) aperture.

The most interesting variable in Ophiuchus is RS, which is a recurrent nova and the only member of the class, apart from the 'Blaze Star', T Coronae, which can flare up to naked-eye visibility – as it did in 1898, 1933, 1958, 1967 and 1987; the usual magnitude is rather below 12. It is at least 3000 light-years away, and is worth monitoring, as a new outburst may occur at any time. Ophiuchus was the site of the last galactic supernova, Kepler's Star of 1604, which for a while outshone Mars and Jupiter. It is a pity that it appeared before telescopes came into common use.

Another interesting object in the constellation is Munich 15040, better known as Barnard's Star because it was discovered, in 1916, by the American astronomer Edward Emerson Barnard. It is not easy to locate, because its magnitude is below 9. It is the closest of all stars apart from the members of the α Centauri system, and has the greatest proper motion known, so that in fact in only 190 years or so it will shift against its background by a distance equal to the apparent diameter of the full moon. It is an extremely feeble red dwarf, and irregularities in its motion have led to the belief that it is attended by at least one companion which is of planetary rather than stellar mass, though definite proof is still lacking. The guide star to it is 66 Ophiuchi, magnitude 4.6. Incidentally, this was the region where an astronomer named Poczobut, in 1777, tried to introduce a new constellation, Taurus Poniatowski, or Poniatowski's Bull; it also included 70 Ophiuchi (a well-known but rather close binary), together with 67 and 68. Not surprisingly, the little Bull has been deleted from current maps.

Ophiuchus contains no less than seven globular clusters in Messier's list. All of them are reasonably bright, so that they are favourite objects for users of binoculars or wide-field telescopes. M2 is interesting because it is less condensed than most globulars, and is therefore easier to resolve; it may be compared with its neighbour M10, which is much more concentrated.

Mythologically, Ophiuchus is identified with Aesculapius, son of Apollo and Coronis, whose skill in

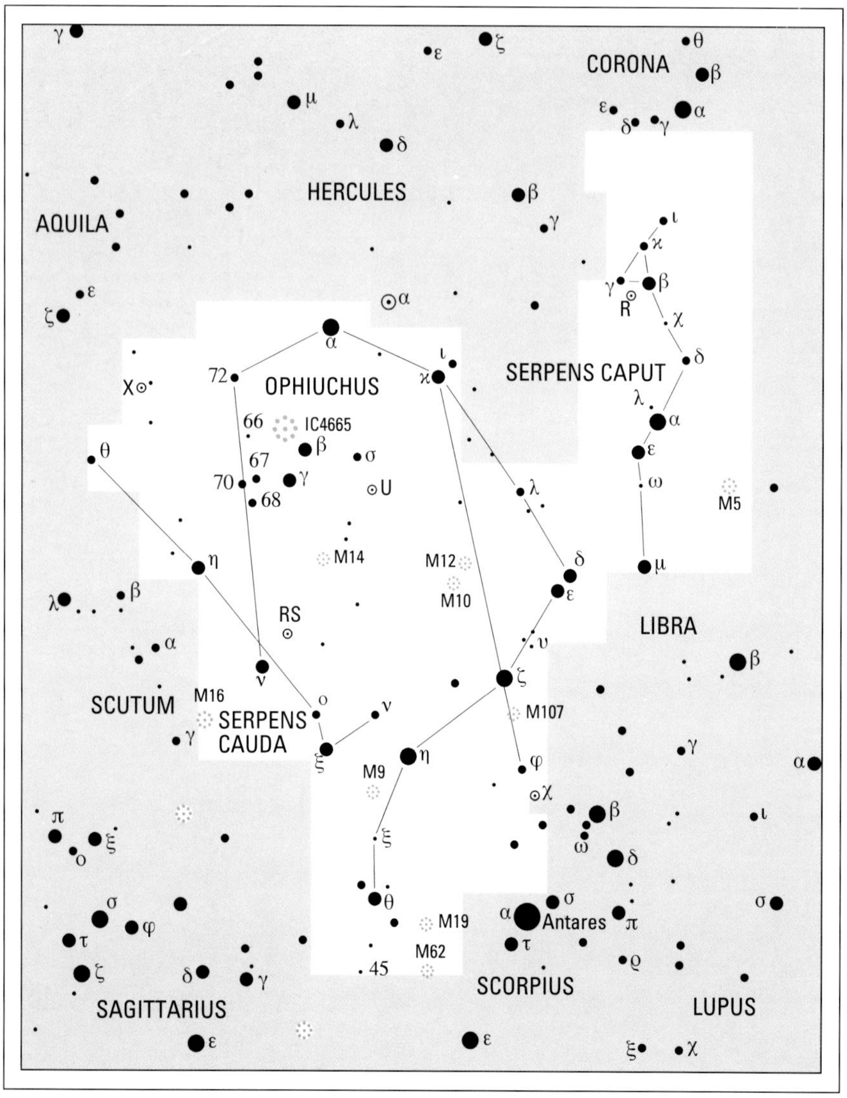

◀ ***These constellations*** *are relatively hard to identify, particularly as they are so confused. In mythology Ophiuchus was the Serpent-bearer (a former name for it was Serpentarius) and Serpens was the reptile with which he was struggling – and which he has apparently pulled in half! The only bright star in the region is α Ophiuchi (Rasalhague). Ophiuchus extends into the Zodiac between Scorpius and Sagittarius, so that the major planets can pass through it.*

▶ ***The globular cluster M12*** *in Ophiuchus, photographed by John Fletcher using a 25-cm (10-inch) reflector. M12 is easier to resolve into individual stars than other globular clusters such as M10.*

medicine was legendary. He was able even to restore the dead to life, and this angered the God of the Underworld, whose realm was starting to become depopulated. Jupiter reluctantly disposed of Aesculapius by striking him with a thunderbolt, but then relented sufficiently to transport him to the sky.

Serpens. Of the two halves of Serpens, the head (Caput) is much the more conspicuous, and there is one fairly bright star, the reddish α or Unukalhai: magnitude 2.65, distance 88 light-years, luminosity 90 times that of the Sun. The actual head is formed by a little triangle of stars; ϰ (magnitude 4.09), β (3.67) and γ (3.85). Directly between β and γ is the Mira variable R Serpentis, which can rise to naked-eye visibility when at its best but which becomes very faint at minimum. Like almost all members of its class, it is very red. Its period is only nine days less than a year, so that when it peaks at times when it is above the horizon only during the hours of daylight, the maxima are virtually unobservable for several consecutive years. The date of maximum in 1994 was 25 March.

M5, which lies some way from Unukalhai, is one of the finest globular clusters in the entire sky; only ω Centauri, 47 Tucanae, M13 in Hercules and M22 in Sagittarius are brighter. It is very evident in binoculars, and has been known ever since 1702, when Gotfried Kirch discovered it. M5 is easy to resolve; the distance is 27,000 light-years, and, unlike M13, it is particularly rich in variable stars.

The Serpent's body (Cauda) is less prominent, and the brightest star, η or Alava, is only of magnitude 3.26 (distance is 52 light-years, luminosity 17 times that of the Sun). However, Cauda does contain two objects of note. One is M16, the Eagle Nebula, in which is embedded the cluster NGC6611. It is on the fringe of the constellation, adjoining Scutum, and in fact the guide star to it is γ Scuti, magnitude 4.7. The Eagle is a large, diffuse area of nebulosity, while the cluster is reasonably well marked. The two are not difficult to locate, but of course photographs are needed to bring out their full glory; there is a mass of detail, and there are areas of dark nebulosity together with small, circular 'globules', which will eventually condense into stars.

The other main feature of Cauda is θ (Alya), a particularly wide and easy double. The components are identical twins, each of magnitude 4.5 and of spectral type A5. With the naked eye, θ appears as a single star of magnitude 3.4, but good binoculars will show both components. Of course they make up a genuine binary system, but they are a long way apart – perhaps 900 times the distance between the Earth and the Sun, so that the revolution period is immensely long. The distance from us is just over 100 light-years. It may be said that θ Serpentis is one of the best 'demonstration doubles' in the entire sky. To find it, follow through the line of θ, η and δ Aquilae (Star Map 8); if this is prolonged for an equal distance beyond, it will reach θ Serpentis.

OPHIUCHUS

BRIGHTEST STARS

No.	Star	R.A. h	m	s	Dec. °	′	″	Mag.	Spectrum	Proper name
55	α	17	34	56	+12	33	36	2.08	A5	Rasalhague
35	η	17	10	22	−15	43	30	2.43	A2	Sabik
13	ζ	16	37	09	−10	34	02	2.56	O9.5	Han
1	δ	16	14	21	−03	41	39	2.74	M1	Yed Prior
60	β	17	43	28	+04	34	02	2.77	K2	Cheleb
27	ϰ	16	57	40	+09	22	30	3.20	K2	
2	ε	16	18	19	−04	41	33	3.24	G8	Yed Post
42	θ	17	22	00	−24	59	58	3.27	B2	
64	ν	17	59	01	−09	46	25	3.34	K0	

Also above magnitude 4.3: 72 (3.73), γ (3.75), λ (Marfik) (3.82), 67 (3.97), 70 (4.03), φ (4.28), 45 (4.29).

VARIABLES

Star	R.A. h	m	Dec. °	′	Range (mags)	Type	Period (d)	Spectrum
χ	16	27.0	−18	27	4.2–5.0	Irregular	–	B
U	17	16.5	+01	13	5.9–6.6	Algol	1.68	B+B
X	18	38.3	+08	50	5.9–9.2	Mira	334	M+K
RS	17	50.2	−06	43	5.3–12.3	Recurrent nova	--	O+M

DOUBLES

Star	R.A. h	m	Dec. °	′	P.A. °	Sep. ″	Mags	
ϱ	16	25.6	−23	27	344	3.1	5.3, 6.0	
η	17	10.4	−15	43	247	0.5	3.0, 3.5	Binary, 64 years Difficult test

CLUSTERS AND NEBULAE

M	NGC	R.A. h	m	Dec. °	′	Mag.	Dimensions ′	Type
9	6333	17	19.2	−18	31	7.9	9.3	Globular cluster
10	6254	16	57.1	−04	06	6.6	15.1	Globular cluster
12	6218	16	47.2	−01	57	6.6	14.5	Globular cluster
14	6402	17	37.6	−03	15	7.6	11.7	Globular cluster
19	6273	17	02.6	−26	16	7.1	13.5	Globular cluster
62	6266	17	01.2	−30	07	6.6	14.1	Globular cluster
107	6171	16	32.5	+03	13	8.1	10.0	Globular cluster
	6633	18	27.7	+06	34	4.6	27	Open cluster
	IC4665	17	46.3	+05	43	4.2	41	Open cluster

SERPENS

BRIGHTEST STARS

No.	Star	R.A. h	m	s	Dec. °	′	″	Mag.	Spectrum	Proper name
(Caput)										
24	α	15	44	16	+06	25	32	2.65	K2	Unukalhai
(Cauda)										
58	η	18	21	18	−02	53	56	3.26	K0	Alava
63	θ	18	56	13	+04	12	13	3.4 (combined)	A5+A5	Alya

Also above magnitude 4.3: Caput: μ (3.54), β (3.67), ε (3.71), δ (3.80), γ (3.85), ϰ (4.09); Cauda: ξ (3.54), ο (4.26).

VARIABLE (Caput)

Star	R.A. h	m	Dec. °	′	Range (mags)	Type	Period (d)	Spectrum
R	15	50.7	+15	08	5.1–14.4	Mira	356	M

DOUBLES

Star	R.A. h	m	Dec. °	′	P.A. °	Sep. ″	Mags	
δ	16	34.8	+10	32	177	4.4	4.1, 5.2	Binary, 3168y (Caput)
θ	18	56.2	+04	12	104	22.3	4.5, 4.5	(Cauda)

CLUSTERS AND NEBULAE

M	NGC	R.A. h	m	Dec. °	′	Mag.	Dimensions ′	Type
5	5904	15	18.6	+02	05	5.8	17.4	Globular cluster (Caput)
16	6611	18	18.8	−13	47		35 × 28	Nebula and cluster (Cauda) Eagle Nebula and cluster NGC6611

Scorpius, Sagittarius, Corona Australis

Magnitudes
-1
0
1
2
3
4
5
Variable star
Galaxy
Planetary nebula
Gaseous nebula
Globular cluster
Open cluster

Scorpius is often, incorrectly, referred to as Scorpio. The leader is Antares, which is sufficiently close to the celestial equator to reach a reasonable altitude over Britain and the northern United States – though the extreme southern part of the Scorpion does not rise in these latitudes; the southernmost bright star, θ or Sargas, has a declination of almost −43 degrees. Antares is generally regarded as the reddest of the first-magnitude stars, though its colour is much the same as that of Betelgeux. It is interesting to compare these two red supergiants. Antares is 330 light-years away, and 7500 times as luminous as the Sun, so that it has only about half the power of Betelgeux; it is slightly variable, but the fluctuations, unlike those of Betelgeux, are too slight to be noticed with the naked eye.

Antares has a companion which looks slightly greenish by comparison. Both are enveloped in a huge cloud of very rarefied material, detected at infra-red wavelengths. The brightest part of the cloud associated with ρ Ophiuchi lies less than four degrees to the north-north-west.

The long chain of stars making Scorpius is striking; it ends in the 'sting', where there are two bright stars close together – λ (Shaula) and υ (Lesath). They give the impression of being a wide double, but there is no true association, because Lesath is 1570 light-years away, Shaula only 275. Lesath is extremely luminous, and could match 15,000 Suns, so that it is far superior to Antares; Shaula – only just below the first magnitude as seen from Earth – is a mere 1300 Sun-power. Both Shaula and Lesath are hot and bluish-white. Antares is flanked by τ and σ, both above the third magnitude. μ and ζ, further south, look like naked-eye doubles – again a line-of-sight effect. The separation between the two stars of ζ is nearly 7 minutes of arc; the fainter of the pair is 2500 light-years away, more remote than its brighter, orange neighbour.

The scorpion's head is made up of β (Graffias or Akrab), ν and ω. β is a fine double, so wide virtually any telescope splits it; the primary is a spectroscopic binary.

Scorpius is crossed by the Milky Way, and there are many fine star fields. There are also four clusters in Messier's list. M6 and M7 are among the most spectacular open clusters in the sky, though they are inconveniently far south from Britain and the northern United States. Both are easily seen with the naked eye, and can be resolved with binoculars; M7, the brighter of the two, was described by Ptolemy as 'a nebulous cluster following the sting of Scorpius'. Because of its large size, it is best seen with a very low magnification. M6, the Butterfly, is also very prominent, and is further away: 1300 light-years, as against 800 light-years for M7. Another bright open cluster is NGC6124, which forms a triangle with the μ and ζ pairs. It can be detected with binoculars without difficulty.

The other two Messier objects are globulars. M4 is very easy to locate, because it is in the same binocular field with Antares and less than two degrees to the west. It is just visible with the naked eye, and binoculars show it well; it is one of the closest of all globulars. No more than 7500 light-years away, it is very rich in variable stars. M80 is not so prominent, but can be located easily between Antares and β. It is 36,000 light-years away. More remote than M4, it looks much smaller; it is also more compact, with a diameter of perhaps 50 light-years. It is relatively poor in variable stars, but in 1860 a bright nova was seen in it, rising to the seventh magnitude; in the lists it is given as T Scorpii. It soon faded away, but in case it is a recurrent nova, M80 is worth monitoring.

Sagittarius is exceptionally rich, and the glorious star clouds hide our view of that mysterious region at the centre of the Galaxy. Since Sagittarius is the southernmost of the Zodiacal constellations, it is never well seen from Britain or the northern United States; part of it never rises at all. The brightest stars are ε (Kaus Australis) and σ (Nunki), with α and β only just above the fourth magnitude.

β or Akrab, in the far south of the constellation, is an easy double, and makes up a naked-eye pair with its

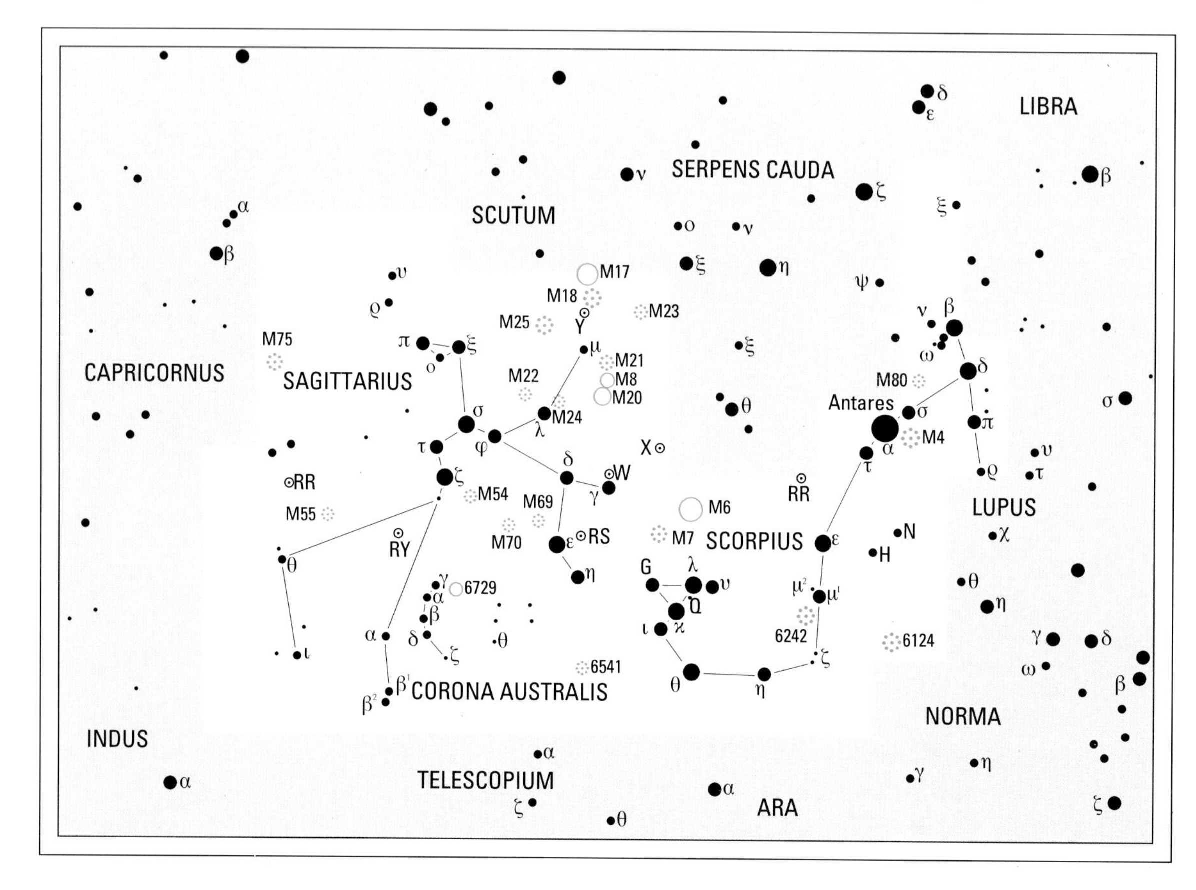

◀ ***The two southernmost constellations** of the Zodiac. From the British Isles or the northern United States, parts of Scorpius and Sagittarius never rise. Antares can be seen, and is at its best during summer evenings. From southern countries Scorpius passes overhead, and rivals Orion for the title of the most glorius constellation in the sky, while the star clouds in Sagittarius hide our view of the centre of the Galaxy. Scorpius adjoins Libra, which was once known as the Scorpion's Claws, and the star formerly known as γ Scorpii has been given a free transfer, so that it is now called σ Librae.*

neighbour. ζ or Ascella is a very close, difficult binary. The components are almost equal, and the revolution period is 21 years; the separation is only about 0.3 of a second of arc, so that a telescope of at least 38-centimetre (15-inch) size is needed to resolve the pair. Users of telescopes of around this size will find it a useful test object.

There are comparatively few bright variables, but RY Sagittarii, again in the southern part of the constellation, is an R Coronae star; usually it is of magnitude 6, but falls to 15. Unfortunately, its southern declination makes it an awkward target for British and North American observers.

Sagittarius abounds in clusters and nebulae. There are three superb galactic nebulae. M20 (the Trifid) and M8 (the Lagoon) are not far from λ and μ; they can be seen as whitish patches in binoculars. Photography brings out their vivid clouds. Close by is the open cluster M21, easy to resolve. M17, in the northern section, and known variously as the Omega, Swan or Horseshoe Nebula, is magnificent. Of the globular clusters, M22 in particular is very fine, and was the first member of the class to be discovered (by Abraham Ihle, as long ago as 1665).

There is a great deal to be seen in Sagittarius, and one has to be systematic about it; for example, once μ has been identified it is not hard to move on to M25, M17, M21, M20 and M18, though care must be taken not to confuse them. Incidentally, M24 is not a true nebular object at all, but merely a star cloud in the Milky-Way – though it does contain an open cluster, NGC6603, which lies in its northern part. When the star clouds in the area are high above the horizon, sweeping along them with binoculars or a low-power telescope will give breathtaking views.

Corona Australis, or Corona Austrinus, is one of Ptolemy's original constellations, but no legends appear to be attached to it. It is small, with no stars above the fourth magnitude, but the little semi-circle consisting of γ, α, β, δ and θ is distinctive enough, close to the relatively obscure α Sagittarii. γ is a close binary, and makes a useful test object. NGC6541 is a globular cluster, just detectable with binoculars; it lies some way from the main pattern, between θ Coronae and θ Scorpii. The variable nebula NGC6729 surrounds the erratic variable R Coronae Australis (do not confuse it with R Coronae Borealis). The changes in the nebula mimic the fluctuations of the star, but this is faint, below the range of small telescopes.

SCORPIUS

BRIGHTEST STARS

No.	Star	R.A. h	m	s	Dec. °	′	″	Mag.	Spectrum	Proper name
2	α	16	29	24	−26	25	55	0.96	M1	Antares
35	λ	17	33	36	−37	06	14	1.63	B2	Shaula
	θ	17	37	19	−42	59	52	1.87	F0	Sargas
26	ε	16	50	10	−34	17	36	2.29	K2	Wei
7	δ	16	00	20	−22	37	18	2.32	B0	Dschubba
	ϰ	17	42	29	−39	01	48	2.41	B2	Girtab
8	β	16	05	26	−19	48	19	2.64	B0+B2	Graffias
34	υ	17	30	46	−37	17	45	2.69	B3	Lesath
23	τ	16	35	53	−28	12	58	2.82	B0	
20	σ	16	21	11	−25	35	34	2.85	B1	Alniyat
6	π	15	58	51	−26	06	50	2.89	B1	
	ι^1	17	47	35	−40	07	37	3.03	F2	
	μ^1	16	51	52	−38	02	51	3.04	B1	
	G	17	49	51	−37	02	36	3.21	K2	
	η	17	12	09	−43	14	21	3.33	F2	

Also above magnitude 4.3: μ^2 (3.57), ζ^2 (3.62), ϱ (3.88), ω^1 (3.96), ν (4.00), ξ (4.16), H (4.16), N (4.23), Q (4.29).

VARIABLE

Star	R.A. h	m	Dec. °	′	Range (mags)	Type	Period (d)	Spectrum
RR	16	55.6	−30	35	5.0–12.4	Mira	279	M

DOUBLES

Star	R.A. h	m	Dec. °	′	P.A. °	Sep. ″	Mags	
ξ	16	04.4	−11	22	051	7.6	4.8, 7.3	A is double
β	16	05.4	−19	48	021	13.6	2.6, 4.9	A is double
σ	16	21.2	−25	36	273	20.0	2.9, 8.5	
α	16	29.4	−26	26	273	2.7	1.2, 5.4	Binary, 878y, red, green

CLUSTERS AND NEBULAE

M	NGC	R.A. h	m	Dec. °	′	Mag.	Dimensions ′	Type
4	6121	16	23.6	−26	32	5.9	26.3	Globular cluster.
6	6405	17	40.1	−32	13	4.2	50	Open cluster (Butterfly)
7	6475	17	53.9	−34	49	3.3	80	Open cluster
80	6093	16	17.0	−22	59	7.2	8.9	Globular cluster
	6124	16	25.6	−40	40	5.8	29	Open cluster
	6242	16	55.6	−39	30	6.4	9	Open cluster

SAGITTARIUS

BRIGHTEST STARS

No.	Star	R.A. h	m	s	Dec. °	′	″	Mag.	Spectrum	Proper name
20	ε	18	24	10	−34	23	05	1.85	B9	Kaus Australia
34	σ	18	55	16	−26	17	48	2.02	B3	Nunki
38	ζ	19	02	37	−29	52	49	2.59	A2	Ascella
19	δ	18	20	59	−29	49	42	2.70	K2	Kaus Meridonalis
22	λ	18	27	58	−25	25	18	2.81	K2	Kaus Borealis
41	π	19	09	46	−21	01	25	2.89	F2	Albaldah
10	γ	18	05	48	−30	25	26	2.99	K0	Alnasr
	η	18	17	37	−36	45	42	3.11	M3	
27	φ	18	45	39	−26	59	27	3.17	B8	
40	τ	19	05	56	−27	40	13	3.32	K1	

Also above magnitude 4.3: ξ^2 (3.51), ο (3.77), μ (Polis) (3.86), ϱ (3.93), β^1 (Arkab) (3.93), α (Rukbat) (3.97), ι (4.13), β^2 (4.29).

VARIABLES

Star	R.A. h	m	Dec. °	′	Range (mags)	Type	Period (d)	Spectrum
X	17	47.6	−27	50	4.2–4.8	Cepheid	7.01	F
W	18	05.0	−29	35	4.3–5.1	Cepheid	7.59	F–G
RS	18	17.6	−34	06	6.0–6.9	Algol	2.41	B–A
Y	18	21.4	−18	52	5.4–6.1	Cepheid	5.77	F
RY	19	16.5	−33	31	6.0–15	R Coronæ	–	Gp
RR	19	55.9	−29	11	5.6–14.0	Mira	335	M

DOUBLES

Star	R.A. h	m	Dec. °	′	P.A. °	Sep. ″	Mags	
η	18	17.6	−36	46	105	3.6	3.2,7.8	
β^1	19	22.6	−44	28	077	28.3	3.9,8.0	Wide naked-eye pair with β^2

CLUSTERS AND NEBULAE

M	NGC	R.A. h	m	Dec. °	′	Mag.	Dimensions ′	Type
8	6523	18	03.8	−24	23	6.0	90 x 40	Nebula (Lagoon)
17	6618	18	20.8	−16	11	7.0	46 x 37	Nebula (Omega)
18	6613	18	19.0	−17	08	6.9	9	Open cluster
20	6514	18	02.6	−23	02	7.5	29 x 27	Nebula (Trifid)
21	6531	18	04.6	−22	30	5.9	13.0	Open cluster
22	6656	18	36.4	−23	54	5.1	24.0	Globular cluster
23	6494	17	56.8	−19	01	5.5	27	Open cluster
24	6603	18	16.9	−18	29	4.5	90	Star cloud
25	IC4725	18	31.6	−19	15	4.6	31.0	Open cluster round
28	6626	18	24.5	−24	52	6.9	11.0	Globular cluster
54	6715	18	55.1	−30	29	7.7	9.1	Globular cluster
55	6809	19	40.0	−30	58	6.9	19.0	Globular cluster
69	6637	18	31.4	−32	21	7.7	7.1	Globular cluster
70	6681	18	43.2	−32	18	8.1	7.8	Globular cluster
75	6864	20	06.1	−21	55	8.6	6.0	Globular cluster

CORONA AUSTRALIS

The brightest stars are α (Meridiana) and β, each 4.11; the position of α is R.A. 19h 09m 28s.2, dec. −37° 54′ 16″. The other stars making up the little semi-circle are γ (4.21), δ (4.59) and ζ (4.75).

DOUBLES

Star	R.A. h	m	Dec. °	′	P.A. °	Sep. ″	Mags	
ϰ	18	33.4	−38	44	359	21.6	5.9, 5.9	
γ	19	06.4	−37	04	109	1.3	4.8, 5.1	Binary, 12y, good test

CLUSTERS AND NEBULAE

M	NGC	R.A. h	m	Dec. °	′	Mag.	Dimensions ′	Type
	6541	18	08.0	−43	42	6.6	13.1	Globular cluster
	6729	19	01.9	−36	57	var	1 (var)	Variable nebula round R Coronae Australis

Andromeda, Triangulum, Aries, Perseus

Magnitudes

–1

0

1

2

3

4

5

Variable star

Galaxy

Planetary nebula

Gaseous nebula

Globular cluster

Open cluster

Andromeda is a large, prominent northern constellation, commemorating the beautiful princess who was chained to a rock on the seashore to await the arrival of a monster, though fortunately the dauntless hero Perseus was first on the scene. Andromeda adjoins Perseus to one side and Pegasus to the other; why Alpheratz was transferred from the Flying Horse to the Princess remains a mystery.

The three leading stars of Andromeda are all of magnitude 2.1. Their individual names are often used; α is Alpheratz, β is Mirach and γ is Almaak. Their distances are respectively 72, 88 and 121 light-years; their luminosities 96, 115 and 95 times that of the Sun. Alpheratz is an A-type spectroscopic binary; Mirach is orange-red, with colour that is very evident in binoculars. It has been suspected of slight variability. Almaak is a particularly fine double, with a K-type orange primary and a hot companion which is said to look slightly blue-green by contrast. The pair can be resolved with almost any telescope, and the companion is a close binary, making a useful test for a telescope of about 25-centimetre (10-inch) aperture. δ, between Alpheratz and Mirach, is another orange star of type K.

R Andromedae, close to the little triangle of θ (4.61), σ (4.62) and ϱ (5.18), is a Mira variable which can at times rise above the sixth magnitude, and is readily identifiable because it is exceptionally red. The trick is to locate it when it is near maximum, so that the star field can be memorized and the variable followed down to its minimum – though if you are using a small telescope you will lose it for a while, since it drops down to almost the 15th magnitude.

Of course the most celebrated object in Andromeda is the Great Spiral, M31. It can just be seen with the naked eye when the sky is dark and clear, and the Arab astronomer Al-Sûfi called it 'a little cloud'. It lies at a narrow angle to us, which is a great pity; if it were face-on it would indeed be glorious. The modern value for its distance is 2.2 million light-years, though if the Cepheid standard candles have been slightly under-estimated, as is possible, this value may have to be revised slightly upwards. It is a larger system than ours, and has two dwarf elliptical companions, M32 and NGC205, which are easy telescopic objects.

It has to be admitted that M31 is not impressive when seen through a telescope, and photography is needed to bring out its details. Novae have been seen in it, and there has been one supernova, S Andromedae of 1885, which reached the sixth magnitude – though it was not exhaustively studied, simply because nobody was aware of its true nature; at that time it was still believed that M31, like other spirals, was a minor feature of our own Galaxy.

The open cluster NGC752, between γ Andromedae and β Trianguli, is within binocular range, though it is scattered and relatively inconspicuous. It is worth seeking out the planetary nebula NGC7662, close to the triangle made up of γ, κ and τ; a 25-centimetre (10-inch) telescope shows its form, though the hot central star is still very faint.

Triangulum is one of the few constellations which merits its name; the triangle made up of α, β and γ is distinctive even though only β is as bright as the third magnitude. There is one reasonably bright Mira star, R Trianguli, some way from γ, but the main object of interest is the Triangulum Spiral, M33, which lies some way from α in the direction of Andromeda, and is just south of a line joining α Trianguli to β Andromedae. It is looser than M31, but placed at a better angle to us. Some observers claim to be able to see it with the naked eye; binoculars certainly show it, but it can be elusive telescopically, because its surface brightness is low. It is much less massive than our Galaxy.

Aries. According to legend, this constellation honours a flying ram which had a golden fleece, and was sent by

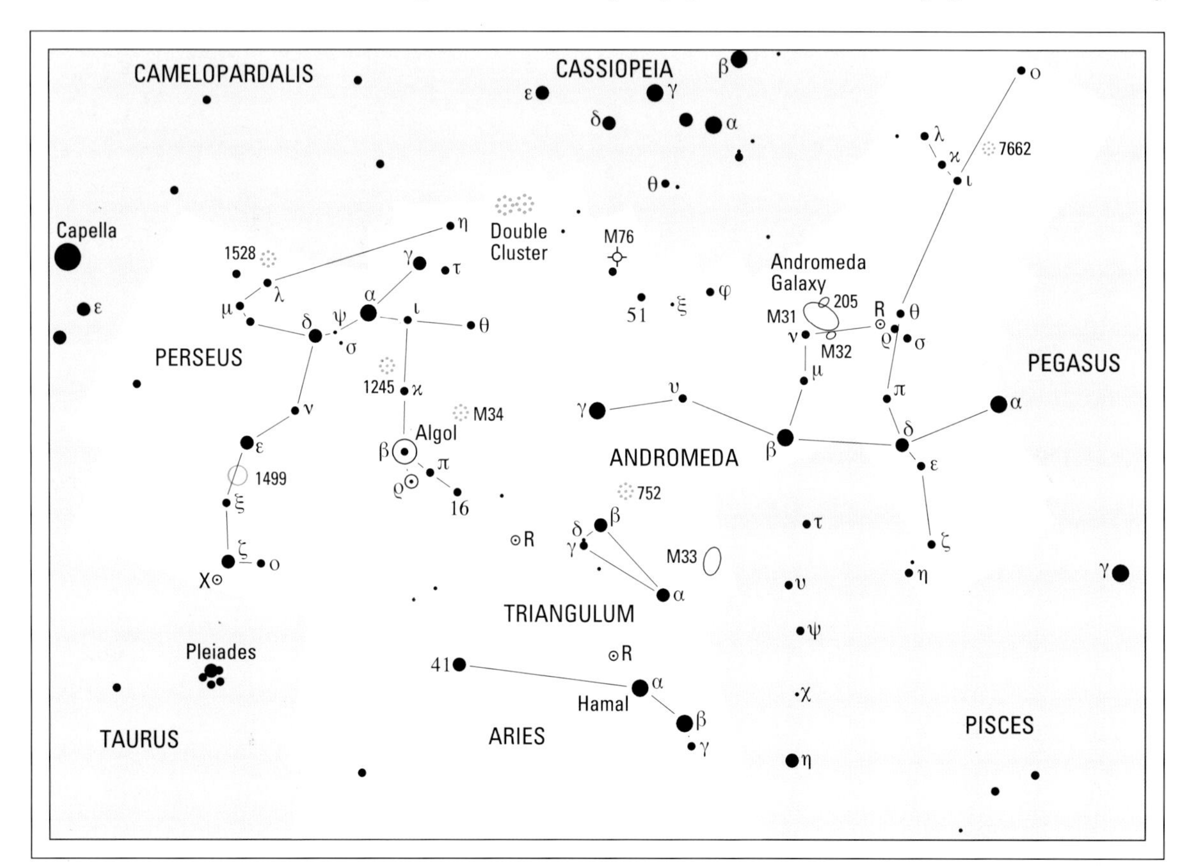

◀ ***The constellations*** *in this map are best seen during evenings in northern autumn (southern spring), though it is true that the northernmost parts of Perseus and Andromeda – as well as Capella, in Auriga – are circumpolar from the British Isles or the northern United States and are always very low from Australia and New Zealand. Andromeda adjoins the Square of Pegasus, and indeed Alpheratz (α Andromedae) is one of the four stars of the Square. Aries is, of course, in the Zodiac, though precession has now carried the vernal equinox across the border of Pisces.*

Mercury to rescue the two daughters of the King of Thebes, who were about to be assassinated by their wicked stepmother. Aries is fairly distinctive, with a small trio of stars (α, β, γ) of which α, or Hamal, is reddish and of the second magnitude. γ, or Mesartim, is a wide, easy double with equal components. Binoculars will not split it, but almost any small telescope will do so.

Perseus. The gallant hero is well represented in the sky, and has an easily-identified shape. The constellation is immersed in the Milky Way, and is very rich. There are no first-magnitude stars, but the leader, α or Mirphak, is not far below. It is of type F, 620 light-years away and 6000 times as luminous as the Sun.

β, or Algol, is the prototype eclipsing binary, and one of the most famous stars in the sky; it lies in the head of Medusa, the Gorgon, who had been decapitated by Perseus but whose glance could still turn any living creature to stone. Algol's period is 2 days 20 hours 48 minutes 56 seconds; the primary eclipse is only about 72 per cent total, but is enough to drop the apparent magnitude from 2.1 to 3.4. The secondary minimum, when the fainter component is hidden, amounts to less than a tenth of a magnitude.

The main component (A) is of type B, and is a white star 100 times as luminous as the Sun, with a diameter of 4 million kilometres (2.5 million miles). The secondary (B) is of type G; it is about three times as luminous as the Sun, and is about 5.5 million kilometres (3.4 million miles) across, so that it is larger though less massive than the primary. The true separation is around 10.5 million kilometres (6.5 million miles), so that the components are too close together to be seen separately; there is also a third star in the system, well away from the eclipsing pair.

We can work out a good deal about the evolution of the Algol system. Originally the secondary (B) was more massive than its partner, so that it swelled out and left the Main Sequence earlier. As it expanded, its gravitational grip on its outer layers weakened, and material was captured by the other star (A), which eventually became the more massive of the two. The process is still going on. The system is a source of radio waves, from which we can tell that a stream of material is making its way from B to A. This is what is termed mass transfer, and is of the utmost importance in all studies of binary systems.

The fluctuations of Algol are easy to follow with the naked eye; the times of mimima are given in almanacs or in monthly astronomy periodicals such as *Astronomy Now* and *Sky & Telescope*. Suitable comparison stars are ζ (2.85), ε (2.89) and ϰ (3.80), as well as γ Andromedae (2.14). Mirphak is rather too bright. Avoid using ϱ Persei, which is a red semi-regular variable with a range of magnitude from 3 to 4, and a very rough period which may range between 33 and 55 days.

ζ Persei is highly luminous (15,000 times as powerful as the Sun) and is the senior member of a 'stellar association', made up of a group of hot, luminous stars which are moving outwards from a common centre and presumably had a common origin. In the same binocular field with ζ are o (3.83) and the irregular variable X Persei, which has a range of between magnitudes 6 and 7 and is of special note because it is an X-ray emitter.

M34, an open cluster near Algol, can be seen with binoculars. However, it pales in comparison with NGC 869 and 885, which make up the Sword-Handle. They are easy to locate – γ and δ, in the W of Cassiopeia, point to them – and can be seen with the naked eye; telescopes show a wonderful pair of clusters in the same low-power field. They rank among the most beautiful sights in the stellar sky.

ANDROMEDA

BRIGHTEST STARS

No.	Star	R.A. (h m s)	Dec. (° ′ ″)	Mag.	Spectrum	Proper name
21	α	00 08 23	+29 05 26	2.06	A0p	Alpheratz
43	β	01 09 44	+35 37 14	2.06	M0	Mirach
57	γ	02 03 54	+42 19 47	2.14	K2+A0	Almaak
31	δ	00 39 20	+30 51 40	3.27		

Also above magnitude 4.3: 51 (3.57), σ (3.6v), λ (3.82), μ (3.87), ξ (4.06), υ (4.09) ϰ (4.14), φ (4.25), ι (4.29)

VARIABLES

Star	R.A. (h m)	Dec. (° ′)	Range (mags)	Type	Period (d)	Spectrum
R	00 24.0	+38 35	5.8–14.9		409	S

DOUBLES

Star	R.A. (h m)	Dec. (° ′)	P.A. (°)	Sep. (″)	Mags	
γ	02 03.9	+42 20	063	9.8	2.3, 4.8	B is a binary, 61y; 5.5, 6.3; sep 0″.5

CLUSTERS AND NEBULAE

M	NGC	R.A. (h m)	Dec. (° ′)	Mag.	Dimensions (′)	Type
31	224	00 42.7	+41 16	3.5	178 × 63	Sb galaxy. Great.
32	221	00 42.7	+40 52	8.2	7.6 × 5.8	E2 galaxy. Com to M31.
	205	00 40.4	+41 41	8.0	17.4 × 9.8	E6 galaxy. Com to M31.
	752	01 57.8	+37 41	5.7	50	Open cluster
	7662	23 25.9	+42 33	9.2	20″ × 130″	Planetary nebula

TRIANGULUM

BRIGHTEST STARS

No.	Star	R.A. (h m s)	Dec. (° ′ ″)	Mag.	Spectrum	Proper name
4	β	02 09 32	+34 59 14	3.00	A5	

Also above magnitude 4.3: α (Rasalmothallah) (3.41), γ (4.01).

VARIABLES

Star	R.A. (h m)	Dec. (° ′)	Range (mags)	Type	Period (d)	Spectrum
R	02 37.0	+34 16	5.4-12.6	Mira	266.5	

GALAXY

M	NGC	R.A. (h m)	Dec. (° ′)	Mags	Dimensions (′)	Type
33	598	01 33.9	+30 39	5.7	62 × 39	Sc galaxy

ARIES

BRIGHTEST STARS

No.	Star	R.A. (h m s)	Dec. (° ′ ″)	Mag.	Spectrum	Proper name
13	α	02 07 10	+23 27 45	2.00	K2	Hamal
6	β	01 54 38	+20 48 29	2.64	A5	Sheratan

Also above magnitude 4.3: 41 (c) (Nair al Butain) (3.63), γ (Mesartim) (3.9) (combined magnitude).

VARIABLES

Star	R.A. (h m)	Dec. (° ′)	Range (mags)	Type	Period (d)	Spectrum
R	00 24.0	+38 35	5.8–14.9		409	S

DOUBLES

Star	R.A. (h m)	Dec. (° ′)	P.A. (°)	Sep. (″)	Mags
γ	01 53.5	+19 18	000	7.8	4.8, 4.8

PERSEUS

BRIGHTEST STARS

No.	Star	R.A. (h m s)	Dec. (° ′ ″)	Mag.	Spectrum	Proper name
33	α	03 24 19	+49 51 40	1.80	F5	Mirphak
26	β	03 08 10	+40 57 21	2.12 (max)	B8	Algol
44	ζ	03 54 08	+31 53 01	2.85	B1	Atik
45	ε	03 57 51	+40 00 37	2.89	B0.5	
23	γ	03 04 48	+53 30 23	2.93	G8	
39	δ	03 42 55	+47 47 15	3.01	B5	
25	ϱ	03 05 10	+38 50 25	3.2 (max)	M4	Gorgonea Terti

Also above magnitude 4.3: η (Miram) (3.76), ν (3.77), ϰ (Misam) (3.80), ε (3.83), τ (Kerb) (3.85), υ (Nembus) (4.04), ξ (Menkib) (4.04), ι (4.05), φ (4.07), θ (4.12), μ (4.14), ψ (4.23), 16 (4.23), 16 (4.23), λ (4.29).

VARIABLES

Star	R.A. (h m)	Dec. (° ′)	Range (mags)	Type	Period (d)	Spectrum
β	03 08.2	+40 57	2.1–3.4	Algol	2.87	B+G
ϱ	03 05.2	+38 50	3.2–4.2	Semi-reg.	33/55	M
X	03 55.4	+31 03	6.0–7.0	Irregular	-	X-ray source

DOUBLES

Star	R.A. (h m)	Dec. (° ′)	P.A. (°)	Sep. (″)	Mags
η	02 50.7	+55 54	300	28.3	3.3,8.5

CLUSTERS AND NEBULAE

M	NGC	R.A. (h m)	Dec. (° ′)	Mag.	Dimensions (′)	Type
34	1039	02 42.0	+42 47	5.2	35	Open cluster
76	650-1	01 42.4	+51 34	12.2	65″ × 290″	Planetary nebula

Pegasus, Pisces

Magnitudes

-1
0
1
2
3
4
5

Variable star

Galaxy

Planetary nebula

Gaseous nebula

Globular cluster

Open cluster

Pegasus forms a square – though one of its main stars has been stolen by the neighbouring Andromeda. The stars in the Square of Pegasus are not particularly bright; Alpheratz is of the second magnitude, the others between 2.5 and 3. However, the pattern is easy to pick out because it occupies a decidedly barren region of the sky. On a clear night, try to count the number of stars you can see inside the Square first with the naked eye, and then with binoculars. The answer can be somewhat surprising.

Three of the stars in the Square are hot and white. α Pegasi (Markab) is of type B9, 100 light-years away and 75 times as luminous as the Sun. γ (Algenib), which looks the faintest of the four, is also the most remote (520 light-years) and the most powerful (equal to 1300 Suns); the spectral type is B. The fourth star, β (Scheat), is completely different. It is an orange-red giant of type M, and the colour is evident even with the naked eye, so that binoculars bring it out well, and the contrast with its neighbours is striking. Moreover, it is variable. It has a fairly small range, from magnitude 2.3 to 2.5, but the period – around 38 days – is more marked than with most other semi-regular stars. The changes can be followed with the naked eye, α and β make good comparison stars.

When making estimates of this kind, allowance has to be made for what is termed extinction, the dimming of a star due to atmospheric absorption which naturally increases at lower altitudes above the horizon (see table).

The right ascensions of β and α are about the same, and the difference in declination is about 13 degrees. Suppose that β is at an altitude of 32 degrees; it will be dimmed by 0.2 of a magnitude. If α is directly below (as it may be to northern-hemisphere observers; in southern altitudes the reverse will apply) the altitude will be 32 − 13 = 19 degrees, and the dimming will be 0.5 magnitude. If the two look equal, α will actually be the brighter by 0.3 magnitude, so that β will be 2.7. Try to find a comparison star at an altitude equal to that of the variable. This is unimportant with telescopic variables; extinction will not change noticeably over a telescopic or binocular field of view.

It is also interesting to compare the real luminosities of the stars in the Square. As we have seen, absolute magnitude is the apparent magnitude which a star would have if it could be seen from a standard distance of 10 parsecs, or 32.6 light-years. The values for the four stars are: Alpheratz −0.1, α Pegasi +0.2, β Pegasi −1.4 (rather variable), and γ −3.0, so that γ would dominate the scene.

The other leading star of Pegasus is ε, which is well away from the Square and is on the border of Equuleus. It is a K-type orange star, 520 light-years away and 4500 times as luminous as the Sun. It has been strongly suspected of variability, and naked-eye estimates are worthwhile; α is a good comparison, though in general ε should be slightly but detectably the brighter of the two.

The globular cluster M15, close to ε, was discovered in 1746 by the Italian astronomer Maraldi. To find it, use θ and ε as guides. It is just below naked-eye visibility, but binoculars show it as a fuzzy patch; it has an exceptionally condensed centre, and is very rich in variable stars. It is also very remote, at a distance of over 49,000 light-years. The real diameter cannot be far short of 100 light-years.

Pisces is one of the more obscure Zodiacal constellations, and consists mainly of a line of dim stars running along south of the Square of Pegasus. Mythologically its associations are rather vague; it is sometimes said to represent two fishes into which Venus and Cupid once changed themselves in order to escape from the monster Typhon, whose intentions were anything but honourable.

α, magnitude 3.79, has three proper names: Al Rischa, Kaïtain or Okda. It is a binary, not difficult to split with a small telescope; both components have been suspected of slight variability in brightness and colour, but firm evidence is lacking. Both are of type A, and the distance from us is 100 light-years. ζ is another easy double, and here too slight variability has been suspected.

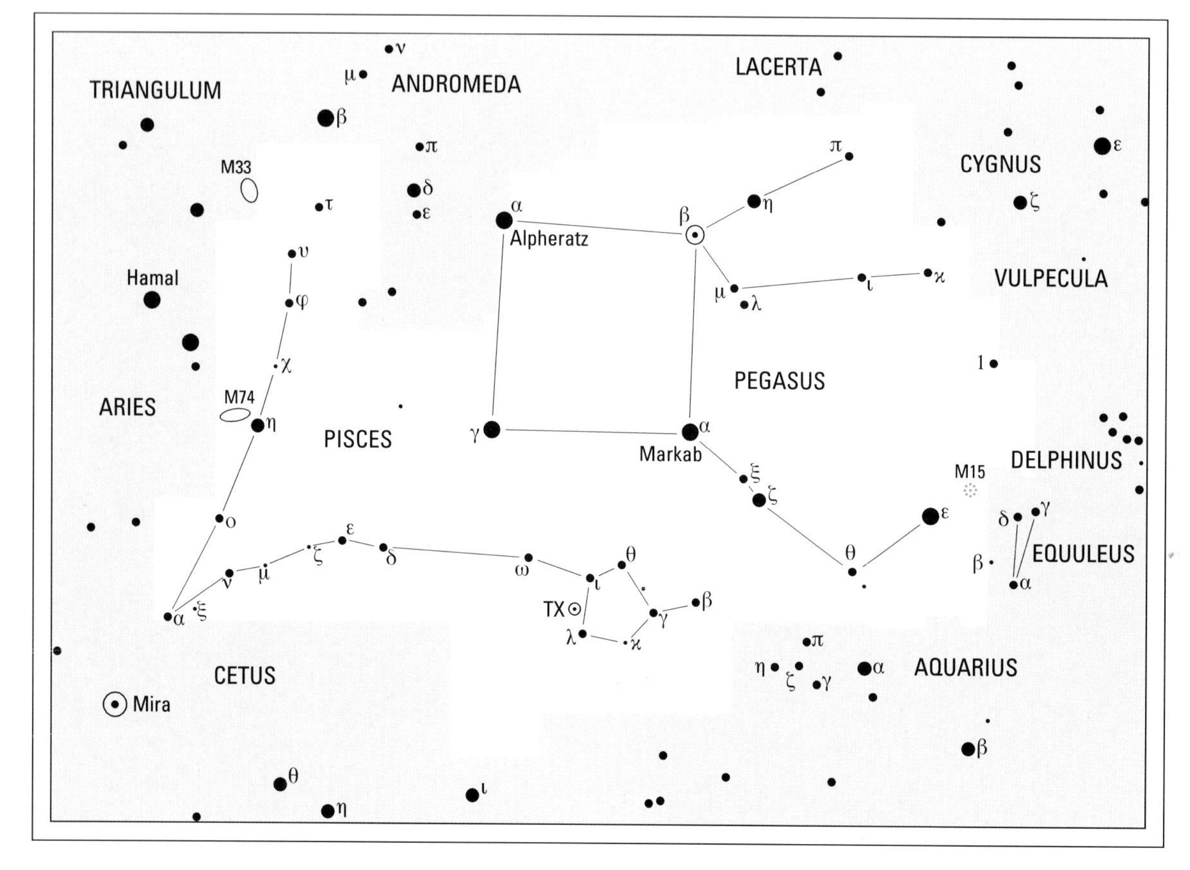

◀ ***Pegasus** is the most prominent constellation of the evening sky during northern autumn (southern spring). The four main stars – one of which has been illogically transferred to Andromeda – make up a square, which is easy enough to identify even though maps tend to make it seem smaller and brighter than it really is. In fact the brightest star in Pegasus, ε, is some way from the Square. Pisces is a very dim Zodiacal constellation occupying the area between Pegasus and Cetus.*

◀ ***The globular cluster M15** in Pegasus, photographed by Bernard Abrams using a 25-cm (10-inch) reflector. It can be found near to ε.*

EXTINCTION TABLE

Altitude above horizon, °	Dimming in magnitude.
1	3.0
2	2.5
4	2.0
10	1.0
13	0.8
15	0.7
17	0.6
21	0.4
26	0.3
32	0.2
43	0.1

Above this altitude, extinction can be neglected.

The exceptionally red N-type semi-regular variable TX (19) Piscium is worth locating. It is easily found, near the 'circlet' made up of ι, θ, γ and λ; as it never falls below magnitude 7.7 it is always within binocular range, and its hue is almost as strong as that of the famous Garnet Star, μ Cephei.

The galaxy M74, discovered by Méchain in 1780, is one of the less massive spirals in Messier's catalogue; it can be seen with a 7.6-centimetre (3-inch) telescope, but can be rather elusive. It lies within a couple of degrees of η Piscium. There is a fairly well-defined nucleus, but the spiral arms are loose and faint, so that even Sir John Herschel mistook it for a globular cluster. The distance is around 26 million light-years.

One object in Pisces which is worth finding, though requiring at least a 25-centimetre (10-inch) telescope, is the white dwarf Wolf 28, better known as Van Maanen's Star; it was discovered in 1917 by the Dutch astronomer Adriaan van Maanen. Its position is R.A. 00h 46m.5, dec. +05 degrees 09 minutes, about two degrees south of δ Piscium. Its visual magnitude is 12.4, and it is one of the dimmest stars known, with a luminosity only about 1/6000 of that of the Sun. The diameter is about the same as that of the Earth, but in mass it is equal to the Sun, so that the density must be about a million times that of water; if you could go there and stand on the surface, you would find that your weight has been increased by about 10 million times. The proper motion amounts to nearly 3 seconds of arc per year, and the distance is less than 14 light-years, so that this is one of the nearest known of all white dwarfs.

PEGASUS

BRIGHTEST STARS

No.	Star	R.A. h	m	s	Dec. °	′	″	Mag.	Spectrum	Proper name
8	ε	21	44	11	+09	52	30	2.38	K2	Enif
53	β	23	03	46	+28	04	58	2.4 (max)	M2	Scheat
54	α	23	04	45	+15	12	19	2.49	B9	Markab
88	γ	00	13	14	+15	11	01	2.83	B2	Algenib
44	η	22	43	00	+30	13	17	2.94	G2	Matar
42	ζ	22	41	27	+10	49	53	3.40	B8	Homan
48	μ	22	50	00	+24	36	06	3.48	K0	Sadalbari

Also above magnitude 4.3: θ (Biham) (3.53), ι (3.76), 1 (4.08), κ (4.13), ξ (Al Suud al Nujam) (4.19), π (4.29). Alpheratz (α Andromedae) was formerly included in Pegasus, as δ Pegasi.

VARIABLE

Star	R.A. h	m	Dec. °	′	Range (mags)	Type	Period (d)	Spectrum
β	23	03.8	+28	05	2.4–2.8	Semi-reg.	38	M

CLUSTER

M	NGC	R.A. h	m	Dec. °	′	Mag.	Dimensions ′	Type
15	7078	21	30.0	+12	10	6.3	12.3	Globular cluster

PISCES

The brightest star is η (Alpherg), R.A. 01h 31m 29s, dec.+15° 20′ 45″ mag. 3.62. Also above magnitude 4.3: γ (3.69), α (Al Rischa) (3.79), ω (4.01), ι (4.13), o (Torcular) (4.26), θ (4.28), ε (4.28).

VARIABLE

Star	R.A. h	m	Dec. °	′	Range (mags)	Type	Period (d)	Spectrum
TX	23	46.4	+03	29	6.9–7.7	Irregular	-	N

DOUBLES

Star	R.A. h	m	Dec. °	′	P.A. °	Sep. ″	Mags	
α	02	02.0	+02	46	279	1.9	4.2, 5.1	Binary, 933y
ζ	01	13.7	+07	35	063	23.0	5.6, 6.5	

GALAXY

M	NGC	R.A. h	m	Dec. °	′	Mag.	Dimension ′	Type
74	628	01	36.7	+15	47	9.2	10.2 × 9.5	Sc Galaxy

Capricornus, Aquarius, Piscis Australis

Magnitudes

−1

0

1

2

3

4

5

Variable star

Galaxy

Planetary nebula

Gaseous nebula

Globular cluster

Open cluster

Capricornus has been identified with the demigod Pan, but the mythological association is decidedly nebulous, and the pattern of stars certainly does not recall the shape of a goat, marine or otherwise. Neither can it be said that there is a great deal of interest here, even though the constellation covers over 400 square degrees of the sky. δ is the only star above the third magnitude; it is about 49 light-years away, and some 13 times as luminous as our own Sun.

β Capricorni is one of the less powerful of the naked-eye stars, and is not much more than twice the luminosity of the Sun, though its distance is not known with any certainty and may be less than the official value given in the Cambridge catalogue. It has a sixth-magnitude companion which is within binocular range and is itself a very close double. The bright star appears to be a spectroscopic triple, so that β Capricorni is a very complex system indeed.

α^1 and α^2 make up a wide pair, easily separable with the naked eye, but there is no genuine association. The brighter star, α^2, is 117 light-years away, while the fainter component, α^1, is very much in the background at a distance of 1600 light-years; both are of type G, but while the more remote star is well over 5000 times as powerful as the Sun, the closer member of the pair could equal no more than 75 Suns. This is a classic example of an optical pair.

There are no notable variables in Capricornus, but there is one Messier object, the globular cluster M30, which lies close to ζ (which is, incidentally, a very luminous G-type giant). M30 was discovered by Messier himself in 1764, and described as 'round; contains no star'. It is in fact a small globular with a brightish nucleus; it is 41,000 light-years away, and has no characteristics of special note.

Aquarius, with an area of almost a thousand square degrees, is larger than Capricornus, but it is not a great deal more conspicuous. It is known as the Water-bearer, but its mythological associations are vague, though it has sometimes been identified with Ganymede, the cup-bearer to the Olympian gods. Its main claim to fame is that it lies in the Zodiac. Most of it is in the southern hemisphere of the sky.

Both α and β are very luminous and remote G-type giants. The most interesting star is ζ, which is a fine binary with almost equal components; both are F-type subgiants about 100 light-years away, with a real separation of at least 15,000 million kilometres (over 9000 million miles). This is an excellent test object for a telescope of around 7.6-centimetre (3-inch) aperture.

There is a distinctive group of stars between Fomalhaut, in Piscis Australis, and α Pegasi. The three stars labelled 'ψ Aquarii' are close together, with χ and φ nearby; several of them are orange, and they have often been mistaken for a very loose cluster, though they are not really associated with each other.

R Aquarii is a symbiotic or Z Andromedae type variable. It is made up of a cool red giant together with a hot subdwarf – both of which seem to be intrinsically variable. The whole system is enveloped in nebulosity, and the smaller star seems to be pulling material away from its larger, less dense companion. R Aquarii is none too easy to locate, but users of larger telescopes will find that it repays study.

M2 is a particularly fine globular cluster, forming a triangle with α and β Aquarii. Some people claim that they can see it with the naked eye; with binoculars it is easy. It was discovered by Maraldi as long ago as 1746, and is very remote, at around 55,000 light-years. Its centre is not so condensed as with most globulars, and the edges are not hard to resolve.

M72 is another globular, discovered by Méchain in 1780; it is 62,000 light-years away, and comparatively 'loose'. It is one of the fainter objects in Messier's list, and

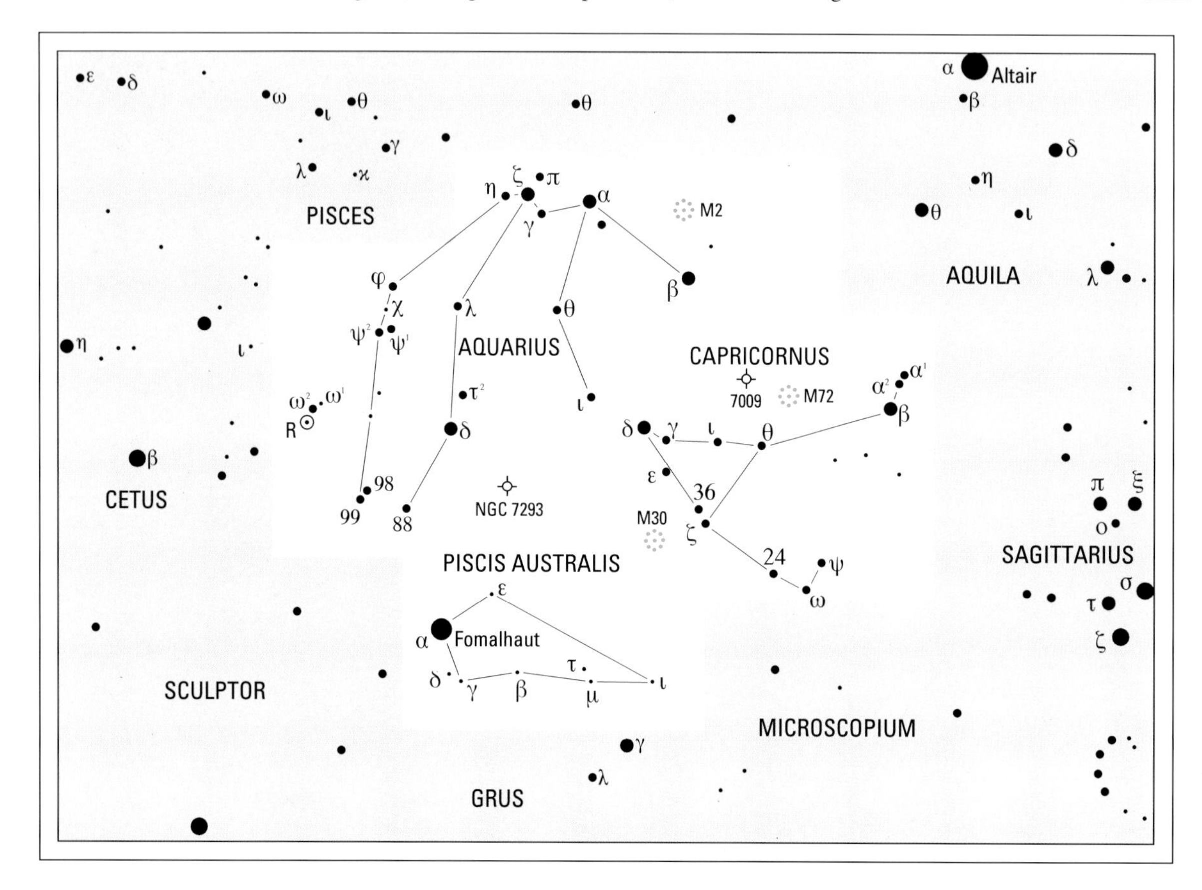

◀ ***The two Zodiacal constellations** of Capricornus and Aquarius occupy a wide area, but contain little of immediate interest, and together with Pisces and Cetus they give this whole region a decidedly barren look. Fomalhaut, in Piscis Australis, is the southernmost of the first-magnitude stars to be visible from the British Isles or the northern United States; northern observers, who never see it high up, do not always appreciate how bright it really is. The celestial equator just passes through the northernmost part of Aquarius.*

◀ *Globular cluster M2 in the constellation Aquarius, photographed by John Fletcher using a 25-cm (10-inch) reflector.*

is none too easy to locate; it lies between θ Capricorni and ε Aquarii (not shown), and is surprisingly difficult to resolve into individual stars.

M73, less than 2 degrees from ν Aquarii (magnitude 4.51), is not a real cluster at all, even though it has been given an NGC number; it is made up of a few disconnected stars below the tenth magnitude.

There are two interesting planetary nebulae in Aquarius. NGC7009, the Saturn Nebula, is about one degree west of ν, and is a beautiful object in large telescopes, with a prominent belt of obscuring material. It is about 3900 light-years away, and around half a light-year in diameter.

NGC7293, the Helix Nebula, is the largest and the brightest of all the planetaries, and is said to be visible in binoculars as a faint patch, though a telescope is needed to show it clearly because it lies so close to ν. When photographed, the Helix is seen to be not unlike the Ring Nebula in Lyra (M57), but the central star is only of the 13th magnitude.

Piscis Australis, or Piscis Austrinus, is a small though ancient constellation, apparently not associated with any myth or legend.

The only star above the fourth magnitude it contains is Fomalhaut, which is the southernmost of the first-magnitude stars visible from the latitudes of the British Isles (from north Scotland it barely rises). It is easy to find by using β and α Pegasi, in the Square, as pointers; but beware of confusing it with Diphda or β Ceti, which is roughly aligned with the other two stars of the Square, Alpheratz and γ Pegasi. However, Diphda is a magnitude fainter than Fomalhaut.

Fomalhaut is a pure white star, 22 light-years away and 13 times more luminous than the Sun; it is therefore one of our closer stellar neighbours. In 1983, the Infra-Red Astronomical Satellite found that it is associated with a cloud of cool matter which may be planet-forming; as with Vega, β Pictoris and other such stars, we cannot certainly claim that a planetary system exists there, but neither can we rule it out, The Southern Fish contains nothing else of particular note, though β is a wide and easy optical double.

CAPRICORNUS

BRIGHTEST STARS

No.	Star	R.A. h	m	s	Dec. °	′	″	Mag.	Spectrum	Proper name
49	δ	21	47	02	−16	07	38	2.87	A5	Deneb al Giedi
9	β	20	21	00	−14	46	53	3.08	F8	Dabih

Also above magnitude 4.3: α^2 (Al Giedi) (3.57), γ (Nashira) (3.68), ζ (Yen) (3.74), θ (4.07), ω (4.11), φ (4.14), α^1 (4.24), ι (4.28).

DOUBLE

Star	R.A. h	m	Dec. °	′	P.A. °	Sep. ″	Mags	
α	20	18.1	−12	33	291	377.7	3.6, 4.2	Naked-eye pair
β	20	21.0	−14	47	267	205.0	3.1, 6.0	

CLUSTER

M	NGC	R.A. h	m	Dec. °	′	Mag.	Dimensions ′	Type
30	7099	21	40.4	23	11	7.5	11.0	Globular cluster

AQUARIUS

BRIGHTEST STARS

No.	Star	R.A. h	m	s	Dec. °	′	″	Mag.	Spectrum	Proper name
22	β	21	31	33	−05	34	16	2.91	G0	Sadalsuud
34	α	22	05	47	−00	19	11	2.96	G2	Sadalmelik
76	δ	22	54	39	−15	49	15	3.27	A2	Scheat

Also above magnitude 4.3: ζ (3.6 combined magnitude), 88 (c^2) (3.66), λ (3.74), ε (Albali), (3.77), γ (3.84), 98 (b^1) (3.97), τ (4.01), η (4.02), θ (Ancha) (4.16), φ^1(4.21), τ (4.22), ι (4.27).

VARIABLE

Star	R.A. h	m	Dec. °	′	Range (mags)	Type	Period (d)	Spectrum
R	23	43.8	−15	17	5.8–12.4	Symbiotic	387	M+pec

DOUBLE

Star	R.A. h	m	Dec. °	′	P.A. °	Sep. ″	Mags	
ζ	2	28.8	−00	01	200	2.0	4.3, 4.5	Binary, 856y

CLUSTERS AND NEBULAE

M	NGC	R.A. h	m	Dec. °	′	Mag.	Dimensions ′	Type
2	7089	21	33.5	−00	49	6.5	12.9	Globular cluster
72	6981	20	53.5	−12	32	9.3	5.9	Globular cluster
	7293	22	29.6	−20	48	6.5	770″	Planetary nebula (Helix Nebula)
	7009	21	04.2	−11	22	8.3	2″.5 × 100″	Planetary nebula (Saturn Nebula)

M73 (NGC 6994), R.A. 20h 58.9, dec, −12° 38′, is an asterism of four stars.

PISCIS AUSTRALIS

BRIGHTEST STAR

No.	Star	R.A. h	m	s	Dec. °	′	″	Mag.	Spectrum	Proper name
24	α	22	57	39	−29	37	20	1.16	A3	Fomalhaut

Also above mag. 4.3: ε (4.17), δ (4.21), β (Fum el Samakah) (4.29).

DOUBLE

Star	R.A. h	m	Dec. °	′	P.A. °	Sep. ″	Mags	
β	22	51.5	−32	21	172	30.3	4.4, 7.9	(Optical)

Cetus, Eridanus (northern), Fornax

Magnitudes

−1

0

1

2

3

4

5

Variable star

Galaxy

Planetary nebula

Gaseous nebula

Globular cluster

Open cluster

Cetus is a vast constellation, covering 1232 square degrees. Mythologically it is said to represent the sea-monster which was sent to devour the Princess Andromeda, but which was turned to stone when Perseus showed it the Gorgon's head.

The brightest star, β (Diphda), can be found by using Alpheratz and γ Pegasi, in the Square, as pointers. It is an orange K-type star, 68 light-years away and 75 times as luminous as the Sun. It has been strongly suspected of variability, and is worth monitoring with the naked eye, though the lack of suitable comparison stars makes it awkward to estimate. θ and η lie close together; θ is white, and η, with a K-type spectrum, rather orange. In the same binocular field there is a faint double star, 37 Ceti.

τ Ceti is of special interest. It is only 11.9 light-years away and about one-third as luminous as the Sun, with a K-type spectrum. It is one of the two nearest stars which can be said to be at all like the Sun (ε Eridani is the other), and it may be regarded as a promising candidate for the centre of a planetary system, so that efforts have been made to 'listen out' for signals from it which might be interpreted as artificial – so far with a total lack of success. The flare star UV Ceti lies less than three degrees south-west of τ.

The 'head' of Cetus is made up of α, γ, μ, ξ and δ. α (Menkar) is an M-type giant, 130 light-years away and 132 times as luminous as the Sun; it too has been suspected of slight variability. It is a binary with a very long revolution period; it is fairly easy to split with a small telescope.

Mira (ο Ceti) is the prototype long-period variable, and has been known to exceed the second magnitude at some maxima, though at others it barely rises above 4. It is visible with the naked eye for only a few weeks every year, but, when at its best, it alters the whole aspect of that part of the sky. It was the first variable star to be identified, and is also the closest of the M-type giants; its distance is 95 light-years, and it is well over 100 times as powerful as the Sun. The average period is 331 days. This is not much more than a month shorter than a year, and so there are spells when maxima occur when Mira is too close to the Sun to be seen. It has a faint companion which has the same motion through space; it is believed to be rather variable, and its closeness to the main star makes it a difficult test object.

M77 is a massive Seyfert spiral galaxy, and is a strong radio source. It lies near δ, and is not hard to locate, but its nucleus is so bright compared with the spiral arms that with low or even moderate magnifications it takes on the guise of a rather fuzzy star. The distance is about 52 million light-years. NGC247, close to β, is fairly large, but has a low surface brightness, and is also placed at an unfavourable angle to us, so that the spiral form is not well displayed. All the other galaxies in Cetus are considerably fainter.

Eridanus. Only the northern part of this immensely long constellation is shown here; the rest sprawls down into the far south of the sky. In mythology Eridanus represents the River Po, into which the reckless youth Phaethon fell after he had obtained permission to drive the Sun-chariot for one day – with the result that the Earth was set on fire, and Jupiter had to call a reluctant halt to the proceedings by striking Phaethon with a thunderbolt.

There is not a great deal to see in the northern part of the River. There are only two stars above the third magnitude; β or Kursa, close to Rigel in Orion (type A, 96 light-years away, 83 times as luminous as the Sun) and γ or Zaurak (type M, 114 light-years away, 120 Sun-power). It is worth looking at the δ–ε pair. δ (Rana) is fairly close, at 29 light-years, and is a K-type star only 2.6 times as luminous as the Sun; next to it is ε, at 10.7 light-years, which, with τ Ceti, is one of the two nearest stars to bear any resemblance to the Sun. the IRAS satellite found that it is associated with cool material, and as a possible

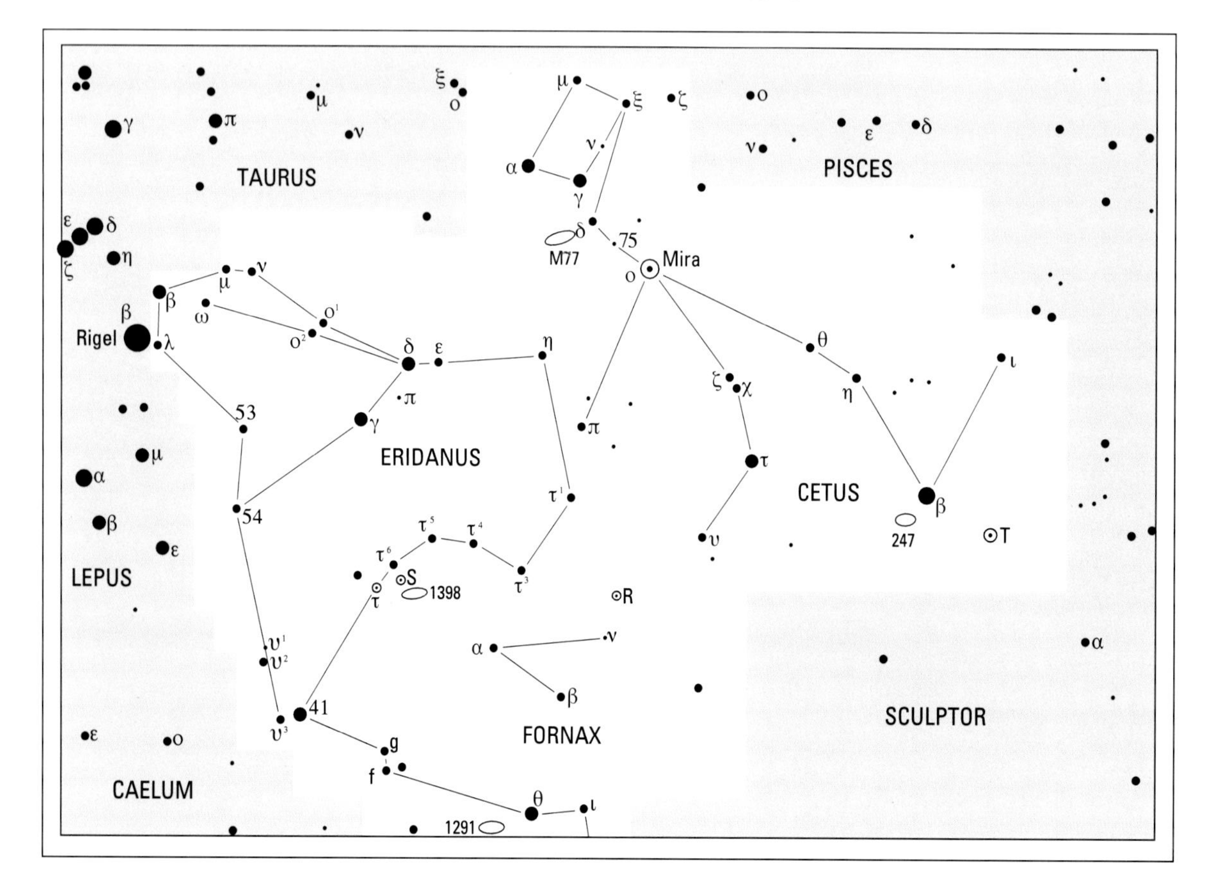

◄ ***The constellations*** *in this map are seen to advantage during evenings in late autumn (northern hemisphere) or late spring (southern). Cetus is large but rather faint, though the 'head', containing α (Menkar) is not hard to identify. Eridanus is so immensely long that not all of it can be conveniently shown on one map; the southern part is contained in Star Map 22, the south polar region – the 'river' ends with the brilliant Achernar, which does not rise from anywhere north of Cairo. Lepus is also shown here, but is described with Orion.*

planetary centre it is decidedly promising. It is smaller and less luminous than τ Ceti, and is in fact the feeblest star visible with the naked eye apart from the far-southern ε Indi. Its absolute magnitude is 6.1, so that from our standard distance of 32.6 light-years it could not easily be seen without optical aid.

The two Omicrons, o^1 (Beid) and o^2 (Keid), lie side by side, but are not connected. Beid is 277 light-years away and well over 150 times as luminous as the Sun, while Keid is a complex system only 16 light-years away. It is a wide, easy binary; the secondary is itself a binary consisting of a feeble red dwarf, of exceptionally low mass (no more than 0.2 that of the Sun) together with a white dwarf whose diameter is about twice that of the Earth and which seems to be associated with a third body of sub-stellar mass.

Fornax is a 'modern' group whose name has been shortened. (There are other cases too: for example, Crux Australis, the Southern Cross, is catalogued officially simply as 'Crux'.) It was added to the sky by Lacaille in 1752; it was originally Fornax Chemica, the Chemical Furnace.

It is marked by a triangle of inconspicuous stars: α (magnitude 3.87), β (4.46) and ν (4.69). It is crowded with galaxies, but all these are inconveniently faint for users of small telescopes. α is a wide double.

▼ ***This photograph** of the stars in the constellations Cetus and Pisces was made in the optical spectrum. The bright, reddish star slightly above and to the right of the centre is Mira Ceti, a variable red giant star which is 400 light-years away from Earth. It is a prototype long-period variable, giving its name to the class. Over 600 million km (380 million miles) in diameter, it varies in brightness over a period of about 331 days. This photograph was taken on 22 October 1979, when it was near to its maximum brightness.*

CETUS

BRIGHTEST STARS

No.	Star	R.A. h	m	s	Dec. °	′	″	Mag.	Spectrum	Proper name
16	β	00	43	35	−17	59	12	2.04	K0	Diphda
92	α	03	02	17	+04	05	23	2.54	M2	Menkar
31	η	01	08	35	−10	10	56	3.45	K2	
86	γ	02	43	18	+03	14	09	3.47	A2	Alkaffaljidhina
52	τ	01	44	04	−15	56	15	3.50	G8	

Also above magnitude 4.3: ι (3.56), θ (3.60), ζ (3.73), υ (4.00), δ (4.07), π (4.25), μ (4.27), ξ^2 (4.28).
The variable Mira (ο Ceti) has been known to rise to magnitude 1.6, but most maxima are much fainter than this.

VARIABLES

Star	R.A. h	m	Dec. °	′	Range (mags)	Type	Period (d)	Spectrum
ο	02	19.3	−02	58	1.6–10.1	Mira	331	M
T	00	21.8	−20	03	5.0–6.9	Semi-reg.	159	M

DOUBLES

Star	R.A. h	m	Dec. °	′	P.A. °	Sep. ″	Mags
χ	01	49.6	−10	41	250	183.8	4.9, 6.9
γ	02	43.3	+03	14	294	2.8	3.8, 7.3
66	02	12.8	−02	24	234	16.5	5.7, 7.5
ο	02	19.3	−02	58	085	0.3	var, 9.5v

CLUSTERS AND NEBULAE

M	NGC	R.A. h	m	Dec. °	′	Mag.	Dimensions ′	Type
77	1068	02	42.7	−00	01	8.8	6.9 × 5.9	SBp galaxy (Seyfert galaxy)
	247	00	47.1	−20	46	8.9	20.0 × 7.4	Spiral galaxy

ERIDANUS

BRIGHTEST STARS

No.	Star	R.A. h	m	s	Dec. °	′	″	Mag.	Spectrum	Proper name
67	β	05	07	51	−05	05	11	2.79	A3	Kursa
34	γ	03	58	02	−13	30	31	2.95	M0	Zaurak

Also above magnitude 4.3: δ (Rana) (3.54), τ^4 (Angetenar) (3.69), ε (3.73), υ^2 (Theemini) (3.82), 53 (Sceptrum) (3.87), η (Azha) (3.89), ν (3.93), μ (4.09), o^1 (4.02), o^3 (Beid) (4.04), λ (4.27); o^2 (Keid), close to o^1, is of magnitude 4.43.
ζ (Zibal), now of magnitude 4.80, is one of the few stars to be strongly suspected of fading during historic times, though the evidence is inconclusive.

DOUBLE

Star	R.A. h	m	Dec. °	′	P.A. °	Sep. ″	Mags	
o^2	04	15.2	−07	39	107	82.8	4.9, 9.5	B is double

FORNAX

The brightest star in Fornax is α: R.A. 03h 12m 04s.2, dec. −28° 59′ 13″, mag. 3.87.
There are no other stars above magnitude 4.3.

DOUBLE

Star	R.A. h	m	Dec. °	′	P.A. °	Sep. ″	Mags	
α	03	12.1	−28	59	298	4.0	4.0, 7.0	Binary 314y

Orion, Canis Major, Canis Minor,

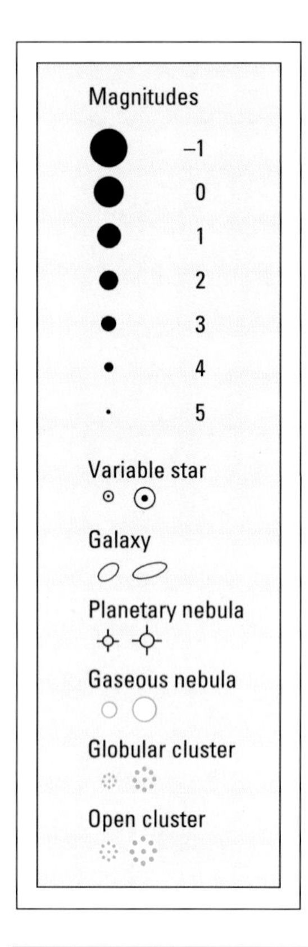

Orion, the Hunter, is generally regarded as the most splendid of all the constellations. The two leaders are very different from each other; though lettered β, Rigel is the brighter, and is particularly luminous, since it could match 60,000 Suns and is some 900 light-years away. If it were as close to us as Sirius, its magnitude would be -10, and it would be one-fifth as brilliant as the full Moon. It has a companion star, which is above magnitude 7, and would be easy to see if it were not so overpowered by Rigel, and even so it has been glimpsed with a 7.6-centimetre (3-inch) telescope under good conditions. The companion is itself a close binary, with a luminosity 150 times that of the Sun. α (Betelgeux) has a official magnitude range of from 0.4 to 0.9, but it seems definite that at times it can rise to 0.1, almost equal to Rigel. Good comparison stars are Procyon and Aldebaran, but allowance must always be made for extinction. The apparent diameter of Betelgeux is greater than for any other star beyond the Sun, and modern techniques have enabled details to be plotted on its surface.

The other stars of the main pattern are γ (Bellatrix), κ (Saiph) and the three stars of the Belt, δ (Mintaka), ε (Alnilam) and ζ (Alnitak). Bellatrix is 2200 times as luminous as the Sun; all the others outshine the Sun by more than 20,000 times, and are over 1000 light-years away. Indeed, Saiph is not much less powerful than Rigel, but is even more remote, at 2200 light-years. Mintaka is an eclipsing binary with a very small range (magnitude 2.20 to 2.35), while both it and Alnitak have companions which are easy telescopic objects.

σ, in the Hunter's Sword, is a famous multiple, and of course θ, the Trapezium, is responsible for illuminating the wonderful nebula M42. M43 (an extension of M42) and M78 (north of the Belt) are really only the brightest parts of a huge nebular cloud which extends over almost the whole of Orion. Other easy doubles are ι and λ.

The red semi-regular variable W Orionis is in the same binocular field with π^6 (magnitude 4.5), the southernmost member of a line of stars which, for some strange reason, are all lettered π. It has an N-type spectrum, and is always within binocular range; its colour makes it readily identifiable, and it is actually redder than Betelgeux, though the hue is not so striking because the star is much fainter. U Orionis, on the border of Orion and Taurus, is a Mira star which rises to naked-eye visibility at maximum; it is a member of a well-marked little group lying between τ Tauri and η Geminorum.

Canis Major, Orion's senior Dog, is graced by the presence of Sirius, which shines as much the brightest star in the sky even though it is only 26 times as luminous as the Sun; it is a mere 8.6 light-years away, and is the closest of all the brilliant stars apart from α Centauri. Though it is pure white, with an A-type spectrum, the effects of the Earth's atmosphere make it flash various colours. All stars twinkle to some extent, but Sirius shows the effect more than any others simply because it is so bright. The white dwarf companion would be easy to see if it were not so overpowered; the revolution period is 50 years. It is smaller than the planet Neptune, but is as massive as the Sun.

ε (Adhara), δ (Wezea), η (Aludra) and o^2 are all very hot and luminous; Wezea, indeed, could match 130,000 Suns, and is over 3000 light-years away. It is not easy to appreciate that of all the bright stars in Canis Major, Sirius is much the least powerful. Adhara, only just below the official 'first magnitude', has a companion which is easy to see with a small telescope.

There are two fine open clusters in Canis Major. M41 lies in the same wide field with the reddish ν^2, forming a triangle with ν^2 and Sirius; it is a naked-eye object, and can be partly resolved with binoculars. NGC2362, round the hot, luminous star τ (magnitude 4.39), is 3500 light-years away, and seems to be a very young cluster; with a low power it looks almost stellar, but higher magnification soon resolves it. In the same low-power field is the β Lyrae eclipsing binary UW Canis Majoris, which is an exceptionally massive system. According to one estimate the masses of the two components are 23 and 19 times that of the Sun, so that they rank as cosmic heavyweights. The total luminosity of the system is at least 16,000 times that of the Sun.

Canis Minor, the Little Dog, includes Procyon, 11.4 light-years away and 10 times as luminous as the Sun. Like Sirius, it has a white dwarf companion, but the dwarf is so faint and so close-in that it is a very different object. The

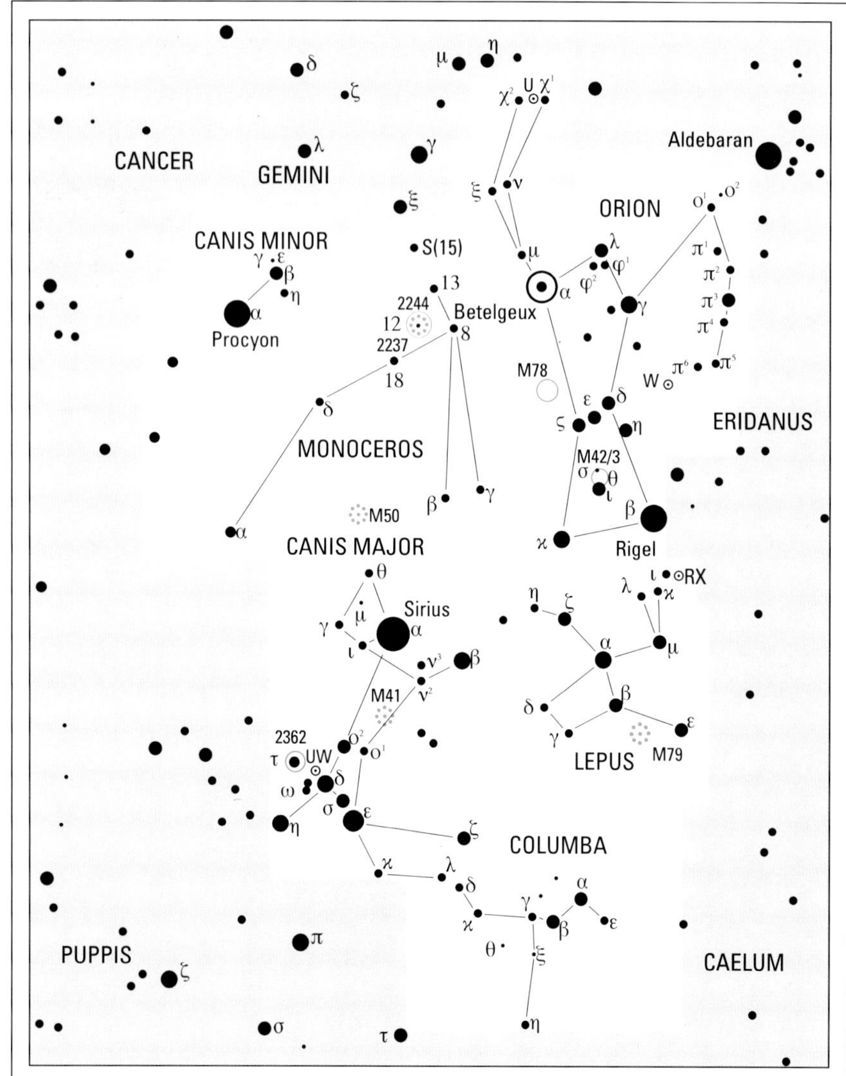

◀ ***Orion*** *is probably the most magnificent of all the constellations, and since it is crossed by the celestial equator it is visible from every inhabited country (though from the proposed observatory at the South Pole, Rigel will be permanently above the horizon and Betelgeux never!). Orion is a superb guide to other groups; the Belt stars point southwards to Sirius and northwards to Aldebaran. Orion and his retinue dominate the evening sky all through northern winter (southern summer). The stars in the southernmost part of this map do not rise over Britain.*

Monoceros, Lepus, Columba

ORION

BRIGHTEST STARS

Star	R.A.			Dec.			Mag.	Spectrum	Proper name
	h	m	s	°	′	″			
19 β	05	14	32	−08	12 06	0.	12	B8	Rigel
58 α	05	55	10	+07	24	26	0.1–0.9	M2	Betelgeux
24 γ	05	25	08	+06	20	59	1.64	B2	Bellatrix
46 ε	05	36	13	−01	12	07	1.70	B0	Alnilam
50 ζ	05	50	45	−01	56	34	1.77	O9.5	Alnitak
53 κ	05	47	45	−09	40	11	2.06	B0	Saiph

VARIABLES

Star	R.A.		Dec.		Range	Type	Period	Spectrum
	h	m	°	′	(mags)		(d)	
U	05	55.8	+20	10	4.8–12.6	Mira	372	M
W	05	05.4	+01	11	5.9–7.7	Semi-reg	12	N

DOUBLES

Star	R.A.		Dec.		P.A.	Sep.	Mags
	h	m	°	′	°	″	
λ	05	36.1	+09	56	043	4.4	3.6,5.5
ι	05	35.4	−05	55	141	11.3	2.8,6.9
β	05	14.5	−08	12	202	9.5	0.1,6.8

M	NGC	R.A.		Dec.		Mag.	Dimensions	Type
		h	m	°	′		′	
M42	1976	05	35.4	−05	27	5	66 × 60	Great Nebula
M43	1982	05	35.6	−05	16	7	20 × 15	Extension of M42
M78	2068	05	46.7	+00	03	8	8 × 8	Nebula

CANIS MAJOR

BRIGHTEST STARS

Star	R.A.			Dec.			Mag.	Spectrum	Proper name
	h	m	s	°	′	″			
9 α	06	45	09	−16	42	58	−1.46	A1	Sirius
21 ε	06	58	38	−28	58	20	1.50	B2	Adhara
25 δ	07	08	23	−26	23	36	1.86	F8	Wezea
2 η	07	24	06	−29	18	11	2.44	B5	Aludra

VARIABLES

Star	R.A.		Dec.		Range	Type	Period	Spectrum
	h	m	°	′	(mags)		(d)	
UW	07	18.4	−24	34	4.0–5.3	β Lyrae	4.39	07

DOUBLE

Star	R.A.		Dec.		P.A.	Sep.	Mags
	h	m	°	′	°	″	
α	06	45.1	−16	43	005	4.5	−1.5, 8.5

CLUSTERS AND NEBULAE

M	NGC	R.A.		Dec.		Mag.	Dimensions	Type
		h	m	°	′		′	
M41	2287	06	47.0	−20	44	4.5	38	Open cluster
	2362	07	17.8	−24	57	4	8	Open cluster

CANIS MINOR

BRIGHTEST STARS

Star	R.A.			Dec.			Mag.	Spectrum	Proper name
	h	m	s	°	′	″			
10 α	07	39	18	+05	13	30	0.38	F5	Procyon
3 β	07	27	09	+08	17	21	2.90	B8	Gomeisa

MONOCEROS

The brightest star is β: R.A. 06h 28m 49s, dec. −07° 01′ 58″, mag. 3.7.

DOUBLE

Star	R.A.		Dec.		P.A.	Sep.	Mags
	h	m	°	′	°	″	
ε	06	23.8	+04	36	027	13.4	4.5, 6.5
S (15)	06	41.0	+09	54	AB 213	2.8	4.7v, 7.5

CLUSTERS AND NEBULAE

M	NGC	R.A.		Dec.		Mag.	Dimensions	Type
		h	m	°	′		′	
50	2323	07	03.2	−08	20	5.9	16	Open cluster
	2237	06	32.3	+05	03	~6	80 × 60	Nebula
	2244	06	32.4	+04	52	5	24	Open cluster

LEPUS

BRIGHTEST STARS

Star	R.A.			Dec.			Mag.	Spectrum	Proper name
	h	m	s	°	′	″			
11 α	05	32	44	−17	49	20	2.58	F0	Arneb
9 β	05	28	15	−20	45	35	2.84	G2	Nihal
2 μ	05	12	56	−16	12	20	3.31	B9	

VARIABLES

Star	R.A.		Dec.		Range	Type	Period	Spectrum
	h	m	°	′	(mags)		(d)	
RX	05	11.4	−11	51	5.0–7.0	Irregular	–	M

DOUBLE

Star	R.A.		Dec.		P.A.	Sep.	Mags
	h	m	°	′	°	″	
κ	05	13.2	−12	56	358	2.6	4.5,7.4
β	05	28.2	−20	46	330	2.5	2.8,7.3
γ	05	44.5	−22	27	350	96.3	3.7,6.3

CLUSTERS AND NEBULAE

M	NGC	R.A.			Dec.		Mag.	Dimensions	Type
		h	m	s	°	′			
M79	1904	05	24	30	−24	33	9.9	8.7	Globular cluster

COLUMBA

BRIGHTEST STARS

Star	R.A.			Dec.			Mag.	Spectrum	Proper name
	h	m	s	°	′	″			
α	05	39	39	−34	04	27	2.64	B8	Phakt
β	05	50	57	−35	46	06	3.12	K2	Wazn

Also above magnitude 4.3: δ (3.85), ε (3.87), η (3.96).

revolution period is 40 years. The only other brightish star in Canis Major is β, which makes a pretty little group with the much fainter ε, η and γ.

Monoceros is not an original constellation; it was created by Hevelius in 1690, and although it represents the fabled unicorn there are no legends attached to it. Much of it is contained in the large triangle bounded by Procyon, Betelgeux and Saiph. There are no bright stars, but there are some interesting doubles and nebular objects, and the constellation is crossed by the Milky Way. β is a fine triple; William Herschel, who discovered it in 1781, called it 'one of the most beautiful sights in the heavens'. S Monocerotis is made up of a whole group of stars, together with the Cone Nebula, which is elusive but not too hard to photograph. The open cluster NGC2244, round the star 12 Monocerotis (magnitude 5.8), is easy to find with binoculars; surrounding it is the Rosette Nebula, NGC2237, which is 2600 light-years away and over 50 light-years across. Photographs show the dark dust-lanes and globules which give it such a distinctive appearance. M50 is an unremarkable open cluster near the border between Monoceros and Canis Major.

Lepus, the Hare, is placed here because it represents an animal which Orion is said to have been particularly fond of hunting. Of the two leaders, α (Arneb) is an F-type supergiant, 950 light-years away and 6800 times as luminous as the Sun; β (Nihal) is of type G, 316 light-years away and 600 Sun-power. γ is a wide, easy double. R Leporis, nicknamed the Crimson Star, is a Mira variable making a triangle with κ (4.36) and μ; it can reach naked-eye visibility, and can be followed with binoculars for parts of its cycle. It is cool by stellar standards – hence its strong red colour – but is 1000 light-years away, and at least 500 times more powerful than the Sun.

M79, discovered by Méchain in 1780, is a globular cluster at a distance of 43,000 light years; it lies in line with α and β. It is not too easy to find with binoculars, but a small telescope will show it clearly.

Columba (originally Columba Noae, Noah's Dove) contains little of immediate interest, but the line of stars south of Orion, of which α and β are the brightest members, makes it easy to identify. μ, magnitude 5.16, is one of three stars which seem to have been 'shot out' of the Orion nebulosity, and are now racing away from it in different directions; the other two are 53 Arietis and AE Aurigae. μ Columbae is of spectral type O9.5, so that it is certainly very young; it has the high proper motion of 0.025 of a second of arc per year.

Taurus, Gemini

Taurus is a large and conspicuous Zodiacal constellation, representing the bull into which Jupiter once changed himself for thoroughly discreditable reasons. It has no well-defined pattern, but it does contain several objects of special interest.

α (Aldebaran), in line with Orion's Belt, is an orange-red star of type K0, 68 light-years away and 100 times as luminous as the Sun. It looks very similar to Betelgeux, though it is not nearly so remote or powerful; it makes a good comparison for Betelgeux, though generally it is considerably the fainter of the two. The stars of the Hyades cluster extend from it in a sort of V-formation, but there is no true association; Aldebaran is not a cluster member, and merely happens to lie about halfway between the Hyades and ourselves – which is rather a pity, since its brilliant orange light tends to drown the fainter stars. The leading Hyades are γ (3.63), ε (3.54), δ (3.76) and θ (3.42). The cluster was not listed by Messier, presumably because there was not the slightest chance of confusing it with a comet.

Because the Hyades are so scattered, they are best seen with binoculars. σ consists of two dim stars close to Aldebaran; δ makes up a wide pair with the fainter star 64 Tauri, of magnitude 4.8; and θ is a naked-eye double, made up of a white star of magnitude 3.4 and a K-type orange companion of magnitude 3.8. The colour contrast is striking in binoculars. Here, too, we are dealing with a line-of-sight effect; the white star is the closer to us by 15 light-years, though undoubtedly the two have condensed out of the same nebula which produced all the rest of the Hyades.

Messier did include the Pleiades in his catalogue, and gave them the number 45. Of course, they have been known since very early times; they are referred to by Homer and Hesiod, and are mentioned three times in the Bible. The leader, η Tauri or Alcyone, is of the third magnitude; then follow Electra, Atlas, Merope, Maia, Taygete, Celaeno, Pleione and Asterope. This makes nine, though the cluster is always nicknamed the Seven Sisters. However, Pleione is close to Atlas, and is an unstable shell star which varies in light, while Celaeno (magnitude 5.4) and Asterope (5.6) are easy to overlook. On the next clear night, see how many separate stars you can see in the cluster without optical aid; if you can manage a dozen, you are doing very well indeed. Binoculars show many more, and the total membership of the cluster amounts to several hundreds. The average distance of the stars is just over 400 light-years.

The Pleiades are at their best when viewed under very low magnification. The leading stars are hot and bluish-white, and the cluster – unlike the Hyades – is certainly very young; there is considerable nebulosity, so that star formation is presumably still going on. This nebulosity is very difficult to see through a telescope, but is surprisingly easy to photograph.

The other nebular object is M1, the Crab, which is the remnant of the supernova of 1054. It can be glimpsed with powerful binoculars, close to the third-magnitude ζ; a telescope shows its form, but photography is needed to bring out its intricate structure. It is expanding, and inside is a pulsar which powerful equipment can record as a faint, flickering object – one of the few pulsars to be optically identified.

λ Tauri is an Algol variable, easy to follow with the naked eye; good comparison stars are γ, ο, ξ and μ. The real separation of the components is of the order of 14 million kilometres (nearly 9 million miles), so that they cannot be seen separately; eclipses of the primary are 40 per cent total. The distance is 326 light-years; λ is much more luminous than Algol, but is also much further away. The only other Algol stars to exceed magnitude 5 at maximum are Algol itself, δ Librae and the far-southern ζ Phoenicis.

Of the other leading stars in Taurus, ζ (Alheka) is a highly luminous B-type giant, 490 light-years away and

Magnitudes
-1
0
1
2
3
4
5
Variable star
Galaxy
Planetary nebula
Gaseous nebula
Globular cluster
Open cluster

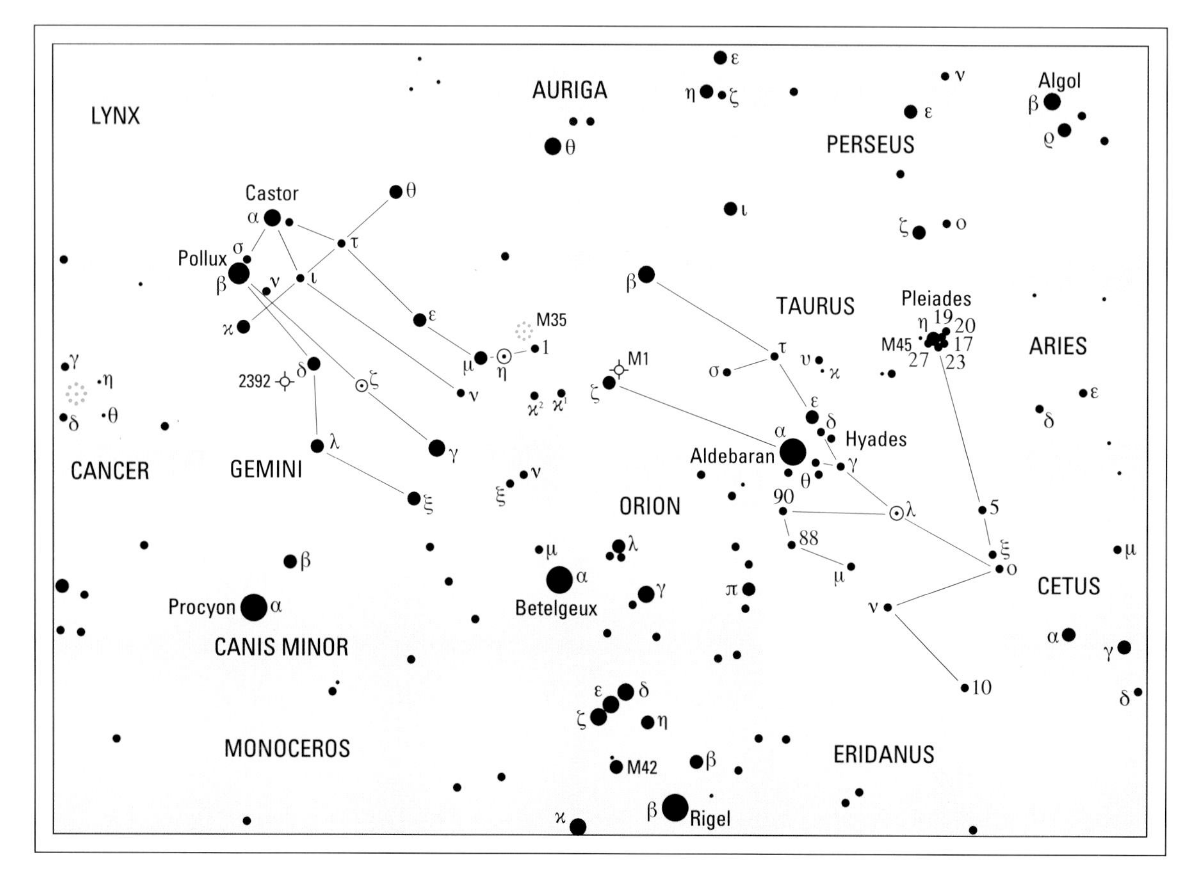

◀ ***These two large**, important Zodiacal constellations form part of Orion's retinue, and are thus best seen during evenings in northern winter (southern summer). Taurus contains the two most famous open clusters in the sky, the Pleiades and the Hyades, while the 'Twins', Castor and Pollux, make an unmistakable pair. The Milky Way flows through Gemini, and there are many rich star fields. Canis Minor is shown here, but is described with Map 16.*

1300 times as powerful as the Sun. β (Alnath) is very prominent, and has been transferred from Auriga to Taurus – which seems illogical, as it belongs much more naturally to the Auriga pattern. It is 130 light-years from us, and can equal 470 Suns.

Gemini. The Heavenly Twins, Castor and Pollux, make up a striking pair. Pollux is the brighter; it is 36 light-years away as against 46 light-years for Castor, and it is an orange K-type star, outshining the Sun by some 60 times. Castor is a fine binary with a revolution period of 420 years; though the separation is less than it used to be a century ago, it is still a suitable target for small telescopes. Each component is a spectroscopic binary, and there is a third member of the system, YY Geminorum, which is an eclipsing binary.

There are two notable variables in Gemini. ζ is a typical Cepheid, with a period of 10.15 days; this is almost twice the period of δ Cephei itself, and ζ Geminorum is correspondingly the more luminous, since at its peak it is well over 5000 times as luminous as the Sun. η, or Propus, is a red semi-regular with an extreme range of magnitude 3.1 to 3.9, and a rough period of around 233 days; a good comparison star is μ, which is of the same spectral type (M3) and the same colour. Also in the Twins is U Geminorum, the prototype dwarf nova. Stars of this type are known either as U Geminorum stars or as SS Cygni stars; it is true that U Geminorum is much the fainter of the two, since its 'rest' magnitude is only 14.9 and it never reaches magnitude 8. The average interval between outbursts is just over 100 days.

M35 is a very conspicuous cluster close to η and μ. It is 2850 light-years away, and was discovered by de Chéseaux in 1746; Messier called it 'a cluster of very small stars'. It is worth seeking out NGC2392, the Eskimo Nebula, which is a planetary lying between ϰ and λ; the central star is of the tenth magnitude. The Eskimo is decidedly elusive, but photographs taken with larger telescopes show its curious 'face'. Like all planetaries it is expanding, and has now reached a diameter of more than half a light-year. It was William Herschel who first called these objects 'planetary nebulae', because he thought that their disks made them look like planets – but the name could hardly be less appropriate.

▼ ***The Pleiades cluster*** *in Taurus, photographed by Bernard Abrams using a 25-cm (10-inch) reflector. Known since ancient times, Messier included them as 45.*

TAURUS

BRIGHTEST STARS

No.	Star	R.A. h	m	s	Dec. °	′	″	Mag.	Spectrum	Proper name
87	α	04	35	55	+16	30	33	0.85	K5	Aldebaran
112	β	05	26	17	+28	36	27	1.65	B7	Al Nath
25	η	03	47	29	+24	06	18	2.87	B7	Alcyone
123	ζ	05	37	39	+21	08	33	3.00	B2	Alheka
35	λ	04	00	41	+12	29	15	3.4 (max)	B3	
78	θ^2	04	28	40	+15	52	15	3.42	A7	

Also above magnitude 4.3: ε (Ain) (3.54), ο (3.60), 27 (Atlas) (3.63), γ (Hyadum Primus) (3.63), 17 (Electra) (3.70), ξ (3.74), δ (3.76), θ^1 (3.85), 20 (Maia) (3.88), ν (3.91), 5 (4.11), 23 (Merope) (4.18), ϰ (4.22), 88 (4.25), 90 (4.27), 10 (4.28), μ (4.29), υ (4.29), 19 (Taygete) (4.30), τ (4.28), δ^3 (4.30). β (Al Nath) was formerly included in Auriga, as γ Aurigæ.

VARIABLES

Star	R.A. h	m	Dec. °	′	Range (mags)	Type	Period (d)	Spectrum
λ	04	00.7	+12	29	3.3–3.8	Algol	3.95	B+A
BU (Pleione)	03	49.2	+24	08	4.8–5.5	Irregular	–	Bp
T	04	22.0	+19	32	8.4–13.5	T Tauri	–	G–K
SU	05	49.1	+19	04	9.0–16.0	R Coronæ	–	G0p

DOUBLES

Star	R.A. h	m	Dec. °	′	P.A. °	Sep. ″	Mags	
θ	04	28.7	+15	32	346	337.4	3.4, 3.8	Naked-eye
σ	04	39.3	+15	55	193	431.2	4.7, 5.1	Naked-eye
K+67	04	25.4	+22	18	173	339	4.2, 5.3	Naked-eye

CLUSTERS AND NEBULAE

M	NGC	R.A. h	m	Dec. °	′	Mag.	Dimensions ′	Type
1	1952	05	34.5	+22	01	10	6.4	Supernova remnant (Crab)
45	1432/5	03	47.0	+24	07	3	110	Open cluster (Pleiades)
		04	27	+16	00	1	330	Open cluster (Hyades)

GEMINI

BRIGHTEST STARS

No.	Star	R.A. h	m	s	Dec. °	′	″	Mag.	Spectrum	Proper name
78	β	07	45	19	+28	01	34	1.14	K0	Pollux
66	α	07	34	36	+31	53	18	1.58	A0	Castor
24	γ	06	37	43	+16	23	57	1.93	A0	Alhena
13	μ	06	22	58	+22	30	49	2.88	M3	Tejat
27	ε	06	43	56	+25	07	52	2.98	G8	Mebsuta
7	η	06	14	53	+22	30	24	3.1 (max)	M3	Propus
31	ξ	06	45	17	+12	53	44	3.36	F5	Alzirr

Also above magnitude 4.3: δ (Wasat) (3.53), ϰ (3.57), λ (3.58), θ (3.60), ζ (Mekbuda) (3.7 max), ι (3.79), υ (4.06), ν (4.15), 1 (4.16), ϱ (4.18), σ (4.28).

VARIABLES

Star	R.A. h	m	Dec. °	′	Range (mags)	Type	Period (d)	Spectrum
η	06	14.9	+22	30	3.1–3.9	Semi-regular	+233	M
ζ	07	04.1	+20	34	3.7–4.1	Cepheid	10.15	F–G

DOUBLES

Star	R.A. h	m	Dec. °	′	P.A. °	Sep. ″	Mags	
η	06	14.9	+22	30	266	1.4	3v, 8.8	Binary, 470y
α	07	34.6	+31	53	AB 088	2.5	1.9, 2.9	Binary 420y
					AC 164	72.5	8.8	

CLUSTERS AND NEBULAE

M	NGC	R.A. h	m	Dec. °	′	Mag.	Dimensions ′	Type
35	2168	06	08.9	+24	20	5	28	Open cluster
	2392	07	29.2	+20	55	10	13″ × 44″	Planetary nebula (Eskimo Nebula)

Auriga, Lynx

Magnitudes

- –1
- 0
- 1
- 2
- 3
- 4
- 5

Variable star

Galaxy

Planetary nebula

Gaseous nebula

Globular cluster

Open cluster

Auriga, the Charioteer, is a brilliant northern constellation, led by Capella. In mythology it honours Erechthonius, son of Vulcan, the blacksmith of the gods; he became King of Athens, and invented the four-horse chariot.

Capella is the sixth brightest star in the entire sky, and is only 0.05 of a magnitude inferior to Vega. It and Vega are on opposite sides of the north celestial pole, so that when Capella is high up Vega is low down, and vice versa; from Britain, neither actually sets, and Capella is near the zenith or overhead point during evenings in winter. It can be seen from almost all inhabited countries, though it is lost from the extreme southern tip of New Zealand.

Capella is yellow, like the Sun, but is a yellow giant rather than a dwarf – or, rather, two giants, because it is a very close binary. One component is 90 times as luminous as the Sun, and the other 70 times; the distance between them is not much more than 100 million kilometres (60 million miles).

The distance from us is 42 light-years. The second star of Auriga, β (Menkarlina), is also a spectroscopic binary, and is actually an eclipsing system with a very small magnitude range. The components are more or less equal, and a mere 12 million kilometres (7.5 million miles) apart; both are of type A.

Of course, the most intriguing objects in Auriga are the two eclipsing binaries ε and ζ, which have been described earlier. It is sheer chance that they lie side by side, because they are at very different distances – 520 light-years for ζ; as much as 4600 light-years for ε. The third member of the trio of the Haedi or Kids, η Aurigae, is a useful comparison; the magnitude is 3.17.

It is worth keeping a close watch on ε, because even during long intervals between eclipses it seems to fluctuate slightly. The catalogues give its normal magnitude as 2.99, in which case it appears very slightly but perceptibly brighter than η. All three Kids are in the same low-power binocular field, and this is probably the best way to make estimates of ε; ζ is much fainter, and the only really useful comparison star is ν, of magnitude 3.97.

Of the other main stars of the Charioteer, ι and the rather isolated δ, are reddish, with K-type spectra. θ is white, and has two companions; the closer pair makes up a slow binary system, while the more remote member of the group, of magnitude 10.6, merely lies in almost the same line of sight.

Auriga is crossed by the Milky Way, and there are several fine open clusters, of which three are in Messier's list. M36 and M38 were both discovered by Guillaume Legentil in 1749, and M37 by Messier himself in 1764; but no doubt all had been recorded earlier, because all are bright.

M36 is easy to resolve, and is 3700 light-years away. M37, at about the same distance, is in the same lower-power field as θ, which is a very good way of identifying it; the brightest stars in the cluster form a rough trapezium. M38 is larger and looser, and rather less bright. It lies slightly away from the mid-point of a line joining θ to ι, and within half a degree of it is a much smaller and dimmer cluster, NGC1907.

Note also the Flaming Star Nebula round the irregular variable AE Aurigae – one of the 'runaway stars' which seem to have been ejected from the Orion nebulosity (the others are 53 Arietis and μ Columbae). AE Aurigae illuminates the diffuse nebulosity, which is elusive telescopically though photographs show intricate structure. The distance is of the order of 1600 light-years.

Lynx is a very ill-defined and obscure northern constellation, created by Hevelius in 1790; it has no mythological associations, and it has been said that only a lynx-eyed observer can see anything there at all. In fact there is one brightish star, α (magnitude 3.13), which is decidedly isolated, and forms an equilateral triangle with Regulus

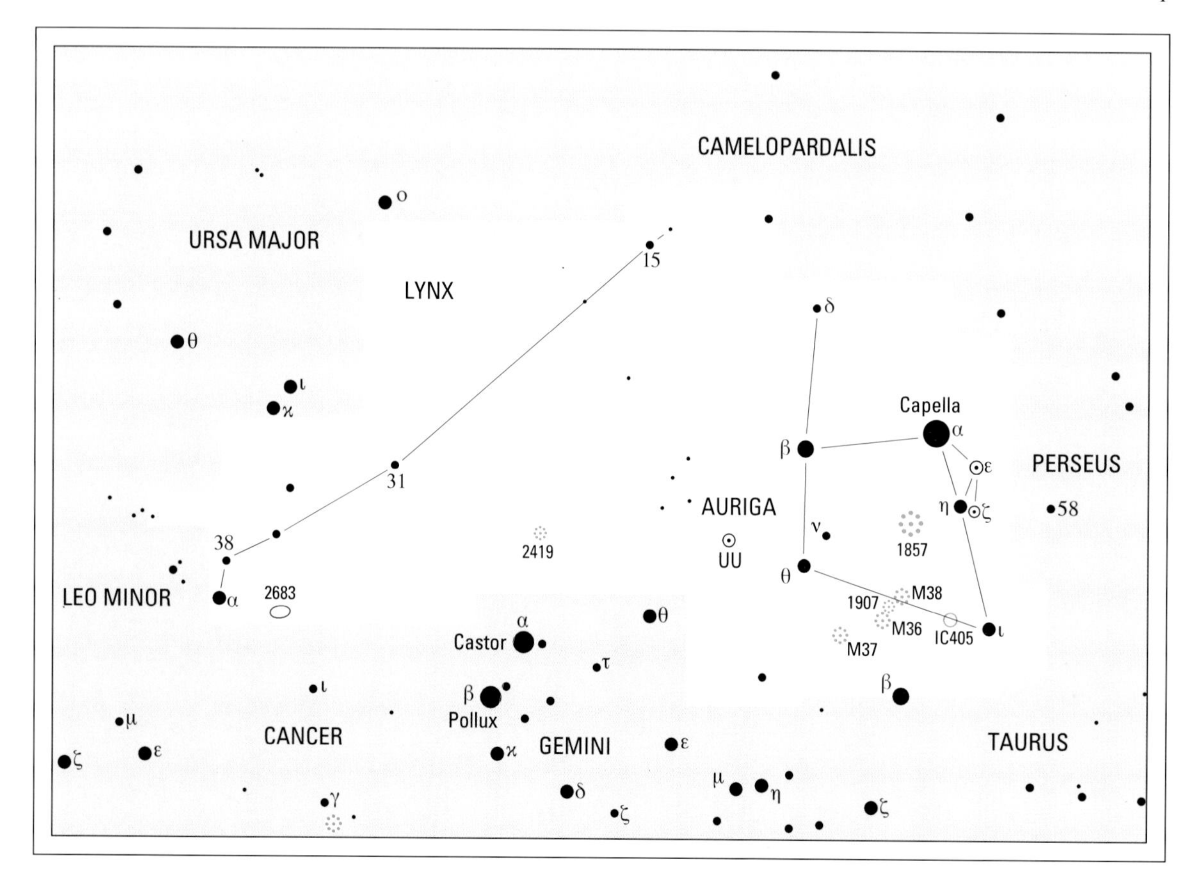

◄ ***Capella,*** *the brightest star in Auriga – and the sixth brightest star in the entire sky – is near the zenith or overhead point during evenings in winter, as seen from the northern hemisphere; this is the position occupied by Vega during summer evenings. From Britain or the northern United States, Capella does not set, though at its lowest it skims the horizon. The Auriga quadrilateral is very easy to identify; a fifth bright star, Alnath, which seems logically to belong to the Auriga pattern, has been transferred to Taurus, and is now β Tauri instead of γ Aurigae.*

and Pollux. It is of type M, and obviously red; its distance is 166 light-years, and it is 120 times as luminous as the Sun. None of the other stars in Lynx have been given Greek letters, though one of them, 31 Lyncis, has been dignified with a proper name: Alsciaukat.

The globular cluster NGC2419, about 7 degrees north of Castor, is faint and none too easy to identify. This is not because it is feeble – on the contrary it is exceptionally large, and must be around 400 light-years across – but because it is so far away.

The distance has been estimated at around 300,000 light-years, and though this may be rather too great it is clear that the cluster is at the very edge of the Milky Way system. It may even be escaping altogether, in which case it will become what is termed an intergalactic tramp. It is very rich, and, predictably, its leading stars are red and yellow giants.

We have little direct knowledge of the isolated star systems that lie between the galaxies. There is every reason to believe they exist, but since they will be so much less luminous than full-scale galaxies they will be far less easy to detect.

Indeed, galaxies of very low surface brightness may also be very elusive. Modern electronic techniques used with large telescopes may be able to track these isolated objects, but at the moment we do not know how many of them there are. At least it seems unlikely that NGC2419 will become the only intergalactic tramp.

AURIGA

BRIGHTEST STARS

No.	Star	R.A. h m s	Dec. ° ′ ″	Mag.	Spectrum	Proper name
13	α	05 16 41	+45 59 53	0.08	G8	Capella
34	β	05 59 32	+44 56 51	1.90	A2	Menkarlina
37	θ	05 59 43	+37 12 45	2.62	A0p	
3	ι	04 56 59	+33 09 58	2.69	K3	Hassaleh
7	ε	05 01 58	+43 49 24	2.99v	F0	Almaaz
10	η	05 06 31	+41 14 04	3.17	B3	

Also above magnitude 4.3: δ (3.72), ζ (Sadatoni) (3.75) (max).

VARIABLES

Star	R.A. h m	Dec. ° ′	Range (mags)	Type	Period (d)	Spectrum
ε	05 02.0	+43 49	3.0–3.8	Eclipsing	9892	F
ζ	05 02.5	+41 05	3.7–4.1	Eclipsing	972	K+B
UU	06 36.5	+38 27	5.1–6.8	Semi-reg.	234	N

DOUBLE

Star	R.A. h m	Dec. ° ′	P.A. °	Sep. ″	Mags
θ	05 59.7	+37 13	AB 313	3.6	2.6,7.1
			AC 297	50.0	10.6

CLUSTERS AND NEBULAE

M	NGC	R.A. h m	Dec. ° ′	Mag.	Dimensions ′	Type
36	1960	05 36.1	+34 08	6.0	12	Open cluster
37	2099	05 52.4	+32 33	5.6	24	Open cluster
38	1912	05 28.7	+35 50	6.4	21	Open cluster
	1857	05 20.2	+39 21	7.0	6	Open cluster
	IC405	05 16.2	+34 16	var	30 × 19	Nebula: Flaming Star Nebula, round AE Aurigæ

LYNX

BRIGHTEST STAR

No.	Star	R.A. h m s	Dec. ° ′ ″	Mag.	Spectrum	Proper name
40	α	09 21 03	+34 23 33	3.13	MO	

Also above magnitude 4.3: 38 (3.92), 31 (Alsciaukat) (4.25).

CLUSTERS AND NEBULAE

M	NGC	R.A. h m	Dec. ° ′	Mag.	Dimensions ′	Type
	2419	07 38.2	+38 53	10.4	41	Globular cluster
	2683	08 52.7	+33 25	9.7	9.3 × 2.5	Sc galaxy

▼ ***Open clusters** M38 and NGC1907 in the constellation Auriga, taken by Bernard Abrams using a 25-cm (10-inch) reflector. They lie within half a degree of each other; NGC1907 is much the smaller and dimmer.*

Carina, Vela, Pyxis, Antlia, Pictor,

Carina, the Keel. We now come to the main constellations of the southern sky, most of which are inaccessible from the latitudes of Britain, Europe or most of the mainland United States. The brightest part of the old Argo is the Keel, which contains Canopus, the second brightest star in the sky. It looks half a magnitude fainter than Sirius, but this is only because it is so much more remote. According to the Cambridge catalogue it is 200,000 times as luminous as the Sun, and therefore well over 7500 times as luminous as Sirius. The spectral type is F, and this means that in theory it should look slightly yellowish, but to most observers it appears pure white. Its declination is 53 degrees S. Over parts of Australia and South Africa it sets briefly, but it is circumpolar from Sydney, Cape Town and the whole of New Zealand.

The second brightest star in the Keel is β or Miaplacidus, of type A and 85 times as luminous as the Sun. ε and ι Carinae, together with ϰ and δ Velorum, make up the False Cross, which is of much the same shape as the Southern Cross and is often confused with it, even though it is larger and not so brilliant. As with the Southern Cross, three of its stars are hot and bluish-white while the fourth – in this case ε Carinae – is red; ε is of type K, 202 light-years from us and 600 Sun-power. ι Carinae is of type F, very luminous (6800 times more so than the Sun) and over 800 light-years away. Its proper name is Tureis, but it has also been called Aspidske.

ZZ Carinae is a bright Cepheid, and R Carinae is one of the brightest of all Mira stars, rising to magnitude 3.9 at some maxima. However, the most interesting variable is η, which has been described earlier. For a while during the 19th century it outshone even Canopus; today it is just below naked-eye visibility, but it may brighten again at any time. The associated nebula can be seen with the naked eye; it contains a famous dark mass nicknamed the Keyhole. Telescopically, η looks quite unlike a normal star, and its orange hue is very pronounced. In the future – perhaps tomorrow, perhaps not for a million years – it will explode as a supernova, and it will then provide us with a truly magnificent spectacle.

The cluster IC2602, round θ Carinae, is very fine; it forms a triangle with β and ι. Also imposing is NGC 2516, which lies in line with δ Velorum and ε Carinae in the False Cross; NGC2867, between ι Carinae and ϰ Velorum, is a planetary nebula which is just within binocular range. The whole of Carina is very rich, and there are a great many spectacular star fields.

Vela, the Sails of Argo, are also full of interest, though less striking than the Keel. The brightest star is γ (Regor), which is a Wolf–Rayet star of spectral type W, and is very hot and unstable. It is a fine, easy double, and there are three fainter companions nearby. δ Velorum, in the False Cross, has a fifth-magnitude companion which is visible in a very small telescope, and in the same binocular field lies the open cluster NGC2391, round the 3.6-magnitude star o Velorum. In a low-power telescope, or even in binoculars, the cluster has a vaguely cruciform appearance. Another naked-eye cluster is NGC2547, near Regor.

Pyxis (originally Pyxis Nautica, the Mariner's Compass). A small constellation north of Vela. The only object of immediate interest is the recurrent nova T Pyxidis, which is normally of about the 14th magnitude, but has flared up to near naked-eye visibility on several occasions. It makes a triangle with α and γ, but in its usual state it is not at all easy to identify.

Antlia, originally Antlia Pneumatica, was added to the sky by Lacaille in 1752, and seems to be one of the totally unnecessary constellations. It adjoins Vela and Pyxis, and is entirely unremarkable.

Pictor (originally Equuleus Pictoris, the Painter's Easel) is another of Lacaille's constellations. It lies near Canopus; there are no bright stars, but β – which has no individual name – has become famous because of the associated cloud of cool material which may be planet-forming. It is

Magnitudes
-1
0
1
2
3
4
5
Variable star
Galaxy
Planetary nebula
Gaseous nebula
Globular cluster
Open cluster

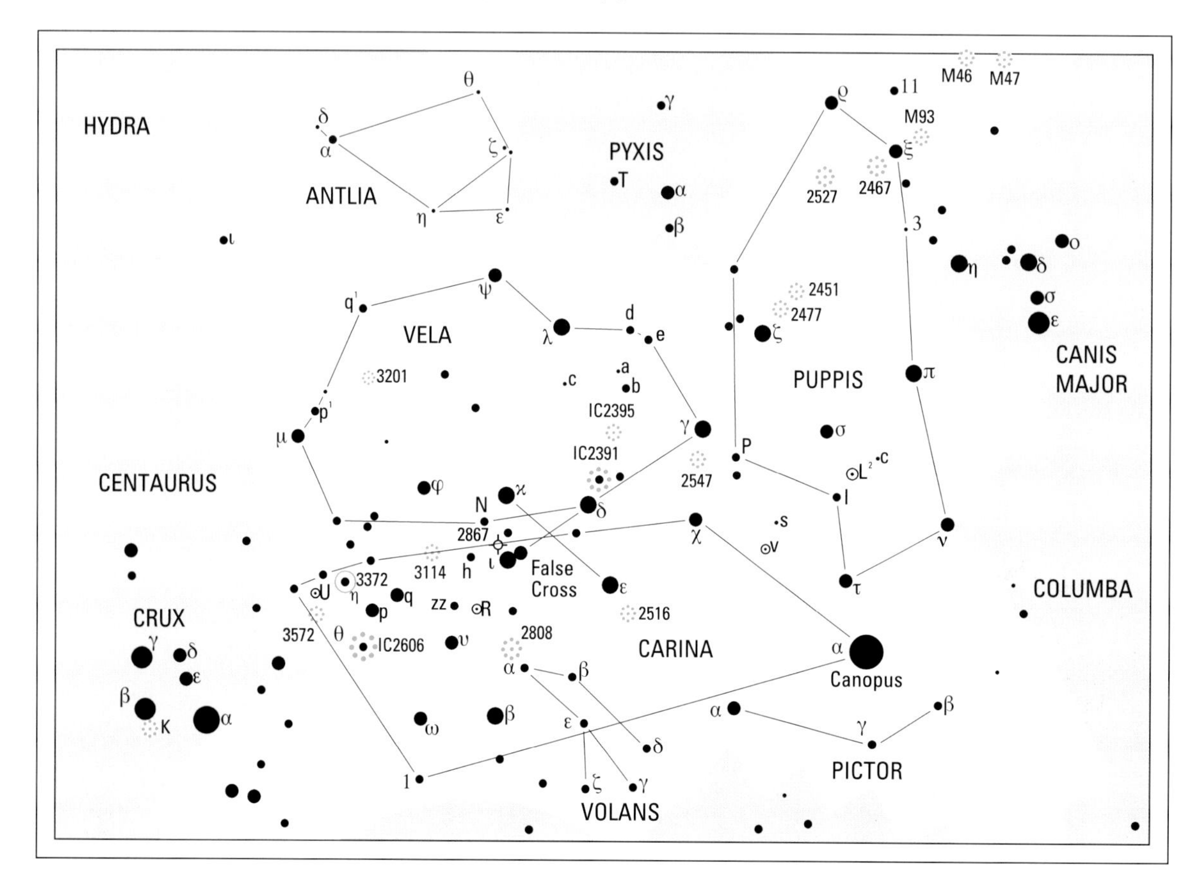

◀ ***This region*** *is well south of the equator, and most of it is invisible from Britain or the northern United States, though part of Puppis can be seen. From southern countries such as Australia, Canopus – the second brightest star in the sky – is near the zenith during evenings around February; it rises from Alexandria, but not from Athens – an early proof that the Earth is not flat. Carina, Vela and Puppis were once combined as Argo Navis, the Ship Argo; another section formed when Argo was dismembered was Malus (the Mast), part of which survives as Pyxis. The whole region, particularly Carina, is very rich.*

Volans, Puppis

78 light-years away, and also 78 times as luminous as the Sun. In 1925 a bright nova, RR Pictoris, flared up in the Painter, and remained fairly prominent for some time before fading back to obscurity.

Volans (originally Piscis Volans, the Flying Fish). A small constellation which, rather confusingly, intrudes into Carina between Canopus and Miaplacidus. It contains little of interest, though γ is a wide, easy double.

Puppis. The Argo's poop, part of which is sufficiently far north to rise in British latitudes though the brightest star, ζ, cannot do so. ζ is a very hot 0-type star, 63,000 times as luminous as the Sun and therefore the equal of Rigel in Orion; it is 2400 light-years away. L^2 is a semi-regular variable with a range of magnitude from 3.4 to just below 6; V Puppis is of the β Lyrae type, with a range of about half a magnitude.

There are only three Messier objects in Puppis, because the rest of the constellation never rises over France, where Messier spent all his life. All three are open clusters. M46 and M47 are neighbours, more or less in line with β Canis Majoris and Sirius. M93, in the binocular field with ξ Puppis, is fairly bright and condensed. Admiral Smyth, the well-known last-century amateur astronomer, commented that the arrangement of the brighter stars in M93 reminded him of a starfish. The distance is 3600 light-years.

CARINA

BRIGHTEST STARS

Star	R.A. (h m s)	Dec. (° ' ")	Mag.	Spectrum	Proper name
α	06 23 57	−52 41 44	−0.72	F0	Canopus
β	09 13 12	−69 43 02	1.68	A0	Miaplacidus
ε	08 22 31	−59 30 34	1.86	K0	Avior
ι	09 17 05	−59 16 31	2.25	F0	Tureis
θ	10 42 57	−64 23 39	2.76	B0	
υ	09 47 06	−65 04 18	2.97	A0	
1(ZZ)	09 45 15	−62 30 28	3.3 (max.)	G0	
ϱ	10 32 01	−61 41 07	3.32	B3	
ω	10 13 44	−70 02 16	3.32	B7	
w	10 17 05	−61 19 56	3.40	K5	
q	09 10 58	−58 58 01	3.44	B0	
x	07 56 47	−52 58 56	3.47	B2	

Also above magnitude 4.3: u (3.78), c (3.84), R (3.9 max.), x (3.91), 1 (4.00), h (4.08).

VARIABLES

Star	R.A. (h m)	Dec. (° ')	Range (mags)	Type	Period (d)	Spectrum
η	10 45.1	−59 41	−0.8–7.9	Irregular	–	Pec
ZZ	09 45.2	−62 30	3.3–4.2	Cepheid	35.5	F-K
R	09 32.2	−62 47	3.9–10.5	Mira	309	M
U	10 57.8	−59 44	5.7–7.0	Cepheid	38.8	F-G

DOUBLES

Star	R.A. (h m)	Dec. (° ')	P.A. (°)	Sep. (")	Mags
υ	09 47.1	−65 04	127	5.0	3.1, 6.1

CLUSTERS AND NEBULAE

M	NGC	R.A. (h m)	Dec. (° ')	Mag.	Dimensions (')	Type
	IC2602	10 43.2	−64 24	2	50	Open cluster, round θ
	2516	07 58.3	−60 52	3.8	30	Open cluster
	3114	10 02.7	−60 07	4.2	35	Open cluster
	3572	11 10.4	−60 14	6.6	7	Open cluster
	2808	09 12.0	−64 52	6.3	14	Globular cluster
	3372	10 43.8	−59 52	6		Nebula, round η
	2867	09 21.4	−58 19	9.7	11"	Planetary nebula

VELA

BRIGHTEST STARS

Star	R.A. (h m s)	Dec. (° ' ")	Mag.	Spectrum	Proper name
γ	08 09 32	−47 20 12	1.78	WC7	Regor
δ	08 44 42	−54 42 30	1.96	A0	Koo She
λ	09 08 00	−43 25 57	2.21	K5	Al Suhail al Wazn
κ	09 22 07	−55 00 38	2.50	B2	Markeb
μ	10 46 46	−49 25 12	2.69	G5	
N	09 31 13	−57 02 04	3.13	K5	

Also above magnitude 4.3: φ (3.54), ψ (3.60), o (3.62), c (3.75), p (3.84),b (3.84), q (3.85), a (3.91), 4 (4.14), x (4.28).

DOUBLES

Star	R.A. (h m)	Dec. (° ')	P.A. (°)	Sep. (")	Mags	
δ	08 44.7	−54 43	153	2.6	2.1, 5.1	
μ	10 46.8	−49 25	055	2.3	2.7, 6.4.	Binary, 116y
γ	08 09.5	−47 20	AB 220	41.2	1.9, 4.2	
			AC 151	62.3	8.2	
			AD 141	93.5	9.1	
			DE 146	1.8	12.5	

CLUSTERS AND NEBULAE

M	NGC	R.A. (h m)	Dec. (° ')	Mag.	Dimensions (')	Type
IC	2391	08 40.2	−53 04	2.5	50	Open cluster (o Velorum)

CLUSTERS AND NEBULAE

M	NGC	R.A. (h m)	Dec. (° ')	Mag.	Dimensions (')	Type
IC	2395	08 41.1	−48 12	4.6	8	Open cluster
	2547	08 10.7	−49 16	4.7	20	Open cluster
	3201	10 17.6	−46 25	6.7	18	Globular cluster

PYXIS

The brightest star is α: R.A. 08h 43m 35s.5, dec. −33° 11′ 11″, mag. 3.68.
Also above magnitude 43: β (3.97), γ (4.01)

VARIABLE

Star	R.A. (h m)	Dec. (° ')	Range (mags)	Type	Period (d)	Spectrum
T	09 04.7	−32 23	6.3–14.0	Recurrent nova	–	

ANTLIA

The only star brighter than magnitude 4.3 is α: R.A. 10h 27m 09s, dec. −31° 04′ 14″, mag. 4.25.

PICTOR

BRIGHTEST STARS

Star	R.A. (h m s)	Dec. (° ' ")	Mag.	Spectrum	Proper name
α	06 48 11	−61 56 29	3.27	A5	–

The only other star above magnitude 4.3 is β (3.85). This is the star now known to be associated with a disk of material which may be planet-forming.

VOLANS

The brightest star is γ: R.A. 07h 08m 42s.3, dec. −70° 29′ 50″, combined magnitude 3.6.
Also above magnitude 4.3: β (3.77), ς (3.95), δ (3.98), α (4.00).

DOUBLES

Star	R.A. (h m)	Dec. (° ')	P.A. (°)	Sep. (")	Mags
γ	07 08.8	-70 30	300	13.6	4.0, 5.9

PUPPIS

BRIGHTEST STARS

Star	R.A. (h m s)	Dec. (° ' ")	Mag.	Spectrum	Proper name
ζ	08 03 35	−40 00 12	2.25	O5.8	Suhail Hadar
π	07 17 09	−37 05 51	2.70	K5	
ϱ	08 07 33	−24 18 15	2.81	F6	Turais
τ	06 49 56	−50 36 53	2.93	K0	
υ	06 37 45	−43 11 45	3.17	B8	
σ	07 29 14	−43 18 05	3.25	K5	
ε	07 49 18	−24 51 35	3.34	G3	Asmidiske
ξ	07 13 13	−45 10 59	3.4 (max)	M5	

Also above magnitude 4.3: c (3.59), s (3.73), α (3.82), 3 (3.96), P (4.11), 11 (4.20).

VARIABLES

Star	R.A. (h m)	Dec. (° ')	Range (mags)	Type	Period (d)	Spectrum
L^2	07 13.5	−44 39	3.4–6.2	Semi-reg.	140	M
V	07 58.2	−49 15	4.7–5,2	β Lyræ	1.45	B+B

CLUSTERS AND NEBULAE

M	NGC	R.A. (h m)	Dec. (° ')	Mag.	Dimensions (')	Type
46	2437	07 41.8	−14 49	6.1	27	Open cluster
47	2422	07 36.6	−14 30	4.4	30	Open cluster
93	2447	07 44.6	−23 52	6.2	22	Open cluster
	2477	07 52.3	−38 33	5.8	27	Open cluster
	2451	07 45.4	−37 58	2.8	45	Open cluster
	2527	08 05.3	−28 10	6.5	22	Open cluster
	2467	07 52.5	−26 24		14 × 32	Open cluster

Centaurus, Crux Australis, Triangulum Australe, Circinus,

Magnitudes

–1

0

1

2

3

4

5

Variable star

Galaxy

Planetary nebula

Gaseous nebula

Globular cluster

Open cluster

Centaurus was one of Ptolemy's original 48 groups. α and β, are the Pointers to the Southern Cross; α, the brightest star in the sky apart from Sirius and Canopus, has been known as Toliman, Rigel Kentaurus and Rigel Kent, but astronomers refer to it simply as α Centauri. It is the nearest of the bright stars, and only slightly further away than its dim red dwarf companion Proxima, which is only of the 11th magnitude and is difficult to identify; it lies two degrees from α, and is a feeble flare star.

α itself is a magnificent binary, with components of magnitudes 0.0 and 1.2. The primary is a G-type yellow star rather more luminous than the Sun; the K-type secondary is the larger of the two, but has less than half the Sun's luminosity. The revolution period is 80 years. The apparent separation ranges from 2 to 22 seconds of arc, so that the pair is easy to resolve with a small telescope.

β, known as Agena or Hadar, is a B-type star, 460 light-years away and 10,500 times the luminosity of the Sun. γ is a binary with almost equal components, but the separation is less than 1.5 seconds of arc, so that at least a 10-centimetre (4-inch) telescope is needed to resolve it. The Mira variable R Centauri lies between α and β. At its best it reaches naked-eye visibility.

ω Centauri is much the finest globular cluster in the sky. To the naked eye, it is a hazy patch in line with Agena and second-magnitude ε Centauri. It is one of the nearer globulars at around 17,000 light-years. It probably contains over a million stars, concentrated near the centre of the system within no more than a tenth of a light-year.

There are several bright open clusters in Centaurus, notably the two near λ. There is also a remarkable galaxy, NGC5128, which is crossed by a dark dust-lane, and is a fairly easy telescopic object. In 1986 a bright supernova was discovered in it by an Australian amateur astronomer, Robert Evans, using his 32-centimetre (13-inch) reflector.

Crux Australis. There can be few people who cannot identify the Southern Cross, though it was not accepted as a separate constellation until 1679. One of the four stars, δ, is more than a magnitude fainter than the rest, which rather spoils the symmetry; neither is there a central star to make an X, as with Cygnus in the far north. α, β and δ are hot and bluish-white, while γ is a red giant of type M. α (Acrux) is a wide double, and there is a third star in the same telescopic field. β, of type B, is slightly variable.

Triangulum Australe. The three leaders, α, β and γ, form a triangle; α is identifiable because of its orange-red hue. It is 55 light-years away, and 96 times as luminous as the Sun. The globular cluster NGC6025 lies near β and is not far below naked-eye visibility; binoculars show it well.

Circinus was one of Lacaille's additions, lying between the Pointers and Triangulum Australe. α is a wide double; γ is a close binary.

Ara lies between θ Scorpii and α Trianguli Australis. Three of its leading stars, β, ζ and η, are orange K-type giants; R Arae, in the same binocular field with ζ and η, is an Algol-type eclipsing binary which never becomes as faint as the seventh magnitude. Ara contains several brightish clusters, of which the most notable is the globular NGC6397, close to the β–γ pair. It seems to be no more than 8200 light-years away – probably the closest globular cluster. NGC6352, near α, is considerably brighter even though it is further away.

Telescopium is a small, dim constellation near Ara. The only object of note is the variable RR Telescopii, less than four degrees from α Pavonis. It is unusually very faint indeed, but has to flared up to the seventh magnitude.

Norma is another obscure constellation formed by Lacaille, and once known as Quadra Euclidis, or Euclid's Quadrant. It adjoins Ara and Lupus, and contains two fairly bright open clusters; NGC6067, not far from γ, and the adjacent NGC6087, round the Cepheid variable S Normae.

Lupus is an original constellation. It contains a number of brightish stars, though there is no well-marked pattern. NGC5722, close to ζ, is an open cluster within binocular

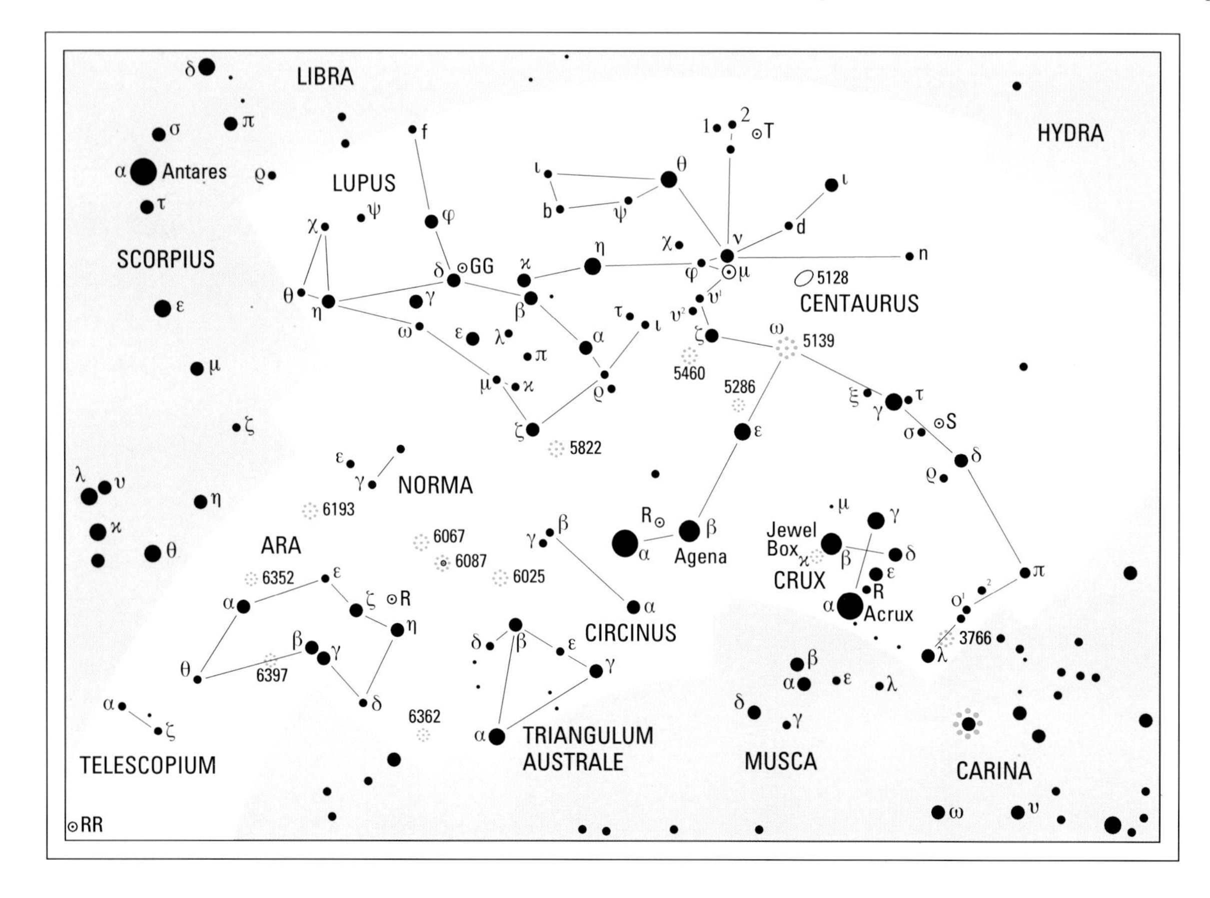

SOUTHERN CROSS

The distances and luminosities of the four main stars in Crux Australis:

	Distance, light-years	Luminosity, Sun = 1
α	360	3200–2000
β	460	8200
γ	88	160
δ	260	1300

◀ ***The Southern Cross**, Crux Australis, is the smallest constellation in the sky, but one of the most conspicuous, even if it is shaped more like a kite than an X. It is almost surrounded by Centaurus, and the brilliant Pointers, α and β Centauri, show the way to it. Most of Centaurus is too far south to be seen from Europe; from New Zealand, Crux is circumpolar. It is highest during evenings in southern autumn. Centaurus is an imposing constellation; it contains the finest of all globular clusters, ω Centauri.*

Ara, Telescopium, Norma, Lupus

range, and κ is an easy double. In 1006 a supernova flared up here, and became almost as bright as the quarter-moon.

The Jewel Box open cluster, round κ Crucis, is one of the loveliest in the sky; its main stars form a triangle, around a striking red supergiant. It is 7700 light-years away, and the cluster is about 25 light-years across; it is believed to be no more than a few million years old, so that by cosmic standards it is a true infant. Close by it is the dark nebula known as the Coal Sack, which can be detected with the naked eye as an almost starless region.

CENTAURUS

BRIGHTEST STARS

Star	R.A. h m s	Dec. ° ′ ″	Mag.	Spectrum	Proper name
α	14 39 37	−60 50 02	−0.27	G2+K1	
β	14 03 49	−60 22 22	0.61	B1	Agena
θ	14 06 41	−38 22 12	2.06	K0	Haratan
γ	12 21 31	−48 57 34	2.17	A0	Menkent
ε	13 39 53	−53 27 58	2.30	B1	
η	14 35 30	−42 09 28	2.31	B3	
ζ	13 55 32	−47 17 17	2.55	B2	Al Nair al Kentaurus
δ	12 08 21	−50 43 20	2.60	B2	
ι	13 20 36	−36 42 44	2.75	A2	
μ	13 49 37	−42 28 25	3.04 max	B3	
κ	14 59 10	−42 06 15	3.13	B2	Ke Kwan
λ	11 35 47	−63 01 11	3.13	B9	
ν	13 49 30	−41 41 16	3.41	B2	

Also above magnitude 4.3: φ (3.83), τ (3.86), υ (3.87), d (3.88), π (3.89), σ (3.91), 65G (4.11), 1 (4.23), n (4.27), 2 (4.19), ξ² (4.27).

VARIABLES

Star	R.A. h m	Dec. ° ′	Range (mags)	Type	Period (d)	Spectrum
R	14 16.6	−59 55	5.3–11.8	Mira	546	M
μ	13 49.6	−42 28	3.0–3.5	Irregular	–	B
T	13 41.8	−33 36	5.5–9.0	Semi-reg.	60	K–M
S	12 24.6	−49 26	6.0–7.0	Semi-reg.	65	N

DOUBLES

Star	R.A. h m	Dec. ° ′	P.A. °	Sep. ″	Mags	
α	14 39.6	−60 50	215	19.7	0.0, 1.2	Binary, 80y
γ	12 41.5	−48 58	353	1.4	2.9, 2.9	Binary, 84y

CLUSTERS AND NEBULAE

M	NGC	R.A. h m	Dec. ° ′	Mag.	Dimensions ′	Type
	IC2944	11 36.6	−63 02	4.5	15	Open cluster (λ Centauri)
	3766	11 36.1	−61 37	5.3	12	Open cluster
	5460	14 07.6	−48 19	5.6	25	Open cluster
	5139	13 25.8	−47 29	3.6	36	Globular cluster (ω Centauri)
	5286	13 46.4	−51 22	7.6	9	Globular cluster
	5128	13 25.5	−43 01	7.0	18.2 × 14.3	SOp galaxy (Centaurus A)

CRUX AUSTRALIS

BRIGHTEST STARS

Star	R.A. h m s	Dec. ° ′ ″	Mag.	Spectrum	Proper name
α	12 26 26	−63 05 56	0.83	B1+B3	Acrux
β	12 47 43	−59 41 19	1.25	B0	
γ	12 31 10	−57 06 47	1.63	M3	
δ	12 15 09	−58 44 55	2.80	B2	

Also above magnitude 4.3: ε (3.59), μ¹ (4.03), ξ (4.04), η (4.15).

DOUBLES

Star	R.A. h m	Dec. ° ′	P.A. °	Sep. ″	Mags
α	12 26.6	−63 06	115	4.4	1.4, 1.9
			202	90.1	1.0, 4.9
γ	12 31.2	−57 07	031	110.6	1.6, 6.7
			082	155.2	9.5
μ¹	12 54.6	−57 11	017	34.5	4.0, 5.2

CLUSTERS AND NEBULAE

M	NGC	R.A. h m	Dec. ° ′	Mag.	Dimensions ′	Type
	4755	12 53.6	−60 20	4	10	Open cluster κ Crucis (Jewel Box)
		12 53	−63	–	400 × 300	Dark nebula (Coal Sack)

TRIANGULUM AUSTRALE

BRIGHTEST STARS

Star	R.A. h m s	Dec. ° ′ ″	Mag.	Spectrum	Proper name
α	16 48 40	−69 01 39	1.92	K2	Atria
β	16 55 08	−63 25 50	2.85	F5	
γ	15 18 54	−68 40 46	2.89	A0	

Also above magnitude 4.3: δ (3.85), ε (4.03), ε (4.11).

CLUSTERS AND NEBULAE

M	NGC	R.A. h m	Dec. ° ′	Mag.	Dimensions ′	Type
	6025	16 03.7	−60 30	5.1	12	Globular cluster

CIRCINUS

BRIGHTEST STARS

Star	R.A. h m s	Dec. ° ′ ″	Mag.	Spectrum	Proper name
α	14 42 28	-64 58 43	3.19	F0	

Also above magnitude 4.3: β (4.07).

DOUBLES

Star	R.A. h m	Dec. ° ′	P.A. °	Sep. ″	Mags
α	14 42.5	−64 59	232	15.7	3.2, 8.6

ARA

BRIGHTEST STARS

Star	R.A. h m s	Dec. ° ′ ″	Mag.	Spectrum	Proper name
β	17 25 18	−55 31 47	2.85	K3	
α	17 31 50	−49 52 34	2.95	B3	Choo
ζ	16 58 37	−55 59 24	3.13	K5	
γ	17 25 23	−56 22 39	3.34	B1	

Also above magnitude 4.3: δ (3.62), θ (3.66), η (3.76), ε¹ (4.06).

VARIABLE

Star	R.A. h m	Dec. ° ′	Range (mags)	Type	Period (d)	Spectrum
R	16 39.7	−57 00	6.0–6.9	Algol	4.42	B

CLUSTERS AND NEBULAE

M	NGC	R.A. h m	Dec. ° ′	Mag.	Dimensions ′	Type
	6193	16 41.3	−48 46	5.2	15	Open cluster
	6352	17 25.5	−48 25	8.1	7.1	Globular cluster
	6362	17 31.9	−67 03	8.3	10.7	Globular cluster
	6397	17 40.7	−53 40	5.6	25.7	Globular cluster

TELESCOPIUM

The brightest star is α: R.A. 18h 26m 58s.2, dec. −45° 58′ 06″, mag. 3.51.
Also above magnitude 4.3: ζ (4.13).

VARIABLES

Star	R.A. h m	Dec. ° ′	Range (mags)	Type	Period (d)	Spectrum
RR	20 04.2	−55 43	6.5–16.5	Z Andromedae	–	F5p

NORMA

The only star in Norma above magnitude 4.3 is γ²: R.A. 16h 19m 50s, dec. −50° 09′ 20″, mag. 4.02.

VARIABLES

Star	R.A. h m	Dec. ° ′	Range (mags)	Type	Period (d)	Spectrum
S	16 18.9	−57 54	6.1-6.8	Cepheid	9.75	F–G

CLUSTERS AND NEBULAE

M	NGC	R.A. h m	Dec. ° ′	Mag.	Dimensions ′	Type
	6067	16 18.9	−54 13	5.6	13	Open cluster
	6087	16 18.9	−57 54	5.4	12	Open cluster (S Normae cluster)

LUPUS

BRIGHTEST STARS

Star	R.A. h m s	Dec. ° ′ ″	Mag.	Spectrum	Proper name
α	14 41 56	−47 23 17	2.30	B1	Men
β	14 58 32	−43 08 02	2.68	B2	Ke Kouan
γ	15 35 08	−41 10 00	2.78	B3	
δ	15 21 22	−40 38 51	3.22	B2	
ε	15 22 41	−44 41 21	3.37	B3	
ζ	15 12 17	−52 05 57	3.41	G8	
η	16 00 07	−38 23 48	3.41	B2	

Also above magnitude 4.3: φ (3.56), κ (3.72), π (3.89), χ (3.95), ρ (4.05), λ (4.05), θ (4.23), μ (4.27).

VARIABLE

Star	R.A. h m	Dec. ° ′	Range (mags)	Type	Period (d)	Spectrum
GG	15 18.9	−40 47	5.4–6.0	β Lyrae	2.16	B+A

DOUBLES

Star	R.A. h m	Dec. ° ′	P.A. °	Sep. ″	Mags
κ	15 11.9	−48 44	144	26.8	3.9, 5.8

CLUSTERS AND NEBULAE

M	NGC	R.A. h m	Dec. ° ′	Mag.	Dimensions ′	Type
	5822	15 16.8	−45 39	7	40	Open cluster

Grus, Phoenix, Tucana, Pavo, Indus,

Magnitudes

–1

0

1

2

3

4

5

Variable star

Galaxy

Planetary nebula

Gaseous nebula

Globular cluster

Magellanic Cloud

Grus, the Crane, is much the most prominent of the four Southern Birds; one way to identify it is to continue the line from α and β Pegasi, in the Square, through Fomalhaut. The line of stars running from γ through β and on to ε and ζ really does give some impression of a bird in flight. The little pairs making up δ and μ give the impression of being wide doubles, though both are due to nothing more than line-of-sight effects.

Of the two leaders of the Crane, α (Alnair) is a bluish-white B-star, 68 light-years away and 230 times as luminous as the Sun. β (Al Dhanab) is an M-type giant, 228 light-years away and 750 Sun-power. The two are almost equally bright, and the contrast between the steely hue of Alnair and the warm orange of Al Dhanab is striking in binoculars – or even with the naked eye. Grus contains a number of faint galaxies, but there is not much of interest here for the user of a small telescope.

Phoenix was the mythological bird which periodically burned itself to ashes, though this did not perturb it in the least and it soon recovered. α (Ankaa) is the only bright star; it is of type K, decidedly orange, lying at a distance of 78 light-years. It is 75 times as luminous as the Sun. It makes up a triangle with Achernar in Eridanus and Al Dhanab in Grus, which is probably the best way to identify it.

The main object of interest is ζ Phoenicis, which is a typical Algol eclipsing binary with a range from magnitude 3.6 to 4.4; the variations are easy to follow with the naked eye, and there are suitable comparison stars in β (3.31), δ (3.95) and η (4.36). Both components are of type B. This is actually the brightest of all stars of its kind apart from Algol itself and λ Tauri.

The interesting variable SX Phoenicis lies less than seven degrees west of Ankaa. It is a pulsating star of the δ Scuti type, with the remarkably short period of only 79 minutes, during which time the magnitude ranges between 7.1 and 7.5 – though the amplitude is not constant from one cycle to another. The spectrum, too, is variable, sometimes being of type A and at others more like type F. The distance is no more than 150 light-years, and the luminosity is roughly twice that of the Sun. Stars of this type are sometimes known as dwarf Cepheids. SX itself forms a triangle with Ankaa and ι (4.71), but the field is not very easy to identify without a telescope equipped with good setting circles.

Tucana, the Toucan. Though the dimmest of the Southern Birds, Tucana is graced by the presence of the Small Magellanic Cloud and two superb globular clusters. The brightest star is α, which is of type K and is decidedly orange; β is a wide double in a fine binocular field. The fainter component is a close binary.

The Small Cloud is very prominent with the naked eye; it is further away than the Large Cloud, but the two are connected by a 'bridge' of material, and are no more than 80,000 light-years apart. The Small Cloud contains objects of all kinds, including many short-period variables – in fact it was by studying these, in 1912, that Henrietta Leavitt was able to establish the period–luminosity relationship which has been so invaluable to astronomers. It has been suggested that the Small Cloud may be of complex form, and that we are seeing it almost 'end-on'.

Almost silhouetted against the Cloud is NGC104 (47 Tucanae), the brightest of all globular clusters apart from ω Centauri. It has even been claimed that 47 Tucanae is the more spectacular of the two, because it is small enough to be fitted into the same moderate-power telescopic field. It is surprisingly poor in variable stars, but there are several of the 'blue stragglers' referred to earlier; photographs taken with the Hubble Space Telescope resolve the cluster right through to its centre. It is about 15,000 light-years away. Telescopically, or even with binoculars, it is evident that its surface brightness is much greater than that of the Small Cloud.

NGC362 is another globular cluster in the same region. It is close to naked-eye visibility, and telescopically it is

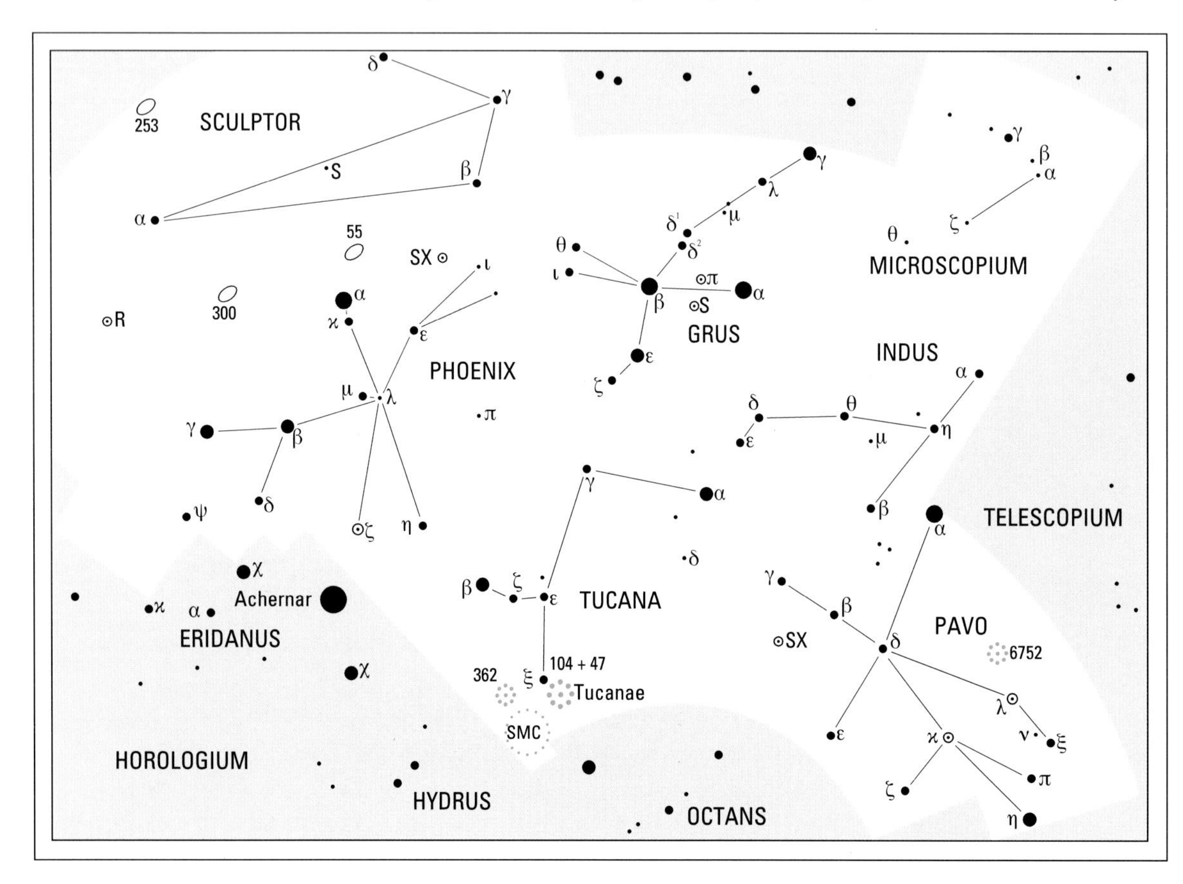

◀ ***The region of the 'Southern Birds'*** *is apt to be somewhat confusing, because only Grus is distinctive, and the other Birds are comparatively ill formed – though in Tucana we find the Small Cloud of Magellan together with the splendid globular cluster 47 Tucanae. However, the fact that Achernar lies nearby is a help in identification. The other constellations in this map – Indus, Microscopium and Sculptor – are very obscure.*

Microscopium, Sculptor

not greatly inferior to 47 Tucanae, though it is less than half the size.

Pavo, the Peacock, has one bright star, α, which can be found by using α Centauri and α Trianguli Australe as pointers; indeed, this is perhaps the best way of leading into the region of the Birds. α has no proper name, and is rather isolated from the rest of the constellation; it is of type B, 230 light-years away and 700 times as luminous as the Sun.

κ Pavonis is a short-period variable with a range of from magnitude 3.9 to 4.7 and a period of just over 9 days; suitable comparison stars are ε (3.96), γ (4.22), ζ (4.01), ξ (4.36) and ν (4.64). (Avoid λ, which is itself a variable of uncertain type; the same comparison stars can be used.) κ Pavonis is of the W Virginis type and much the brightest member of the class. It was once known as a Type II Cepheid, but a W Virginis star is much less luminous than a classical Cepheid with the same period, and κ is no more than about four times as luminous as the Sun; its distance is 75 light-years.

The fine globular cluster NGC6752 lies not far from λ. It is easy to see with binoculars, and is moderately condensed; the distance is about 20,000 light-years. It seems to have been discovered by J. Dunlop in 1828.

Indus is a small constellation created by Bayer in 1603; its brightest star, α, forms a triangle with α Pavonis and Alnair in Grus. There is nothing here of immediate interest for the telescopic observer, but it is worth noting that ε Indi, of magnitude 5.69, is one the nearest stars – just over 11 light-years away – and has only one-tenth the luminosity of the Sun, so that it is actually the feeblest star which can be seen with the naked eye.

If it could be observed from our standard distance of 10 parsecs (32.6 light-years), its apparent magnitude would be 7, and it would be invisible without optical aid. It is of type K, and orange in colour. Despite its low luminosity, it may be regarded as a fairly promising candidate for the centre of a planetary system.

Microscopium is a very dim constellation adjoining Grus and Piscis Australis. It was formerly included in the Southern Fish, so that γ was known a 1 Piscis Australis and ε as 4 Piscis Australis. It contains nothing of special note. α is an easy double (magnitudes 5.0 and 10.0, separation 20.5 seconds, position angle 166 degrees). There are also two Mira variables which can reach binocular visibility at maximum. U ranges from magnitude 7 to 14.4 in a period of 334 days, while S ranges between 7.8 and 14.3 in 209 days. Like most Mira stars they are of spectral type M, and are obviously orange-red.

Microscopium was one of the numerous constellations introduced by Nicolas-Louis de Lacaille in his famous maps of the southern sky in 1752, but frankly it seems unworthy of a separate identity.

Sculptor is another one of Lacaille's groups; originally Apparatus Sculptoris, the Sculptor's Apparatus. It occupies the large triangle bounded by Fomalhaut, Ankaa in Phoenix, and Diphda in Cetus, but the only objects of interest are the various galaxies.

NGC253 lies almost edgewise on to us, and lies not far from α, close to the border between Sculptor and Cetus; it is a favourite photographic target. NGC55 lies near Ankaa on the border between Sculptor and Phoenix, and seems to be one of the nearest galaxies beyond the Local Group, lying at no more than 8 million light-years from us. Like NGC253 it is a spiral, seen almost edge-on; it is easy to identify and attractive to photograph. The south galactic pole lies in Sculptor, and the whole region is noticeably lacking in bright stars.

GRUS

BRIGHTEST STARS

Star	R.A. h m s	Dec. ° ′ ″	Mag.	Spectrum	Proper name
α	22 08 14	−46 57 40	1.74	B5	Alnair
β	22 42 40	−46 53 05	2.11	M3	Al Dhanab
γ	21 53 56	−37 21 54	3.01	B8	
ε	22 48 33	−51 19 01	3.49	A2	

Also above magnitude 4.3: ι (3.90), δ[1] (3.97), δ[2] (4.11), ζ (4.12), θ (4.28)

VARIABLES

Star	R.A. h m	Dec. ° ′	Range (mags)	Type	Period (d)	Spectrum
π[1]	22 22.7	−45 57	5.4–6.7	Semi-reg.	150	S
S	22 26.1	−48 26	6.0–15.0	Mira	401	M

PHOENIX

BRIGHTEST STARS

Star	R.A. h m s	Dec. ° ′ ″	Mag.	Spectrum	Proper name
α	00 26 17	−42 18 22	2.39	K0	Ankaa
β	01 06 05	−46 43 07	3.31	G8	
γ	01 28 22	−43 19 06	3.41	K5	

Also above magnitude 4.3: ζ (3.6, max), ε (3.88), κ (3.94), δ (3.95).

VARIABLES

Star	R.A. h m	Dec. ° ′	Range (mags)	Type	Period (d)	Spectrum
ζ	01 08.4	−55 15	3.6–4.4	Algol	1.67	B+B
SX	23 46.5	−41 35	6.8–7.5	δ Scuti	0.055	A–F

TUCANA

BRIGHTEST STARS

Star	R.A. h m s	Dec. ° ′ ″	Mag.	Spectrum	Proper name
α	22 18 30	−60 15 35	2.86	K3	

Also above magnitude 4.3: β (3.7, combined), γ (3.99), ξ (4.23).

DOUBLES

Star	R.A. h m	Dec. ° ′	P.A. °	Sep. ″	Mags
β	00 31.5	−62 58	169	27.1	4.4, 4.8; B is a close binary (444y)
κ	01 15.8	−68 53	336	5.4	5.1, 7.3

CLUSTERS AND NEBULAE

M	NGC	R.A. h m	Dec. ° ′	Mag.	Dimensions ′	Type
		00 53	−72 50	2.3	280 × 160	Galaxy; Small Cloud of Magellan
	104	00 24.1	−72 05	4.0	30.9	Globular cluster; 47 Tucanæ
	362	01 03.2	−70 51	6.6	12.9	Globular cluster

PAVO

BRIGHTEST STARS

Star	R.A. h m s	Dec. ° ′ ″	Mag.	Spectrum	Proper name
α	20 25 39	−56 44 06	1.94	B3	
β	20 44 57	−66 12 12	3.42	A5	
λ	18 52 13	−62 11 16	3.4 (max)	B1	

Also above magnitude 4.3: δ (3.56), η (3.62), κ (3.9 max), ε (3.96), ζ (4.01), γ (4.22).

VARIABLES

Star	R.A. h m	Dec. ° ′	Range (mags)	Type	Period (d)	Spectrum
κ	18 56.9	−67 14	3.9–4.7	W Virginis	9.09	F
λ	18 52.2	−62 11	3.4–4.3	Irregular	–	B
SX	21 28.7	−69 30	5.4–6.0	Semi-reg.	50	M

CLUSTERS AND NEBULAE

M	NGC	R.A. h m	Dec. ° ′	Mag.	Dimensions ′	Type
	6752	19 10.9	−59 59	5.4	20.4	Globular cluster

INDUS

BRIGHTEST STARS

Star	R.A. h m s	Dec. ° ′ ″	Mag.	Spectrum	Proper name
α	20 37 34	−47 17 29	3.11	K0	Persian

Also above magnitude 4.3: β (3.65).

MICROSCOPIUM

The brightest star is γ: R.A. 21h 01m 17s.3, dec. −32° 5′ 28″, mag. 4.67. It was formerly known as 1 Piscis Australis.

SCULPTOR

The brightest star is α: R.A. 00h 58m 36s.3, dec. −29° 21′ 27″, mag. 4.31.

VARIABLES

Star	R.A. h m	Dec. ° ′	Range (mags)	Type	Period (d)	Spectrum
S	00 15.4	−32 03	5.5–13.6	Mira	365.3	M
R	01 27.0	−32 33	5.8–7.7	Semi-reg.	370	N

CLUSTERS AND NEBULAE

M	NGC	R.A. h m	Dec. ° ′	Mag.	Dimensions ′	Type
	55	00 14.9	−39 11	8.2	32.4 × 6.5	SB galaxy
	253	00 47.6	−25 17	7.1	25.1 × 7.4	Sc galaxy
	300	00 54.9	−37 41	8.7	20.0 × 14.8	Sd galaxy

Eridanus (southern), Horologium, Caelum, Dorado, Reticulum,

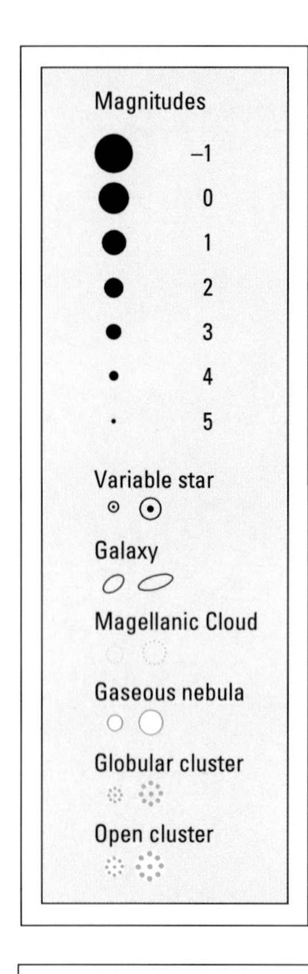

Eridanus, the River, stretches down into the far south, ending at Achernar, which is the ninth brightest star in the sky; it is 85 light-years away, and 750 times as luminous as the Sun. It can be seen from anywhere south of Cairo; from New Zealand it is circumpolar.

There is a minor mystery attached to Acamar, or θ Eridani. Ptolemy ranked it as of the first magnitude, and seems to have referred to it as 'the last in the River', but it is now little brighter than magnitude 3. It is not likely to have faded, and it is just possible that Ptolemy had heard reports of Achernar, which is not visible from Alexandria – though Acamar can be seen, low over the horizon. Acamar, 55 light-years away, is a splendid double, with one component rather brighter than the other; both are white, of type A, and are respectively 50 and 17 times more luminous than the Sun.

Horologium is one of Lacaille's obscure constellations, bordering Eridanus. The only object of any note is the red Mira variable R Horologii, which can rise to magnitude 4.7 at maximum. It is rather isolated, but χ and φ Eridani, near Achernar, point more or less to it.

Caelum is another Lacaille addition; he seems to have had a fondness for sculpture, since Caelum was originally Caela Sculptoris, the Sculptor's Tools. There is nothing of interest here; the constellation borders Columba and Dorado.

Dorado, the Swordfish, once commonly known as Xiphias. It lies between Achernar and Canopus. The most notable star is β, which is a bright Cepheid variable; it has a period of over 9 days, and is therefore considerably more luminous than δ Cephei itself. If its distance is correctly given in the Cambridge catalogue, it is 7500 light-years away, with a peak luminosity 200,000 times that of the Sun.

Most of the Large Magellanic Cloud lies in Dorado, and here we have the superb nebula 30 Doradûs, probably the finest in the sky. The Large Cloud, 169,000 light-years away, was once classed as an irregular galaxy, but shows clear indications of barred spirality. It remains visible with the naked eye even in moonlight, and is of unique importance to astronomers – which is partly why so many of the latest large telescopes have been sited in latitudes from which the Cloud is accessible. There have been various novae in it, and the spectacular supernova of 1987.

Reticulum was originally Reticulus Rhomboidalis, the Rhomboidal Net. It is a small but compact group bordering Eridanus and Hydrus, not far from Achernar and the Large Cloud. Of its leading stars, β (3.85), γ (4.51), δ (4.56) and ε (4.44) are all orange, with K- or M-type spectra, so that they are quite distinctive. The Mira variable R Reticuli, with a magnitude range of from 6.5 to 14, lies in the same wide field with α, and when near maximum is a binocular object. Close beside it is R Doradûs, just across the boundary of the Swordfish, which is a red semi-regular star always within binocular range.

Hydrus, the Little Snake, is easy enough to find, though it is far from striking. α and β are relatively nearby stars;

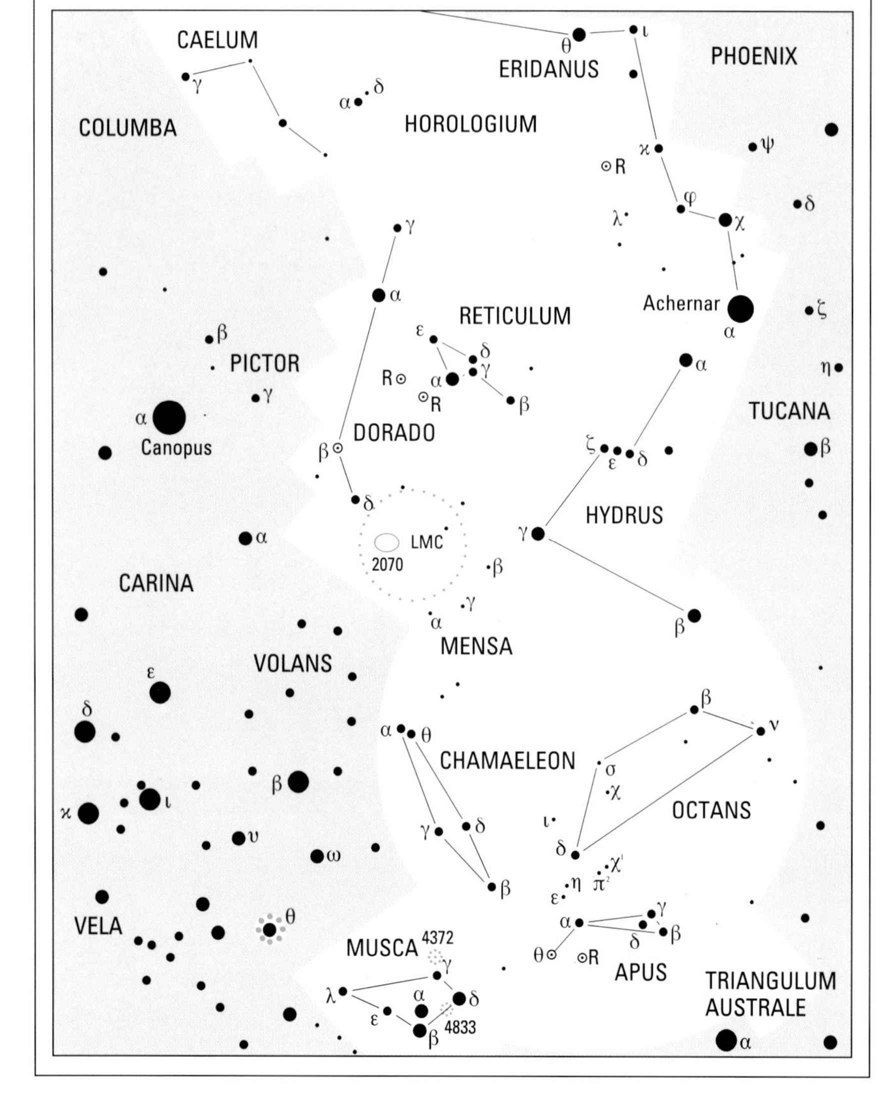

◀ ***The region of the south celestial pole*** *is decidedly barren; if there is any mist, for example, the whole area of the sky will appear completely blank, and even against a dark sky the south polar star, σ Octantis, is none too easy to identify. The nearest reasonably bright star to the pole is β Hydri, but the distance is almost 13°. The polar area is divided up into small constellations, few of which are easy to locate, but the Large Cloud of Magellan is present, mainly in Dorado but extending into Mensa. It is also worth noting that Achernar lies fairly close to α Hydri.*

▲ ***Large Magellanic Cloud*** *(top left) and Small Magellanic Cloud (right) are the nearest notable galaxies: the LMC is 169,000 light-years away, the SMC slightly further.*

Hydrus, Mensa, Chamaeleon, Musca, Apus, Octans

α is at a distance of 36 light-years, and β, at less than 21 light-years, is even closer. Since β is a G-type star, only 2½ times as luminous as the Sun, it may well have a system of planets, though we have no proof. It is a mere 12 degrees away from the south celestial pole.

Mensa is yet another Lacaille creation, originally Mons Mensae, the Table Mountain. It has the unenviable distinction of being the only constellation with no star as bright as the fifth magnitude, but at least a small part of the Large Magellanic Cloud extends into it.

Chamaeleon. Another dim group. The best way to find it is to follow a line from ι Carinae, in the false cross, through Miaplacidus (β Carinae) and extend it for some distance. The four leading stars of Chamaeleon, α (4.07), β (4.26), γ (4.11) and δ (4.45), are arranged in a diamond pattern; β lies roughly between Miaplacidus and α Trianguli Australe.

Musca Australis, the Southern Fly, generally known simply as Musca (there used to be a Musca Borealis, in the northern hemisphere, but this has now disappeared from our maps; no doubt somebody has swatted it). There are two bright globular clusters, NGC4833 near δ and NGC4372 near γ. They are not easy to locate with binoculars, but are well seen in a small telescope.

Apus was added to the sky by Bayer in 1603, originally under the name of Avis Indica, the Bird of Paradise. To find it, take a line from α Centauri through α Circini and continue until you come to α Apodis; the other main stars of the constellation – γ, δ and ε – make up a small triangle. δ is a red M-type star, and has a K-type companion at a separation of 103 seconds of arc. θ, in the same wide field with α, is a semi-regular variable which is generally within binocular range; also in the field is R Apodis, which is below magnitude five and suspected of variability.

Octans lies nearest to the pole. The brightest star in the southernmost constellation is ν, which is a K-type orange giant 75 times as luminous as the Sun. The south polar star, σ Octantis, is only of magnitude 5.5, and is not too easy to locate at first glance. A good method, using 7-power binoculars, is as follows:

Identify α Apodis, as given above. In the same field as α Apodis are two faint stars, ε Apodis (5.2) and η Apodis (5.0). These point straight to the orange δ Octantis (4.3), which has two dim stars, π^1 and π^2 Octantis, close beside it. Now put δ Octantis at the edge of the field, and continue the line from Apus. χ Octantis (5.2) will be on the far side of the field; centre it, and you will see two more stars of about the same brightness, σ and ι. These three are in the same field, and make up a triangle. The south polar star, σ, is the second in order from δ. Using 12-power binoculars, the three are in the same field with υ (5.7).

σ Octantis is of type F, and is less than seven times as luminous as the Sun, so that it pales in comparison with the northern Polaris. The pole is moving slowly away from it, and the separation will have grown to a full degree by the end of the century.

ERIDANUS

BRIGHTEST STARS

Star	R.A. h m s	Dec. ° ′ ″	Mag.	Spectrum	Proper name
α	01 37 43	−57 14 12	0.46	B5	Achernar
θ	02 58 16	−40 18 17	2.92	A3+A2	Acamar

Also above magnitude 4.3: υ^4 (3.56), φ (3.56), χ (3.70), υ^2 (3.82), υ^3 (3.96), ϰ (4.25), ι (4.11), e (4.27), g (4.27).

DOUBLE

Star	R.A. h m	Dec. ° ′	P.A. °	Sep. ″	Mags
θ	02 58.3	−40 18	088	8.2	3.4, 4.5

HOROLOGIUM

The brightest star is α; R.A. 04h 14m 00.0s, dec. −42° 17′ 40″, mag. 3.186.

VARIABLE

Star	R.A. h m	Dec. ° ′	Range (mags)	Type	Period (d)	Spectrum
R	02 53.9	−49 53	4.7–14.3	Mira	404	M

CAELUM

The brightest star is α: R.A. 04h 40m 33s.6, dec. −41° 51′ 50″, mag. 4.45.

DORADO

BRIGHTEST STAR

Star	R.A. h m s	Dec. ° ′ ″	Mag.	Spectrum	Proper name
α	04 34 00	−55 02 42	3.27	A0	

Also above magnitude 4.3: β (3.7 max), γ (4.25).

VARIABLES

Star	R.A. h m	Dec. ° ′	Range (mags)	Type	Period (d)	Spectrum
β	05 33.6	−62 29	3.7–4.1	Cepheid	9.84	F
R	04 36.8	−62 05	4.8–6.6	Semi-reg.	338	M

CLUSTERS AND NEBULAE

M	NGC	R.A. h m	Dec. ° ′	Mag.	Dimensions ′	Type
	–	05 24	−69 45	0	650 × 550	Galaxy; Large Cloud of Magellan
	2070	05 38.7	−69 06	3	40 × 25	Nebula 30 Dorad*s in Large Cloud of Magellan

RETICULUM

BRIGHTEST STARS

Star	R.A. h m s	Dec. ° ′ ″	Mag.	Spectrum	Proper name
α	04 14 25	−62 28 26	3.35	G6	

Also above magnitude 4.3: β (3.85).

HYDRUS

BRIGHTEST STARS

Star	R.A. h m s	Dec. ° ′ ″	Mag.	Spectrum	Proper name
β	00 25 46	−77 15 15	2.80	G1	
α	01 58 46	−61 34 12	2.86	F0	
γ	03 47 14	−74 14 20	3.24	M0	

Also above magnitude 4.3: δ (4.09), ε (4.11).

MENSA

The brightest star is α: R.A. 6h 10m 14s.6, dec. −74° 45′ 11″, mag. 5.09.
A small part of the Large Cloud of Magellan extends into Mensa.

CHAMAELEON

The brightest star is α: R.A. 08h 18m 31s.7, dec. −76° 55′ 10″, mag. 4.07.
Also above magnitude 4.3: γ (4.11).

MUSCA

BRIGHTEST STARS

Star	R.A. h m s	Dec. ° ′ ″	Mag.	Spectrum	Proper name
α	12 37 11	−69 08 07	2.69	B3	
β	12 46 17	−68 06 29	3.05	B3	

Also above magnitude 4.3: δ (3.62), λ (3.64), γ (3.87), ε (4.11).

DOUBLE

Star	R.A. h m	Dec. ° ′	P.A. °	Sep. ″	Mags	
β	12 46.3	−68 06	014	1.4	4.7, 5.1	Binary; period many centuries

CLUSTERS AND NEBULAE

M	NGC	R.A. h m	Dec. ° ′	Mag.	Dimensions ′	Type
	4833	13 00	−70 53	7.3	13.5	Globular cluster
	4372	12 25.8	−72 40	7.8	18.6	Globular cluster

APUS

The brightest star is α: R.A. 14h 47m 51s.6, dec. −79° 02′ 41″, mag. 3.83.
Also above magnitude 4.3: γ (3.89), β (4.24).

VARIABLES

Star	R.A. h m	Dec. ° ′	Range (mags)	Type	Period (d)	Spectrum
θ	14 05.3	−76 48	6.4-8.6	Semi-reg.	119	M

DOUBLE

Star	R.A. h m	Dec. ° ′	P.A. °	Sep. ″	Mags
δ	16 20.3	−78 41	012	102.9	4.7, 5.1

OCTANS

The brightest star is γ: R.A. 21h 41m 29s dec. −77° 23′ 24″, mag. 3.76.
The south polar star is σ: R.A. 20h 15m 1s dec. −89° 08′, mag. 5.46.

VARIABLES

Star	R.A. h m	Dec. ° ′	Range (mags)	Type	Period (d)	Spectrum
δ	22 20.0	−80 26	4.9–5.4	Semi-reg.	55	M

THE PRACTICAL ASTRONOMER

◀ ***The author's*** *22-cm (8½-inch) telescope in its weather-proof housing at his home in Selsey – a practical proposition for most amateur astronomers.*

The Beginner's Guide to the Sky

Most people take at least a passing interest in astronomy; after all, the skies are all around us, and not even the most myopic observer can fail to appreciate the Sun, the Moon and the stars! But astronomy as a serious hobby is quite another matter.

Let it be said at the outset that astronomy as a hobby, and astronomy as a career are two very different things. The professional astronomer must have a science degree, and there is no short cut, but the amateur needs nothing but interest and enthusiasm, and astronomy is still the one science in which amateurs can, and do, carry out really valuable research.

One popular misconception is that a large, expensive telescope is necessary. This is quite wrong. Much can be done with very limited equipment, or even none at all. So let us begin at the very beginning.

The first step is to do some reading, and absorb the basic facts. Next, obtain an outline star map and learn your way around the night sky. If tackled systematically, it takes a surprisingly short time; because the stars do not move perceptibly in relation to each other, a constellation can always be found again after it has been initially identified. The best procedure is to select one of two constellations which are glaringly obvious, such as Orion, the Great Bear or (in the southern hemisphere) the Southern Cross, and use them as guides to the less prominent groups. Remember, too, that there are only a few thousand naked-eye stars, and the main patterns stand out clearly, while the planets can soon be tracked down; Venus and Jupiter are far brighter than any star, while Mars is distinguished by its strong red hue. Only Saturn can look confusingly stellar.

Cameras can be introduced at an early stage. Any camera capable of giving a time-exposure will do; pictures of star trails, for instance, can be really spectacular, particularly if taken against a dramatic background. You may also pick up a meteor, or an artificial satellite which crawls across the field of view while the exposure is being made.

The naked-eye observer can do some valuable work. Meteor studies are important, both visually and photographically, and so are observations of aurorae, though admittedly these are limited to people who live at fairly high latitudes. Some variable stars are well within naked-eye range; Betelgeux in Orion and γ Cassiopeiae, the middle star of the W Pattern, are two examples, and it is fascinating to watch the steady fading and subsequent brightening of eclipsing binaries such as Algol.

However, sooner or later the question of buying optical equipment will arise. The essential here is to avoid the temptation to go straight round to the nearest camera shop and spend a few tens of pounds or dollars (or even more than a hundred) on a very small telescope. It may look nice, but it is not likely to be of much use, and the obvious alternative is to invest in binoculars, which have most of the advantages of a small telescope apart from sheer magnification and few of the drawbacks. They can also, of course, be used for other more mundane activities such as bird-watching.

The main disadvantage of binoculars is that the magnification is generally fixed. Zoom pairs, with variable magnification, are obtainable, but on the whole, it is probably better to accept the fixed-power limitation. With increased aperture and magnification, the field of view

▶ ***A 'neck' attachment*** *can be bought or made which will make it possible to hold the binoculars steady. James Savile demonstrates.*

▶ ***A converted camera tripod*** *will also serve quite satisfactorily as a binocular mount.*

▼ ***Light pollution.*** *This picture of Dublin taken at midnight from a hill overlooking the city shows the effect of unshielded light on the sky.*

becomes smaller and the binoculars become heavier. Beyond a magnification of about ×12, some sort of a mounting or tripod is desirable. A word of warning, too: every time you pick up the binoculars, even for a moment, loop the safety-cord round your neck. Fail to do so, and it is only a question of time before the binoculars are dropped, with disastrous results.

The Moon is a constant source of enjoyment to the binocular-user; the mountains, craters, valleys and rays are beautifully brought out, and it takes very little time to learn the main features. The Sun is emphatically to be avoided (never use binoculars to look at it, even with the addition of a dark filter), but there is plenty to see among the stars. Binoculars bring out the diverse colours really well; there are clusters, groups, rich fields, and nebulae. And there is always the chance that we will be treated to the spectacle of a bright comet. This has not happened often in recent years, though of course the brilliant comets Hyakutake (1996) and Hale–Bopp (1997) were spectacular.

These, then, are the first steps in home astronomy. If your interest is maintained, it will then be time to consider obtaining a telescope.

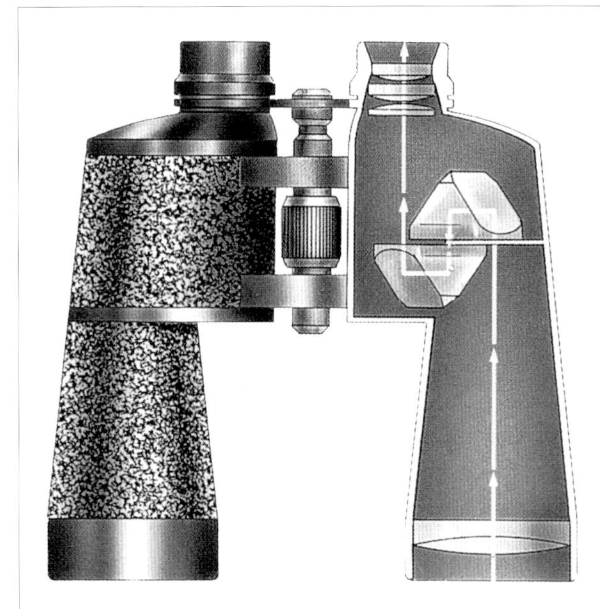

Binoculars
Binoculars are graded according to their magnification and their aperture, which is always given in millimetres. Thus a 7 × 50 pair yields a magnification of seven, with each object-glass 50 mm in diameter. If only one pair is to be obtained, this is probably a wise choice, because binoculars of this type have a wide field and are lightweight enough to be 'handy'.

◀ ***Orion,*** *photographed over a saguaro cactus near Tucson, Arizona, USA. This photograph was taken by David Cortner using a simple hand-driven mounting which tracks the stars by turning a screw at the correct rate.*

▼ ***Star trails****. Using an ordinary camera, a standard film and a long exposure, it is possible to make pictures of star trails, showing the apparent movement of the heavens. I took this picture at La Silla, in Chile, in 1990. The prominent trail to the far left is Jupiter. The trail above and to the right of the dome is Betelgeux.*

Choosing a Telescope

Astronomical telescopes are of two basic types. First there is the refractor, which collects its light by means of a glass lens (or combination of lenses) known as an objective or object-glass. Secondly there is the reflector, in which light is collected by a curved mirror. The aperture of the telescope is determined by the diameter of the object-glass (for a refractor) or the main mirror (for a reflector). In each case the actual magnification is done by a smaller lens known as an eyepiece or ocular. Obviously, the larger the aperture of the telescope, the more light can be collected, and the higher the magnification which can be used.

Each type of instrument has its own advantages, and also its own drawbacks. Aperture for aperture, the refractor is the more effective, and it also needs comparatively little maintenance; but it is much more expensive than a reflector of equivalent light-grasp, and it is less portable.

There are various forms of reflectors, of which the most common is the Newtonian; here the main mirror is parabolic, and the secondary mirror is flat. The main problem is that the mirrors need periodical re-coating with some reflective substance, usually aluminium, and they are always liable to go out of adjustment. Compound telescopes such as Schmidt-Cassegrains are becoming very popular, and have the advantage of being more portable than Newtonians, but unfortunately they are very costly.

In choosing a telescope, much depends upon the main interests of the observer; for example, anyone who intends to concentrate on the Sun will be wise to select a refractor, while the deep-sky enthusiast will prefer a reflector. Moreover, there is always the temptation to begin with a very small telescope – say a 5-centimetre (2-inch) refractor, or a 7.5-centimetre (3-inch) Newtonian – which will cost a relatively small sum. This is emphatically not to be recommended. A telescope of this kind may look nice, but the mounting will probably be unsteady, and the field of view will be small. Moreover, there is the question of magnification. In general, it is true to say that the maximum useful power for a telescope of good optical quality is × 20 per centimetre of aperture (× 50 per inch) – so that, for example, a 7.5-centimetre (3-inch) reflector will bear no more than a power of 150. If you use too high a power, the image will be so faint that it will be completely useless. If you see a telescope which is advertised by its magnifying power only, avoid it; it is the aperture which matters, at least 7.5 centimetres (3 inches) for a refractor and 15 centimetres (6 inches) for a reflector for serious work.

Take care with the choice of eyepieces. At least three will be necessary: one giving low power (wide views), one moderate power (general views) and one high power (for more detailed views, particularly of the Moon).

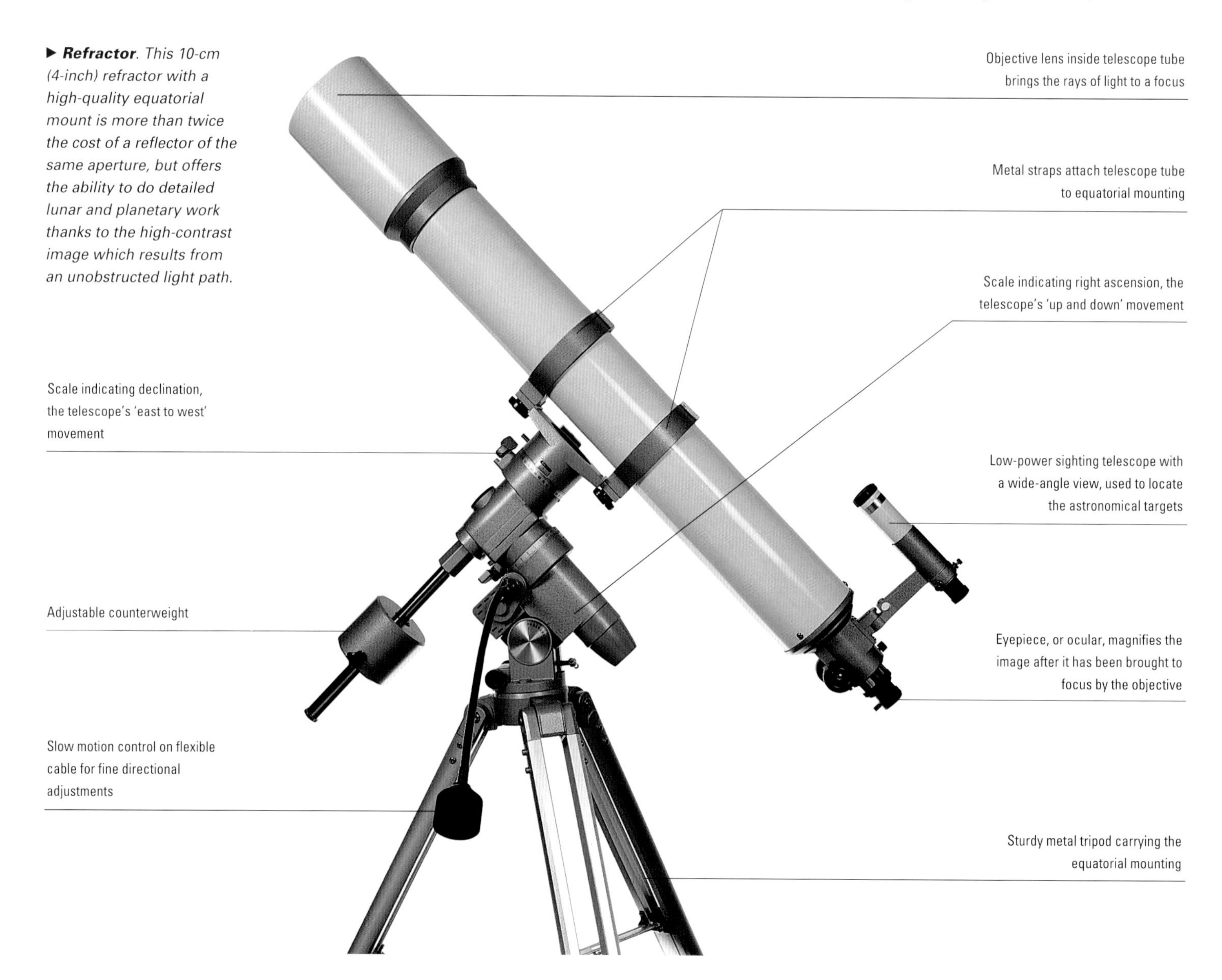

▶ ***Refractor**. This 10-cm (4-inch) refractor with a high-quality equatorial mount is more than twice the cost of a reflector of the same aperture, but offers the ability to do detailed lunar and planetary work thanks to the high-contrast image which results from an unobstructed light path.*

▲ ***Seen with naked eye***, *the Full Moon will be covered by the thickness of a pencil held at arm's length. The seas may be observed, but it is difficult to see more detailed features.*

▶ ***Through a small telescope*** *or binoculars far more detail on the Moon's surface becomes visible.*

▼ ***Dobsonian***. *Lacking all frills such as fine adjustments for declination and right ascension, it offers the maximum power for minimum outlay. For about the same price as an 11.5-cm (4½-inch) Newtonian reflector, significantly more power for deep sky observations is available with this 15-cm (6-inch) Dobsonian. It cannot be mechanically guided and is unsuitable for lunar and planetary observations.*

▼ ***Newtonian reflector***. *Reflectors are readily available and relatively cheap. However, reflectors with objectives less than 15 cm (6 inches) should be avoided by those intending to undertake serious observing.*

▼ ***Schmidt-Cassegrain***. *A 25-cm (10-inch) Schmidt-Cassegrain with automatic high-speed slewing and go-to facilities controlled by a handset. Modern instruments such as this offer a high degree of sophistication even to beginners who can programme the telescope to seek out many deep-sky objects in rapid succession.*

Home Observatories

Small telescopes are portable; larger ones are not. It may well be that the serious amateur observer will want to set up his telescope in a permanent position, and this means building an observatory, which is not nearly so difficult as might be imagined, and is well within the scope of most people.

The simplest form is the run-off shed. Here, the shed is run on rails, and is simply moved back when the telescope is to be used. It is wise to make the shed in two parts which meet in the middle; if the shed is a single construction, it has to have a door, which must be either hinged or removable. If hinged, it flaps; if it is removable, there are problems when trying to replace it in the dark with a wind blowing, as it tends to act as an effective sail! The main disadvantage is that the user is unprotected during observation, stray light can be a nuisance, and any strong breeze can shake the telescope.

For a refractor, a run-off roof arrangement is suitable. The slidable portion can either be the top half of the shed, or merely the actual roof, and an arrangement involving bicycle-chains and a hand crank is relatively simple to construct. This is excellent for a refractor, but less so for a Newtonian or Cassegrain reflector, because here we have to contend with a restricted view of the sky.

A 'dome' need not be a graceful construction; it can even be square (a contradiction in terms!) and mounted on a circular rail, so that the entire building revolves and there is a removable section of the roof. This is suitable for a relatively small instrument, though larger 'total rotators' become so heavy that they tend to stick.

◀ ***The author's*** *run-off shed for his 32-cm (12.5-inch) reflector. The shed is in two parts which run back on rails in opposite directions.*

▼ ***The author's*** *39-cm (15-inch) Newtonian reflector at Selsey. The mounting is of the fork type.*

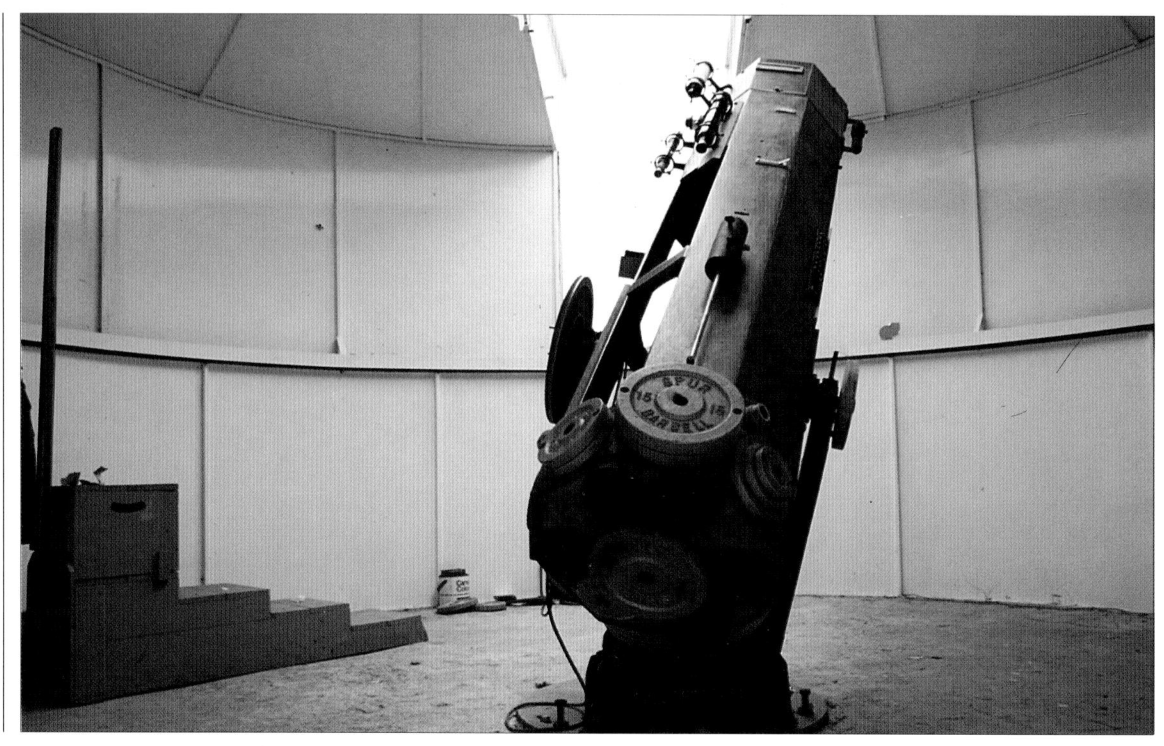

One form of dome is what can be called the wedding-cake pattern. The walls can be permanent, so that the upper part rotates on a rail; the viewing portion of the roof can be hinged, so that it is simply swung back (this is easier than completely removing it). A true dome looks much more decorative, though the hemispherical section is much more difficult to make.

Certainly there is a great deal to be said for observing from inside a dome rather than in the open air, but there are a few points to be borne in mind. First, take care about the siting; make sure that you remain as clear as possible from inconvenient trees and nearby lights. Secondly, make sure that you cannot be obstructed. An observatory which is not anchored down comes into the category of a portable building, and is not subject to planning permission, which is important to remember if you happen to have an awkward local council. Finally, ensure that everything is secure. In the 1990s, when law and order has broken down so completely, it is essential to take all possible precautions – something which was much less pressing 20 years ago, and which we hope will again be less pressing 20 years hence.

▲ ***Auckland Observatory***, *New Zealand. This contains a 51-cm (20-inch) reflector, and is a fine example of an amateur-built and amateur-run observatory which produces work of full professional standard.*

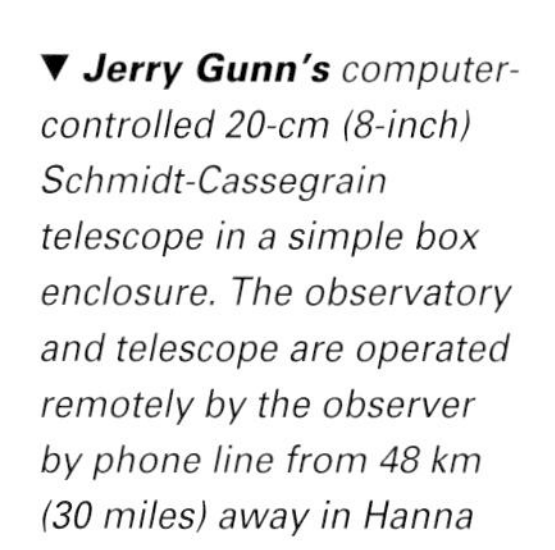

▼ ***Jerry Gunn's*** *computer-controlled 20-cm (8-inch) Schmidt-Cassegrain telescope in a simple box enclosure. The observatory and telescope are operated remotely by the observer by phone line from 48 km (30 miles) away in Hanna City, Illinois, USA. The observatory lid opens by remote control, and a CCD camera attached to the telescope sends images back to Gunn in his basement in Peoria. He uses the system to monitor the brightness changes in variable stars.*

◀ ***The Mountain Skies Observatory*** *built by Curtis MacDonald, near Laramie, Wyoming, USA. It houses a 31-cm (12.5-inch) Newtonian reflector with a 12-cm (5-inch) refractor on the same mounting. The dome has a wooden framework covered with Masonite and painted with exterior latex. This photograph was taken by moonlight with Comet Hale-Bopp in the background.*

Glossary

A

Aberration of starlight: The apparent displacement of a star from its true position in the sky due to the fact that light has a definite velocity (299,792.5 kilometres per second). The Earth is moving around the Sun, and thus the starlight seems to reach it 'at an angle'. The apparent positions of stars may be affected by up to 20.5 seconds of arc.

Absolute magnitude: The apparent magnitude that a star would have if it were observed from a standard distance of 10 parsecs, or 32.6 light-years. The absolute magnitude of the Sun is +4.8.

Absolute zero: The lowest limit of temperature: −273.16 degrees C. This value is used as the starting point for the Kelvin scale of temperature, so that absolute zero = 0 Kelvin.

Absorption of light in space: Space is not completely empty, as used to be thought. There is appreciable material spread between the planets, and there is also material between the stars; the light from remote objects is therefore absorbed and reddened. This effect has to be taken into account in all investigations of very distant objects.

Absorption spectrum: A spectrum made up of dark lines against a bright continuous background. The Sun has an absorption spectrum; the bright background or continuous spectrum is due to the Sun's brilliant surface (**photosphere**), while the dark absorption lines are produced by the solar atmosphere. These dark lines occur because the atoms in the solar atmosphere absorb certain characteristic wavelengths from the continuous spectrum of the photosphere.

Acceleration: Rate of change of velocity. Conventionally, increase of velocity is termed acceleration; decrease of velocity is termed deceleration, or negative acceleration.

Aerolite: A meteorite whose composition is stony.

Aeropause: A term used to denote that region of the atmosphere where the air-density has become so slight as to be disregarded for all practical purposes. It has no sharp boundary, and is merely the transition zone between 'atmosphere' and 'space'.

Airglow: The faint natural luminosity of the night sky due to reactions in the Earth's upper atmosphere.

Air resistance: Resistance to a moving body caused by the presence of **atmosphere**. An artificial satellite will continue in orbit indefinitely only if its entire orbit is such that the satellite never enters regions where air resistance is appreciable.

Airy disk: The apparent size of a star's disk produced even by a perfect optical system. Since the star can never be focused perfectly, 84 per cent of the light will concentrate into a single disk, and 16 per cent into a system of surrounding rings.

Albedo: The reflecting power of a planet or other non-luminous body. A perfect reflector would have an albedo of 100 per cent.

Altazimuth mount: A telescope mounting in which the instrument can move freely in both **altitude** and **azimuth**. Modern computers make it possible to drive telescopes of this sort effectively, and most new large telescopes are on altazimuth mountings.

Altitude: The angular distance of a celestial body above the horizon, ranging from 0 degrees at the horizon to 90 degrees at the **zenith**.

Ångström unit: The unit for measuring the wavelength of light and other electromagnetic vibrations. It is equal to 100 millionth part of a centimetre. Visible light ranges from about 7500 Å (red) down to about 3900 Å (violet).

Antenna: A conductor, or system of conductors, for radiating or receiving radio waves. Systems of antennae coupled together to increase sensitivity, or to obtain directional effects, are known as antenna arrays. or as radio telescopes when used in **radio astronomy**.

Apastron: The point in the orbit of a **binary** system where the stars are at their furthest from each other. The closest point is known as the **periastron**.

Aphelion: The orbital position of a planet or other body when it is furthest from the Sun. The closest point is known as the **perihelion**.

Apogee: The point in the orbit of the Moon or an artificial satellite at which the body is furthest from the Earth. The closest point is known as the **perigee**.

Arc, degree of: One 360th part of a full circle (360°).

Arc minute, arc second: One 60th part of a degree of arc. One minute of arc (1') is in turn divided into 60 seconds of arc (60").

Ashen light: The faint luminosity of the night side of the planet Venus, seen when Venus is in the crescent stage. It is probably a genuine phenomenon rather than a contrast effect, but its cause is not certainly known.

Asteroids: The minor planets, most of which move around the Sun between the orbits of Mars and Jupiter. Several thousands of asteroids are known; much the largest is Ceres, whose diameter is 1003 kilometres. Only one asteroid (Vesta) is ever visible with the naked eye.

Astrology: A pseudo-science which claims to link the positions of the planets with human destinies. It has no scientific foundation.

Astronomical unit: The distance between the Earth and the Sun. It is equal to 149,597,900 kilometres, usually rounded off to 150 million kilometres.

Astrophysics: The application of the laws and principles of physics to all branches of astronomy. It has often been defined as 'the physics and chemistry of the stars'.

Atmosphere: The gaseous mantle surrounding a planet or other body. It can have no definite boundary, but merely thins out until the density is no greater than that of surrounding space.

Atom: The smallest unit of a chemical element which retains its own particular character. (Of the 92 elements known to occur naturally, hydrogen is the lightest and uranium is the heaviest.)

Aurorae (polar lights): Aurora Borealis in the northern hemisphere, Aurora Australis in the southern. They are glows in the upper atmosphere, due to charged particles emitted by the Sun. Because the particles are electrically charged, they tend to be attracted towards the magnetic poles, so that aurorae are seen best at high latitudes.

Azimuth: The horizontal direction or bearing of a celestial body, reckoned from the north point of the observer's horizon. Because of the Earth's rotation, the azimuth of a body is changing all the time.

B

Background radiation: Very weak microwave radiation coming from space, continuously from all directions and indicating a general temperature of 3 degrees above **absolute zero**. It is believed to be the last remnant of the Big Bang, in which the universe was created about 15,000 million years ago. The Cosmic Background Explorer satellite (COBE) has detected slight variations in it.

Baily's Beads: Brilliant points seen along the edge of the Moon's disk at a total solar eclipse, just before totality and again just after totality has ended. They are due to the Sun's light shining through valleys between mountainous regions on the limb of the Moon.

Barycentre: The centre of gravity of the Earth–Moon system. Because the Earth is 81 times more massive than the Moon, the barycentre lies within the terrestrial globe.

Binary star: A star made up of two components that are genuinely associated, and are moving around their common centre of gravity. They are very common. With some binaries the separations are so small that the components are almost touching each other, and cannot be seen separately, although they can be detected by means of spectroscopy (see **spectroscope**). See also **eclipsing binary**; **spectroscopic binary**.

Black hole: A region of space surrounding a very massive collapsed star, or 'collapsar', from which not even light can escape.

Bode's Law: An empirical relationship between the distances of the planets from the Sun, discovered by J. D. Titius in 1772 and made famous by J. E. Bode. The law seems to be fortuitous, and without any real scientific basis.

Bolide: A brilliant meteor, which may explode during its descent through the Earth's atmosphere.

Bolometer: A very sensitive radiation detector, used to measure slight quantities of radiation over a very wide range of wavelengths.

C

Carbon-nitrogen cycle: The stars are not 'burning' in the usual sense of the word; they are producing their energy by converting hydrogen into helium, with release of radiation and loss of mass. One way in which this conversion takes place is by a whole series of reactions, involving carbon and nitrogen as catalysts. It used to be thought that the Sun shone because of this process, but modern work has shown that another cycle, the so-called proton-proton reaction, is more important in stars of solar type. The only stars which do not shine because of the hydrogen-into-helium process are those which are at a very early or relatively late stage in their evolution.

Cassegrain reflector: A type of reflecting telescope (see **reflector**) in which the light from the object under study is reflected from the main mirror to a convex secondary, and thence back to the eyepiece through a hole in the main mirror.

Celestial sphere: An imaginary sphere surrounding the Earth, concentric with the Earth's centre. The Earth's axis indicates the positions of the celestial poles; the projection of the Earth's equator on to the celestial sphere marks the celestial equator.

Centrifuge: A motor-driven apparatus with long arm, at the end of which is a cage. When people (or animals) are put into the cage, and revolved and rotated at high speeds, it is possible to study effects comparable with the accelerations experienced in spacecraft. All astronauts are given tests in a centrifuge during their training.

Cepheid: An important type of variable star, Cepheids have short periods of from a few days to a few weeks. and are regular in their behaviour. It has been found that the period of a Cepheid is linked with its real luminosity: the longer the period, the more luminous the star. From this it follows that once a Cepheid's period has been measured, its distance can be worked out. Cepheids are luminous stars, and may be seen over great distances; they are found not only in our Galaxy, but also in external galaxies. The name comes from Delta Cephei, the brightest and most famous member of the class.

Charge-Coupled Device (CCD): An electronic imaging device which is far more sensitive than a photographic plate, and is now replacing photography for most branches of astronomical research.

Chromatic aberration: A defect found in all lenses, resulting in the production of 'false colour'. It is due to the fact that light of all wavelengths is not bent or refracted equally; for example, blue light is refracted more strongly than red, and so is brought to focus nearer the lens. With an astronomical telescope, the object-glass is compound – i.e. made up of several lenses composed of different kinds of glass. In this way chromatic aberration may be reduced, although it can never be entirely cured.

Chromosphere: The part of the Sun's atmosphere lying above the bright surface or **photosphere**, and below the outer **corona.** It is visible with the naked eye only during total solar **eclipses**, when the Moon hides the photosphere; but by means of special instruments it may be studied at any time.

Circular velocity: The velocity with which an object must move, in the absence of air resistance, in order to describe a circular orbit around its primary.

Circumpolar star: A star which never sets, but merely circles the celestial pole and remains above the horizon.

Clusters, stellar: A collection of stars which are genuinely associated. An *open cluster* may contain several hundred stars, usually together with gas and dust; there is no particular shape to the cluster. *Globular clusters* contain thousands of stars, and are regular in shape; they are very remote, and lie near the edge of the Galaxy. Both open and globular clusters are also known in external galaxies. *Moving clusters* are made up of widely separated stars moving through space in the same direction and at the same velocity. (For example, five of the seven bright stars in the Great Bear are members of the same moving cluster.)

Collimator: An optical arrangement for collecting light from a source into a parallel beam.

Colour index: A measure of a star's colour and hence of its surface temperature. The ordinary or visual **magnitude** of a star is a measure of the apparent brightness as seen with the naked eye; the photographic magnitude is obtained by measuring the apparent size of a star's image on a photographic plate. The two magnitudes will not generally be the same, because in the old standard plates red stars will seem less prominent than they appear to the eye. The difference between visual and photographic magnitude is known as the colour index. The scale is adjusted so that for a white star, such as Sirius, colour index = 0. A blue star will have negative colour index; a yellow or red star will have positive colour index.

Colures: Great circles on the celestial sphere. The equinoctial colure, for example, is the great circle which passes through both celestial poles and also the **First Point of Aries** (vernal equinox), i.e. the point where the ecliptic intersects the celestial equator.

Coma: (1) The hazy-looking patch surrounding the nucleus of a **comet**. (2) The blurred haze surrounding the images of stars on a photographic plate, due to optical defects in the equipment.

Comet: A member of the Solar System, moving around the Sun in an orbit which is generally highly eccentric. It is made up of relatively small particles (mainly ices) together with tenuous gas: the most substantial part of the comet is the nucleus, which may be several kilometres in diameter. A comet's tail always points more or less away from the Sun, due to the effects of **solar wind**. There are many comets with short periods, all of which are relatively faint; the only bright comet with a period of less than a century is Halley's. The most brilliant comets have periods so long that their return cannot be predicted. See also **sun-grazers**.

Conjunction: The apparent close approach of a planet to a star or to another planet; it is purely a line-of-sight effect, since the planet is very much closer to us than the star. An *inferior conjunction*, for Mercury and Venus, is the position when the planet has the same **right ascension** as the Sun (see **inferior planets**.) A *superior conjunction* is the position of a planet when it is on the far side of the Sun with respect to the Earth.

Constellation: A group of stars named after a living or a mythological character, or an inanimate object. The names are highly imaginative, and have no real significance. Neither is a constellation made up of stars that are genuinely associated with one another; the individual stars lie at very different distances from the Earth, and merely happen to be in roughly the same direction in space. The International Astronomical Union currently recognizes 88 separate constellations.

Corona: The outermost part of the Sun's atmosphere; it is made up of very tenuous gas at a very high temperature, and is of great extent. It is visible to the naked eye only during total solar **eclipses**.

Coronagraph: A type of **telescope** designed to view the solar **corona** in ordinary daylight; ordinary telescopes are unable to do this, partly because of the sunlight scattered across the sky by the Earth's atmosphere, and partly because of light which is scattered inside the telescope – mainly by particles of dust. The coronagraph was invented by the French astronomer B. Lyot.

Cosmic rays: High-velocity particles reaching the Earth from outer space. The heavy cosmic-ray primaries are broken up when they enter the top part of the Earth's atmosphere, and only the secondary particles reach ground-level. There is still much doubt whether cosmic radiation will prove to be a major hazard in long-term spaceflights.

Cosmology: The study of the universe as a whole; its nature, origin, evolution, and the relations between its various parts.

Counterglow: See **Gegenschein**.

Crab Nebula: The remnant of a **supernova** observed in 1054; an expanding cloud of gas, approximately 6000 **light-years** away, according to recent measurements. It is important because it emits not only visible light but also radio waves and **X-rays**. Much of the radio emission is due to **synchrotron radiation** (that is, the acceleration of charged particles in a strong magnetic field). The Crab Nebula contains a **pulsar**, the first to be identified with an optical object.

Culmination: The time when a star or other celestial body reaches the observer's **meridian**, so that it is at its highest point (*upper culmination*). If the body is circumpolar, it may be observed to cross the meridian again 12 hours later (*lower culmination*). With a non-circumpolar object, lower culmination cannot be observed as, at that point, the object is then below the horizon.

Cybernetics: The study of methods of communication and control that are common both to machines and also to living organisms.

Glossary

D

Day: In everyday language, a day is the amount of time it takes for the Earth to spin once on its axis. A *sidereal day* (see **sidereal time**) is the rotation period measured with reference to the stars (23 hours 56 minutes 4.091 seconds). A *solar day* is the time interval between two successive noons; the length of the mean solar day is 24 hours 3 minutes 56.555 seconds – rather longer than the sidereal day, since the Sun is moving eastwards along the **ecliptic**. The *civil day* is, of course, taken to be 24 hours.

Declination: The angular distance of a celestial body north or south of the celestial equator. It may be said to correspond to latitude on the surface of the Earth.

Density: The mass of a substance per unit volume. Taking water as one, the density of the Earth is 5.5.

Dichotomy: The exact half-phase of Mercury, Venus or the Moon.

Diffraction rings: Concentric rings surrounding the image of a star as seen in a **telescope**. They cannot be eliminated, since they are due to the wave-motion of light. They are most evident in small instruments.

Direct motion: Bodies which move around the Sun in the same sense as the Earth are said to have *direct motion*. Those which move in the opposite sense have *retrograde motion*. The term may also be applied to satellites of the planets. No planet or asteroid with retrograde motion is known, but there are various retrograde satellites and comets. The terms are also used with regard to the apparent movements of the planets in the sky. When moving eastwards against the stars, the planet has direct motion; when moving westwards, it is retrograding.

Diurnal motion: The apparent daily rotation of the sky from east to west. It is due to the real rotation of the Earth from west to east.

Doppler effect: The apparent change in the wavelength of light caused by the motion of the observer. When a light-emitting body is approaching the Earth, more light-waves per second enter the observer's eye than would be the case if the object were stationary; therefore, the apparent wavelength is shortened, and the light seems 'too blue'. If the object is receding, the wavelength is apparently lengthened, and the light is 'too red'. For ordinary velocities the actual colour changes are very slight, but the effect shows up in the spectrum of the object concerned. If the dark lines are shifted towards the red or long-wave, the object must be receding; and the amount of the shift is a key to the velocity of recession. Apart from the galaxies in our **Local Group**, all external systems show red shifts, and this is the observational proof that the universe is expanding. The Doppler principle also applies to radiations at radio wavelengths.

Double star: A star which is made up of two components. Some doubles are *optical*; that is to say, the components are not truly associated, and simply happen to lie in much the same direction as seen from Earth. Most double stars, however, are physically associated or **binary** systems.

E

Earthshine: The faint luminosity of the night hemisphere of the Moon, due to light reflected on to the Moon from the Earth.

Eclipses: These are of two kinds: solar and lunar. (1) A *solar eclipse* is caused by the Moon passing in front of the Sun. By coincidence, the two bodies appear almost equal in size. When the alignment is exact, the Moon covers up the Sun's bright disk for a brief period, either totally or partially (never more than about eight minutes; usually much less). When the eclipse is total the Sun's surroundings – the **chromosphere**, **corona** and **prominences** – may be seen with the naked eye (though you should never look directly at the Sun). If the Sun is not fully covered, the eclipse is *partial,* and the spectacular phenomena of totality are not seen. If the Moon is near its greatest distance from the Earth (see **apogee**) it appears slightly smaller than the Sun, and at central alignment a ring of the Sun's disk is left showing around the body of the Moon; this is an *annular* eclipse, and again the phenomena of totality are not seen. (2) A *lunar eclipse* is caused when the Moon passes into the shadow cast by the Earth; it may be either total or partial. Generally, the Moon does not vanish, as some sunlight is refracted on to it by way of the ring of atmosphere surrounding the Earth.

Eclipsing binary (or Eclipsing variable): A **binary star** made up of two components moving around their common centre of gravity at an angle such that, as seen from the Earth, the components mutually eclipse each other. In the case of the eclipsing binary Algol, one component is much brighter than the other; every $2\frac{11}{12}$ days the fainter star covers up the brighter, and the star seems to fade by more than a magnitude.

Ecliptic: The projection of the Earth's orbit on to the celestial sphere. It may also be defined as 'the apparent yearly path of the Sun against the stars', passing through the constellations of the **Zodiac.** Since the plane of the Earth's orbit is inclined to the equator by 23.5 degrees, the angle between the ecliptic and the celestial equator must also be 23.5 degrees.

Ecosphere: The region around the Sun in which the temperatures are neither too hot nor too cold for life to exist under suitable conditions. Venus lies near the inner edge of the ecosphere; while Mars is near the outer edge. The ecospheres of other stars will depend upon the luminosities of the stars concerned.

Electromagnetic spectrum: The full range of what is termed electromagnetic radiation: **gamma-rays**, **X-rays**, **ultra-violet radiation**, visible light, **infra-red radiation** and radio waves. Visible light makes up only a very small part of the whole electromagnetic spectrum. Of all the radiations, only visible light and some of the radio waves can pass through the Earth's atmosphere and reach ground-level.

Electron: A fundamental particle carrying unit negative charge of electricity; the orbital components of the **atom**.

Electron density: The number of free electrons in unit volume of space. A free electron is not attached to any particular atom, but is moving independently.

Element: A substance which cannot be chemically split up into simpler substances; 92 elements are known to exist naturally on the Earth and all other substances are made up from these fundamental 92. Various extra elements have been made artificially, all of which are heavier than uranium (number 92 in the natural sequence) and most of which are very unstable.

Elongation: The apparent angular distance of a planet from the Sun, or of a satellite from its primary planet.

Emission spectrum: A spectrum consisting of bright lines or bands. Incandescent gases at low density yield emission spectra.

Ephemeris: A table giving the predicted positions of a moving celestial body, such as a planet or a comet.

Epoch: A date chosen for reference purposes in quoting astronomical data. For instance, some star catalogues are given for 'epoch 1950'; by the year 2000 the given positions will have changed slightly because of the effects of **precession**.

Equation of time: The Sun does not move among the stars at a constant rate, because the Earth's orbit is not circular. Astronomers therefore make use of a *mean* sun, which travels among the stars at a speed equal to the average speed of the real Sun. The interval by which the real Sun is ahead of or behind the mean sun is termed the *equation of time.* It can never exceed 17 minutes; four times every year it becomes zero.

Equator, celestial: The projection of the Earth's equator on to the **celestial sphere** divides the sky into two equal hemispheres.

Equatorial mount: A telescope mounting in which the instrument is set upon an axis which is parallel to the axis of the Earth; the angle of the axis must be equal to the observer's latitude. This means that to keep an object in view the telescopes were equatorially mounted, but with the aid of modern computers it has become possible to make effective drives for altazimuth telescopes; altazimuth is now the favoured mounting.

Equinox: Twice a year the Sun crosses the celestial equator, once when moving from south to north (about 21 March) and once when moving from north to south (about 22 September). These points are known respectively as the vernal equinox, or **First Point of Aries**, and the autumnal equinox, or **First Point of Libra**. (The equinoxes are the two points at which the ecliptic cuts the celestial equator.)

Escape velocity: The minimum velocity at which an object must move in order to escape from the surface of a planet, or other body, without being given extra propulsion and neglecting any air resistance. The escape velocity of the Earth is 11 kilometres per second, or about 40,200 kilometres per hour; for the Moon it is only 2.4 kilometres per second; for Jupiter, as much as 60 kilometres per second.

Exosphere: The outermost part of the Earth's atmosphere. It is very rarefied, and has no definite upper boundary, since it simply 'thins out' into surrounding space.

F

Faculae: Bright temporary patches on the surface of the Sun, usually (although not always) associated with **sunspots**. Faculae frequently appear in a position near which a spot group is about to appear, and may persist for some time in the region of a group which has disappeared.

First Point of Aries: The vernal equinox. The **right ascension** of the vernal equinox is taken as zero, and the right ascensions of all celestial bodies are referred to it. See **equinox**.

First Point of Libra: The autumnal equinox. See **equinox**.

Flares, solar: Brilliant outbreaks in the outer part of the Sun's atmosphere, usually associated with active **sunspot** groups. They send out electrified particles which may later reach the Earth, causing magnetic storms and aurorae (see **aurora**); they are also associated with strong outbursts of solar radio emission. It has been suggested that the particles emitted by flares may present a hazard to astronauts who are in space or on the unprotected surface of the Moon.

Flare stars: Faint red dwarf stars which may brighten up by several magnitudes over a period of a few minutes, fading back to their usual brightness within an hour or so. It is thought that this must be due to intense flare activity in the star's atmosphere. Although the energies involved are much higher than for solar **flares**, it is not yet known whether the entire stellar atmosphere is involved, or only a small area, as in the case of flares on the Sun. Typical flare stars are UV Ceti and AD Leonis.

Flash spectrum: Just before the Moon completely covers the Sun at a total solar eclipse, the Sun's atmosphere is seen shining by itself, without the usual brilliant background of the **photosphere**. The dark lines in the spectrum then become bright, producing what is termed the flash spectrum. The same effect is seen just after the end of totality.

Flocculi: Patches on the Sun's surface, observed by instruments based on the principle of the **spectroscope**. Bright flocculi are composed of calcium; dark flocculi are made up of hydrogen.

Focal length: The distance between a lens (or mirror) and the point at which the image of an object at infinity is brought to focus. The focal length divided by the aperture of the mirror or lens is termed the *focal ratio.*

Fraunhofer lines: The dark absorption lines in the Sun's spectrum. named in honour of the German optician J. von Fraunhofer, who first studied and mapped them in 1814.

Free fall: The normal state of motion of an object in space under the influence of the gravitational pull of a central body; thus the Earth is in free fall around the Sun, while an artificial satellite moving beyond the atmosphere is in free fall around the Earth. While no thrust is being applied, a lunar probe travelling between the Earth and the Moon is in free fall; the same applies for a probe in a transfer orbit between the Earth and another planet. While a vehicle is in free fall, an astronaut will have no apparent 'weight', and will be experiencing zero gravity or weightlessness.

Fringe region: The upper part of the **exosphere**. Atomic particles in the fringe region have little chance of collision with one another, and to all intents and purposes they travel in free orbits, subject to the Earth's gravitation.

G

g: Symbol for the force of gravity at the Earth's surface. The acceleration due to gravity is 9.75 metres per second per second at sea level.

Galaxies: Systems of stars; our Galaxy contains about 100,000 million stars, but is not exceptional in size. Galaxies are of various shapes; some are spiral, some elliptical, some irregular. The most remote galaxies known are at least 18,000 million light-years away; all, apart from those of our Local Group, are receding from us, so that the entire universe is expanding.

Galaxy, the: The Galaxy of which our Sun is a member.

Gamma-rays: Extremely short-wavelength electromagnetic radiations. Cosmic gamma-ray sources have to be studied by space research methods.

Gegenschein (or counterglow): A very faint glow in the sky, exactly opposite to the Sun; it is very difficult to observe, and has never been satisfactorily photographed. It is due to tenuous matter spread along the main plane of the **Solar System**, so that it is associated with the **Zodiacal Light**.

Geocentric: Relative to the Earth as a centre – or as measured with respect to the centre of the Earth.

Geocorona: A layer of very tenuous hydrogen surrounding the Earth near the uppermost limit of the atmosphere.

Geodesy: The science which deals with the Earth's form, dimensions, elasticity, mass, gravitation and allied topics.

Geophysics: The science dealing with the physics of the Earth and its environment. Its range extends from the interior of the Earth out to the limits of the magnetosphere. In 1957–8 an ambitious international programme, the International Geophysical Year (IGY). was organized to undertake intensive studies of geophysical phenomena at the time of a sunspot maximum. It was extended to 18 months, and was so successful that at the next sunspot minimum a more limited but still extensive programme was organized, the International Year of the Quiet Sun (IQSY).

Gibbous: A phase of the Moon or a planet which is more than half, but less than full.

Gravitation: The force of attraction which exists between all particles of matter in the universe. Particles attract one another with a force which is directly proportional to the product of their masses and inversely proportional to the square of the distance between them.

Great circle: A circle on the surface of a sphere (such as the Earth, or the celestial sphere) whose plane passes through the centre of the sphere. Thus a great circle will divide the sphere into two equal parts.

Green flash (or green ray): When the Sun is setting, the last visible portion of the disk may flash brilliant green for a very brief period. This is due to effects of the Earth's atmosphere, and is best observed over a sea horizon. Venus has also been known to show a green flash when setting.

Greenwich Mean Time (GMT): The time reckoned from the Greenwich Observatory in London, England. It is used as the standard throughout the world. Also known as Universal Time (UT).

Greenwich Meridian: The line of longitude which passes through the Airy Transit Circle at Greenwich Observatory. It is taken as longitude zero degrees, and is used as the standard throughout the world.

Gregorian reflector: A type of reflecting telescope (see **reflector**) in which the incoming light is reflected from the main mirror on to a small concave mirror placed outside the focus of the main mirror; the light then comes back through a hole in the main mirror and is brought to focus. Gregorian reflectors are not now common.

H

H I and H II regions: Clouds of hydrogen in the **Galaxy**. In *H I regions* the hydrogen is neutral, and the clouds cannot be seen, but they may be studied by radio telescopes by virtue of their characteristic emission at a wavelength of 21 centimetres. In *H II regions* the hydrogen is ionized (see **ion**), generally in the presence of hot stars. The recombination of the ions and free electrons to form neutral atoms gives rise to the emission of light, by which the H II regions can be seen.

Halation ring: A ring sometimes seen around a star image on a photograph. It is purely a photographic effect.

Halo: (1) A luminous ring around the Sun or Moon, due to ice crystals in the Earth's upper atmosphere. (2) The *galactic halo*: The spherical-shaped star cloud around the main part of the **Galaxy**.

Hertzsprung-Russell Diagram (H/R Diagram): A diagram in which stars are plotted according to spectral type and luminosity. It is found that there is a well-defined band known as the **Main Sequence** which runs from the upper left of the Diagram (very luminous bluish stars) down to the lower right (faint red stars); there is also a giant branch to the upper right, while the dim, hot **white dwarfs** lie to the lower left. H/R Diagrams have been of the utmost importance in studies of stellar evolution. If colour index is used instead of spectrum, the diagram is known as a colour-magnitude diagram.

Hohmann orbit: See **transfer orbit**.

Glossary

Hour angle: The time which has elapsed since a celestial body crossed the meridian of the observer.

Hour circle: A great circle on the celestial sphere which passes through both poles of the sky. The zero hour circle corresponds to the observer's meridian.

Hubble Constant: The relationship between the distance of a galaxy and its recessional velocity. Its value is still uncertain, it may be of the order of 60 kilometres per second per megaparsec.

I

Inferior planets: Mercury and Venus, whose orbits lie closer to the Sun than does that of the Earth. When their **right ascensions** are the same as that of the Sun, so that they are approximately between the Sun and the Earth, they reach *inferior conjunction*. If the **declination** is also the same as that of the Sun, the result will be a transit of the planet.

Infra-red radiation: Radiation with wavelengths longer than that of red light, but shorter than microwaves. Infra-red sources in the sky are studied either from high-altitude observatories (as at Mauna Kea) or with space techniques. In 1983 the Infra-Red Astronomical Satellite (IRAS) carried out a full survey of the sky in infra-red.

Ion: An atom which has lost or gained one or more electrons; it has a corresponding positive or negative electrical charge, since in a complete atom the positive charge of the nucleus is balanced out by the combined negative charge of the electrons. The process of producing an ion is termed *ionization*.

Ionosphere: The region above the **stratosphere**, from about 65 up to about 800 kilometres. Ionization of the atoms in this region (see **ion**) produces layers which reflect radio waves, making long-range communication over the Earth possible. Solar events have effects upon the ionosphere, and produce ionospheric storms; on occasion, radio communication is interrupted.

Irradiation: The effect which makes brightly lit or self-luminous bodies appear larger than they really are. For example, the Moon's bright crescent appears larger in diameter than the Earth-lit part of the disk.

J

Julian day: A count of the days, starting from 12 noon on 1 January 4713 BC. The system was introduced by Scaliger in 1582. The 'Julian' is in honour of Scaliger's father, and has nothing to do with Julius Caesar or the Julian Calendar. Julian days are used by **variable star** observers, and for reckonings of phenomena which extend over very long periods of time.

K

Kepler's Laws of Planetary Motion: The three important laws announced by J. Kepler between 1609 and 1618 . They are:

(1) The planets move in elliptical orbits, the Sun being located at one focus of the ellipse, while the other focus is empty.

(2) The radius vector, or imaginary line joining the centre of the planet to the centre of the Sun, sweeps out equal areas in equal times.

(3) The squares of the sidereal periods of the planets are proportional to the cubes of their mean distances from the Sun (Harmonic Law).

Kiloparsec: 1000 **parsecs**, or 3260 **light-years**.

Kirkwood gaps: Regions in the belt of **asteroids** between Mars and Jupiter in which almost no asteroids move. The gravitational influence of Jupiter keeps these zones 'swept clear'; an asteroid which enters a Kirkwood region will be regularly perturbed by Jupiter until its orbit has been changed. They were first noted by the American mathematician Daniel Kirkwood.

L

Laser (Light Amplification by the Simulated Emission of Radiation): A device which emits a beam of light made up of rays of the same wavelength (coherent light) and in phase with one another. It can be extremely intense. Laser beams have already been reflected off the Moon.

Latitude, celestial: The angular distance of a celestial body from the nearest point on the **ecliptic**.

Librations, lunar: Although the Moon's rotation is captured with respect to the Earth, there are various effects, known as librations, which enable us to examine 59 per cent of the total surface instead of only 50 per cent, although no more than 50 per cent can be seen at any one time. There are three librations: in longitude (because the Moon's orbital velocity is not constant), in latitude (because the Moon's equator is inclined by 6 degrees to its orbital plane), and diurnal (due to the rotation of the Earth).

Light-year: The distance travelled by light in one year. It is equal to 9.46 million million million kilometres.

Limb: The edge of the visible disk of the Sun. Moon, a planet, or the Earth (as seen from space).

Local Group of galaxies: The group of which our Galaxy is a member. There are more than two dozen systems, of which the most important are the Andromeda Spiral, our Galaxy, the Triangulum Spiral and the two Clouds of Magellan.

Longitude, celestial: The angular distance from the vernal equinox to the foot of a perpendicular drawn from a celestial body to meet the **ecliptic**. It is measured eastwards along the ecliptic from zero degrees to 360 degrees.

Lunation (synodical month): The interval between successive new moons: 29 days 12 hours 44 minutes. See also **synodic period**.

Lyot filter (monochromatic filter): A device used for observing the Sun's prominences and other features of the solar atmosphere, without the necessity of waiting for a total **eclipse**. It was invented by the French astronomer B. Lyot.

M

Mach number: The velocity of a vehicle moving in an atmosphere divided by the velocity of sound in the same region. Near the surface of the Earth, sound travels at about 1200 kilometres per hour; so Mach 2 would be $2 \times 1200 = 2400$ kilometres per hour.

Magnetic storm: A sudden disturbance of the Earth's magnetic field, shown by interference with radio communication as well as by variations in the compass needle. It is due to charged particles sent out from the Sun, often associated with solar **flares**. A *magnetic crochet* is a sudden change in the Earth's magnetic field due to changing conditions in the lower **ionosphere**. The crochet is associated with the flash phase of the flare, and commences with it; the storm is associated with the particles, which reach the Earth about 24 hours later.

Magnetohydrodynamics: The study of the interactions between a magnetic field and an electrically conducting fluid. The Swedish scientist H. Alfven is regarded as the founder of magnetohydrodynamics.

Magnetosphere: The region round a body in which that body's magnetic field is dominant. In the **Solar System**, Jupiter has the largest magnetosphere; the other giants, as well as the Earth and Mercury, have pronounced magnetic fields, but the Moon, Venus and Mars do not.

Magnitude: This is really a term for 'brightness', but there are several different types. (1) *Apparent* or *visual magnitude*: the apparent brightness of a celestial body as seen with the eye. The brighter the object, the lower the magnitude. The planet Venus is of about magnitude −4 ; Sirius, the brightest star, −1.4; the Pole Star, +2; stars just visible with the naked eye, +6; the faintest stars that can be recorded with the world's largest telescopes, below +20. A star's apparent magnitude is no reliable key to its luminosity. (2) *Absolute magnitude*: the apparent magnitude that a star would have if seen from a standard distance of 10 **parsecs** (32.6 **light-years**). (3) *Photographic magnitude*: the magnitude derived from the size of a star's image on a photographic plate. (4) *Bolometric magnitude*: this refers to the total radiation sent out by a star, not merely to visible light.

Main Sequence: The well-defined band from the upper left to lower right of a **Hertzsprung-Russell Diagram**. The Sun is typical Main Sequence star.

Maser (Microwave Amplification by Simulated Emission of Radiation): The same basic principle as that of the laser, but applied to radio wavelengths rather than to visible light.

Mass: The quantity of matter that a body contains. It is not the same as weight, which depends upon local gravity; thus on the Moon an Earthman has only one-sixth of his normal weight, but his mass remains unaltered.

Meridian, celestial: The **great circle** on the celestial sphere which passes through the **zenith** and both celestial poles. The meridian cuts the observer's horizon at the exact north and south points.

Messier numbers: Numbers given by the 18th-century French astronomer Charles Messier to various nebulous objects. including open and globular clusters, gaseous nebulae and galaxies. Messier's catalogue contained slightly over a hundred objects. His numbers are still used; thus the Andromeda Spiral is M31, the Orion Nebula M42, the Crab Nebula M1, and so on.

Meteor: Cometary debris; a small particle which enters the Earth's upper atmosphere and burns away, producing the effect known as a shooting star.

Meteorite: A larger body, which is able to reach ground-level without being destroyed. There is a fundamental difference between meteorites and **meteors**; a meteorite seems to be more nearly related to an **asteroid** or minor planet. Meteorites may be stony (*aerolites*), iron (*siderites*) or of intermediate type. In a few cases meteorites have produced craters; the most famous example is the large crater in Arizona, which is almost 1.5 kilometres in diameter and was formed in prehistoric times.

Meteoroids: The collective term for meteoritic bodies. It was once thought that they would present a serious hazard to spacecraft travelling outside the Earth's atmosphere, but it now seems that the danger is very much less than was feared, even though it cannot be regarded as entirely negligible.

Micrometeorite: An extremely small particle, less than 0.01016 centimetres in diameter, moving around the Sun. When a micrometeorite enters the Earth's atmosphere, it cannot produce a shooting-star effect, as its mass is too slight. Since 1957, micrometeorites have been closely studied from space probes and artificial satellites.

Micron: A unit of length equal to one thousandth of a millimetre. There are 10,000 **Ångströms** to one micron. The usual symbol is Ì

Midnight Sun: The Sun seen above the horizon at midnight. This can occur for some part of the year anywhere inside the Arctic and Antarctic Circles.

Milky Way: The luminous band stretching across the night sky. It is due to a line-of-sight effect; when we look along the main plane of the **Galaxy** (that is, directly towards or away from the galactic centre) we see many stars in roughly the same direction. Despite appearances, the stars in the Milky Way are not closely crowded together. The term used to be applied to the Galaxy itself, but is now restricted to the appearance as seen in the night sky.

Millibar: The unit which is used as a measure of atmospheric pressure. It is equal to 1000 dynes per square centimetre. The standard atmospheric pressure is 1013.25 millibars (75.97 centimetres of mercury).

Minor planets: See **asteroids**.

Molecule: A stable association of atoms; a group of atoms linked together. For example, a water molecule (H_2O) is made up of two hydrogen atoms and one atom of oxygen.

Month: (1) *Calendar month*: the month in everyday use. (2) *Anomalistic month*: the time taken for the Moon to travel from one **perigee** to the next. (3) *Sidereal month*: the time taken for the Moon to complete one journey around the **barycentre**, with reference to the stars.

Multiple star: A star made up of more than two components physically associated, which orbit their mutual centre of gravity.

N

Nadir: The point on the celestial sphere immediately below the observer. It is directly opposite to the overhead point or **zenith**.

Nebula: A mass of tenuous gas in space together with what is loosely termed 'dust'. If there are stars in or very near the nebula, the gas and dust will become visible, either because of straightforward reflection or because the stellar radiation excites the material to self-luminosity. If there are no suitable stars, the nebula will remain dark, and will betray its presence only because it will blot out the light of stars lying beyond it. Nebulae are regarded as regions in which fresh stars are being formed out of the interstellar material.

Neutrino: A fundamental particle which has no mass and no electric charge – which makes them extremely difficult to detect.

Neutron: A fundamental particle whose mass is equal to that of a **proton**, but which has no electric charge. Neutrons exist in the nuclei of all atoms apart from that of hydrogen.

Neutron star: A star made up principally or completely of **neutrons**, so that it will be of low luminosity but almost incredibly high density. Theoretically, a neutron star should represent the final stage in a star's career. It is now thought probable that the remarkable radio sources known as **pulsars** are in fact neutron stars.

Newtonian reflector: The common form of astronomical **reflector**. Incoming light is collected by a mirror, and directed on to a smaller flat mirror placed at 45 degrees. The light is then sent to the side of the tube, where it is brought to a focus and the eyepiece is placed. Most small and many large reflectors are of Newtonian type.

Noctilucent clouds: Rare, strange clouds in the **ionosphere**, best seen at night when they continue to catch the rays of the Sun, after it has set. They lie at altitudes of greater than 80 kilometres, and are noticeably different from normal clouds. It is possible that they are produced by meteoritic dust in the upper atmosphere.

Nodes: The points at which the orbit of a planet, a comet or the Moon cuts the plane of the **ecliptic**, either as the body is moving from south to north (*ascending node*) or from north to south (*descending node*). The line joining these two points is known as the *line of nodes*.

Nova: A star which undergoes a sudden outburst, flaring up to many times its normal brilliancy for a while before fading back to obscurity. A nova is a binary system in which one component is a white dwarf; it is the white dwarf which is responsible for the outbursts.

Nutation: A slight, slow 'nodding' of the Earth's axis, due to the fact that the Moon is sometimes above and sometimes below the ecliptic, and therefore does not always pull on the Earth's equatorial bulge in the same direction as the Sun. The result is that the position of the celestial pole seems to 'nod' by about 9 seconds of arc to either side of its mean position with a period of 18 years 220 days. Nutation is superimposed on the more regular shift of the celestial pole caused by precession.

O

Object-glass (objective): The main lens of a refracting telescope (see **refractor**).

Obliquity of the ecliptic: The angle between the ecliptic and the celestial equator. Its value is 23 degrees 26 minutes 54 seconds. It may also be defined as the angle by which the Earth's axis is tilted from the perpendicular to the orbital plane.

Occultation: The covering up of one celestial body by another. Thus the Moon may pass in front of a star or (occasionally) a planet; a planet may occult a star; and there have been cases when one planet has occulted another – for instance, Venus occulted Mars in 1590. Strictly speaking, solar **eclipses** are occultations of the Sun by the Moon.

Opposition: The position of a planet when it is exactly opposite the Sun in the sky, and so lies due south at midnight. At opposition, the Sun, the Earth and the planet are approximately aligned, with the Earth in the mid position. Obviously, the **inferior planets** (Mercury and Venus) can never come to opposition.

Orbit: The path of an artificial or natural celestial body. See also **transfer orbit**.

Ozone: Triatomic oxygen (0_3). The ozone layer in the Earth's upper atmosphere absorbs many of the lethal short-wavelength radiations coming from space. Were there no ozone layer, it is unlikely that life on Earth could ever have developed.

P

Parallax, trigonometrical: The apparent shift of a body when observed from two different directions. The separation of the two observing sites is called the baseline. The Earth's orbit provides a baseline 300 million kilometres long (since the radius of the orbit is 150 million kilometres); therefore, a nearby star observed at a six-monthly interval will show a definite parallax shift relative to the more distant stars. It was in this way that Bessel, in 1838, made the first measurement of the distance of a star (61 Cygni). The method is useful out to about 300 **light-years**, beyond which the parallax shifts become too small to detect.

Parsec: The distance at which a star would show a parallax of one second of arc. It is equal to 3.26 **light-years**, 206,265 astronomical units, or 30.8 million million million kilometres. (Apart from the Sun, no star lies within one parsec of us.)

Glossary

Penumbra: (1) The comparatively light surrounding parts of a **sunspot**. (2) The area of partial shadow lying to either side of the main cone of shadow cast by the Earth. During lunar **eclipses**, the Moon must move through the penumbra before reaching the main shadow (or **umbra**). Some lunar eclipses are penumbral only.

Periastron: The point of the orbit of a member of a **binary** system in which the stars are at their closest to each other. The most distant point is termed **apastron**.

Perigee: The point in the orbit of the Moon or an artificial satellite at which the body is closest to the Earth. The most distant point is the **apogee**.

Perihelion: The point in the orbit of a member of the **Solar System** in which the body is at its closest to the Sun. The most distant point is the **aphelion**. The Earth reaches perihelion in early January.

Periodic times: See **sidereal period**.

Perturbations: The disturbances in the orbit of a celestial body produced by the gravitational pulls of others.

Phases: The apparent changes in shape of the Moon and some planets depending upon the amount of the sunlit hemisphere turned towards us. The Moon, Mercury and Venus show complete phases, from new (invisible) to full. Mars can show an appreciable phase, since at times less than 90 per cent of its sunlit face is turned in our direction. The phases of the outer planets are insignificant.

Photometry: The measurement of the intensity of light. The device now used for accurate determinations of star magnitudes is the *photoelectric photometer*, which consists of a photoelectric cell used together with a **telescope**. (A photoelectric cell is an electronic device. Light falls upon the cell and produces an electric current; the strength of the current depends on the intensity of the light.)

Photosphere: The bright surface of the Sun.

Planet: A non-luminous body moving round a star. It is likely that other stars have planetary systems similar to that of the Sun, but as yet there is no definite proof.

Planetarium: An instrument used to show an artificial sky on the inner surface of a large dome, and to reproduce celestial phenomena of all kinds. A planetarium projector is extremely complicated, and is very accurate. The planetarium is an educational device, and has become very popular in recent years. Planetaria have been set up in many large cities all over the world, and are also used in schools and colleges.

Planetary nebula: A faint star surrounded by an immense 'shell' of tenuous gas. More than 300 are known in our Galaxy. They are so called because their telescopic appearance under low magnification is similar to that of a planet.

Plasma: A gas consisting of ionized atoms (see **ion**) and free electrons, together with some neutral particles. Taken as a whole, it is electrically neutral, and is a good conductor of electricity.

Poles, celestial: The north and south points of the **celestial sphere**.

Populations, stellar: There are two main types of star regions. *Population I* areas contain a great deal of interstellar material, and the brightest stars are hot and white; it is assumed that star formation is still in progress. The brightest stars in *Population II* areas are red giants, well advanced in their evolutionary cycle; there are almost no hot, white giant stars, and there is little interstellar material, so that star formation has apparently ceased. Although no rigid boundaries can be laid down, it may be said that the arms of spiral galaxies are mainly of Population I; the central parts of spirals, as well as elliptical galaxies and globular clusters, are mainly of Population II.

Position angle: The apparent direction of one object with reference to another measured from the north point of the main object through east (90 degrees), south (180 degrees) and west (270 degrees).

Precession: The apparent slow movement of the celestial poles. It is caused by the pull of the Moon and the Sun upon the Earth's equatorial bulge. The Earth behaves rather in the manner of a top which is running down and starting to topple, but the movement is very gradual; the pole describes a circle on the celestial sphere, centred on the pole of the ecliptic, which is 47 degrees in diameter and takes 25,800 years to complete. Because of precession, the celestial equator also moves. and this in turn affects the position of the **First Point of Aries** (vernal equinox), which shifts westwards along the **ecliptic** by 50 seconds of arc each year. Since ancient times, this motion has taken the vernal equinox out of Aries into the adjacent constellation of Pisces (the Fishes). Our present Pole Star will not retain its title indefinitely. In AD 12,000, the north polar star will be the brilliant Vega, in Lyra.

Prism: A glass block having flat surfaces inclined to one another. Light passing through a prism will be split up, since different colours are refracted by different amounts.

Prominences: Masses of glowing gas, chiefly hydrogen, above the Sun's bright surface. They are visible with the naked eye only during total solar **eclipses**, but modern equipment allows them to be studied at any time. They are of two main types, eruptive and quiescent.

Proper motion: The individual motion of a star on the **celestial sphere**. Because the stars are so remote, their proper motions are slight. The greatest known is that of Barnard's Star (a red dwarf at a distance of 6 **light-years**); this amounts to one minute of arc every six years, so that it will take 180 years to move by an amount equal to the apparent diameter of the Moon. The proper motions of remote stars are too slight to be measured at all.

Proton: A fundamental particle with unit positive electrical charge. The nucleus of a hydrogen atom consists of one proton. See also **neutron**.

Pulsar: A **neutron star** radio source which does not emit continuously, but in rapid, very regular pulses. Their periods are short (often much less than one second).

Purkinje effect: An effect inherent in the human eye, which makes it less sensitive to light of longer wavelength when the general level of intensity is low. Consider two lights, one red and one blue, which are of equal intensity. If the intensity of both are reduced by equal amounts, the blue light will appear to be the brighter of the two.

Q

Quadrature: The position of the Moon or a planet when at right angles to the Sun as seen from Earth. Thus the Moon is in quadrature when it is seen at half-phase.

Quantum: The smallest amount of light-energy which can be transmitted at any given wavelength.

Quasar: A very remote immensely luminous object, now known to be the core of a very active galaxy – possibly powered by a massive black hole inside it. Quasars are also known as QSOs (Quasi-Stellar Objects). *BL Lacertae objects* are of the same type, though less important.

R

Radar astronomy: The technique of using radar pulses to study astronomical objects. Most planets and some asteroids have been contacted by radar, and the radar equipment carried in space probes such as Magellan has provided us with detailed maps of the surface of Venus.

Radial velocity: The towards-or-away movement of a celestial body, measured by the **Doppler effect** in its spectrum. If the spectral lines are red-shifted, the object is receding; if the shift is to the blue, the object is approaching. Conventionally, radial velocity is said to be positive with a receding body, negative with an approaching body.

Radiant: The point in the sky from which the **meteors** of any particular shower appear to radiate (for example, the August shower has its radiant in Perseus, so that the meteors are known as the Perseids). The meteors in a shower are really moving through space in parallel paths, so that the radiant effect is due merely to perspective.

Radio astronomy: Astronomical studies carried out in the long-wavelength region of the **electromagnetic spectrum**. The main instruments used are known as *radio telescopes*; they are of many kinds, ranging from 'dishes', such as the 76-metre (250-foot) paraboloid at Jodrell Bank (Cheshire), to long lines of aerials.

Radio galaxies: Galaxies which are extremely powerful emitters of radio radiation.

Red shift: The **Doppler** displacement of spectral lines towards the red or long-wave end of the spectrum, indicating a velocity of recession. Apart from the members of the **Local Group,** all galaxies show red shifts in their spectra.

Reflector: A telescope in which the light is collected by means of a mirror.

Refraction: The change in direction of a ray of light when passing from one transparent substance into another.

Refractor: A telescope which collects its light by means of a lens. The light passes through this lens (**object-glass**) and is brought to focus; the image is then magnified by an eyepiece.

Resolving power: The ability of a **telescope** to separate objects which are close together; the larger the telescope the greater its resolving power. Radio telescopes (see **radio astronomy**) have poor resolving power compared with optical telescopes.

Retardation: The difference in the time of moonrise between one night and the next. It may exceed one hour, or it may be as little as a quarter of an hour.

Retrograde motion: In the **Solar System**, movement in a sense opposite to that of the Earth in its orbit; some comets, notably Halley's, have retrograde motion. The term is also used with regard to the apparent movements of planets in the sky; when the apparent motion is from east to west, relative to the fixed stars, the direction is retrograde. The term may be applied to the rotations of planets. Since Uranus has an axial inclination of more than a right angle, its rotation is technically retrograde; Venus also has retrograde axial rotation.

Reversing layer: The gaseous layer above the bright surface or **photosphere** of the Sun. Shining on its own, the gases would yield bright spectral lines; but as the photosphere makes up the background, the lines are reversed, and appear as dark absorption or **Fraunhofer lines**. Strictly speaking, the whole of the Sun's **chromosphere** is a reversing layer.

Right ascension: The right ascension of a celestial body is the time which elapses between the culmination of the **First Point of Aries** and the culmination of the body concerned. For example, Aldebaran in Taurus culminates 4h 33m after the First Point of Aries has done so; therefore the right ascension of Aldebaran is 4h 33m. The right ascensions of bodies in the **Solar System** change quickly. However, the right ascensions of stars do not change, apart from the slow cumulative effect of **precession**.

Roche limit: The distance from the centre of a planet, or other body, within which a second body would be broken up by gravitational distortion. This applies only to an orbiting body which has no appreciable structural cohesion, so that strong, solid objects, such as artificial satellites, may move safely well within the Roche limit for the Earth. The Roche limit lies at 2.44 times the radius of the planet from the centre of the globe, so that for the Earth it is about 9170 kilometres above ground-level. Saturn's ring system lies within the Roche limit for Saturn.

RR Lyrae variables: Regular **variable stars** whose periods are very short (between about 1¼ hours and about 30 hours). They seem to be fairly uniform in luminosity; each is around 100 times as luminous as the Sun. They can therefore be used for distance measures, in the same way as **Cepheids**. Many of them are found in star clusters, and they were formerly known as cluster-Cepheids. No RR Lyrae variable appears bright enough to be seen with the naked eye.

S

Saros: A period of 18 years 11.3 days, after which the Earth, Moon. and Sun return to almost the same relative positions. Therefore, an **eclipse** of the Sun or Moon is liable to be followed by a similar eclipse 18 years 11.3 days later. The period is not exact, but is good enough for predictions to be made – as was done in ancient times by Greek philosophers.

Satellite: A secondary body orbiting a primary. The Earth has one satellite (the Moon); Jupiter has 16, Saturn 18, Uranus 15, Neptune eight and Pluto one, while Mercury and Venus are unattended.

Schmidt telescope (or Schmidt camera): A type of **telescope** which uses a spherical mirror and a special glass correcting plate. With it, relatively wide areas of the sky may be photographed with a single exposure; definition is good all over the plate. In its original form, the Schmidt telescope can be used only photographically. The largest Schmidt in use is the 122-centimetre instrument at Palomar.

Scintillation: Twinkling of stars. It is due entirely to the effects of the Earth's atmosphere; a star will scintillate most violently when it is low over the horizon, so that its light is passing through a thick layer of atmosphere. A planet, which shows up as a small disk rather than a point, will generally twinkle much less than a star.

Seasons: Effects on the climate due to the inclination of the Earth's axis. The fact that the Earth's distance from the Sun is not constant has only a minor effect upon our seasons.

Second of arc: One 360th of a degree. See **arc minute, arc second**.

Secular acceleration: Because of friction produced by the tides, the Earth's rotation is gradually slowing down; the 'day' is becoming longer. The average daily lengthening is only 0.00000002 seconds, but over a sufficiently long period the effect becomes detectable. The lengthening of terrestrial time periods gives rise to an apparent speeding-up of the periods of the Sun, Moon and planets. Another result of these tidal phenomena is that the Moon is receding from the Earth slowly.

Seeing: The quality of the steadiness and clarity of a star's image. It depends upon conditions in the Earth's atmosphere. From the Moon, or from space, the 'seeing' is always perfect.

Seismometer: An earthquake recorder. Very sensitive seismometers were taken to the Moon by the Apollo astronauts, and provided interesting information about seismic conditions there.

Selenography: The study of the Moon's surface.

Sextant: An instrument used for measuring the altitude of a celestial body above the horizon.

Seyfert galaxies: Galaxies with small, bright nuclei. Many of them are radio sources, and show evidence of violent disturbances in their nuclei.

Shooting-star: The luminous appearance caused by a meteor falling through the Earth's atmosphere.

Sidereal period: The time taken for a planet or other body to make one journey around the Sun (365.2 days in the case of the Earth). The term is also used for a satellite in orbit around a planet. Also known as periodic time.

Sidereal time: The local time reckoned according to the apparent rotation of the celestial sphere. It is zero hours when the **First Point of Aries** crosses the observer's meridian. The sidereal time for any observer is equal to the right ascension of an object which lies on the meridian at that time. Greenwich sidereal time is used as the world standard (this is, of course, merely the local sidereal time at Greenwich Observatory).

Solar apex: The point on the celestial sphere towards which the Sun is apparently travelling. It lies in the constellation Hercules; the Sun's velocity towards the apex is 19 kilometres per second. The point directly opposite in the sky to the solar apex is termed the solar *antapex*. This motion is distinct from the Sun's rotation around the centre of the Galaxy, which amounts to about 320 kilometres per second.

Solar constant: The unit for measuring the amount of energy received on the Earth's surface by solar radiation. It is equal to 1.94 calories per minute per square centimetre. (A calorie is the amount of heat needed to raise the temperature of 1 gram of water by 1 degree C.)

Solar flares: See **flares, solar**.

Solar parallax: The trigonometrical parallax of the Sun. It is equal to 8.79 seconds of arc.

Solar System: The system made up of the Sun, the planets, satellites, comets, asteroids, meteoroids and interplanetary dust and gas.

Solar time, apparent: The local time reckoned according to the Sun. Noon occurs when the Sun crosses the observer's meridian, and is therefore at its highest in the sky.

Solar wind: A steady flow of atomic particles streaming out from the Sun in all directions. It was detected by means of space probes, many of which carry instruments to study it. Its velocity in the neighbourhood of the Earth exceeds 965 kilometres per second. The intensity of solar wind is enhanced during solar storms.

Solstices: Times when the Sun is at its northernmost point in the sky (declination 23½°N, around 22 June), or at its southernmost point (23½°S, around 22 December). The dates of the solstices vary somewhat, because of the calendar irregularities due to leap years.

Glossary

Spacesuit: Equipment designed to allow an astronaut to operate outside the atmosphere.

Specific gravity: The density of any substance compared with that of an equal volume of water.

Spectroheliograph: An instrument used for photographing the Sun in the light of one particular wavelength only. If adapted for visual use, it is known as a *spectrohelioscope.*

Spectroscope: An instrument used to analyse the light on a star or other luminous object. Astronomical spectroscopes are used in conjunction with telescopes. Without them our knowledge of the nature of the universe would still be very rudimentary.

Spectroscopic binary: A **binary star** whose components are too close together to be seen separately, but whose relative motions cause opposite **Doppler** shifts which are detectable spectroscopically.

Speculum: The main mirror of reflecting telescope (see **reflector**). Older mirrors were made of speculum metal; modern ones are generally of glass.

Spherical aberration: The blurred appearance of an image as seen in a telescope, due to the fact that the lens or mirror does not bring the rays falling on its edge and on its centre to exactly the same focus. If the spherical aberration is noticeable, the lens or mirror is of poor quality, and should be corrected.

Spicules: Jets up to 16,000 kilometres in diameter, in the solar chromosphere. Each lasts for 4–5 minutes.

Spiral nebula: A now obsolete term for a spiral galaxy.

Star: A self-luminous gaseous body. The Sun is a typical star.

Steady-state theory: A theory according to which the universe has always existed, and will exist for ever. The theory has now been abandoned by almost all astronomers.

Stratosphere: The layer in the Earth's atmosphere lying above the **troposphere**. It extends from about 11 to about 64 kilometres above sea-level.

Sublimation: The change of a solid body to the gaseous state without passing through a liquid condition. (This may well apply to the polar caps on Mars.)

Sundial: An instrument used to show the time, by using an inclined style, or gnomon, to cast a shadow on to a graduated dial. The gnomon points to the celestial pole. A sundial gives apparent time; to obtain mean time, the value shown on the dial must be corrected by applying the **equation of time**.

Sun-grazers: Comets which at **perihelion** make very close approaches to the Sun. All the sun-grazers are brilliant comets with extremely long periods.

Sunspots: Darker patches on the solar photosphere; their temperature is about 4000 degrees C (as against about 6000 degrees C for the general photosphere), so that they are dark only by contrast; if they could be seen shining on their own, their surface brilliance would be greater than that of an arc-light. A large sunspot consists of a central darkish area or umbra, surrounded by a lighter area or penumbra, which may be very extensive and irregular. Sunspots tend to appear in groups, and are associated with strong magnetic fields; they are also associated with **faculae** and with solar **flares**. They are most common at the time of solar maximum (approximately every 11 years). No sunspot lasts for more than a few months at most.

Supergiant stars: Stars of exceptionally low density and great luminosity. Betelgeux in Orion is a typical supergiant.

Superior conjunction: The position of a planet when it is on the far side of the Sun as seen from Earth.

Superior planets: The planets beyond the orbit of the Earth in the Solar System: that is to say, all the principal planets apart from Mercury and Venus.

Supernova: A colossal stellar outburst. A Type I supernova involves the total destruction of the white dwarf component of a **binary** system; a Type II supernova is produced by the collapse of a very massive star. At its peak, a supernova may exceed the combined luminosity of all the other stars of an average galaxy.

Synchronous satellite: An artificial satellite moving in a west-to-east equatorial orbit in a period equal to that of the Earth's axial rotation (approximately 24 hours): as seen from Earth the satellite appears to remain stationary, and is of great value as a communications relay. Many synchronous satellites are now in orbit.

Synchrotron radiation: Radiation emitted by charged particles moving at relativistic velocities in a strong magnetic field. Much of the radio radiation coming from the **Crab Nebula** is of this type.

Synodic period: The interval between successive oppositions of a **superior planet**. For an **inferior planet**, the term is taken to mean the interval between successive conjunctions with the Sun.

Syzygy: The position of the Moon in its orbit when at new or full phase.

T

Tektites: Small, glassy objects which are aerodynamically shaped, and seem to have been heated twice. It has been suggested that they are meteorite, but it is now generally believed that they are of terrestrial origin.

Telemetry: The technique of transmitting the results of measurements and observations made on instruments in inaccessible positions (such as unmanned probes in orbit) to a point where they can be used and analysed.

Telescope: The main instrument used to collect the light from celestial bodies, thereby producing an image which can be magnified. There are two main types: the reflector and the refractor. All the world's largest telescopes are reflectors, because a mirror can be supported by its back, whereas a lens has to be supported around its edge – and if it is extremely large, it will inevitably sag and distort under its own weight, thereby rendering itself useless.

Terminator: The boundary between the day and night hemispheres of the Moon or a planet. Since the lunar surface is mountainous, the terminator is rough and jagged, and isolated peaks may even appear to be detached from the main body of the Moon. Mercury and Venus, which also show lunar-type phases, seem to have almost smooth terminators, but this is probably because we cannot see them in such detail (at least in the case of Mercury, whose surface is likely to be as mountainous as that of the Moon). Mars also shows a smooth terminator, although it is now known that the surface of the planet is far from being smooth and level. Photographs of the Earth taken from space or from the Moon show a smooth terminator which appears much 'softer' than that of the Moon, because of the presence of atmosphere.

Thermocouple: An instrument used for measuring very small quantities of heat. When used in conjunction with a large telescope, it is capable of detecting remarkably feeble heat-sources.

Tides: The regular rise and fall of the ocean waters, due to the gravitational pulls of the Moon and (to a lesser extent) the Sun.

Time dilation effect: According to relatively theory, the 'time' experienced by two observers in motion compared with each other will not be the same. To an observer moving at near the velocity of light, time will slow down; also, the observer's mass will increase until at the actual velocity-of-light time will stand still and mass will become infinite! The time and mass effects are entirely negligible except for very high velocities, and at the speeds of modern rockets they may be ignored completely.

Transfer orbit (or Hohmann orbit): The most economical orbit for a spacecraft which is sent to another planet. To carry out the journey by the shortest possible route would mean continuous expenditure of fuel, which is a practical impossibility. What has to be done is to put the probe into an orbit which will swing it inwards or outwards to the orbit of the target planet. To reach Mars, the probe is speeded up relative to the Earth, so that it moves outwards in an elliptical orbit; calculations are made so that the probe will reach the orbit of Mars and rendezvous with the planet. To reach Venus, the probe must initially be slowed down relative to the Earth, so that it will swing inwards towards the orbit of Venus. With a probe moving in a transfer orbit, almost all the journey is carried out in free fall, so that no propellant is being used. On the other hand, it means that the distances covered are increased, so that the time taken for the journey is also increased.

Transit: (1) The passage of a celestial body, or a point on the celestial sphere, across the observer's meridian; thus the **First Point of Aries** must transit at 0 hours **sidereal time.** (2) Mercury and Venus are said to be in transit when they are seen against the disk of the

Sun at inferior conjunction. Transits of Mercury are quite frequent (e.g. one in 1973 and the next in 1986), but the next transit of Venus will not occur until 2004; the last took place in 1882. Similarly, a satellite of a planet is said to be in transit when it is seen against the planet's disk. Transits of the four large satellites of Jupiter may be seen with small telescopes; also visible are shadow transits of these satellites, when the shadows cast by the satellites are seen as black spots on the face of Jupiter.

Transit instrument: A **telescope** which is specially mounted; it can move only in elevation, and always points to the meridian. Its sole use is to time the moments when stars cross the meridian, so providing a means of checking the time. The transit instrument set up at Greenwich Observatory by Sir George Airy, in the 19th century, is taken to mark the Earth's prime meridian (longitude zero degrees). Although still in common use it is likely that they will become obsolete before long.

Trojans: Asteroids which move around the Sun at a mean distance equal to that of Jupiter. One group of Trojans keeps well ahead of Jupiter and the other group well behind, so that there is no danger of collision. More than a dozen Trojans are now known.

Troposphere: The lowest part of the Earth's atmosphere, reaching to an average height of about 11 kilometres above sea-level. It includes most of the mass of the atmosphere, and all the normal clouds lie within it. Above, separating the troposphere from the **stratosphere**, is the *tropopause*.

Twilight, astronomical: The state of illumination of the sky when the Sun is below the horizon, but by less than 18 degrees.

Twinkling: Common term for **scintillation**.

U

Ultra-violet radiation: Electromagnetic radiation which has a wavelength shorter than that of violet light, and so cannot be seen with the naked eye. The ultra-violet region of the **electromagnetic spectrum** lies between visible light and X-radiation. The Sun is a very powerful source of ultra-violet, but most of this radiation is blocked out by layers in the Earth's upper atmosphere – which is fortunate for us, since in large quantities ultra-violet radiation is lethal. Studies of the ultra-violet radiations emitted by the stars have to be carried out by means of instruments sent up in rockets or artificial satellites.

Umbra: (1) The dark inner portion of a sunspot. (2) The main cone of shadow cast by a planet or the Moon.

Universal time: The same as Greenwich Mean Time.

V

Van Allen Zones (or Van Allen Belts): Zones around the Earth in which electrically charged particles are trapped and accelerated by the Earth's magnetic field. They were detected by J. Van Allen and his colleagues in 1958, from results obtained with the first successful US artificial satellite, Explorer 1. Apparently there are two main belts. The outer, made up mainly of electrons, is very variable, since it is strongly affected by events taking place in the Sun; the inner zone, composed chiefly of protons, is more stable. On the other hand, it may be misleading to talk of two separate zones; it may be that there is one general belt whose characteristics vary according to distance from the Earth. The Van Allen radiation is of great importance in all geophysical research, and probably represents the major discovery of the first years of practical astronautics.

Variable stars: Stars which fluctuate in brightness over short periods of time.

Variation: (1) An inequality in the motion of the Moon, due to the fact that the Sun's pull on it throughout its orbit is not constant in strength. (2) Magnetic variation: the difference, in degrees, between magnetic north and true north. It is not the same for all places on the Earth's surface, and it changes slightly from year to year because of the wandering of the magnetic pole.

Vernal Equinox: See **First Point of Aries**.

Vulcan: The name given to a hypothetical planet once believed to move around the Sun at a distance less than that of Mercury. It is now certain that Vulcan does not exist.

W

White dwarf: A very small, extremely dense star. The atoms in it have been broken up and the various parts packed tightly together with almost no waste space, so that the density rises to millions of times that of water; a spoonful of white dwarf material would weigh many tonnes. Evidently a white dwarf has used up all its nuclear 'fuel'; it is in the last stages of its active career, and has been aptly described as a bankrupt star. **Neutron stars** are even smaller and denser than white dwarfs.

Widmanstätten patterns: If an iron **meteorite** is cut, polished and then etched with acid, characteristic figures of the iron crystals appear. These are known as Widmanstätten patterns. They are never found except in meteorites.

Wolf-Rayet stars: Exceptionally hot, greenish-white stars whose spectra contain bright emission lines as well as the usual dark absorption lines. Their surface temperature may approach 100,000 degrees C, and they seem to be surrounded by rapidly expanding envelopes of gas. Attention was first drawn to them in 1867 by the astronomers Wolf and Rayet, after whom the class is named. Recently, it has been found that many of the Wolf-Rayet stars are **spectroscopic binaries**.

X

X-rays: Electromagnetic radiations of very short wavelength. There are many X-ray sources in the sky; studies of them must be undertaken by space research methods.

X-ray astronomy: X-rays are very short electromagnetic radiations, with wavelengths of from 0.1 to 100 **Ångströms.** Since X-rays from space are blocked by the Earth's atmosphere, astronomical researches have to be carried out by means of instruments taken up in rockets. The Sun is a source of X-rays; the intensity of the X-radiation is greatly enhanced by solar **flares**. Sources of X-rays outside the **Solar System** were first found in 1962 by American astronomers, who located two sources, one in Scorpius and the other in Taurus; the latter has now been identified with the Crab Nebula. Since then, various other X-ray sources have been discovered, some of which are variable.

Y

Year: The time taken for the Earth to go once around the Sun; in everyday life it is taken to be 365 days (366 days in Leap Year). (1) *Sidereal year*: The true revolution period of the Earth: 365.26 days, or 365 days 6 hours 9 minutes 10 seconds. (2) *Tropical year*: The interval between successive passages of the Sun across the **First Point of Aries**. It is equal to 365.24 days, or 365 days 5 hours 48 minutes 45 seconds. The tropical year is about 20 minutes shorter than the sidereal year because of the effects of precession, which cause a shift in the position of the First Point of Aries. (3) *Anomalistic year*: The interval between successive **perihelion** passages of the Earth. It is equal to 365.26 days, or 365 days 6 hours 13 minutes 53 seconds. It is slightly longer than the sidereal year because the position of the perihelion point moves by about 11 seconds of arc annually. (4) *Calendar year*: The mean length of the year according to the Gregorian calendar. It is equal to 365.24 days, or 365 days 5 hours 49 minutes 12 seconds.

Z

Zenith: The observer's overhead point (altitude 90 degrees).

Zenith distance: The angular distance of celestial body from the observer's zenith.

Zodiac: A belt stretching right round the sky, 8 degrees to either side of the **ecliptic**, in which the Sun, Moon and bright planets are always to be found. It passes through 13 constellations, the 12 commonly known as the Zodiacal groups plus a small part of Ophiuchus (the Serpent-bearer).

Zodiacal constellations: The 12 constellations used in **astrology**. They are Aquarius, Aries, Cancer, Capricornus, Gemini. Leo, Libra, Pisces, Sagittarius, Scorpius, Taurus and Virgo.

Zodiacal light: A cone of light rising from the horizon and stretching along the ecliptic. It is visible only when the Sun is a little way below the horizon, and is best seen on clear, moonless evenings or mornings. It is thought to be due to small particles scattered near the main plane of the **Solar System**. A fainter extension along the ecliptic is known as the *Zodiacal band.*

Index

A

B

C

Index

Index

N

O

P

Index

Index

T

ACKNOWLEDGEMENTS

Many people have helped in the preparation of this book. My special thanks go to Professor Sir Arnold Wolfendale for providing a Foreword, and to Paul Doherty for his excellent artwork. Among those who have provided many photographs are Commander Henry Hatfield, Terry Moseley, Don Trombino, Alan Heath, H. J. P. Arnold, Bernard Abrams, John Fletcher, and the Honourable Adrian Berry. Finally, I am most grateful to Robin Rees, of Messrs Philip's, for all his help and encouragement.

Photographic Credits

Where one of a number of photographs appears over two pages, it is credited on the first page only.

Abbreviations used are:
t top; c centre; b bottom; l left; r right
NASA National Aeronautics and Space Administration
JPL Jet Propulsion Laboratory
PM Patrick Moore Collection
STScl Space Telescope Science Institute

1 European Southern Observatory; **2** J. Hester, P. Scowen (Arizona State University)/NASA; **6** PM; **8** J. Hester (Arizona State University)/NASA; **10–11** NASA; **12**t PM, **12**c The Royal Society/PM, **12**bl The National Portrait Gallery/PM, **12**bc PM; **13** PM; **15** PM; **16–17** PM; **18**l PM, **18**r European Southern Observatory; **20**tl PM, **20**tr PM, **20**bl PM, **20**br Dr Seth Shostak/Science Photo Library; **21**t PM; **22–23** PM; **24**tl Novosti, **24**cl Popperphoto, **24**bl NASA, **24**r NASA; **25**r PM; **26**tl Novosti, **26**cNASA, **26**bl PM, **26**r NASA; **27**t NASA/Woodmansterne, **27**c PHOTRI/ZEFA, **27**b NASA/PM; **28**t NASA, **28**br Hubble Heritage Team (AURA/STScl/NASA); **29**tl & **29**tr C. Struck, P. Appleton (Iowa State University), K. Borne (Hughes STX Corp.), R. Lucas (STScl)/NASA, **29**br NASA; **30–31** NASA; **34** NASA; **35**t F. Damm/ZEFA, **35**b PM; **38** PM/T. J. C. A. Moseley; **39** A. Watson/PM; **40** Paris Observatory/Royal Astronomical Society (1, 2, 3, 4, 5); **41**bl Lick Observatory/Royal Astronomical Society (6), Royal Greenwich Observatory/ Royal Astronomical Society (7), **41**bc A. H. Mikesell; **42** H. R. Hatfield; **43** PM; **44–5** PM; **46** NASA; **47** Woodmansterne/NASA; **48** NASA; **49**t NASA, **49**c NASA, **49**b NASA; **50**t PM, **50**b US Naval Research Laboratory; **51**t M. Robinson (USGS)/NASA, **51**l & **51**r (7) Clementine NASA Science Team/NASA; **61** David Strange/PM; **62** NASA; **63** NASA; **64–5** NASA; **68**b Alan Heath; **69**t NASA, **69**b Novosti; **70–1** NASA; **72**l NASA, **72**r NASA/Science Photo Library; **73** NASA/Science Photo Library; **74** Charles Capen (Lowell Observatory, Arizona); **75** PM; **76–7** NASA; **78–9** NASA/JPL; **82** P. James (University of Toledo), S. Lee (University of Colorado)/NASA; 84–5 NASA; **83** D. Crisp & the WFPC-2 Science Team (JPL/California Institute of Technology)/NASA; **84–5** NASA; **86–7** NASA/PM; **88**l PM, **88**r NASA; **89t** B. Zellner/NASA, **89**bl NASA, **89**br B. Zellner, A. Sorrs, HSTI/NASA; **90** PM; **91** European Southern Observatory; **92**t PM, **92**b Charles Capen (Lowell Observatory, Arizona); **93** NASA; **95** NASA; **96** NASA/JPL; **97** NASA/JPL; **98** HST Comet Team/NASA; **99**l (t, c, b) Dr H. Weaver & T. Ed Smith (STScl)/NASA, **99**tr PM, **99**br PM; **100** NASA/JPL; **101** NASA; **102–3** NASA; **107**t Charles Capen (Lowell Observatory, Arizona), **107**b NASA; **108–9** NASA/JPL; **110**tl NASA, **110**bl NASA, **110**r NASA/JPL; **111** NASA/JPL; **112–13** NASA; **114–15** NASA; **118–19** NASA; **120** D. A. Allen Anglo-Australian Telescope; **121** NASA; **122** NASA/JPL; **123**t JPL/California Institute of Technology, **123**b E. Karkoschka (University of Arizona)/ NASA; **124**t NASA, **124**bl NASA, **124**br JPL/California Institute of Technology; **125** NASA; **128** NASA/JPL; **129**t L. Sromovsky (University of Wisconsin-Madison)/NASA, **129**b NASA/JPL; **130–1** NASA; **132** PM; **133**t PM, **133**c NASA, **133**b PM; **134** A. Stern (Southwest Research Institute), M. Buie (Lowell Observatory)/ESA/NASA; **135**t D. L. Rabinowitz (Kitt Peak National Observatory, Arizona), **135**c (3) A. Smette, C. Vanderriest (La Silla); **136** PM; **138**t Peter Carrington/PM, **138**c (7) PM, **138**l British Aerospace Dynamics Group, **138**br Max Planck Institute for Radio Astronomy; **139** Max Planck Institute for Radio Astronomy; **140–1** PM; **142**t Kent Blackwell, **142**c Kent Blackwell, **142**b PM; **143**t Martin Mobberley, **143**b PM; **144** PM; **145**tl John Fletcher, **145**tr PM, **145**c PM, **145**b D. F. Trombino; **146–7** PM; **148**tl D. F. Trombino, **148**bl PM, **148**r PM; **149**t PM, **149cr** Vic Urban; **150–1** D. F. Trombino; **153**bl PM, **153**r Brookhaven National Laboratory; **154**tr Kitt Peak National Observatory, Arizona, **154**br H. J. P. Arnold; **155**tr (3) Commonwealth Science and Industrial Research Organization, Sydney, Australia; **156**cl PM, **156**br D. F. Trombino; **157**tr PM; **159** tr Henry Brinton, **159**c PM, **159**bl D. F. Trombino, **159**br NASA; **160** Institute of Space and Astronautical Science, Japan; **161**t NASA, **161**bl D. F. Trombino, **161**br Paul Doherty; **162–3** Commonwealth Science and Industrial Research Organization, Sidney, Australia; **164–5** PM; **167**b PM; **168**b PM; **170** PM; **172**b European Southern Observatory; **175** S. Andrew; **177** STScl/NASA; **178** PM; **179** NASA; **180**t PM, **180**b Hale Observatories; **181**tl R. W. Arbour, **181**tr G. Sonneborn (GSFC)/NASA and J. Pun (NOAO)/SINS Collaboration; **182** David Malin/Anglo-Australian Observatory from original negative by the UK 1.2m Schmidt Telescope © Royal Observatory, Edinburgh, 1980; **183**b Hubble Heritage Team (AURA/STScl/NASA); **184** Hale Observatories; **185** R. Elson and R. Sword (Cambridge, UK)/NASA and J. Westphal (Caltech); **186**t NASA, **186**bl John Fletcher, **186**br H. R. Hatfield; **187** European Southern Observatory; **188** R. O'Dell, K. P. Handron (Rice University, Houston, Texas)/NASA; **189**t Hubble Heritage Team (AURA/STScl/NASA), **189**b A. Caulet (ST-ECF, ESA)/NASA; **190–1** National Optical Astronomical Observatories/Science Photo Library; **192** Hale Observatories; **193**t NASA, **193**br ESA/NASA; **195**t NASA/PM, **195**b Lund Observatory; **196**tr John Fletcher, **196**l Photolabs/Royal Observatory, Edinburgh from original negatives by the UK 1.2m Schmidt Telescope © Royal Observatory Edinburgh; **197** John Fletcher; **198**bl Lick Observatory; **198**bc Hale Observatories; **198**br Hale Observatories; **199**t Hale Observatories (3), **199**bl Hale Observatories; **199**bc Lick Observatory, **199**br Hale Observatories; **200** PM; **201**tr J. Bahcall (Institute for Advanced Study, Princeton)/NASA, **201**tl & **201**bl J. Bahcall (Institute for Advanced Study, Princeton), M. Disney (University of Wales)/NASA, **201**cr NASA/ESA; **202** NASA; **203** R. Windhorst, S. Pascarelle (Arizona State University)/NASA; **204–5** NASA; **206** NASA; **208–9** E. T. Archive; **219** J. P. Harrington and K. J. Borkowski (University of Maryland)/NASA; **227** Space Frontiers; **229** B. Whitmore (STScl)/NASA; **234–5** John Fletcher; **241** Bernard Abrams; **243** John Fletcher; **245** John Sanford/Science Photo Library; **249** Bernard Abrams; **251** Bernard Abrams; **258** Fred Espenak/Science Photo Library; **260–1** PM; **262**tl PM, **262**bl PM, **262**r Derek St Romaine; **263**l David Cortner/Galaxy Picture Library, **263**r PM; **264** Derek St Romaine; **265**tl Space Frontiers, **265**tr Space Frontiers, **265**b (3) Derek St Romaine; **266** PM; **267**t PM, **267**l Curtis MacDonald/Galaxy Picture Library, **267**r Jerry Gunn/Galaxy Picture Library.

Artwork Credits
Paul Doherty: 14, 32–3, 39, 62, 66–7, 68, 75, 80–1, 93, 104–5, 107, 116–17, 126–7, 128, 133br, 152, 153, 154cl, 155b, 172–3, 174, 181, 183t, 263.

Credit is also due to Mick Saunders, David Hardy, Ed Stuart, Richard Lewis, David Mallott and Kuo Kang Chen.

Mapping Credits
Moon Maps © George Philip Limited
Whole Sky Maps: Wil Tirion
Star Maps prepared by Paul Doherty and produced by Louise Griffiths